THERMODYNAMICS

NEW YORK · JOHN WILEY & SONS, INC.
LONDON ·

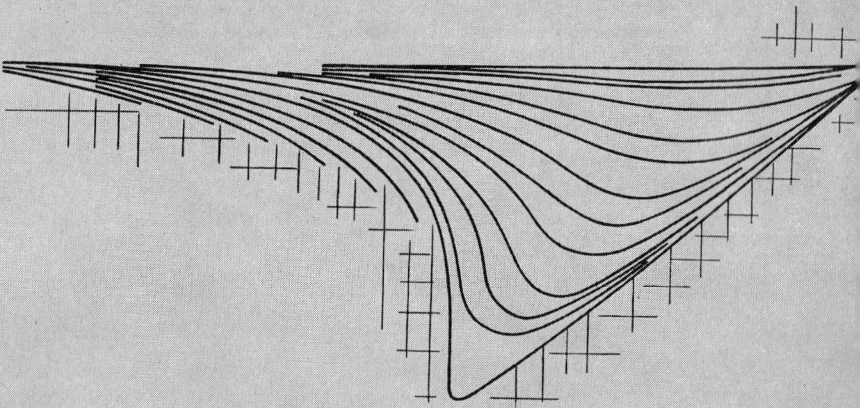

THERMODYNAMICS

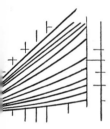

Fifth Printing with Supplementary Problems

Gordon J. Van Wylen

Chairman,
Department of Mechanical Engineering
University of Michigan

FIFTH PRINTING, MARCH, 1963
(with Supplementary Problems)

Library of Congress Catalog Card Number: 59-9356

Printed in the United States of America

To my wife

Margaret DeWitt Van Wylen

Preface

The fact that so many have recently undertaken the writing of thermodynamics textbooks for engineering students compels anyone undertaking the writing of another one to ask, at least of himself, the *raison d'être* for his book. In this case I can cite two reasons. The first is a frankly personal one. Several years ago I began writing supplementary notes for my students. On finding this a challenging and stimulating experience, I decided to expand this effort into a textbook, and have found that the personal rewards justified the effort and discipline required to bring the book to completion. The second reason is that I believe there is room for a book that combines three features, namely, an approach directed toward students, a rigorous treatment of fundamentals, and a distinctly engineering perspective. This book represents an effort to fill these needs. Thus, the objective has been to give the student an appreciation and understanding of both the physical and the mathematical aspects of thermodynamics.

The experience of most people is that they do not gain a thorough understanding of thermodynamics until they have reviewed the subject several times. Often this additional study is undertaken in graduate courses. I fully realize that I have taken several graduate courses before writing this book, and that the temptation is to put much material that I learned as a graduate student in a textbook designed for undergraduates. However, the topics in this book have been so selected that I am convinced that the undergraduate is capable of mastering all the material in it, and that one who does master it will be qualified to tackle the majority of engineering problems that require a thermodynamic analysis.

It is intended that this textbook be covered in two 15-week semesters. The suggested work for the first semester is Chapters 1 through 10, and for the second semester, the remainder of the book. The suggested division for three 10-week terms is Chapters 1 through 8, 9 through 12, and 13 through 16.

In writing a book one has to make a number of decisions regarding terminology and symbols. The most difficult decision for me was whether to use the terms closed and open system or the corresponding terms system and control volume. There are certain advantages to each approach. I finally decided to use closed and open system in the body of the text and to introduce the concept of the control volume in the Appendix. The reason for this is a pedagogical one based on personal experience in teaching thermodynamics.

In deciding what symbols to use for various units, the American Standard Association booklet entitled *Letter Symbols for Heat and Thermodynamics* was the primary guide. However, in order to emphasize the matter of units I decided to use such symbols as lbf/in.^{2} for pressure (rather than psi) and ft^3/lbm for specific volume (instead of cu ft/lb). The reason for this was, first of all, to distinguish between pound mass and pound force, and second, to emphasize the necessity for consistent units for each term of a given equation. The second reason also led to the decision not to carry the factor J in the equations, but rather to emphasize that the only requirement is consistent units. Logically this should also have led to the decision to leave the constant g_c out of all equations. However, in view of the difficulty undergraduates seem to have with this matter I decided to include this. No doubt the individual instructor can, if he so desires, encourage students to leave this constant out of equations also, and introduce it only when it is necessary to obtain dimensionally correct solutions.

The macroscopic point of view has been adopted throughout this text, though occasionally a few amplifying remarks are made from the microscopic point of view.

There are many possible sequences that may be followed when writing a thermodynamics textbook, and the propriety of any one can be judged only from the standpoint of taste or pedagogical experience. One ground rule I have followed is to present new concepts and definitions only in the context in which they are meaningful and relevant. This is particularly true with regard to thermodynamic properties. The properties specific volume, pressure, and temperature are defined in Chapter 2. The pure substance is considered in Chapter 3, and the student is introduced to thermodynamic tables (such as the steam tables), but only to the extent of specific volume, pressure, and temperature. After introducing the first law for a closed system, the internal energy is defined, and the steam tables are used for values of internal energy (which is the only readily available table that gives values of internal energy). Enthalpy is introduced in connection with the open system where its significance is quite evident, and reference is then made to enthalpy values in the tables of thermodynamic properties. Entropy is introduced after the second law and the inequality of Clausius have been discussed. The Helmholtz function and the Gibbs function are introduced in relationship to availability in Chapter 10. Throughout the book I have attempted to follow this pattern of introducing new terms only as they are relevant.

A few comments might well be made on the content of the chapters. Frequently students have very little concept of actual physical equipment, and for this reason I included Chapter 1, which is a brief descriptive chapter. It can be omitted with no lack of continuity. Chapters 2, 3, and 4 cover basic definitions and concepts. Some additional material on temperature measurement might be desirable, but I felt that this could probably be best introduced by the individual instructor, possibly using some demonstrations to supplement the discussion, or perhaps in a laboratory course. Chapter 5 is devoted to the first law and is somewhat longer than many chapters on this subject. I believe that frequently the first law is treated too lightly, particularly as it involves flow into and out of open systems. Conservation of mass is also introduced in this chapter, and an attempt is made to show the relationship of $E = mc^2$ to macroscopic thermodynamics.

The acid test of any thermodynamics textbook probably lies in the treatment of the second law. In this regard I followed the classical approach. Chapter 6 deals with the Kelvin-Planck and Clausius statements, the Carnot cycle, and the thermodynamic scale of temperature. This leads to Chapter 7 where the inequality of Clausius is introduced and entropy is defined. The principle of the increase of entropy is also considered in this chapter. This would logically lead to the matter of availability. However, I felt that the ideal gas and ideal-gas mixture should be covered first. Thus, Chapter 8 deals with the ideal gas and Chapter 9 deals with the mixture of ideal gases and vapors. The *Gas Tables* by Keenan and Kaye are introduced in Chapter 8, and frequent use of these tables is made throughout the rest of the book.

Chapter 10 deals with availability and irreversibility. Though the material is not new in any way, the development and arrangement of the material is somewhat different from that of most textbooks. The concepts of availability and irreversibility are very powerful tools for analysis, and the increasing number of technical papers dealing with them is evidence that more engineers are appreciating their utility and importance.

In the last half of this book I have deviated somewhat from the traditional approach in that I have considered processes in reciprocating machinery in Chapter 10 and cycles in Chapter 11. Thus, I do not have separate chapters on refrigeration, compression of gases, vapor power cycles, etc. The reason for this is twofold. First of all, cycles are frequently overemphasized, and by devoting a single chapter to cycles I attempted to bring the matter into proper focus. Second, since thermodynamics is an engineering science, I thought it best to have a general treatment of many topics. For example, such concepts as volumetric efficiency and mean effective pressure are common to all reciprocating machines, and are therefore best treated in a general way.

It is difficult to know the optimum number of fluid-flow topics to include in a textbook on thermodynamics. I have selected a number that are not only important in themselves, but will also assist in bridging the gap between a student's courses in thermodynamics and fluid flow. The topics covered are one-dimensional flow through nozzles (including a discussion of the normal shock) and orifices, and axial flow through blade passages (both impulse and reaction). This material is covered in Chapter 13.

Chapter 14 deals with Maxwell relations, equations of state, and generalized charts. I believe that students gain a better understanding of the mathematical treatment if they have first gained familiarity with the physical aspects of the subject. I also attempted to give some basic ideas of how tables of thermodynamic properties are compiled. Perhaps this was added to satisfy an unanswered question I had as an undergraduate student.

Chapter 15 deals with chemical reactions with an emphasis on combustion. Certain aspects of this chapter, which were included in order to provide a fairly complete coverage, can be omitted by students who are familiar with chemical principles. The enthalpy of formation is introduced in this chapter and I have also used the terms enthalpy and internal energy of combustion, which I believe are better terms than constant-pressure and constant-volume heating value. The third law of thermodynamics is introduced in order to give meaning to absolute entropy, and this in turn is utilized in the second-law analysis of combustion processes.

The final chapter deals with equilibrium, with particular emphasis on chemical equilibrium of ideal-gas mixtures. The chemical potential and the Gibbs phase rule are also introduced in this chapter.

In general, I have not attempted to give specific references, but have included a selected list of general references at the end of the text.

It is very difficult to give specific acknowledgment in a text such as this. First of all, the number to whom acknowledgment is due is very large, because many have contributed to my knowledge on the subject, either directly or through their writings. In the second place, although on some matters I specifically recall the professor or textbook involved, on other matters I have long since forgotten the occasion on which I learned a given point. This is true also of some of the problems that are included in the text, because I have included those that were of definite educational value to me, though in many cases I am not certain of the original source. Therefore, although specific mention is not possible, I do acknowledge the help of many in the writing of this book.

However, I would like to acknowledge my indebtedness to three professors who stimulated my interest in the subject and taught me much of what I know. These are the late Professor C. F. Kessler of the University of Michigan, Professor J. S. Doolittle, formerly of The Pennsylvania State University, and Professor J. H. Keenan of the Massachusetts Institute of Technology. I also owe a real debt of gratitude to my colleagues at the University of Michigan who offered many valuable suggestions in the course of the development and writing of this text. Many valuable comments were received from my colleagues in other universities who used the preliminary edition of this book, and this help is acknowledged with gratitude. A good number of students, student assistants, and secretaries were also of considerable help. Also, I would be remiss without acknowledging the help and grace of God, the Creator of the universe and the author of life.

I wish the reader a profitable and enjoyable journey through the book. Your comments, criticisms, and suggestions will be much appreciated.

GORDON J. VAN WYLEN

Ann Arbor, Michigan
March, 1959

CONTENTS

Symbols

a	acceleration
a, A	specific Helmholtz function and total Helmholtz function
A	area
AF	air-fuel ratio
C_p	constant-pressure specific heat
C_v	constant-volume specific heat
C_{po}	zero-pressure constant-pressure specific heat
$C_{v\infty}$	zero-pressure constant-volume specific heat
c	velocity of sound
E	a thermodynamic property that represents the total energy of a system
f	fugacity
F	force
FA	fuel-air ratio
g	acceleration due to gravity
g, G	specific Gibbs function and total Gibbs function
g_c	a constant that relates force, mass, length, and time
h, H	specific enthalpy and total enthalpy
i, I	specific irreversibility and total irreversibility
I	specific impulse
J	proportionality factor to relate units of work to units of heat
k	specific-heat ratio: C_p/C_v
K_P	equilibrium constant involving ideal gas
KE	kinetic energy
lbf	pound force

lbm	pound mass
lb mole	pound mole
lw, LW	lost work per unit mass and total lost work
m	mass
$\dot{m}$	mass rate of flow
M	molecular weight
$\boldsymbol{M}$	Mach number
mf	mass fraction
n	number of mols
n	polytropic expone:
p, P	partial pressure and total pressure
PE	potential energy
P_r	relative pressure as used in gas tables
q, Q	heat transfer per unit mass and total heat transfer
$\dot{Q}$	rate of heat transfer
Q_H, Q_L	heat transfer with high-temperature body and heat transfer with low-temperature body; sign to be determined from context
R	gas constant
$\bar{R}$	universal gas constant
s, S	specific entropy and total entropy
t, T	temperature in degrees F or degrees C and absolute temperature in degrees R or degrees K
u, U	specific internal energy and total internal energy
v, V	specific volume and total volume
vf	volume fraction
$\bar{V}$	velocity
V_r	relative volume as defined in gas tables
w, W	work per unit mass and total work
$\dot{W}$	work per unit time, or power
w_{rev}	reversible work between two states assuming heat transfer with surroundings only
x	quality
x	mole fraction
Z	elevation

Subscripts

f	properties of saturated liquid
g	properties of saturated vapor
i	properties of saturated solid
fg	difference in property for saturated vapor and saturated liquid
ig	difference in property for saturated vapor and saturated solid
i	properties of a fluid entering a system
e	properties of a fluid leaving a system
c	property at the critical point

r	reduced property
o	property of the surroundings
0	stagnation properties

Superscripts

$-$	bar over symbol means properties on a molal basis
$h^\circ, u^\circ, g^\circ$	enthalpy, internal energy, and Gibbs function of formation
s°	absolute entropy

Script Letters

$\mathcal{E}$	electrical potential
$\mathcal{R}$	reheat factor
$\mathcal{S}$	surface tension
$\mathcal{T}$	tension

Greek Letters

α	volume expansivity
β_S	adiabatic compressibility
β_T	isothermal compressibility
η	efficiency
μ	degree of saturation
μ	chemical potential
μ_J	Joule-Thomson coefficient
ρ	density
τ	time
σ	denotes a change within the boundary of a system
ϕ	a property associated with entropy as defined in the gas tables
ϕ	relative humidity
ϕ	availability function for a closed system
ψ	availability function for an open system
ω	humidity ratio or specific humidity

CHAPTER 1

SOME DESCRIPTIVE CONSIDERATIONS

In this chapter a number of pieces of equipment that involve thermodynamics are described. The purpose of this chapter is to enable the student to have some understanding of the physical equipment he will be considering in a number of thermodynamic problems. There are at least two dangers in writing a chapter such as this. The first is that the student will have a limited view of the equipment which requires a thermodynamic analysis, and the second is that he might not realize that the scope of thermodynamics is very broad and involves metallurgy, chemical reactions, liquefaction and separation of air, and a host of other applications. Therefore this chapter will be of value only if the student considers it a very incomplete and elementary introduction.

A number of terms will be used in this chapter which are not defined until later chapters. This is done deliberately with the intention of introducing the student to the correct usage of the terms before actually defining them.

The pieces of equipment considered in this chapter are the steam power plant, the refrigerator, the gas turbine, and the liquid-propellant rocket. It is presumed that the student is already familiar with the general arrangement of a reciprocating internal-combustion engine and the reciprocating air compressor.

1.1 The Simple Steam Power Plant

A schematic diagram of a simple steam power plant is shown in Fig. 1.1. High-pressure superheated steam leaves the boiler, which is also referred to as a steam generator, and enters the turbine. The steam expands in the turbine and in doing so, does work, which enables

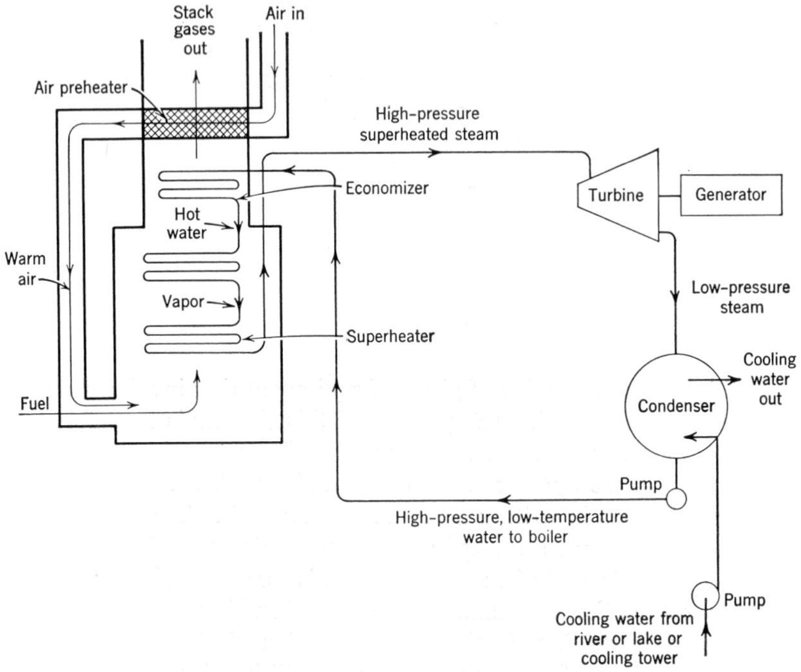

Fig. 1.1 Schematic diagram of a steam power plant.

the turbine to drive the electric generator. The low-pressure steam leaves the turbine and enters the condenser, where heat is transferred from the steam (causing it to condense) to the cooling water. Since large quantities of cooling water are required, power plants are frequently located near rivers or lakes. When the supply of cooling water is limited, a cooling tower may be used. The pressure of the condensate leaving the condenser is increased in the pump, thus enabling the condensate to flow into the boiler.

In many boilers an economizer is used. As a result of heat transfer from the combustion gases, the temperature of the feedwater is increased in the economizer, but it remains in the liquid phase. In the boiler further heat transfer from the combustion gases results in

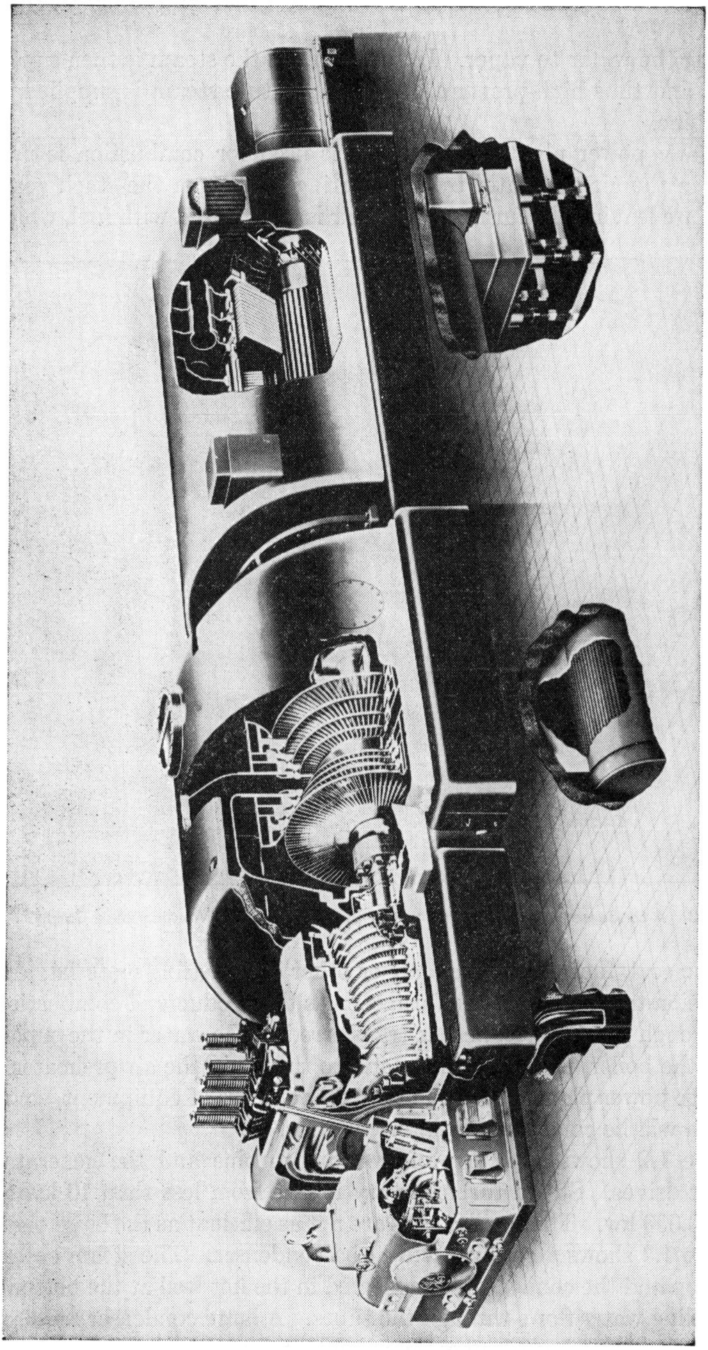

Fig. 1.2 A large steam turbine. (Courtesy General Electric Corp.)

changing the water to vapor. In most boilers the steam is then superheated, and thus high-pressure, high-temperature steam is supplied to the turbine.

In many power plants the air that is used for combustion is preheated in the air preheater by transferring heat from the stack gases as they are leaving the furnace. This air is then mixed with fuel, which

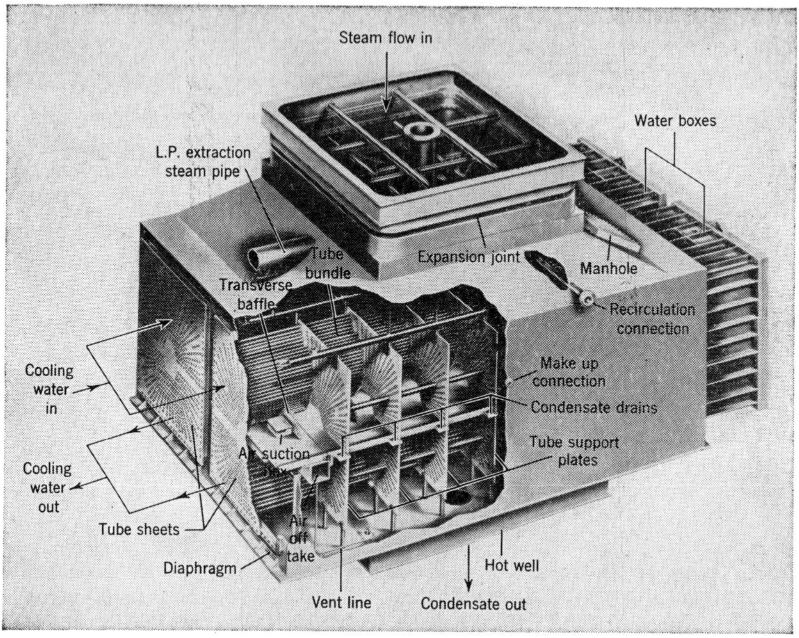

Fig. 1.3 A condenser used in a large power plant. (Courtesy Westinghouse Corp.)

might be coal, fuel oil, natural gas, or other combustible material, and combustion takes place in the furnace. As the products of combustion pass through the furnace, heat is transferred to the water in the superheater, the boiler, the economizer, and to the air in the air preheater.

A large power plant will have many other pieces of equipment, some of which will be considered in later chapters.

Figure 1.2 shows a picture of a steam turbine and the generator which it drives. Steam turbines vary in size from less than 10 kw to over 400,000 kw. The smaller sizes are not as efficient as the large ones.

Figure 1.3 shows a cutaway view of a condenser. The steam enters at the top and the condensate is collected in the hot well at the bottom. The cooling water flows through the tubes. A large condenser requires a tremendous number of tubes, as is evident from Fig. 1.3.

Figure 1.4 shows a large steam generator. The flow of air and products of combustion are indicated. The condensate, also called boiler feedwater, enters at the economizer inlet and the superheated steam leaves at the superheater outlet.

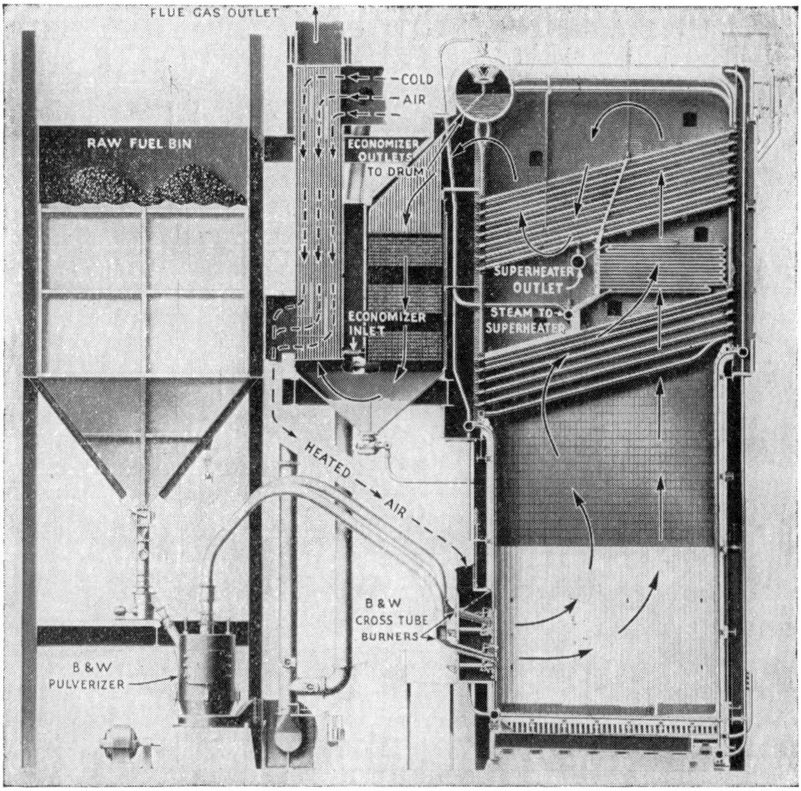

Fig. 1.4 A large steam generator. (Courtesy Babcock and Wilcox Corp.)

Much attention is now being given to the development of nuclear reactors for power generators. In some power plants the reactor essentially replaces the boiler, the energy source being the fissionable material instead of the combustible material of a conventional power plant.

1.2 The Refrigerator

A simple refrigeration cycle is shown schematically in Fig. 1.5. The refrigerant enters the compressor as a low-pressure, slightly superheated vapor. It is then compressed, and leaves the compressor and enters the condenser as a vapor at some elevated pressure. The refrigerant is condensed as a result of heat transfer to cooling water or to the surroundings. The refrigerant leaves the condenser as a high-pressure liquid. The pressure of the liquid is decreased as it flows through the expansion valve, and as a result some of the liquid flashes into vapor. The remaining liquid, now at a low pressure, is vaporized in the evaporator as a result of heat transfer from the refrigerated space. This vapor then enters the compressor.

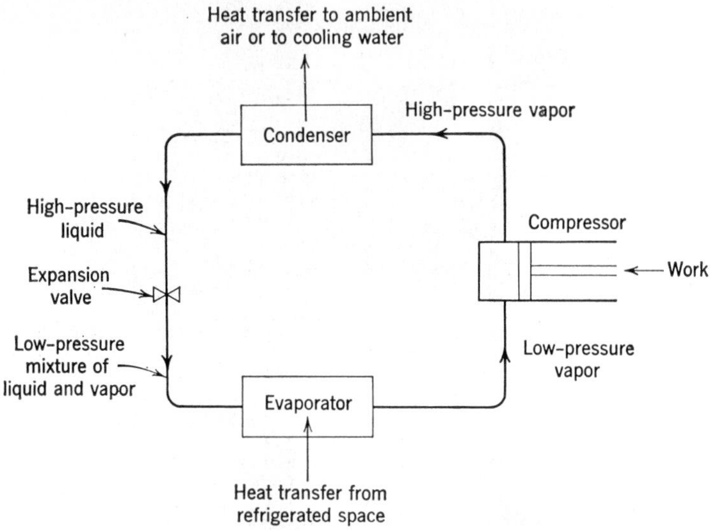

Fig. 1.5 Schematic diagram of a simple refrigeration cycle.

In a typical home refrigerator the compressor is usually located in the rear near the bottom. The compressors are usually hermetically sealed; that is, the motor and compressor are mounted in a sealed housing, and the electric leads for the motor pass through this housing. This is done to prevent leakage of the refrigerant. The condenser is also located at the back of the refrigerator and is so arranged that the air in the room flows past the condenser by natural convection. The expansion valve takes the form of a long capillary tube and the evaporator is located around the outside of the freezing compartment inside the refrigerator.

Figure 1.6 shows a large centrifugal unit which is used to provide refrigeration for an air-conditioning unit. In this unit, brine is cooled, and this in turn is circulated to provide cooling where needed. The expansion valve is not shown, for it is hidden from view by the compressor.

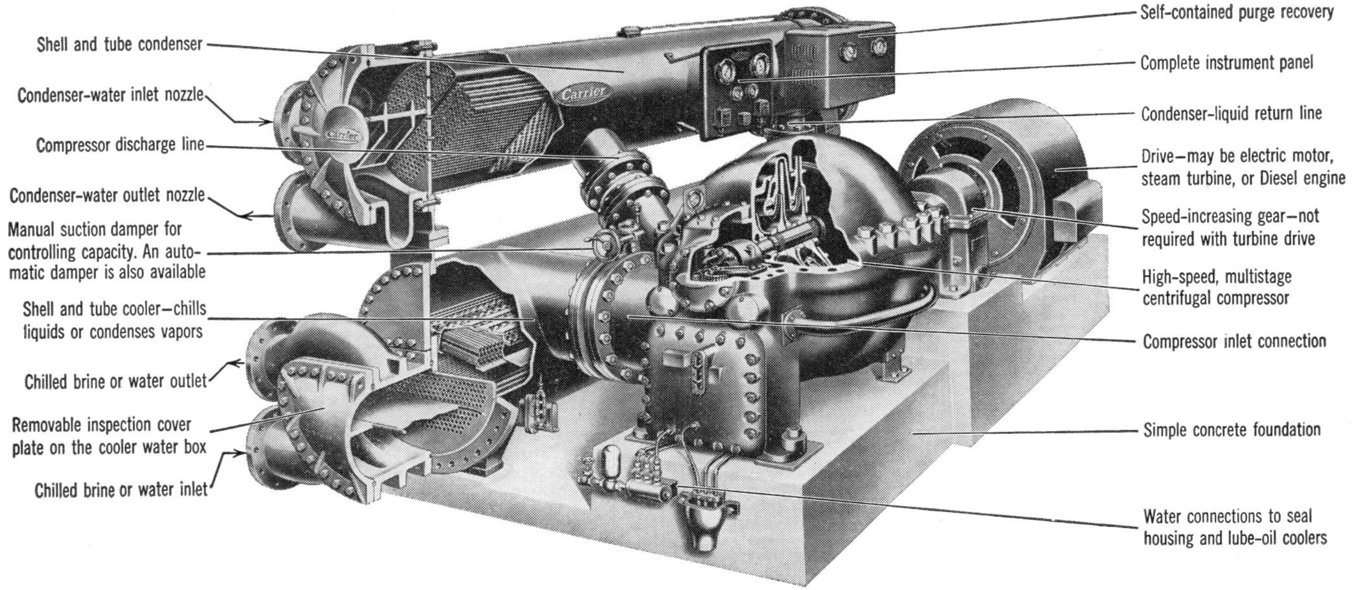
Shell and tube condenser
Condenser-water inlet nozzle
Compressor discharge line
Condenser-water outlet nozzle
Manual suction damper for controlling capacity. An automatic damper is also available
Shell and tube cooler—chills liquids or condenses vapors
Chilled brine or water outlet
Removable inspection cover plate on the cooler water box
Chilled brine or water inlet
Self-contained purge recovery
Complete instrument panel
Condenser-liquid return line
Drive—may be electric motor, steam turbine, or Diesel engine
Speed-increasing gear—not required with turbine drive
High-speed, multistage centrifugal compressor
Compressor inlet connection
Simple concrete foundation
Water connections to seal housing and lube-oil coolers

Fig. 1.6 A refrigeration unit for an air-conditioning system. (Courtesy Carrier Corp.)

1.3 The Gas Turbine

Gas turbines are most commonly used for jet aircraft. However, some use of gas turbines is being made in power generation and ship and automobile propulsion. A schematic diagram of a gas turbine is shown in Fig. 1.7. The essential elements are the compressor, the combustion chamber, and the turbine. The relatively high-pressure air

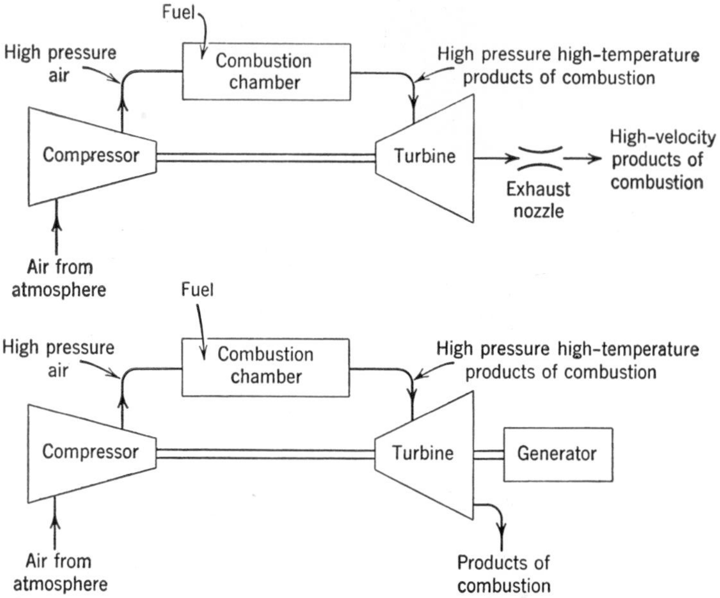

Fig. 1.7 Schematic diagram of a gas turbine.

leaving the compressor is utilized in burning the fuel, and the high-pressure, high-temperature products of combustion enter the turbine and expand, thus doing work. In a jet engine, the products of combustion are only partially expanded in the turbine (only enough work is done to drive the compressor) and the remainder of the expansion takes place in the nozzle, thus giving rise to a high-velocity jet leaving the nozzle. An after-burner may also be used, in which case additional fuel is burned in the nozzle, thus giving an increase in thrust. If the gas turbine is utilized for power generation, the gas is completely expanded in the turbine, and the turbine not only drives the compressor, but also a generator.

Figure 1.8 shows an aircraft turbojet engine, and Fig. 1.9 shows a unit that might be used for power generation. There are many other

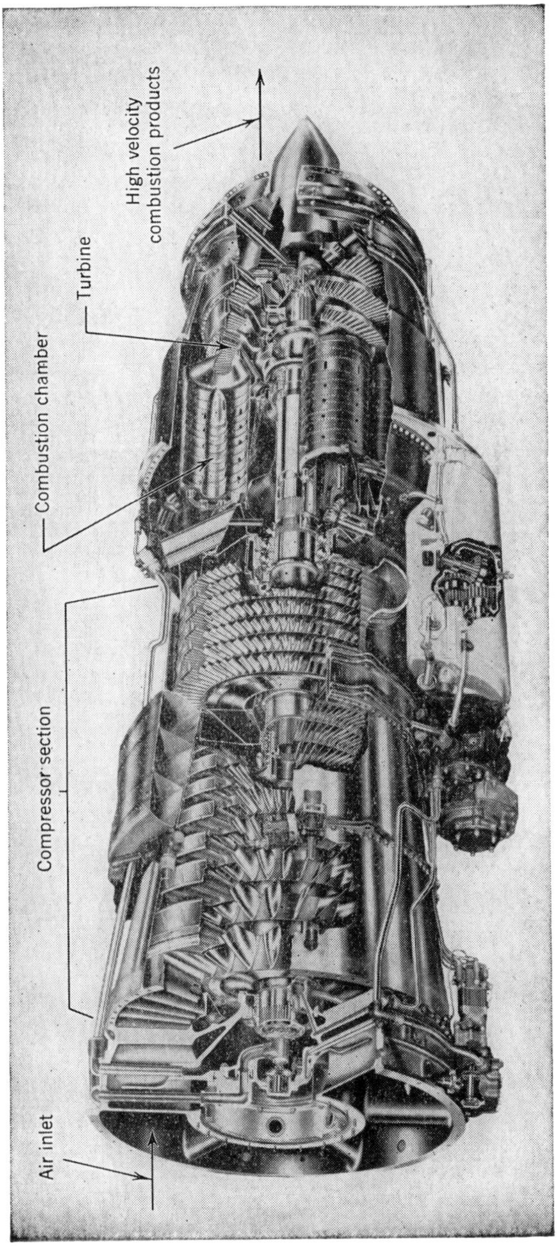

Fig. 1.8 Aircraft jet engine. (Courtesy Pratt Whitney Corp.)

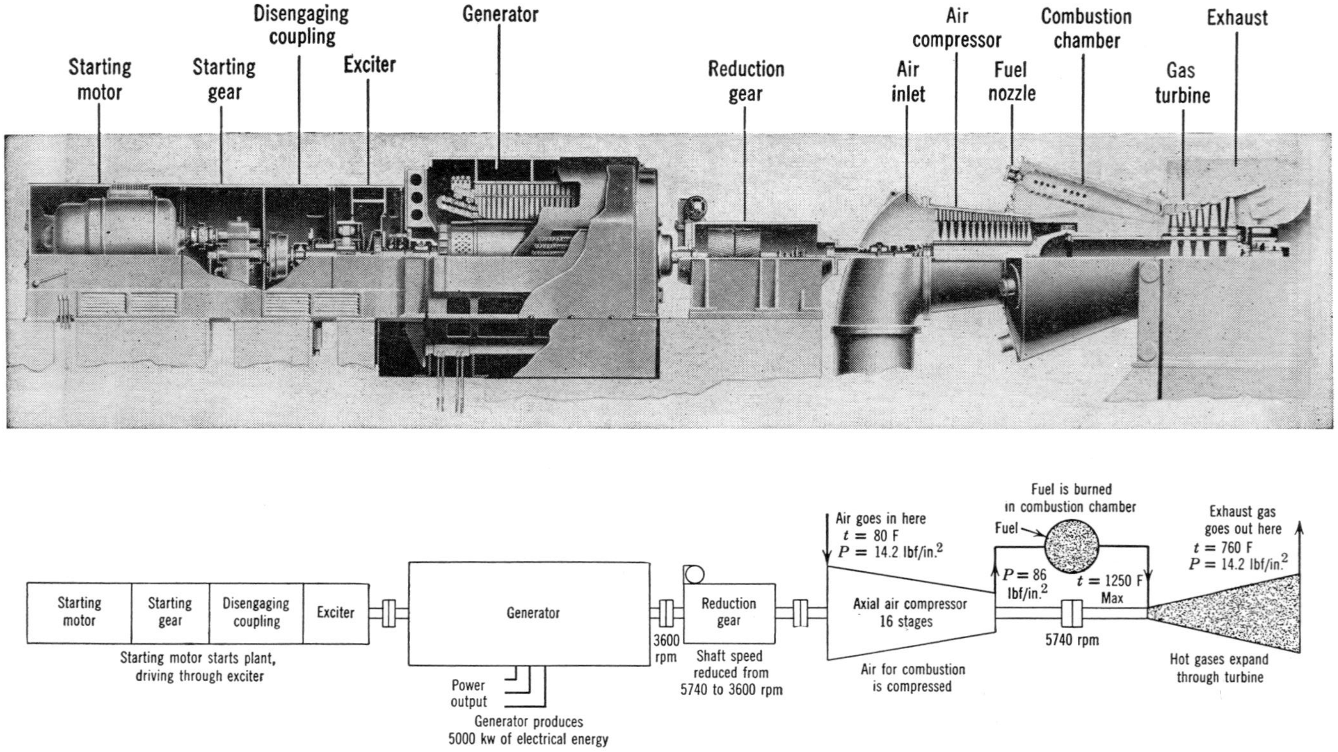

Fig. 1.9 A 5000-kw gas turbine power plant for electric power generation. (Courtesy Westinghouse Corp.)

pieces of equipment that can be incorporated into the cycle, and some of these will be considered in a later chapter.

1.4 The Liquid-Propellant Rocket

The advent of guided missiles and satellites has brought to prominence the rocket engine as a propulsion power plant. Rockets may be classified as either liquid propellant or solid propellant, depending on the fuel used.

Figure 1.10 shows a simplified schematic diagram of a liquid-propellant rocket. The oxidizer and fuel are pumped through the

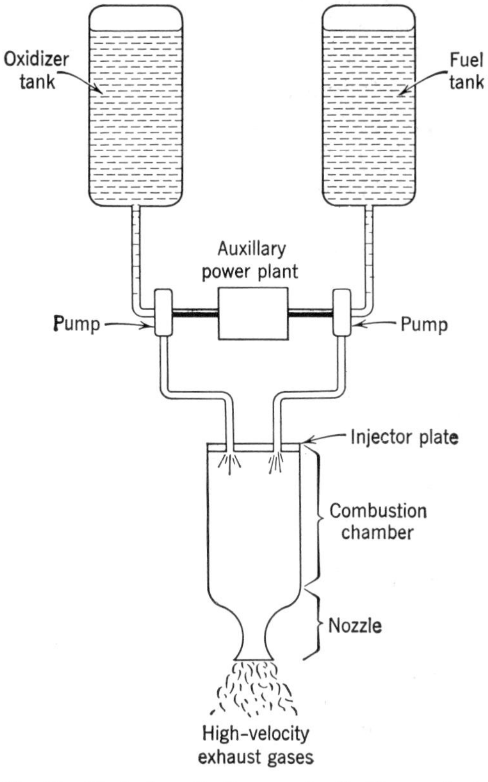

Fig. 1.10 Simplified schematic diagram of a liquid-propellant rocket.

injector plate into the combustion chamber where combustion takes place at a fairly high pressure. The high-pressure, high-temperature products of combustion expand as they flow through the nozzle, and as a result they leave the nozzle with a high velocity. The momentum change associated with this increase in velocity gives rise to the forward thrust.

The oxidizer and fuel must be pumped into the combustion chamber, and some auxiliary power plant is necessary to drive the pumps. In a large rocket this auxiliary power plant must be very reliable and have a relatively high power output, yet it must be light in weight. The oxidizer and fuel tanks occupy the largest part of the volume of an actual rocket, and the range of a rocket is determined largely by the amount of oxidizer and fuel that can be carried. Many different fuels and oxidizers have been considered and tested, and much effort has gone into the development of fuels and oxidizers that will give a higher thrust per unit mass of reactants. Liquid oxygen is frequently used as the oxidizer in liquid-propellant rockets.

Much work has also been done on solid-propellant rockets. They have been very successfully used for jet-assisted take-offs of airplanes, as well as in other military rockets that involve short-duration firing. They are much simpler in regard to both the principle on which they operate and the logistic problems involved in their use by the military.

CHAPTER 2

DEFINITIONS

In the previous chapter several machines and processes involving thermodynamics were described. With this as a background we begin a formal development of thermodynamics. Thermodynamics can be defined as the science that deals with heat and work and those properties of substances that bear a relation to heat and work. Like all sciences, the basis of thermodynamics is experimental observation. In thermodynamics these findings have been formalized into three laws, which are known as the first, second, and third laws of thermodynamics. In addition to these, the zeroth law of thermodynamics, which logically precedes the first law, has been set forth. Newton's second law of motion is also involved in many problems that require a thermodynamic analysis.

In the chapters that follow, these laws are presented and applied to a variety of systems, and the thermodynamic properties are related to these laws of thermodynamics and to Newton's second law. In this course the objective of the student should be to gain a thorough understanding of the fundamentals and an ability to apply these fundamentals to thermodynamic problems. The purpose of the examples and the problems is to further this objective. It should be emphasized that it will not be necessary for the student to memorize equations, for all the problems can be solved by the application of the definitions and laws of thermodynamics.

The remaining sections of this chapter deal with some concepts and definitions basic to thermodynamics.

2.1 Thermodynamic Systems

The first step in every thermodynamic analysis is to specify the system. A *system* is defined as a region in space or a quantity of matter upon which attention is focused for study. Everything external to the system is the surroundings, and the system is separated from the surroundings by the system boundaries. These boundaries may be either movable or fixed.

It is convenient to distinguish three classes of systems, namely, the closed system, the open system, and the isolated system. The *closed system* is a system of fixed mass and identity, whose boundaries are determined by the space the matter occupies. Figure 2.1 illustrates a closed system, for the gas in the cylinder is considered the system. If a Bunsen burner is placed under the cylinder, the temperature of the gas will increase and the piston will rise. As the piston rises, the boundary of the system moves. As we shall see later, heat and work cross the boundary of the system during this process, but the matter that comprises the system is fixed.

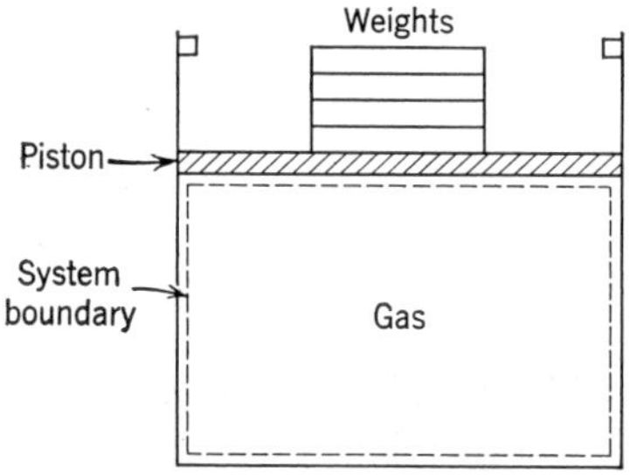

Fig. 2.1 Example of a closed system.

The *open system* is one in which matter crosses the boundary of the system.* Heat and work may also cross the boundary. Figure 2.2, which shows a schematic diagram of an air compressor, illustrates an open system. The important thing to note about the air compressor is that matter crosses the boundary of the system as the low-pressure

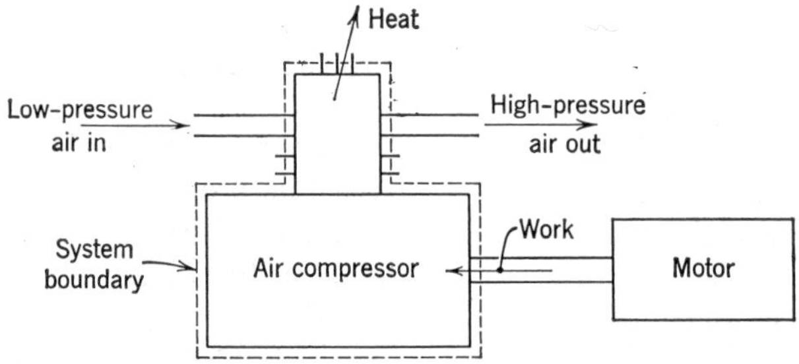

Fig. 2.2 Example of an open system.

* For an alternate definition see Appendix I.

air enters the compressor and the high-pressure air leaves. In addition, work crosses the boundary of the system through the driving shaft and heat is transferred across the system boundary from the cylinder walls. The emphasis in the open system is on a given region in space, and the quantity of matter within the boundary of the system is not fixed.

An *isolated system* is one which is completely uninfluenced by the surroundings. It is of fixed mass and no heat or work cross the boundary of the system.

In summary then, the system is specified by defining certain boundaries, and the matter that comprises the system is the matter within these boundaries. For the closed system this matter is of fixed identity; for the open system it varies with time.

2.2 Macroscopic vs. Microscopic Point of View

An investigation into the behavior of a system might be undertaken from either a microscopic or macroscopic point of view. From the microscopic point of view we would consider matter to be composed of molecules, each molecule having at a given instant a certain position, velocity, and energy and for each molecule these change very frequently as a result of collisions. The behavior of the system could be described by summing up the behavior of each molecule. Further, each molecule could be considered in terms of the fundamental particles of which it is composed and these also could be considered in analyzing the behavior of the system. Such a study is known as statistical mechanics, and it is a most important field of investigation.

In thermodynamics, on the other hand, we adopt the macroscopic point of view. As the name implies, we are concerned with effects of the action of many molecules, and these effects can be perceived by our senses. As an example, consider the pressure a gas exerts on the walls of its container. This pressure results from the change in momentum of the molecules as they strike the wall. We know that each molecule does not have the same change in momentum during the collision with the wall. From a macroscopic point of view, we are not concerned with the action of the individual molecules, but we measure the force on a given area by using, for example, a pressure gage. In fact, these macroscopic observations are completely independent of our assumptions regarding the nature of matter.

All the theory and development in this book will be presented from a macroscopic point of view, though occasionally a few remarks will be made regarding the significance of the microscopic viewpoint. This does not imply that the macroscopic and microscopic viewpoints are not related. The same conclusions must necessarily be reached when a problem is approached from the two viewpoints. It is the technique involved that differs.

A few remarks should be made regarding the continuum. From the macroscropic point of view, we are always concerned with volumes that are very large compared to molecular dimensions, and, therefore, with systems that contain many molecules. Since we are not concerned with the behavior of individual molecules, we can treat the substance as being continuous, disregarding the action of individual molecules, and this we call a continuum. This is, of course, only a convenient assumption, which loses validity when the mean free path of the molecules approaches the order of magnitude of the dimensions of the vessel, as, for example, in high-vacuum technology. In much engineering work the assumption of a continuum is valid and convenient, and goes hand in hand with the macroscopic point of view.

2.3 Properties and State of a Substance

If we consider a given mass of water, we recognize that this water can exist in various forms. If initially it is a liquid, it may on heating become a vapor, or on cooling it may become a solid. And in each phase the water may exist at various pressures and temperatures, or to use the thermodynamic term, in various states. The state may be identified or described by certain observable, macroscopic properties, some familiar ones being temperature, pressure, and density. In later chapters other properties will be introduced. Each of the properties of a substance in a given state has only one definite value, and these properties always have the same value for a given state, regardless of how the substance arrived at that state. In fact, a *property* can be defined as any quantity that depends on the state of the system and is independent of the path by which the system arrived in the given state. Conversely, the state is specified or described by the properties, and later we shall consider the matter of how many independent properties a substance can have, i.e., what is the minimum number of properties that must be specified in order to fix the state of the substance.

Thermodynamic properties can be divided into two general classes, namely, intensive and extensive properties. If a quantity of matter in a given state is divided into two equal parts, each part will have the same value of intensive properties as the original, and half the value of the extensive properties. Pressure, temperature, and density are examples of intensive properties. Mass and total volume are examples of extensive properties. Extensive properties per unit mass, such as specific volume, are intensive properties.

We frequently will refer not only to the properties of a substance but to the properties of a system. When we do so we necessarily imply that the value of the properties has significance for the entire system, and this implies what is called equilibrium. For example, if the gas that comprises the system in Fig. 2.1 is in thermal equilibrium, the temperature will be the same throughout the entire system, and we may speak of the temperature as a property of the system. We may also consider mechanical equilibrium, and this is related to pressure. If a system is in mechanical equilibrium, there is no tendency for the pressure at any point to change with time as long as the system is isolated from the surroundings. There will be a change in pressure with elevation, due to the influence of gravitational forces. However, in many thermodynamic problems this change is so small that it can be neglected. The matter of chemical equilibrium is also important and will be considered in Chapter 16.

2.4 Processes and Cycles

Whenever one or more of the properties of a system change we say that a change in state has occurred. For example, when one of the weights on the piston in Fig. 2.3 is removed, the piston rises and a change in state occurs, for the pressure decreases and the specific volume increases. Similarly, steam that expands through a steam turbine undergoes a change in state. The path of the succession of states through which the system passes is called the *process*.

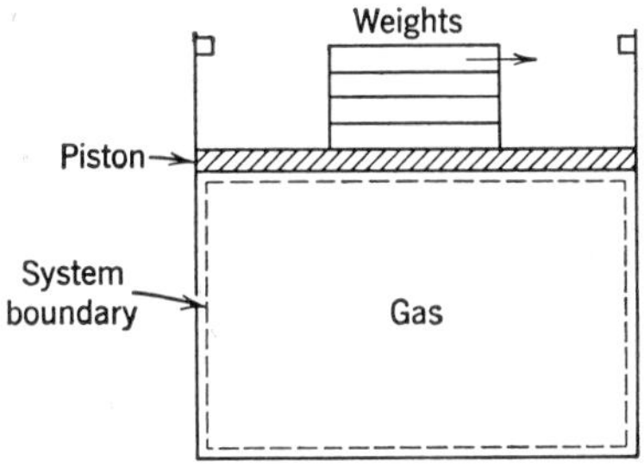

Fig. 2.3 Example of a system that may undergo a quasistatic process.

Let us consider the matter of equilibrium of a system as it undergoes a change in state. The moment the weight is removed from the piston in Fig. 2.3, mechanical equilibrium does not exist and as a result the piston is moved upward until mechanical equilibrium is again restored. The question that arises is this: Since the properties describe the state of a system only when it is in equilibrium, how can we describe the states of a system during a process if the actual process occurs only when equilibrium does not exist. One step in the answer to this question concerns definition of an ideal process, which we call a quasistatic process. A *quasistatic process* is one in which the deviation from thermodynamic equilibrium is infinitesimal, and all the states the system passes through during a quasistatic process may be considered as equilibrium states. Many actual processes closely approach a quasistatic process, and may be so treated with essentially no error. If the weights on the piston in Fig. 2.3 are small and are taken off one by one, the process could be considered quasistatic. On the other hand, if all the weights were removed at once, the piston would rise rapidly until it hit the stops. This would be a nonequilibrium process, and the system would not be in equilibrium at any time during this change of state.

For nonequilibrium processes, we are limited to a description of the system before the process occurred and after the process is completed and equilibrium is restored. We are not able to specify each state through which the system passed, but, as we shall see later, we are able to describe certain over-all effects which occur during the process.

Several processes are described by the fact that one property remains constant. The prefix iso- is used to describe this. An isothermal process is a constant-temperature process, an isobaric (sometimes called isopiestic) process is a constant-pressure process, and an isovolumic process is a constant-volume process. Sometimes the prefix iso- is also used for a process in which the initial and final values of the property are the same. This, however, is not the preferred use.

When a system in a given state goes through a number of different processes and finally returns to its initial state, the system has undergone a *cycle*. Therefore, at the conclusion of a cycle all the properties have the same value they had at the beginning. Steam (water) that circulates through a steam power plant undergoes a cycle.

A distinction should be made between a thermodynamic cycle, which has just been described, and a mechanical cycle. A four-stroke cycle internal-combustion engine goes through a mechanical cycle once every two revolutions. However, the working fluid does not go through a thermodynamic cycle in the engine, since air and fuel are burned and changed to products of combustion which are exhausted to the atmosphere. The term cycle will refer to a thermodynamic cycle unless otherwise designated.

2.5 Units of Force and Mass

In the remaining part of the chapter we will consider a number of properties, beginning in this section with mass and the related concept, force.

It is essential to understand clearly the units of force and mass. Force, mass, length, and time are related by Newton's second law of motion, which states that force on a body of fixed mass is proportional to the product of the mass and acceleration.

$$F \sim ma$$

If we wish to write this as an equality, a constant must be introduced which will have a magnitude and dimensions which depend on the units chosen for force, mass, length, and time. Let this constant be $1/g_c$, so that we write

$$F = \frac{ma}{g_c} \tag{2.1}$$

Let us illustrate the use of this equation by applying it to the units used in this text. Consider a certain platinum cylinder which we say by definition has a mass of 1 lbm. The symbol to be used for pound mass is lbm. (The standard pound mass is actually defined in terms of the kilogram.) Let this one-pound mass be held in the earth's gravitational field at such a location that the acceleration due to gravity is 32.1740 ft/sec^2, which is standard gravitational acceleration. The force exerted on the pound mass at this location due to the gravitational field is defined as a pound force (lbf). With this system of units, let us determine the constant g_c.

$$F = \frac{ma}{g_c}$$

$$1 \text{ lbf} = \frac{1 \text{ lbm} \times 32.174 \text{ ft/sec}^2}{g_c}$$

Therefore,

$$g_c = 32.174 \text{ lbm-ft/lbf-sec}^2$$

It should be emphasized that g_c is a constant and depends only on the units involved, and not on the acceleration due to gravity g at a particular location. To illustrate this, let us calculate the force due to gravity on a pound mass at a location where the acceleration due to gravity is 30.0 ft/sec^2.

$$F = \frac{ma}{g_c}$$

$$F = \frac{1 \text{ lbm} \times 30.0 \text{ ft/sec}^2}{32.174 \text{ lbm-ft/lbf-sec}^2} = 0.933 \text{ lbf}$$

Notice that the answer is dimensionally correct when the units of g_c as well as the magnitude are used, and this procedure is recommended.

The weight of a body is really a measure of the force with which the body is attracted to the earth. In the example above, at the point where g is 30.0 ft/sec^2, the body has a weight of 0.933 lbf, but it still has a mass of 1 lbm. Unless high altitudes or elevations are involved, the variation of g with altitude is not significant for most engineering calculations, and the value of g may be taken as the standard value, $g = 32.17$ ft/sec^2. In this case the weight of a body in pounds force is numerically equal to the mass of the body in pounds mass. Nevertheless, the distinction between the two should always be kept in mind, and the correct units always used. To repeat, the weight of a body is a measure of the force with which it is attracted to the earth and depends on the acceleration due to gravity at the point under consideration, whereas mass is a property of the body and is independent of elevation.

A unit of mass frequently encountered in engineering is the slug. The slug is defined as the mass that will be accelerated at the rate of 1 ft/sec^2 when acted upon by a force of 1 lbf. Let us calculate g_c for this system of units, using Eq. 2.1.

$$1 \text{ lbf} = \frac{1 \text{ slug} \times 1 \text{ ft/sec}^2}{g_c}$$

$$g_c = 1 \text{ slug-ft/lbf-sec}^2$$

In this case the magnitude of g_c is unity, but it has the dimensions indicated. This system is called the British Engineering System.

One other system which should be mentioned, since it is occasionally encountered in the literature, is the English Absolute System. In this system the unit of force is the poundal, which is defined as the force required to accelerate a mass of 1 lbm at the rate of 1 ft/sec^2. Then

$$F = \frac{ma}{g_c}$$

$$1 \text{ poundal} = \frac{1 \text{ lbm} \times 1 \text{ ft/sec}^2}{g_c}$$

$$g_c = 1 \text{ lbm-ft/poundal-sec}^2$$

Again g_c has the magnitude of unity and the dimension indicated above.

An illustration using the metric system might also aid in tying in this concept with systems of units encountered in other courses. The unit of force in the metric system is defined in terms of the unit of mass (the gram), the unit of length (the centimeter), and the unit of time (the second). A dyne is the force required to accelerate a mass of 1 gm at the rate of 1 cm/sec^2.

$$1 \text{ dyne} = \frac{1 \text{ gm} \times 1 \text{ cm/sec}^2}{g_c}$$

$$g_c = 1 \text{ gm-cm/dyne-sec}^2$$

Thus we see that the only system in which g_c has a magnitude other than unity is the system used in this text, although in every system g_c has dimensions. One other fact to note is that the word *pound* by itself should not be used, but rather *pound mass* or *pound force,* whichever is appropriate.

The foregoing discussion is summarized in the following table:

Mass	Length	Time	Force	g_c
lbm	ft	sec	lbf	32.174 lbm-ft/lbf-sec^2
slug	ft	sec	lbf	1 slug-ft/lbf-sec^2
lbm	ft	sec	poundal	1 lbm-ft/poundal-sec^2
gm	cm	sec	dyne	1 gm-cm/dyne-sec^2

The following relations between various units may also be written:

$$1 \text{ lbf} = 32.174 \text{ lbm-ft/sec}^2 = 1 \text{ slug-ft/sec}^2 = 32.174 \text{ poundals}$$
$$= 4.448 \times 10^5 \text{ dynes}$$

$$1 \text{ lbm} = \frac{1}{32.174} \text{ lbf-sec}^2\text{/ft} = \frac{1}{32.174} \text{ slugs} = 453.6 \text{ gm}$$

Frequently it is desirable to use the mole as a quantity of mass. This is the quantity of a substance whose mass is numerically equal to the molecular weight of the substance. The symbol lb mole will designate the mass of a pound mole.

2.6 Specific Volume and Density

The specific volume of a substance is defined as the volume per unit mass, and is given the symbol v. The density of a substance is defined as the mass per unit volume, and is therefore the reciprocal of the specific volume. Density is assigned the symbol ρ. Specific volume and density are intensive properties.

We recognize that the specific volume of a system may vary from point to point. Considering the atmosphere as the system, we recognize that the specific volume increases as the elevation increases. Therefore we must be rather precise in our definition of specific volume.

Consider a small volume δV of a substance in a given state, and let the mass be designated δm. The *specific volume* is defined by the relation

$$v = \lim_{\delta V \to \delta V'} \frac{\delta V}{\delta m}$$

where $\delta V'$ is the smallest volume for which the fluid can be considered a continuum.

Thus in a given system we should speak of the specific volume or density at a point in the system, and recognize that this varies with elevation. However, most of the systems that we consider are relatively small, and the change in specific volume with elevation is not significant, and we can speak of one value of specific volume or density for the entire system.

In this text the specific volume and density will usually be given either on a pound-mass or a pound-mole basis. A bar over the symbol will be used to designate the property on a mole basis. Thus, $\bar{v}$ will designate the molal specific volume and $\bar{\rho}$ will designate the molal density. The most common units used in this text for specific volume are ft^3/lbm and $ft^3/lb\text{-}mole$; for density the corresponding units are lbm/ft^3 and $lb\text{-}mole/ft^3$.

2.7 Pressure

When dealing with liquids and gases we ordinarily speak of pressure; in solids we speak of stresses. The pressure in a fluid at rest at a given point is the same in all directions, and we define pressure as the normal component of force per unit area. More specifically, if δA is a small area, and $\delta A'$ is the smallest area over which we can consider the fluid a continuum, and δF_n is the component of force normal to δA, we define *pressure*, P, as

$$P = \lim_{\delta A \to \delta A'} \frac{\delta F_n}{\delta A}$$

The pressure P is the same in all directions of fluids at rest, but in a viscous fluid in motion there is a variation in the normal force with orientation. However, for high Reynolds numbers, the shearing stresses are small as compared to normal stresses, and this variation of pressure with orientation is small, and we may speak with sufficient accuracy of pressure at a point.

In general, the unit for pressure consistent with the other units used in this text is pounds force per square foot (lbf/ft^2). On the other hand, in common parlance and general experimental work, pressures are often measured in pounds force per square inch ($lbf/in.^2$). Therefore, the student should be careful in numerical calculations to introduce the conversion $144 \text{ in.}^2 = 1 \text{ ft}^2$, as necessary.

In most thermodynamic investigations we are concerned with absolute pressure. Most pressure and vacuum gages, however, read the

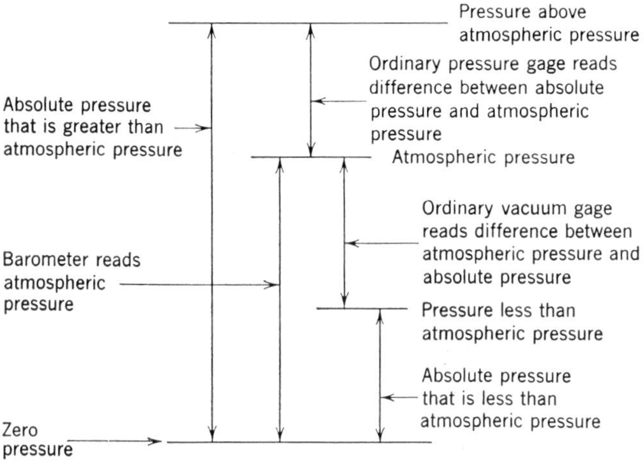

Fig. 2.4 Illustration of terms used in pressure measurement.

difference between the absolute pressure and the atmospheric pressure existing at the gage, and this is referred to as *gage pressure*. This is shown graphically in Fig. 2.4, and the examples given below illustrate the principles involved. Pressures below atmospheric and slightly above atmospheric, and pressure differences (for example, across an orifice in a pipe) are frequently measured with a manometer which contains water, mercury, alcohol, oil, or other fluids. From the principles of hydrostatics the student will recall that for a difference in level of L ft, the pressure difference in pounds per square foot is calculated by the relation

$$\Delta P = \rho \frac{Lg}{g_c}$$

where ρ = density of the fluid in lbm/ft^3. Figure 2.5 illustrates such a manometer.

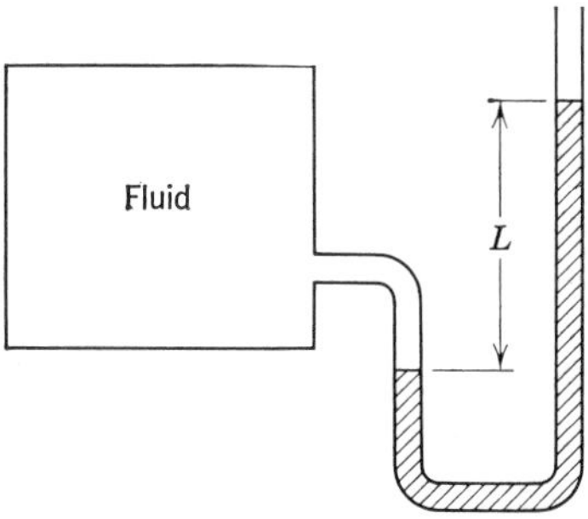

Fig. 2.5 Example of pressure measurement using a column of fluid.

From this relation the student should verify the fact that at ordinary room temperatures

$$1 \text{ in. Hg} = 0.490 \text{ lbf/in.}^2$$

Standard atmospheric pressure is defined as the pressure produced by a column of mercury exactly 760 mm in length, the mercury density being 13.5951 gm/cm^3 and the acceleration due to gravity being standard. Therefore,

$$1 \text{ std atm} = 14.6959 \text{ lbf/in.}^2$$
$$= 1.01324 \times 10^6 \text{ dynes/cm}^2$$

Another common unit of pressure is the atmosphere, which implies a standard atmosphere. Thus, a pressure of 1000 atmospheres is a pressure of 14,696 lbf/in.^{2}

Extremely low pressures (high vacuums) are often measured in microns of mercury (usually only the word micron is used). A micron is one millionth of a meter or 10^{-3} mm. Thus,

$$1 \text{ micron} = 1 \times 10^{-3} \text{ mm}$$

$$1 \text{ micron Hg} = 1 \times 10^{-3} \text{ mm Hg} = 1.933 \times 10^{-5} \text{ lbf/in.}^2$$

In order to distinguish between absolute and gage pressures in this text, the term lbf/in.^{2} or lbf/ft^2 will refer to absolute pressure. The gage pressure will be indicated by lbf/in.^{2} gage or lbf/ft^2 gage. It should also be noted that the symbols psia and psig are much used in technical literature. However, in this text we are emphasizing the matter of units and therefore, except for tables of thermodynamic properties, the symbol lbf/in.^{2} will be used.

2.8 Equality of Temperature

Although temperature is a property with which we are all familiar, an exact definition is difficult. We are aware of temperature first of all as a sense of hotness or coldness when we touch an object. We also learn early in our experience that when a hot body and a cold body are brought into contact, the hot body becomes cooler and the cold body becomes warmer. If these bodies remain in contact for some time, they usually appear to have the same temperature. However, we also realize that our sense of temperature is very unreliable. Sometimes very cold bodies may seem hot, and bodies of different materials which are at the same temperature appear to be at different temperatures.

Because of these difficulties in defining temperature, we define equality of temperature. Consider two blocks of copper, one hot and the other cold, each of which has a thermometer well containing a mercury thermometer. Let these two blocks of copper be brought into thermal communication. We would then observe that the electrical resistance of the hot block decreases with time and for the cold block it increases with time. After a period of time has elapsed, however, no further changes in resistance would be observed. Similarly, when the blocks are first brought in thermal communication, the length of a side of the hot block decreases with time, whereas in the cold block it increases with time. After a period of time, no further change in length of either of the blocks would be perceived. Also, the mercury column of the thermometer in the hot block would drop at first and in the cold block it would rise, but after a period of time no further change in height would be observed. We may say, therefore, that two bodies have *equality of temperature* when no change in any observable property occurs when they are in thermal communication.

2.9 The Zeroth Law of Thermodynamics

Consider now the same two blocks of copper, and also another thermometer. Let one block of copper be brought into contact with the thermometer until equality of temperature is established, and then removed. Then let the second block of copper be brought into contact with the thermometer, and suppose that no change in the mercury level of the thermometer occurs during this operation with the second block. Then we can say that both blocks are in thermal equilibrium with the given thermometer.

The *zeroth law of thermodynamics* states that when two bodies have equality of temperature with a third body, they in turn have equality of temperature with each other. This seems very obvious to us because we are so familiar with this experiment. However, since this fact is not derivable from other laws, and since it logically precedes the first and second laws of thermodynamics, it has been called the zeroth law of thermodynamics. This law is really the basis of temperature measurement, for numbers can be placed on the mercury thermometer, and every time a body has equality of temperature with the thermometer, we can say that the body has the temperature we read on the thermometer. The problem remains, however, of relating the temperatures that we might read on different mercury thermometers, or that we obtain when using different temperature-measuring devices, such as thermocouples and resistance thermometers. This suggests the need for a standard scale for temperature measurements.

2.10 Temperature Scales

There are two common scales for measuring temperature, the Fahrenheit and the Centigrade (also called the Celsius scale). Until 1954 each of these scales was based on two fixed, easily duplicated points, the ice point and the steam point. The temperature of the ice point is defined as the temperature of a mixture of ice and water which is in equilibrium with saturated air at a pressure of 1 atm. The temperature of the steam point is the temperature of water and steam which are in equilibrium at a pressure of 1 atm. On the Fahrenheit scale these two points are assigned the numbers 32 and 212, respectively, and on the Centigrade scale the respective points are numbered 0 and 100. The basis for numbers on the Fahrenheit scale has an interesting background. In searching for an easily reproducible point, Fahrenheit selected the temperature of the human body and assigned it the number 96. He assigned the number 0 to the temperature of a certain mixture of salt, ice, and salt solution. On this scale the ice point was approximately 32. When this scale was slightly revised and fixed in terms of the ice point and steam point, the normal temperature of the human body was found to be 98.6 F.

In this text the letters F and C will denote the Fahrenheit and Centigrade scale, respectively. The usual symbol (°) for degree will not be used, but will rather be implied with the symbol F or C. The symbol t will refer to temperature on either the Fahrenheit or Centigrade scale.

At the Tenth Conference on Weights and Measures in 1954, the Centigrade (or Celsius) scale was redefined in terms of a single fixed point and the ideal-gas temperature scale. The single fixed point is the triple point of water. (The triple point is discussed in Chapter 3 and is defined as the state in which solid, liquid, and vapor of a pure substance can exist in equilibrium.) The magnitude of the degree is defined in terms of the ideal-gas temperature scale, which is discussed in Chapter 8. Thus, the essential features of this new scale are a single fixed point and a definition of the magnitude of the degree. The triple point of water is assigned the value 0.01 C. On this scale the steam point is experimentally found to be 100.00 C. Thus, there is essential agreement between the old and new temperature scales.

It should be noted that we have not yet considered an absolute scale of temperature. The possibility of such a scale rises from the second law of thermodynamics and is discussed in Chapter 6. On the basis of the second law of thermodynamics a temperature scale independent of any thermometric substance can be defined; this absolute scale is usually referred to as the *thermodynamic scale of temperature.*

However, it is very complicated to use this scale directly, and therefore a more practical scale, the *International scale of temperature*, has been adopted, which closely represents the thermodynamic scale.

2.11 The International Temperature Scale

In 1948 the Ninth General Conference of Weights and Measures adopted the International temperature scale, which is described below. This scale is very similar to an earlier one adopted in 1927. It is based on a number of fixed and easily reproducible points which are assigned definite numerical values of temperature, and on specified formulas which relate temperature to the readings on certain temperature-measuring instruments. This scale was so defined that it conforms closely to the thermodynamic temperature scale. The entire scale is given here for the sake of completeness, even though the student will have limited need for it at this point.

The fixed points are as follows, the pressure in each case being exactly 1 atm:

	°C
1. Temperature of equilibrium between liquid and vapor oxygen (oxygen point)	−182.970
2. Temperature of equilibrium between ice and air-saturated water (ice point)	0.000
3. Temperature of equilibrium between liquid water and its vapor (steam point)	100.000
4. Temperature of equilibrium between liquid sulfur and its vapor (sulfur point)	444.600
5. Temperature of equilibrium between solid silver and liquid silver (silver point)	960.800
6. Temperature of equilibrium between solid gold and liquid gold (gold point)	1063.000

The means available for interpolation lead to a division of the scale into four parts:

1. From the ice point to the freezing point of antimony the temperature is determined from measurements on a platinum resistance thermometer using the formula

$$R_t = R_0(1 + At + Bt^2)$$

R_0 is the resistance at the ice point, and the constants A and B are determined by calibration at the steam and sulfur points. The purity and physical condition of the sulfur must be such that R_t/R_0 at 100 C is greater than 1.3910.

2. From the oxygen point to the ice point the temperature is determined from measurements on a platinum resistance thermometer by means of the formula

$$R_t = R_0[1 + At + Bt^2 + C(t - 100)t^3]$$

The constants are determined in the same manner as in 1, and the constant C is determined by calibration at the oxygen point.

3. From the freezing point of antimony to the gold point the temperature is determined from measurements on a platinum vs. platinum-rhodium thermocouple using the relation

$$E = a + bt + ct^2$$

The constants a, b, and c are determined by calibration at the freezing point of antimony, and at the silver and gold points.

4. Above the gold point the temperature is determined by measuring the intensity of radiation, J, and comparing this to the intensity of radiation of the same wave length at the gold point. The formula to determine the temperature is

$$\frac{J_t}{J_{\text{Au}}} = \frac{\exp\,[c_2/\lambda(t_{\text{Au}} + T_0)] - 1}{\exp\,[c_2/\lambda(t + T_0)] - 1}$$

The radiation must be in the visible spectrum, and must be emitted by a black body. The constant c_2 equals 1.438 cm °K, and T_0 is the temperature of the ice point in degrees Kelvin (273.16). λ is the wave length in centimeters.

PROBLEMS

2.1 A 1-kg mass is accelerated with a force of 10 lbf. Calculate the acceleration in ft/sec^2 and cm/sec^2.

2.2 With what force is a mass of 10 slugs attracted to the earth at a point where the gravitational acceleration is 30.6 ft/sec^2? What is its weight in lbf? What is its mass in lbm?

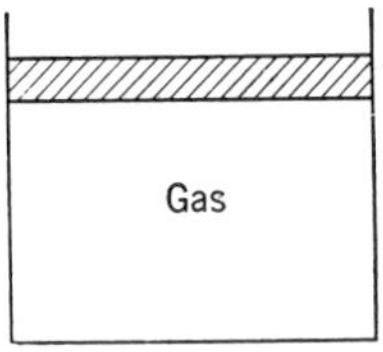

Fig. 2.6 Sketch for Problem 2.3.

2.3 A piston has an area of 1 ft^2. What mass must the piston have if it exerts a pressure of 10 lbf/in.^{2} above atmospheric pressure on the gas enclosed in the cylinder? Assume standard gravitational acceleration.

2.4 A pound mass is "weighed" with a beam balance at a point where $g = 31.0$ ft/sec^2. What reading would be expected? If it is weighed with a spring scale that reads correctly for standard gravity, what reading would be obtained?

2.5 Verify the fact that 1 lbf $= 4.448 \times 10^5$ dynes.

2.6 A pressure gage reads 30.7 lbf/in.^{2}, and the barometer reads 29.7 in. Hg. Calculate the absolute pressure in lbf/in.^{2} and std atm.

2.7 A manometer contains a fluid with a density of 56.0 lbm/ft^3. The difference in level of the two columns is 15 in. What pressure difference is indicated in lbf/in.^{2}?

2.8 A mercury manometer which is used to measure a vacuum reads 29.3 in., and the barometer reads 29.7 in. Hg. Determine the pressure in lbf/in.^{2} and in microns.

2.9 A body of fixed mass is "weighed" at an elevation of 20,000 ft ($g = 32.11$ ft/sec^2) by a spring balance which was calibrated at sea level. The reading on the spring balance is 9.3 pounds. What is the mass of the body?

2.10 In an experimental bomber flying at 40,000 ft ($g = 32.05$ ft/sec^2), the air flow in a piece of apparatus is measured by using a

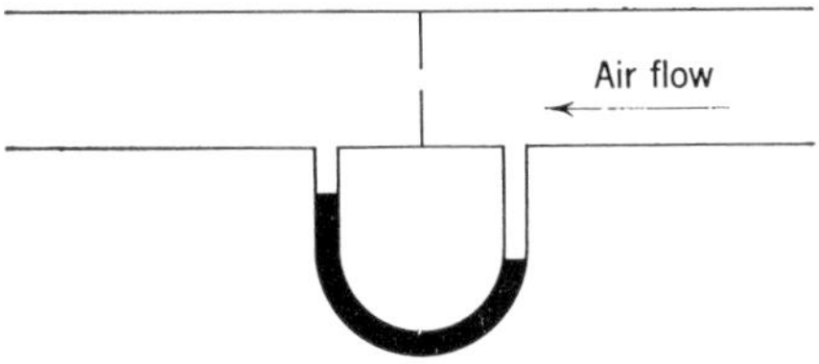

Fig. 2.7 Sketch for Problem 2.10.

mercury manometer. The difference in the level is 30 in. At sea level and the same temperature, mercury has a density of 13.60 gm/cm^3. Determine the pressure drop across the orifice in lbf/in.^{2}

CHAPTER 3

PROPERTIES OF A PURE SUBSTANCE

In the previous chapter we considered three familiar properties of a substance, namely, specific volume, pressure, and temperature. We now turn our attention to pure substances and consider some of the phases in which a pure substance may exist, the number of independent properties a pure substance may have, and methods of presenting thermodynamic properties.

3.1 The Pure Substance

A *pure substance* is one that has a homogeneous and invariable chemical composition even though there is a change of phase. Thus, liquid water, a mixture of liquid water and water vapor (steam), or a mixture of ice and water are all pure substances, for every phase has the same chemical composition. On the other hand, a mixture of liquid air and gaseous air is not a pure substance, since the composition of the liquid phase is different from that of the vapor phase.

Sometimes a mixture of gases such as air is considered a pure substance as long as there is no change of phase. Strictly speaking, this is not true, but rather, as we shall see later, we should say that a mixture of gases such as air exhibits some of the characteristics of a pure substance as long as there is no change of phase.

3.2 Phases of a Pure Substance

Consider as a system 1 lbm of water contained in the piston-cylinder arrangement of Fig. 3.1*a*. Suppose that the piston and weight maintain a pressure of 14.7 lbf/in.^{2} in the cylinder, and that the initial

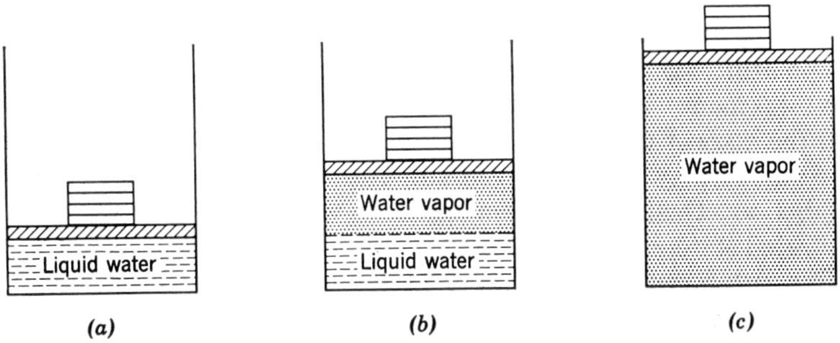

Fig. 3.1 Constant-pressure change from liquid to vapor phase for a pure substance.

temperature is 60 F. As heat is transferred to the water the temperature increases appreciably, the specific volume increases slightly, and the pressure remains constant. When the temperature reaches 212 F, additional heat transfer results in a change of phase (boiling), as indicated in Fig. 3.1*b*. That is, some of the liquid becomes vapor, and during this process both the temperature and pressure remain constant, but the specific volume increases considerably. When the last drop of liquid has vaporized, further transfer of heat results in an increase in both temperature and specific volume of the vapor, Fig. 3.1*c*.

The term *saturation temperature* designates the temperature at which vaporization takes place at a given pressure, this pressure being called the *saturation pressure* for the given temperature. Thus for water at 212 F the saturation pressure is 14.7 lbf/in.^{2}, and for water at 14.7 lbf/in.^{2} the saturation temperature is 212 F. For a pure substance there is a definite relation between saturation pressure and saturation temperature, a typical curve being shown in Fig. 3.2. This is called the vapor-pressure curve.

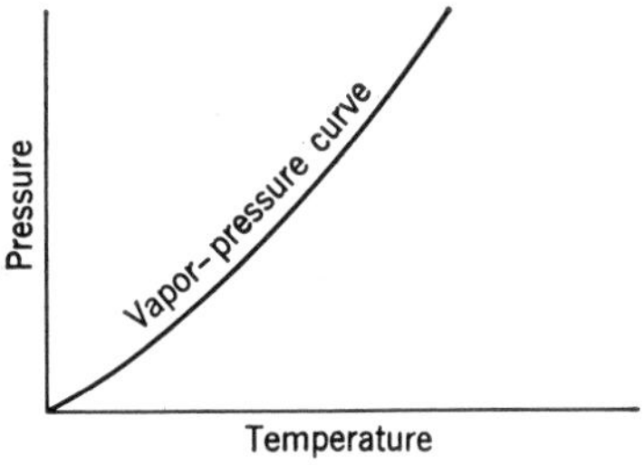

Fig. 3.2 Vapor-pressure curve of a pure substance.

If a substance exists as liquid at the saturation temperature and pressure, it is called *saturated liquid*. If the temperature of the liquid is lower than the saturation temperature for the existing pressure, it is called either a *subcooled liquid* (implying that the temperature is lower than the saturation temperature for the given pressure) or a *compressed liquid* (implying that the pressure is greater than the saturation pressure for the given temperature). Either term may be used, but the latter term will be used in this text.

When a substance exists as part liquid and part vapor at saturation temperature, its *quality* is defined as the ratio of the mass of vapor to the total mass. Thus, in Fig. 3.1*b*, if the mass of the vapor is 0.2 lbm and the mass of the liquid is 0.8 lbm, the quality is 0.2 or 20 per cent. The quality may be considered as an intensive property, and it has the symbol x. Quality has meaning only when the substance is in a saturated state, i.e., at saturation pressure and temperature.

If a substance exists as vapor at the saturation temperature, it is called *saturated vapor*. (Sometimes the term dry saturated vapor is used to emphasize that the quality is 100 per cent.) When the vapor is at a temperature greater than the saturation temperature, it is said to exist as *superheated vapor*. The pressure and temperature of superheated vapor are independent properties, since the temperature may increase while the pressure remains constant. Actually, the substances we call gases are highly superheated vapors.

Consider again Fig. 3.1, and let us plot on the temperature-volume diagram of Fig. 3.3 the constant-pressure line that represents the states through which the water passes as it is heated from the initial state of 14.7 lbf/in.^{2} and 60 F. Let state A represent the initial state and B the saturated-liquid state (212 F). Therefore, line AB represents the process in which the liquid is heated from the initial temperature to the saturation temperature. Point C is the saturated vapor state, and line BC is the constant-temperature process in which the change of phase from liquid to vapor occurs. Line CD represents the process in which the steam is superheated at constant pressure. Temperature and volume both increase during this process.

Now let the process take place at a constant pressure of 100 lbf/in.^{2}, beginning from an initial temperature of 60 F. Point E represents the initial state, the specific volume being slightly less than at 14.7 lbf/in.^{2} and 60 F. Vaporization now begins at point F, where the temperature is 327.8 F. Point G is the saturated-vapor state, and line GH the constant-pressure process in which the steam is superheated.

In a similar manner, a constant pressure of 1000 lbf/in.^{2} is represented by line $IJKL$, the saturation temperature being 544.6 F.

At a pressure of 3206.2 lbf/in.^{2}, represented by line MNO, we find, however, that there is no constant-temperature vaporization process.

Rather, point N is a point of inflection, the slope being zero. This point is called the *critical point*, and at the critical point the saturated-liquid and saturated-vapor states are identical. The temperature, pressure, and specific volume at the critical point are called the *critical*

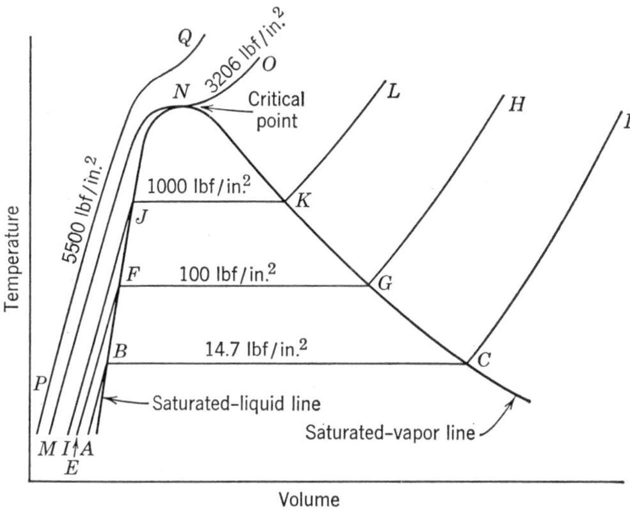

Fig. 3.3 Temperature-volume diagram for water showing liquid and vapor phases. (Not to scale.)

temperature, critical pressure, and *critical volume.* The critical-point data for some substances are given below, and there is a more extensive table in the Appendix.

	Critical Temperature, °F	Critical Pressure, lbf/in.^{2}	Critical Volume, ft^3/lbm
Water	705	3206.2	0.0503
Carbon Dioxide	88	1071	0.0348
Oxygen	−203	735	0.0364
Hydrogen	−400	188	0.534

A constant-pressure process greater than the critical pressure is represented by line PQ. There is no definite change in phase from liquid to vapor, and no definite point at which there is a change from the liquid phase to the vapor phase. For pressures greater than the critical we usually call the substance a liquid when the temperature is less than the critical temperature and a vapor when the temperature is greater than the critical temperature.

In Fig. 3.3 line *NJFB* represents the saturated-liquid line and line *NKGC* represents the saturated-vapor line.

Let us consider one further experiment with the piston-cylinder arrangement. Suppose the cylinder contained 1 lbm of ice at 0 F, 14.7 lbf/in.^{2} When heat is transferred to the ice, the pressure remains constant, the specific volume increases slightly, and the temperature increases until it reaches 32 F, at which point the ice melts while the temperature remains constant. In this state the ice is called *saturated solid*. For most substances the specific volume increases during this melting process, but for water the specific volume of the liquid is less than the specific volume of the solid. When all of the ice has melted, a further heat transfer causes an increase in temperature of the liquid.

If the initial pressure of the ice at 0 F is 0.0505 lbf/in.^{2}, heat transfer to the ice first results in an increase in temperature to 20 F. At this point, however, the ice would pass directly from the solid phase to the vapor phase in the process known as sublimation. Further heat transfer would result in superheating of the vapor.

Finally, consider an initial pressure of the ice of 0.08854 lbf/in.^{2} Again as a result of heat transfer the temperature will increase until it reaches 32 F. At this point, however, further heat transfer may result in some of the ice becoming vapor and some becoming liquid, for at this point it is possible to have the three phases in equilibrium. This is called the *triple point*, which is defined as the state in which three phases may all be present in equilibrium.

This whole matter is best summarized by the diagram of Fig. 3.4, which shows how the solid, liquid, and vapor phases may exist together in equilibrium. Along the sublimation line the solid and vapor phases are in equilibrium, along the fusion line the solid and liquid phases are in equilibrium, and along the vaporization line the liquid and vapor phases are in equilibrium. The only point at which all three phases may exist in equilibrium is the triple point. The vaporization line ends at the critical point because there is no distinct change from the liquid phase to the vapor phase above the critical point.

Thus, when the temperature increases at constant pressure, the pressure being below the triple-point pressure (such as represented by line *AB*), the substance passes directly from the solid to the vapor phase. Along the constant-pressure line *EF*, the substance first passes from the solid to the liquid phase at one temperature, and then from the liquid to the vapor phase at a higher temperature. Constant-pressure line *CD* passes through the triple point, and it is only at the triple point that the three phases may exist together in equilibrium. At a pressure above the critical pressure, such as *GH*, there is no sharp distinction between the liquid and vapor phases. As stated previously, it is considered in the liquid phase when the temperature is below, and in the vapor phase when the temperature is above the critical temperature.

It should be pointed out that there are several different phases in the solid state of most substances. These are associated with various crystalline structures in which the substance may exist. Thus, there

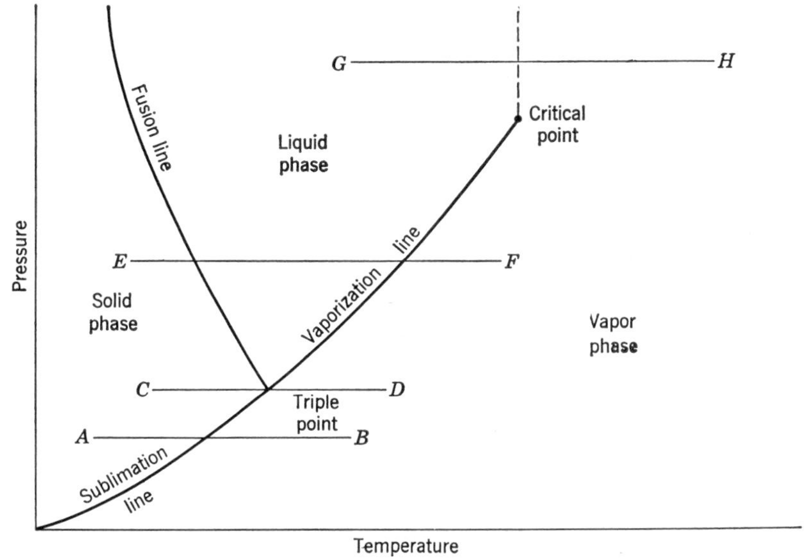

Fig. 3.4 Pressure-temperature diagram for a substance such as water.

might be two phases, both in the solid state, which may exist in equilibrium, and there may be other triple points, representing equilibrium between three phases in the solid state. However, this aspect of thermodynamics is beyond the scope of this text.

3.3 Independent Properties of a Pure Substance

One important reason for introducing the concept of the pure substance is the important fact that a pure substance, in the absence of motion, gravity, surface effects, electricity, and magnetism, has only two independent intensive properties. This conclusion has been reached on the basis of experimental observation. This means that if two independent intensive properties of a pure substance are specified, its state is determined, and there is one definite value for each of the other properties. Thus, if we specify the specific volume and temperature of superheated vapor, the state is fixed and there is one definite value for pressure (and for each of the other properties) in this state. To understand the significance of the term *independent property* consider the saturated-liquid and saturated-vapor states of a pure substance. These two states have the same pressure and same temperature, but are definitely not the same state. In a saturation state, therefore, pressure and temperature are not independent properties. Two independent properties such as pressure and specific volume, or pressure and quality, are required to specify a saturation state of a pure substance.

The reason for mentioning previously that a mixture of gases, such as air, has the same characteristics as a pure substance as long as only one phase is present, concerns precisely this point. The state of air, which is a mixture of gases, is determined by specifying two properties as long as it remains in the gaseous phase, and in this regard air can be treated as a pure substance.

3.4 Equations of State for Vapor Phase

We consider for the moment only the vapor phase, which includes everything we call gas as well as vapor. The relation between pressure, specific volume, and temperature of the vapor phase may be expressed by an equation which is called an *equation of state*. This is, of course, based on the fact that a pure substance has only two independent properties.

To write the equation of state we begin by writing the general relation

$$P = f(v, T) \qquad \text{or} \qquad v = f(P, T)$$

There are several different forms of the equation of state. The simplest is that for an ideal gas.

$$Pv = RT \tag{3.1}$$

The whole matter of ideal gases will be discussed in Chapter 8. However, we point out here that this equation of state for an ideal gas holds for a real gas only as the pressure approaches zero. It also holds with a fair degree of accuracy for highly superheated vapor (air at room temperature is a highly superheated vapor) at somewhat higher pressures.

In order to have an equation of state that is fairly accurate throughout the entire superheated vapor region, more complicated equations have been developed. One of the best known such equations is the Beattie-Bridgeman equation of state. This equation is

$$P = \frac{RT(1 - \epsilon)}{v^2}(v + B) - \frac{A}{v^2} \tag{3.2}$$

where $A = A_0(1 - a/v)$, $B = B_0(1 - b/v)$, $\epsilon = c/vT^3$, and A_0, a, B_0, b, and c are constants for different gases. The values of these constants for various substances are given in Table 3.1.

TABLE 3.1

Constants of the Beattie-Bridgeman Equation of State

Pressure in atmospheres; specific volume in liters per gram mole; temperature in degrees Kelvin:

$\bar{R} = 0.08206$ atm-liters/gm-mole K

Gas	A_0	a	B_0	b	$10^{-4}c$
Helium	0.0216	0.05984	0.01400	0.0	0.0040
Argon	1.2907	0.02328	0.03931	0.0	5.99
Hydrogen	0.1975	−0.00506	0.02096	−0.04359	0.0504
Nitrogen	1.3445	0.02617	0.05046	−0.00691	4.20
Oxygen	1.4911	0.02562	0.04624	0.004208	4.80
Air	1.3012	0.01931	0.04611	−0.001101	4.34
Carbon Dioxide	5.0065	0.07132	0.10476	0.07235	66.00

The matter of equations of state will be discussed further in Chapter 14. The observation to be made here in particular is that an equation of state that accurately describes the relation between pressure, temperature, and specific volume is rather cumbersome and the solution requires considerable time. Therefore, it is much more convenient to tabulate values of pressure, temperature, specific volume, and other thermodynamic properties for various substances. An example of such a table is *Thermodynamic Properties of Steam* by Keenan and Keyes, which we shall ordinarily refer to as the "steam tables." A summary of these tables is given in Table A.1 of the Appendix. The method of developing such a table is to find an equation of state that accurately fits the experimental data, and then to solve the equation of state for the values listed in the table.

3.5 Tables of Thermodynamic Properties

Tables of thermodynamic properties of many substances are available, and in general all these have the same form. Summary tables for a number of different substances are given in the Appendix. In this section we will refer to the steam tables, primarily in order to present the nature of thermodynamic tables. Once the steam tables are understood, all thermodynamic tables can readily be used.

Tables 1 and 2 give the properties of saturated liquid and saturated vapor. In Table 1 these properties are given as a function of saturation temperature, and Table 2 lists the properties as a function of saturation pressure. Since the information contained in these two tables is the same, it is simply a matter of convenience as to which table one uses.

Considering further Table 1, the first column after the temperature gives the corresponding saturation pressure in pounds force per square inch. The next columns give specific volume in cubic feet per pound mass. The first of these gives the specific volume of the saturated liquid, v_f; the second column gives the increase in specific volume when the state changes from saturated liquid to saturated vapor, v_{fg}; the third column gives the specific volume of saturated vapor, v_g. It follows that

$$v_f + v_{fg} = v_g$$

The specific volume of a substance having a given quality can be found by introducing the definition of quality. Quality has already been defined as the ratio of the mass of vapor to total mass of liquid plus vapor when a substance is in a saturation state. Let us consider a mass of 1 lbm having a quality x. The specific volume is the sum of the volume of the liquid and the volume of the vapor. The volume of the liquid is $(1 - x)v_f$. The volume of the vapor is xv_g. Therefore the specific volume v is

$$v = xv_g + (1 - x)v_f \tag{3.3}$$

Since $v_f + v_{fg} = v_g$, Eq. 3.3 can also be written in the following forms

$$v = v_f + xv_{fg} \tag{3.4}$$

$$v = v_g - (1 - x)v_{fg} \tag{3.5}$$

As an example, let us calculate the specific volume of saturated steam at 500 F having a quality of 70 per cent. Using Eq. 3.4

$$v = 0.0204 + 0.7(0.6545) = 0.4785 \text{ ft}^3/\text{lbm}$$

Using Eq. 3.5

$$v = 0.6749 - 0.3(0.6545) = 0.4785 \text{ ft}^3/\text{lbm}$$

When using a slide rule, Eq. 3.5 gives more accuracy for high quality and Eq. 3.4 gives more accuracy for low quality.

In Table 2, the first column after the pressure lists the saturation temperature for each pressure. The next columns list specific volume in a manner similar to Table 1, except that v_{fg} is not listed for low pressures. When necessary, v_{fg} can readily be found by subtracting v_f from v_g.

Table 3 gives the properties of superheated vapor. In the superheat region, pressure and temperature are independent properties, and therefore, for each pressure a large number of temperatures is given. And for each temperature three thermodynamic properties are listed, the first one being specific volume. Thus, the specific volume of steam at a pressure of 100 lbf/in.^{2} and 500 F is 5.589 ft^3/lbm.

Table 4 gives data that enables one to determine the specific volume of a compressed liquid. To demonstrate the use of this table, consider a piston and cylinder, Fig. 3.5, which contains 1 lbm of saturated

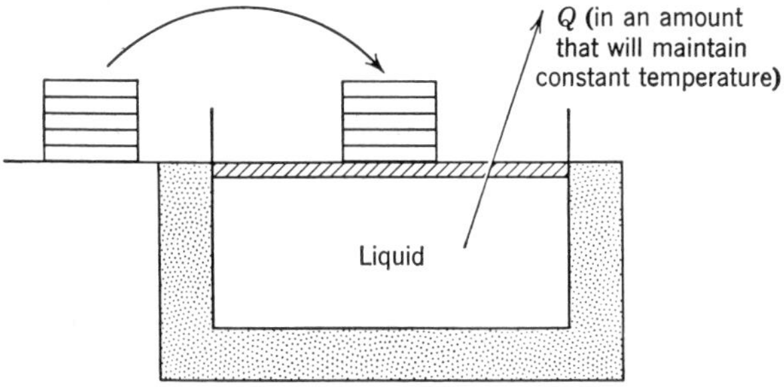

Fig. 3.5 Illustration of compressed liquid state.

liquid at 200 F. Its properties are given in Table 1, and we note that the pressure is 11.53 lbf/in.^{2} Suppose the pressure is increased to 1000 lbf/in.^{2} while the temperature is held constant at 200 F by the necessary transfer of heat, Q. Since water is slightly compressible, we would expect a slight decrease in specific volume during this process. The difference in specific volume between a saturated liquid at a given temperature and a compressed liquid at the same temperature is given in Table 4 of the steam tables. Thus, for the example cited above, at 1000 lbf/in.^{2} and 200 F.

$$(v - v_f)10^5 = -5.4 \text{ ft}^3/\text{lbm} \qquad \text{and} \qquad v_f = 0.016634 \text{ ft}^3/\text{lbm}$$

Therefore

$$v = 0.016634 - 0.000054 = 0.016580 \text{ ft}^3/\text{lbm}$$

One observes that the magnitude of the correction in this case is about 0.3 per cent. For some calculations the accuracy of the data is such that a correction of this order of magnitude is not justified. This is especially true for a liquid that is only slightly compressed, i.e., if the pressure is only 100 or 200 lbf/in.^{2} higher than the saturation pressure. In such a case one can use the specific volume of saturated liquid at the same temperature with only a very small error. Thus, the specific volume of liquid at 100 lbf/in.^{2}, 200 F would be found by using the specific volume of saturated liquid at 200 F. Whether or not to use the correction from Table 4 depends on the accuracy one is interested in. The safe rule is to check the magnitude of the correction and determine whether it should be used.

When interpolating in Table 4, one should interpolate only the correction from Table 4 and obtain from Table 1 the specific volume for saturated liquid at the given temperature. Thus, the specific volume of water at 1000 lbf/in.^{2} and 250 F is found as follows. From Table 4

$$(v - v_f)10^5 = -6.1 \text{ ft}^3/\text{lbm}$$

From Table 1 (at 250 F):

$$v_f = 0.01700 \text{ ft}^3/\text{lbm}$$

$$v = 0.01700 - 0.00006 = 0.01694 \text{ ft}^3/\text{lbm}$$

Table 5 of the steam tables gives the properties of saturated solid and saturated vapor that are in equilibrium. The first column gives the temperature, and the second column gives the corresponding saturation pressure. As would be expected, all these pressures are less than the triple-point pressure. The next two columns give the specific volume of the saturated solid and saturated vapor (note that the tabulated value is $v_g \times 10^{-3}$).

3.6 Thermodynamic Surfaces

The matter discussed in this chapter can be well summarized by a consideration of a pressure-specific-volume-temperature surface. Two such surfaces are shown in Figs. 3.6 and 3.7. Figure 3.6 shows a substance, such as water, in which the specific volume increases during freezing, and Fig. 3.7 shows a substance in which the specific volume decreases during freezing.

In these diagrams the pressure, specific volume, and temperature are plotted on mutually perpendicular coordinates, and all possible equilibrium states are thus represented by a point on the surface. This follows directly from the fact that a pure substance has only two independent intensive properties. All points along a quasistatic process lie on the P-v-T surface, since a quasistatic process always passes through equilibrium states.

The regions of the surface that represent a single phase, namely, the solid, liquid, and vapor phases, are indicated. These surfaces are

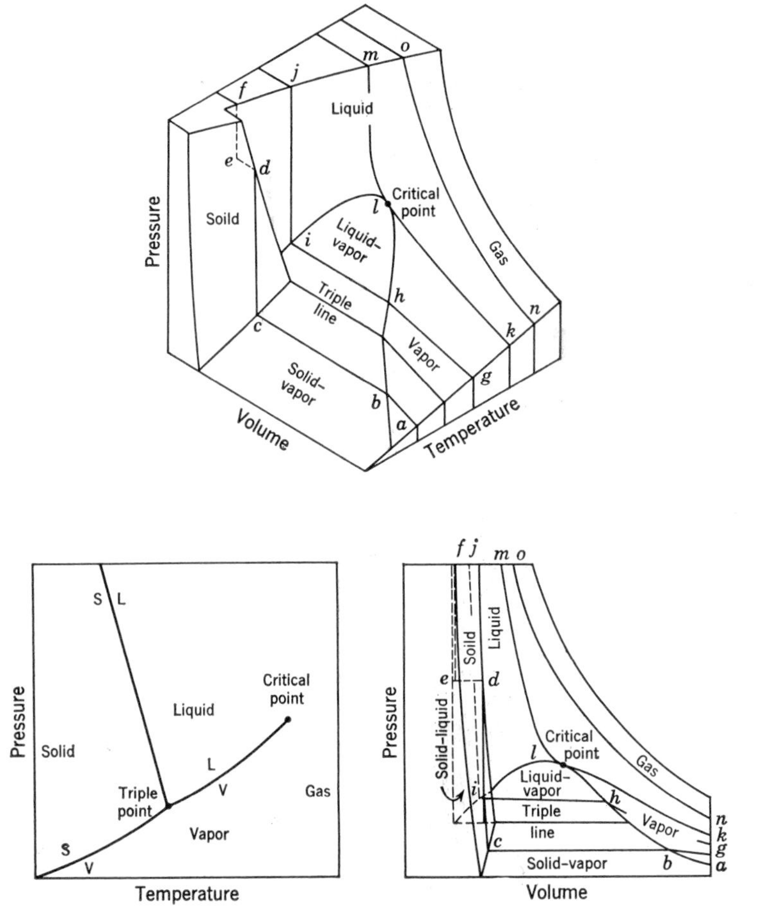

Fig. 3.6 Pressure-volume-temperature surface for a substance that expands on freezing.

curved. The two-phase regions, namely, the solid-liquid, solid-vapor, and liquid-vapor regions, are ruled surfaces. By this we understand that they are made up of straight lines parallel to the specific-volume axis. This, of course, follows from the fact that in the two-phase region, lines of constant pressure are also lines of constant temperature, though the specific volume may change. The triple point actually appears as the triple line on the P-v-T surface, since the pressure and temperature of the triple point are fixed, but the specific volume may vary, depending on the proportion of each phase.

It is also of interest to note the pressure-temperature and pressure-volume projections of these surfaces. We have already considered the

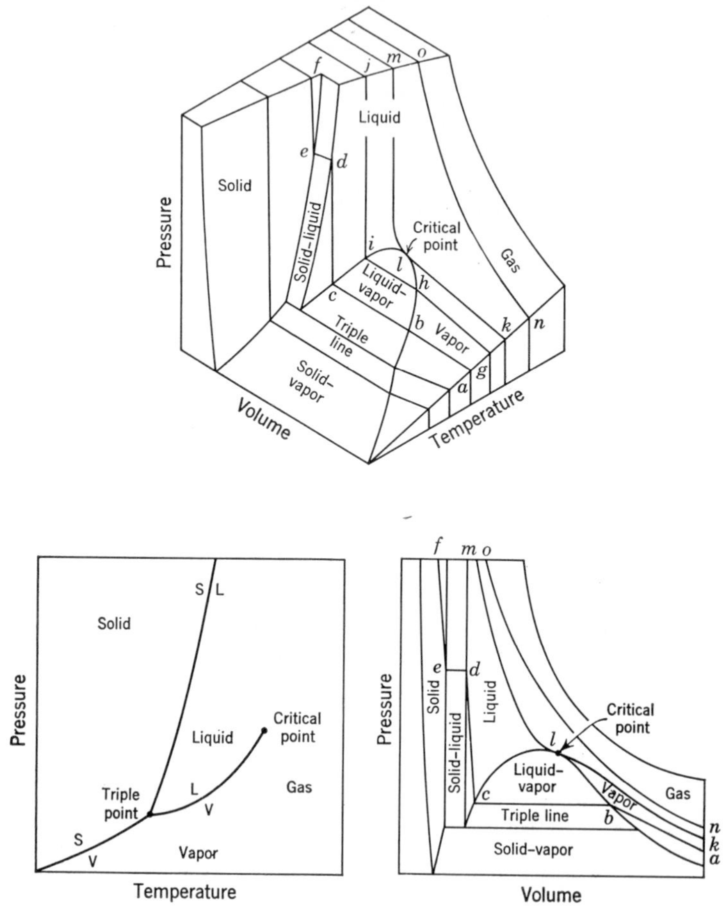

Fig. 3.7 Pressure-volume-temperature surface for a substance that contracts on freezing.

pressure-temperature diagram for a substance such as water. It is on this diagram that we observe the triple point. Various lines of constant temperature are shown on the pressure-volume diagram, and the corresponding constant-temperature sections are lettered identically on the P-v-T surface. The critical isotherm has a point of inflection at the critical point. Various lines of constant temperature are shown on both the P-v-T surface and the P-v diagram, and the corresponding lines are lettered alike.

One notices that with a substance such as water, which expands on freezing, the freezing temperature decreases with an increase in pressure. With a substance that contracts on freezing, the freezing temperature increases as the pressure increases. Thus, as the pressure of vapor is increased along the constant-temperature line *abcdef*, Fig. 3.6, a substance that expands on freezing first becomes solid and then liquid. For the substance that contracts on freezing, the corresponding constant-temperature line, Fig. 3.7, indicates that as the pressure on the vapor is increased, it first becomes liquid and then solid.

Example 3.1 A vessel having a volume of 10 ft^3 contains 3.0 lbm of a liquid water and water vapor mixture in equilibrium at a pressure of 100 $lbf/in.^2$ Calculate

(*a*) The volume and mass of liquid.

(*b*) The volume and mass of vapor.

The specific volume is calculated first:

$$v = \frac{10.0}{3.0} = 3.333$$

The quality can now be calculated, using Eq. 3.5:

$$3.333 = 4.432 - (1 - x)\ 4.414$$

$$(1 - x) = \frac{1.099}{4.414} = 0.249$$

$$x = 0.751$$

Therefore the mass of liquid is

$$3(0.249) = 0.747 \text{ lbm}$$

The mass of vapor is

$$3(0.751) = 2.253 \text{ lbm}$$

The volume of liquid is

$$V_{\text{liq}} = m_{\text{liq}} v_f = 0.747(0.01774) = 0.0133 \text{ ft}^3$$

The volume of the vapor is

$$V_{\text{vap}} = m_{\text{vap}} v_g = 2.253(4.432) = 9.99 \text{ ft}^3$$

Example 3.2 A pressure vessel contains saturated ammonia vapor at 60 F. Heat is transferred to the ammonia until the temperature reaches 200 F. What is the final pressure?

Since the volume does not change during this process, the specific volume also remains constant.

$$v_1 = v_2 = 2.751 \text{ ft}^3/\text{lbm}$$

We know one other property in the final state, namely, the temperature, $t_2 = 200$ F, and therefore the final state is determined.

From the superheat tables, by interpolation

$$P_2 = 142.4 \text{ lbf/in.}^2$$

PROBLEMS

3.1 Plot the following vapor-pressure curves (saturation pressure vs. saturation temperature):

(*a*) Water on Cartesian coordinates, −40 F to 60 F.

(*b*) Water on Cartesian coordinates, 0 to 3500 lbf/in.^{2}

(*c*) Water, Freon-12, and ammonia, on semilog paper (pressure on log scale), 1.0 to 1000 lbf/in.^{2}, −50 F to 300 F.

3.2 Calculate the following specific volumes:

(*a*) Ammonia, 100 F, 80% quality.

(*b*) Freon-12, 10 F, 15% quality.

(*c*) Water, 1500 lbf/in.^{2}, 98% quality.

3.3 Determine the quality (if saturated) or temperature (if superheated) of the following substances in the given states:

(*a*) Ammonia, 80 F, 1.23 ft^3/lbm; 80 lbf/in.^{2}, 4.25 ft^3/lbm.

(*b*) Freon-12, 50 lbf/in.^{2}, 0.6 ft^3/lbm; 50 lbf/in.^{2}, 0.900 ft^3/lbm.

(*c*) Water, 80 F, 2 ft^3/lbm; 1000 lbf/in.^{2}, 0.4 ft^3/lbm.

3.4 Plot a pressure-specific-volume diagram on log log paper (3 × 5 cycles) for water showing the following lines:

(*a*) Saturated liquid.

(*b*) Saturated vapor.

(*c*) The following constant-temperature lines (including the compressed-liquid region): 300 F, 500 F, 700 F, 800 F, 1000 F.

(*d*) The following lines of constant quality: 10%, 50%, 90%.

3.5 Plot a pressure-specific-volume diagram on log log paper (2 × 3 cycles) for Freon-12, showing the following lines:

(*a*) Saturated liquid.

(*b*) Saturated vapor.

(*c*) The following constant-temperature lines: 0 F, 100 F, 230 F, 300 F.

(*d*) The following constant-quality lines: 10%, 50%, 90%.

3.6 The radiator of a heating system has a volume of 2 ft^3 and contains saturated vapor at 20 $lbf/in.^2$ The valves are then closed on the radiator, and as a result of heat transfer to the room the pressure drops to 18 $lbf/in.^2$ Calculate:

(*a*) The total mass of steam in the radiator.
(*b*) The volume and mass of liquid in the final state.
(*c*) The volume and mass of vapor in the final state.

3.7 Steam at the critical state is contained in a rigid vessel. Heat is transferred from the steam until the pressure is 400 $lbf/in.^2$ Calculate the final quality.

3.8 The rigid vessel shown in Fig. 3.8 contains saturated water at 14.7 $lbf/in.^2$ Determine the proportions by volume of liquid and vapor at 14.7 $lbf/in.^2$ necessary to make the water pass through the critical state when heated.

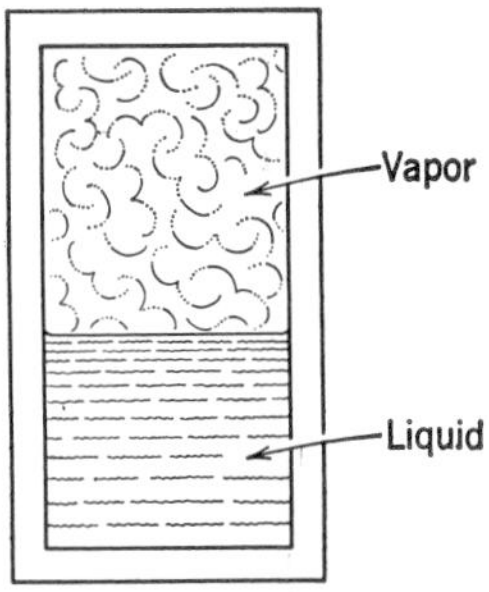

Fig. 3.8 Sketch for Problem 3.8.

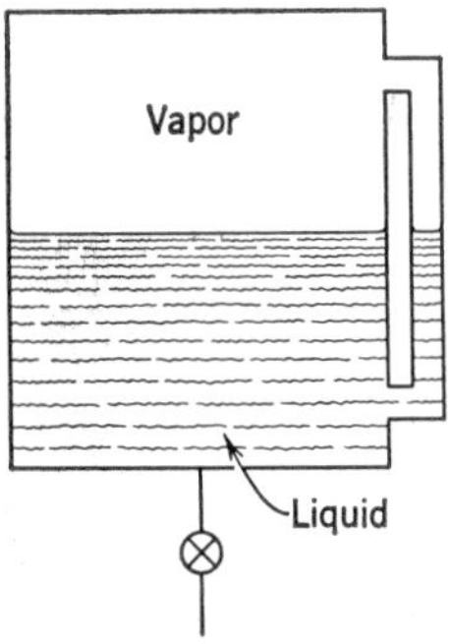

Fig. 3.9 Sketch for Problem 3.10.

3.9 Same as Problem 3.8, but let the contents initially be saturated Freon-12 at 60 F.

3.10 A vessel fitted with a sight glass contains Freon-12 at 80 F. Liquid is withdrawn from the bottom at a slow rate, and the temperature remains constant during the process. If the area of the vessel is 50 $in.^2$ and the level drops 6 in., determine the mass of Freon-12 withdrawn.

3.11 There is a tendency for students to automatically write down that there are 62.4 lbm of liquid water per ft^3. Using the steam tables determine the actual density of water in lbm/ft^3 at the following states:

(*a*) Saturated liquid at 60 F.
(*b*) Liquid at 60 F, 100 $lbf/in.^2$

(*c*) Saturated liquid at 100 lbf/in.^{2}
(*d*) Saturated liquid at 500 F.

3.12 A boiler feed pump delivers 500,000 lbm of water per hr at 1500 lbf/in.^{2}, 460 F. What is the volume rate of flow in ft^3/min? What would be the per cent error if the correction from Table 4 of the steam tables were neglected?

CHAPTER 4

WORK AND HEAT

In this chapter consideration is given to work and heat. It is essential for the student of thermodynamics to understand clearly the definitions of both work and heat, because the correct analysis of many thermodynamic problems depends upon distinguishing between them.

4.1 Definition of Work

Work is usually defined as a force F acting through a displacement dx, the displacement being in the direction of the force. That is,

$$W = \int_1^2 F \cdot dx \tag{4.1}$$

This is a very useful relationship and it enables us to find the work required to raise a weight, to stretch a wire, or move a charged particle through a magnetic field.

However, in view of the fact that we are treating thermodynamics from a macroscopic point of view, it is advantageous to tie in our definition of work with our concepts of systems, properties, and processes. We therefore define work as follows: *Work* is done by a system if the sole effect on things external to the system could be the raising of a weight. Notice that the raising of a weight is in effect a force acting through a distance. Notice, also, that our definition does not say that a weight was actually raised, or that a force actually acted through a given distance. Rather, the definition says that the sole effect external to the system could have been the raising of a weight. Work done by a system is considered positive and work done on a system is considered negative. The symbol W refers to the total work done by a system, and w to the work done per unit mass.

In general, we will speak of work as energy. No attempt will be made to give a rigorous definition of energy. Rather, since the concept is familiar, the term energy will be used as appropriate, and various forms of energy will be identified. Work is the form of energy that fulfills the definition given above.

Let us illustrate this definition of work with a few examples. Consider as a system the battery and motor of Fig. 4.1*a*, and let the motor

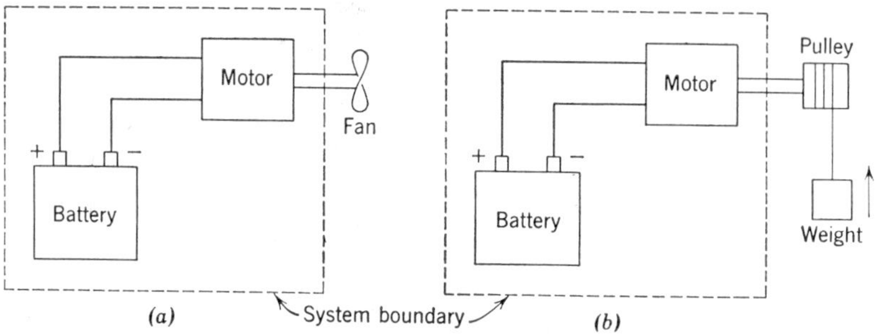

Fig. 4.1 Example of work done at the boundary of a system.

drive a fan. Does work cross the boundary of the system? To answer this question using the definition of work given above, let the fan be replaced with a pulley and weight arrangement shown in Fig. 4.1*b*. As the motor turns, the weight is raised, the sole effect external to the system being the raising of a weight. Thus, for our original system of Fig. 4.1*a*, we conclude that work is crossing the boundary of the system since the sole effect external to the system could be the raising of a weight.

Let the boundaries of the system be changed now to include only the battery shown in Fig. 4.2. Again we ask the question, does work cross the boundary of the system? In answering this question, we will be answering a more general question; namely, does the flow of electrical energy across the boundary of a system constitute work?

The only limiting factor in having the sole external effect the raising of a weight is the inefficiency of the motor. However, as we design a more-and-more efficient motor, with lower bearing and electrical losses, we recognize that we can approach a certain limit, which does meet the requirement of having the only external effect the raising of a weight. Therefore, we can conclude that when there is a flow of electricity across the boundary of a system, as in Fig. 4.2, it is work with which we are concerned.

A final point should be made regarding the work done by an open system: Matter crosses the boundary of the system, and in so doing, a certain amount of energy crosses the boundary of the system. Our definition of work does not include this energy. Thus, in the air compressor of Fig. 2.2, work crosses the boundary of the open system via

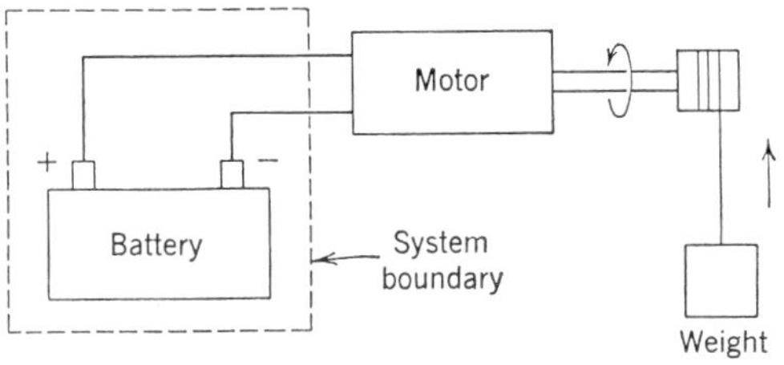

Fig. 4.2 Example of work crossing the boundary of a system because of a flow of an electric current across the system boundary.

the driving shaft on the compressor, but we do not consider any work to cross the boundary of the system with the air flowing into and out of the compressor. This matter will be further amplified in the following chapter.

4.2 Units for Work

We consider work done by a system, such as that done by a steam turbine, as positive, and work done on a system, such as that received by a compressor, as negative. Thus, positive work means that energy leaves the system and negative work means that energy is added to the system.

Our definition of work involves the raising of a weight. The unit of work should therefore be defined in terms of raising a unit weight a given distance at a given location. Let us define our unit of work as the work required to raise a mass of 1 lbm a distance of 1 ft at a location where the acceleration due to gravity is the standard value. This is exactly equivalent to saying that our unit of work is a force of 1 lbf acting through a distance of 1 ft. This unit for work is called the foot-pound force.

Similarly, in the metric system the unit of work is the erg. An erg is the work done by a force of 1 dyne acting through a distance of 1 cm.

This is a very small unit, and for engineering use, a joule is the common unit of work in the metric system. One joule is 10^7 ergs.

Another unit for work which has come into common use as a result of developments in nuclear physics is the electron volt, abbreviated ev. An electron volt is the work required to move an electron through a potential difference of 1 volt. The relation between the electron volt and other units for work is:

$$1 \text{ ev} = 1.608 \times 10^{-12} \text{ erg} = 1.18 \times 10^{-19} \text{ ft-lbf}$$

One million electron volts, abbreviated mev, is also commonly used.

$$1 \text{ mev} = 1.608 \times 10^{-6} \text{ erg} = 1.18 \times 10^{-13} \text{ ft-lbf}$$

Power is the time rate of doing work. That is,

$$\text{power} = \frac{\delta W}{d\tau} = \dot{W}$$

One familiar unit of power is the horsepower (hp).

$$1 \text{ hp} = 33{,}000 \text{ ft-lbf/min}$$

Another familiar unit is the kilowatt (kw). Actually the kilowatt is defined in terms of electrical units, but for our purposes we can define the kilowatt as follows:

$$1 \text{ kw} = 44{,}240 \text{ ft-lbf/min}$$

It follows that

$$1 \text{ hp} = 0.746 \text{ kw}$$

These definitions of power lead to two other units of work, the horsepower-hour (hp-hr) and kilowatt-hour (kw-hr). The horsepower-hour is the work done in 1 hr when the power is 1 hp. Similarly, one kilowatt-hour is the work done in 1 hr when the rate of work is 1 kw.

$$1 \text{ hp-hr} = 33{,}000 \times 60 = 1.98 \times 10^6 \text{ ft-lbf} = 2545 \text{ Btu}$$

$$1 \text{ kw-hr} = 44{,}200 \times 60 = 2.654 \times 10^6 \text{ ft-lbf} = 3412 \text{ Btu}$$

4.3 Work Done at the Moving Boundary of a System in a Quasistatic Process

We have already noted that there are a variety of ways in which work can be done on or by a system. These include work done by a rotating shaft, electrical work, and the work done by the movement of the system boundary, such as the work done in moving the piston in a cylinder. In this section we wish to consider in some detail the work done at the moving boundary of a system during a quasistatic process.

Consider as a system the gas contained in a cylinder and piston, Fig. 4.3. Let one of the small weights be removed from the piston, causing the piston to move upward a distance dl. We can consider this a quasistatic process and calculate the amount of work W done by the system during this process. The total force on the piston is PA, where P is the pressure of the gas and A is the area of the piston. Therefore, the work δW is

$$\delta W = PA\,dl$$

But $A\,dl = dV$, the change in volume of the gas. Therefore,

$$\delta W = P\,dV \tag{4.2}$$

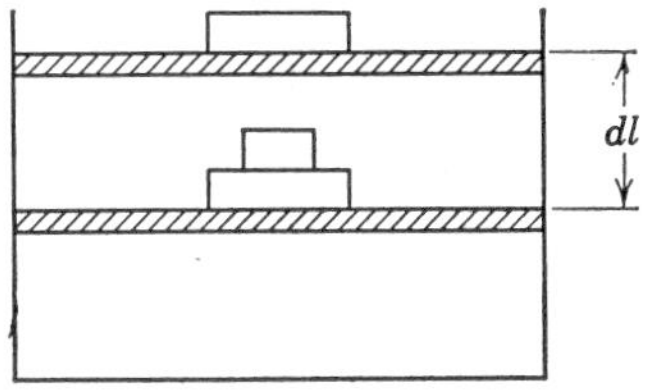

Fig. 4.3. Example of work done at the moving boundary of a system in a quasistatic process.

The work done at the moving boundary during a given quasistatic process can be found by integrating Eq. 4.2. However, this integration can be performed only if we know the relationship between P and V during this process. This relationship might be expressed in the form of an equation, or it might be shown in the form of a graph.

Let us consider a graphical solution first, using as an example a compression process such as the compression of air in the cylinder of a reciprocating air compressor, Fig. 4.4. At the beginning of the process the piston is at position 1, the pressure being relatively low. This

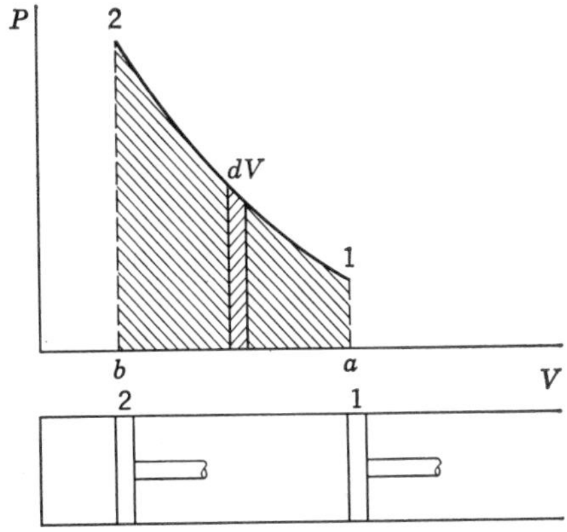

Fig. 4.4 Use of pressure-volume diagram to show work done at the moving boundary of a system in a quasistatic process.

state is represented on a pressure-volume diagram (usually referred to as a P-V diagram) as shown. At the conclusion of the process the piston is in position 2, and the corresponding state of the gas is shown at point 2 on the P-V diagram. Let us assume that this compression was a quasistatic process, and that during the process the system passed through the states shown by the line connecting states 1 and 2 on the P-V diagram. The assumption of a quasistatic process is essential here because each point on line 1–2 represents a definite state, and these states will correspond to the actual state of the system only if the deviation from equilibrium is infinitesimal. The work done on the air during this compression process can be found by integrating Eq. 4.2.

$$ {}_1W_2 = \int_1^2 \delta W = \int_1^2 P\,dV \qquad (4.3) $$

The symbol ${}_1W_2$ is to be interpreted as the work done during the process from state 1 to state 2. It is clear from examining the P-V diagram that the work done during this process, namely, $\int_1^2 P\,dV$ is represented by the area under the curve 1–2, area a–1–2–b–a. In this example the volume decreased, and the area a–1–2–b–a represents work done *on* the system. If the process had proceeded from state 2 to state 1 along the same path, the same area would represent work done *by* the system.

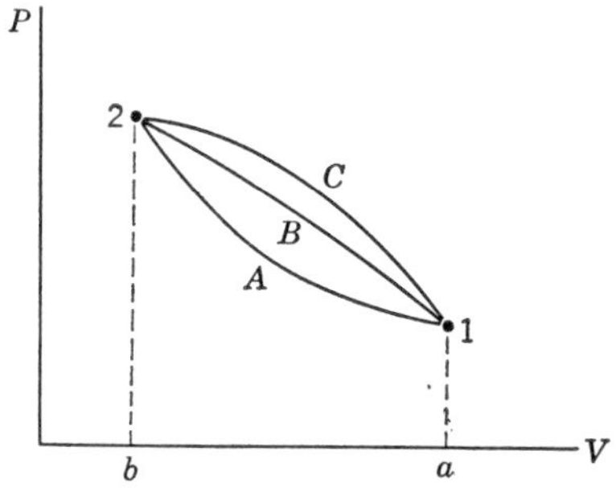

Fig. 4.5 Various quasistatic processes between two given states, indicating that work is a path function.

Further consideration of a P-V diagram, Fig. 4.5, leads to another important conclusion. It is possible to go from state 1 to state 2 along many different quasistatic paths, such as A, B, or C. Since the area underneath each curve represents the work for each process, it is clearly evident that the amount of work involved in each case is not a function of the end state of the process, but rather of the path that is followed in going from one state to another. For this reason work is called a path function, or in mathematical parlance, δW is an inexact differential.

This leads to brief consideration of point and path functions, or, to use another term, exact and inexact differentials. Thermodynamic properties are point functions, a name that arises from the fact that for a given point on a diagram (such as Fig. 4.5) or surface (such as Fig.

3.6) the state is fixed, and thus there is a definite value of each property corresponding to this point. The differentials of point functions are exact differentials, and the integration is simply

$$\int_1^2 dV = V_2 - V_1$$

Thus, we can speak of the volume in state 2 and the volume in state 1, and the change in volume depends only on the initial and final states.

Work, on the other hand, is a path function, for, as has been indicated, the work done in a quasistatic process between two given states depends on the path followed. The differentials of path functions are inexact differentials, and the symbol δ will be used in this text to designate inexact differentials (in contrast to d for exact differentials). Thus, for work we would write

$$\int_1^2 \delta W = {}_1W_2$$

It would be more precise to use the notation ${}_{1A}W_2$, which would indicate the work done during the change from state 1 to 2 along path A. However, implied in the notation ${}_1W_2$ is that the path followed between states 1 and 2 has been specified. It should be noted, we would never speak about the work in the system in state 1 or state 2, and thus we would never write $W_2 - W_1$.

Example 4.1 Consider as a system the gas contained in the cylinder shown in Fig. 4.6, which is fitted with a piston on which a number of small weights are placed. The initial pressure is 20 lbf/in.^{2} and the initial volume of the gas is 1 ft^3.

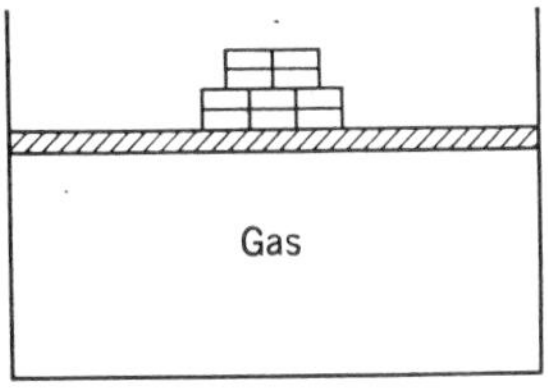

Fig. 4.6 Sketch for Example 4.1

(*a*) Let a Bunsen burner be placed under the cylinder, and let the volume of the gas increase to 3 ft^3 while the pressure remains constant. Calculate the work done by the system during this process

$${}_1W_2 = \int_1^2 P\, dV$$

Since the pressure is constant

$${}_1W_2 = P\int_1^2 dV = P(V_2 - V_1)$$

$$W = 20 \text{ lbf/in.}^2 \times 144 \text{ in.}^2/\text{ft}^2 \times (3 - 1) \text{ ft}^3 = 5760 \text{ ft-lbf}$$

(*b*) Consider the same system and initial conditions, but at the same time that the Bunsen burner is under the cylinder and the piston is rising, let weights be removed from the piston at such a rate that during the process the relation between pressure and volume is given by the expression $PV = \text{constant} = P_1V_1 = P_2V_2$. Let the final volume again be 3 ft^3. Calculate the work done during this process.

We first determine the final pressure.

$$P_2 = \frac{P_1V_1}{V_2} = 20 \times \frac{1}{3} = 6.67 \text{ lbf/in.}^2$$

Again we use Eq. 4.2 to calculate the work.

$${}_1W_2 = \int_1^2 P\,dV$$

We can substitute $P = \text{constant}/V = P_1V_1/V$ into this equation.

$${}_1W_2 = \text{constant}\int_1^2 \frac{dV}{V} = P_1V_1 \ln \frac{V_2}{V_1}$$

$${}_1W_2 = 20 \text{ lbf/in.}^2 \times 144 \text{ in.}^2\text{/ft}^2 \times 1 \text{ ft}^3 \times \ln \tfrac{3}{1} = 3164 \text{ ft-lbf}$$

(*c*) Consider the same system, but let the weights be removed at such a rate during the heat transfer that the expression $PV^{1.3} = $ constant describes the relation between pressure and volume during the process. Again the final volume is 3 ft^3. Calculate the work.

Let us first solve this for the general case of $PV^n = \text{constant}$. Then

$$PV^n = \text{constant} = P_1V_1{}^n = P_2V_2{}^n$$

$$P = \frac{\text{constant}}{V^n} = \frac{P_1V_1{}^n}{V^n} = \frac{P_2V_2{}^n}{V^n}$$

$${}_1W_2 = \int_1^2 P\,dV = \text{constant}\int_1^2 \frac{dV}{V^n} = \text{constant}\left[\frac{V^{-n+1}}{-n+1}\right]_1^2$$

$$= \frac{\text{constant}}{1-n}(V_2{}^{1-n} - V_1{}^{1-n}) = \frac{P_2V_2{}^nV_2{}^{1-n} - P_1V_1{}^nV_1{}^{1-n}}{1-n}$$

$${}_1W_2 = \frac{P_2V_2 - P_1V_1}{1-n}$$

For our problem

$${}_1W_2 = \frac{P_2V_2 - P_1V_1}{1-1.3} = \frac{(4.80 \times 144 \times 3) - (20 \times 144 \times 1)}{1-1.3}$$

$$= 2688 \text{ ft-lbf}$$

(*d*) Consider the same system and initial state as in the three examples above, but let the piston be held by a pin so that the volume remains constant. Also, let heat be transferred from the system until the pressure drops to 10 lbf/in.^{2} Calculate the work.

Since $\delta W = P\,dV$ for a quasistatic process, the work is zero, for in this case there is no change in volume.

The processes for the four examples above are shown on the *P-V* diagram of Fig. 4.7. Process 1–2*a* is a constant-pressure process,

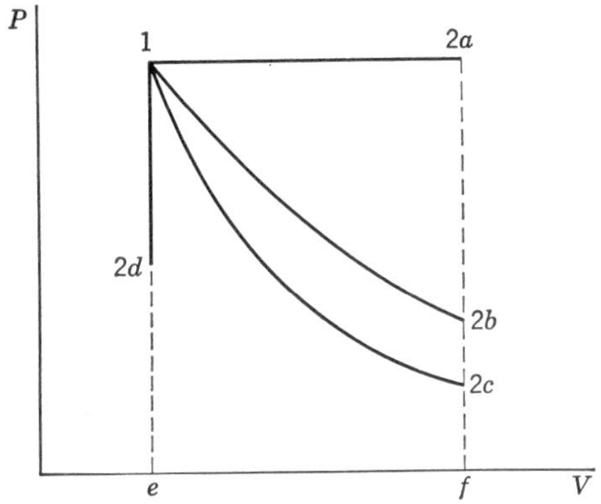

Fig. 4.7 Pressure-volume diagram showing work done in the various processes of Example 4.1.

and area 1–2*a*–*f*–*e*–1 represents the work. Similarly, line 1–2*b* represents the process in which PV = constant, line 1–2*c* the process in which $PV^{1.3}$ = constant, and line 1–2*d* represents the constant-volume process. The student should compare the relative areas under each curve with the numerical results obtained above.

4.4 Some Other Ways in Which Work May Be Done by a System

In the last section we considered the work done by a system at the moving boundary of the system in a quasistatic process. A large number of problems considered in this course will involve this type of process. However, the student should be familiar with other quasistatic processes which involve work, and some of these processes are considered in this section.

The stretched wire Consider as a system a wire which is under a given tension $\mathfrak{I}$. When the length changes by the amount dL, the work done by the system is

$$\delta W = -\mathfrak{I}\,dL$$

The minus sign is necessary because work is done by the system ($+W$) when dL is negative. For a finite process

$$ {}_1W_2 = -\int_1^2 \mathfrak{I}\, dL \tag{4.4} $$

This can be integrated graphically or analytically if the relation between $\mathfrak{I}$ and L is known.

The surface film Consider a system that consists of a liquid film having a surface tension $\mathcal{S}$. When the area of the film changes by an amount dA, the work done by the system is

$$ \delta W = -\mathcal{S}\, dA $$

For finite changes

$$ {}_1W_2 = -\int_1^2 \mathcal{S}\, dA \tag{4.5} $$

The flow of electricity Consider as a system a condenser. Let the potential difference be $\mathcal{E}$, and the amount of electricity stored in the condenser be $\bar{Z}$. When a quantity of electricity $-d\bar{Z}$ leaves the condenser at potential $\mathcal{E}$, the amount of work done by the system is

$$ \delta W = -\mathcal{E}\, d\bar{Z} \tag{4.6} $$

Since the current, i, equals $d\bar{Z}/d\tau$ (where τ = time) we can also write

$$ \delta W = -\mathcal{E}i\, d\tau $$

$$ {}_1W_2 = -\int_1^2 \mathcal{E}i\, d\tau \tag{4.7} $$

Eq. 4.7 may also be written

$$ \frac{\delta W}{d\tau} = -\mathcal{E}i $$

This leads to the definition of a unit of power, the watt. A watt is the power developed by a current of 1 ampere flowing through a potential of 1 volt. This is consistent with our earlier observation that electrical energy is work.

The three processes cited above and the moving boundary of a system have several things in common. First of all, in each case we are concerned with the work in a quasistatic process. Second, in each quasistatic process, the work is given by the integral of the product of an intensive property and the change of an extensive property. These are summarized below:

$$ {}_1W_2 = \int_1^2 P\,dV $$

$$ {}_1W_2 = -\int_1^2 \mathfrak{J}\,dL $$

$$ {}_1W_2 = -\int_1^2 \mathcal{S}\,dA $$

$$ {}_1W_2 = -\int_1^2 \mathcal{E}\,d\bar{Z} $$

Further, in each case the work can be found as the area under a curve on a diagram in which the intensive property is the ordinate and the extensive property the abscissa.

Thus, for a quasistatic process we can write

$$ \delta W = P\,dV - \mathfrak{J}\,dl - \mathcal{S}\,dA - \mathcal{E}\,d\bar{Z} \cdots \tag{4.8} $$

The dots indicate other ways in which work can be done by a system during a quasistatic process. Although the $P\,dV$ type of work is most frequently encountered, the student should bear in mind these other ways in which work can be done.

4.5 Work in Some Other Processes

The main purpose of this section is to emphasize that there are many processes in which work is not equal to $\int P\,dV$. This is a point that frequently needs to be emphasized to the beginning student.

One example is shown in Fig. 4.8, in which a membrane separates a gas from a vacuum. Let the membrane rupture and the gas fill the

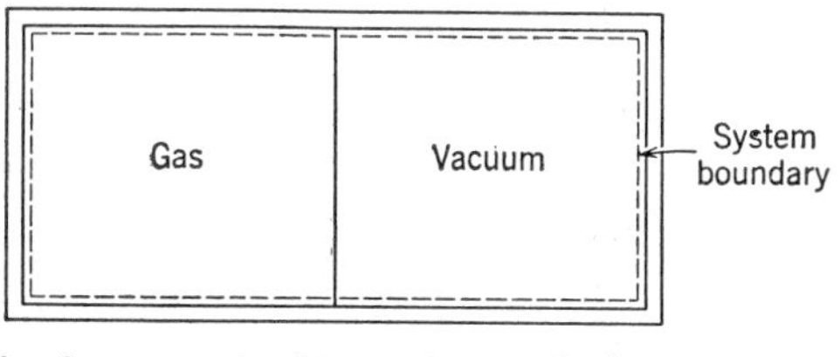

Fig. 4.8 Example of a process involving a change of volume for which the work is zero.

entire volume. Certainly there is a change in pressure and a change in volume of the gas. However, there is no work crossing the boundary of the system, and, therefore, the work for the process is zero. We might attempt to calculate the $\int_1^2 P\,dV$ of the gas, but because this is not a quasistatic process, the $\int_1^2 P\,dV$ is not equal to the work.

The second example, Fig. 4.9, shows a system in which the volume remains constant, but work is done on the gas as the weight is lowered

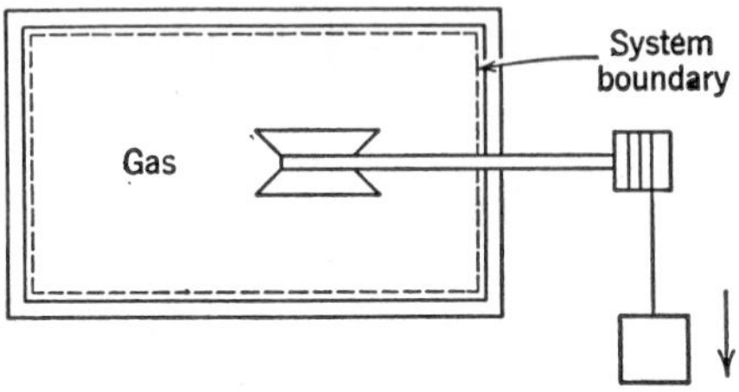

Fig. 4.9 Example of a process for which the change in volume is zero but which does involve work.

and the paddle wheel turns. In this case the work for the process has a certain value, even though the change in volume is zero, and the $\int_1^2 P\,dV = 0$. The reason is that work can cross the boundary of a system in other ways than through the movement of the boundary.

A final example, Fig. 4.10, should be cited, and this involves the same example as Fig. 4.9, except that the boundary of the system is

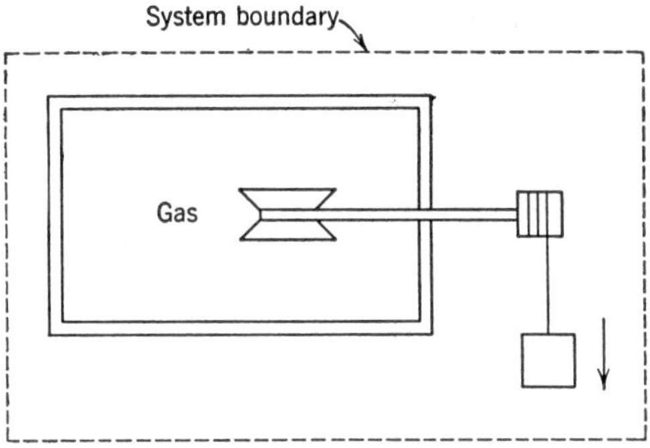

Fig. 4.10 Example of a system so selected that the lowering of a weight does not involve work.

changed to include the weight. In this case all the changes take place within the boundary of the system, and no work crosses the boundary. This emphasizes the fact that work is identified only as it crosses the boundary of the system, and whether or not work is involved depends on what constitutes the system.

The only process for which work is equal to $\int P\,dV$ is a quasistatic process which involves a moving boundary of the system.

4.6 Definition of Heat

The thermodynamic definition of heat is somewhat different from the everyday use of the word. Therefore, it is essential to understand clearly the definition of heat given here, particularly because it is involved in so many thermodynamic problems.

If a block of hot copper is placed in a beaker of cold water, we know from experience that the block of copper cools down and the water warms up until they come to equality of temperature. What causes this decrease in the temperature of the copper and the increase in the temperature of the water? We say that this was caused by the fact that energy was transferred from the copper block to the water. It is out of such a transfer of energy that we arrive at a definition of heat.

Heat is defined as the form of energy that is transferred across a boundary by virtue of a temperature difference or temperature gradient. Implied in this definition is the very important fact that a body never contains heat, but that heat is identified as heat only as it crosses the boundary. Thus, heat is a transient phenomenon. If we consider the hot block of copper as one system and the cold water in the beaker as another system, we recognize that originally neither system contains any heat (they do contain energy, of course). When the copper is placed in the water and the two are in thermal communication, heat is transferred from the copper to the water, until equilibrium of temperature is established. At that point we no longer have heat transfer, since there is no temperature difference. Neither of the systems contains any heat at the conclusion of the process. It also follows that heat is identified at the boundaries of the system, for heat is defined as energy being transferred across the system boundary.

4.7 Units of Heat

As in all other quantities involved in thermodynamics, we must have units for heat. Consider as a system 1 lbm water at 59.5 F, and let a block of hot copper be placed in the water. Further, let the block of copper have such a mass and such a temperature that when thermal equilibrium is established the temperature of the water is 60.5 F. We define as our unit of heat the quantity of heat transferred from the copper to the water, and call the unit of heat the British thermal unit, which is abbreviated Btu. More specifically, this is called the 60-degree Btu, which may be defined as the quantity of heat required to raise 1 lbm of water from 59.5 F to 60.5 F.*

Further, we consider heat transferred to a system positive, and heat transferred from a system negative. Thus, positive heat represents energy transferred to a system, and negative heat represents energy transferred from a system.

The symbol Q is used to represent heat. Heat transfer per unit mass of a closed system is given the symbol q. Thus for a given closed system

$$q = \frac{Q}{m} \tag{4.9}$$

A process in which there is no heat transfer, $Q = 0$, is called an adiabatic process.

4.8 Heat—A Path Function

Heat is a path function. That is, the amount of heat transferred when a system changes from state 1 to state 2 depends on the intermediate states through which the system passes, i.e., its path. Therefore, heat is an inexact differential, and the differential is written δQ. On integrating we write

$$\int_1^2 \delta Q = {}_1Q_2$$

In words, ${}_1Q_2$ is the heat transferred during the given process between state 1 and state 2.

4.9 Comparison of Heat and Work

At this point it is evident that there are many similarities between heat and work, and these are summarized here.

(*a*) Heat and work are both transient phenomena. Systems never possess heat or work, but either or both may occur when a system undergoes a change of state.

(*b*) Both heat and work are boundary phenomena. Both are observed at the boundaries of the system, and both represent energy crossing the boundary of the system.

(*c*) Both heat and work are path functions and inexact differentials.

* Actually the Btu as used today is defined in terms of electrical units. This point is explained in the next chapter.

It should also be noted that in our sign convention, $+Q$ represents heat transferred to the system, and thus is energy added to the system. On the other hand, $+W$ represents work done by the system and thus represents energy leaving the system.

A final illustration may be helpful to indicate the difference between heat and work. Figure 4.11 shows a gas contained in a rigid vessel. Resistance coils are wound around the outside of the vessel. As current flows through the resistance coils the temperature of the gas increases. What crosses the boundary of the system, heat or work?

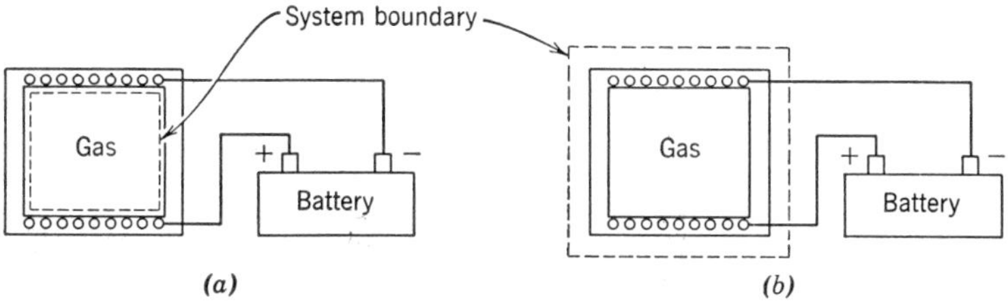

Fig. 4.11 An example showing the difference between heat and work.

In Fig. 4.11*a* we consider only the gas as the system. In this case the energy crosses the boundary of the system because the temperature of the walls is higher than the temperature of the gas. Therefore, we recognize that heat crosses the boundary of the system.

In Fig. 4.11*b* the system includes the vessel and the resistance heater. Electricity crosses the boundary of the system, and as was indicated earlier, this is work.

PROBLEMS

4.1 A cylinder fitted with a piston contains 3 lbm of saturated water vapor at a pressure of 100 lbf/in.^{2} The steam is heated until the temperature is 500 F. During this process the pressure remains constant. Calculate the work done by the steam during the process.

4.2 A cylinder in which the piston is restrained by a spring contains 1 ft^3 of air at a pressure of 15 lbf/in.^{2}, which just balances the atmospheric pressure of 15 lbf/in.^{2} Assume the weight of the piston is negligible. In this initial state, the spring exerts no force on the piston. The gas is then heated until the volume is doubled. The final pressure of the gas is 50 lbf/in.^{2}, and during the process the spring exerts a force which is proportional to the displacement of the piston from the initial position.

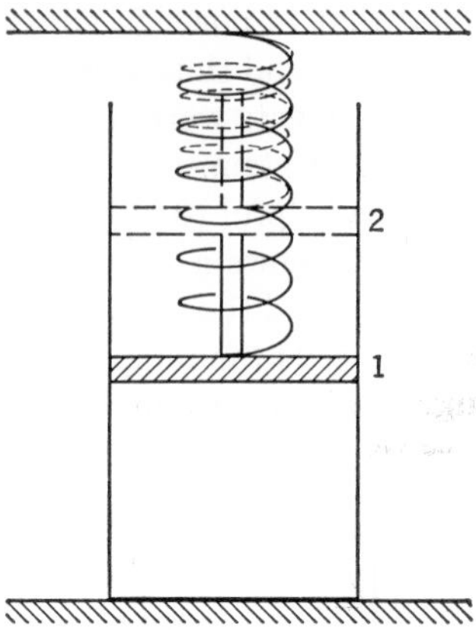

Fig. 4.12 Sketch for Problem 4.2.

(*a*) Show this process on a *P-V* diagram.

(*b*) Considering the gas as the system, calculate the total work done by the system.

(*c*) Of the total work, how much is done against the atmosphere? How much against the spring?

4.3 A balloon which is initially flat is inflated by filling it with compressed air from a tank of compressed air. The final volume of the balloon is 40 ft^3. The barometer reads 29.7 in. Hg. Consider the tank, the balloon, and the connecting pipe as a system. Determine the work for this process.

4.4 A spherical balloon has a diameter of 10 in., and contains air at a pressure of 20 lbf/in.^{2} The diameter of the balloon increases to 12 in. due to heating, and during this process the pressure is proportional to the diameter. Calculate the work done by the air during this process.

4.5 Ammonia is compressed in a cylinder by a piston. The initial temperature is 100 F. The initial pressure is 60 lbf/in.^{2}, and the final pressure is 180 lbf/in.^{2} The following data are available for this process:

Pressure, lbf/in.^{2}	Volume, in.^{3}
60	80.0
80	67.5
100	60.0
120	52.5
140	45.0
160	37.5
180	32.5

(*a*) Determine the work for the process considering the ammonia as the system.

(*b*) What is the final temperature of the ammonia?

4.6 During static tests of rocket motors a compressed gas is often used to force the propellants into the combustion chamber. Consider the arrangement shown in Fig. 4.13. Compressed air is used to expel

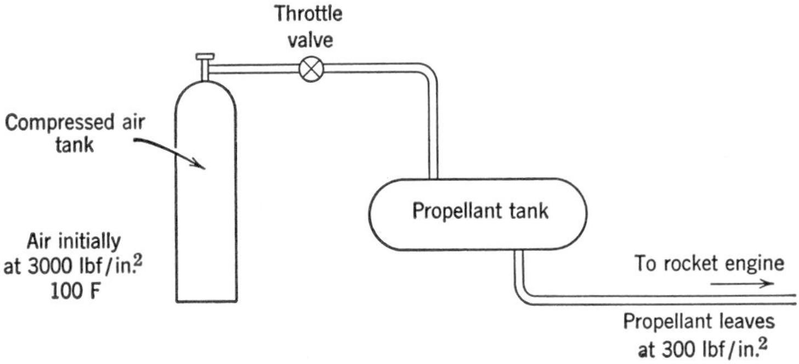

Fig. 4.13 Sketch for Problem 4.6.

the liquid propellant from the propellant tank. The initial pressure of the air is 3000 lbf/in.^{2} and the initial temperature is 100 F. The propellant has a density of 70 lbm/ft^3 and the propellant tank is filled to capacity and contains 2000 lbm of propellant. The propellant leaves at a constant pressure of 300 lbf/in.^{2} Considering the air as the system, determine the work done by the air in forcing the propellant from the propellant tank.

4.7 A room is heated with steam radiators on a winter day. Examine the following systems regarding heat transfer (including sign):

(*a*) The radiator.

(*b*) The room.

(*c*) The radiator and the room.

4.8 Consider a hot-air heating system for a home and examine the following systems for heat transfer:

(*a*) The combustion chamber and combustion gas side of the heat transfer area.

(*b*) The furnace as a whole including the hot and cold air ducts and chimney.

CHAPTER 5

THE FIRST LAW OF THERMODYNAMICS

Having completed our consideration of basic definitions and concepts we are ready to proceed to a discussion of the first law of thermodynamics. Often this law is called the law of the conservation of energy, and as we shall see later, this is essentially true. The procedure will be to first state this law for a closed system, and then extend it to the open system. The law of the conservation of matter will also be considered in this chapter.

5.1 The First Law of Thermodynamics for a Closed System Undergoing a Cycle

The *first law of thermodynamics* states that during any cycle a closed system undergoes, the cyclic integral of the heat is equal to the cyclic integral of the work.

Consider as a system the gas in the container shown in Fig. 5.1. Let this system go through a cycle which is comprised of two processes, the first process being one in which work is done on the system by the paddle which turns as the weight is lowered. Let the system then be returned to its initial state by transferring heat from the system, thus completing the cycle.

We have seen previously that we can measure the work in foot-pounds force and the heat in Btu. Let measurements of work and heat be made during such a cycle for a wide variety of systems and for various amounts of work and heat. When the amount of work and heat is compared, we find that these two are always proportional. This is the first law of thermodynamics, and in equation form it is written

$$\oint \delta W = J \oint \delta Q \tag{5.1}$$

The symbol $\oint$ designates the cyclic integral, and J is a proportionality factor. In words, Eq. 5.1 states that the cyclic integral of the work (i.e., the net work done during the cycle) is proportional to the cyclic integral of the heat (i.e., the net heat transfer during the cycle).

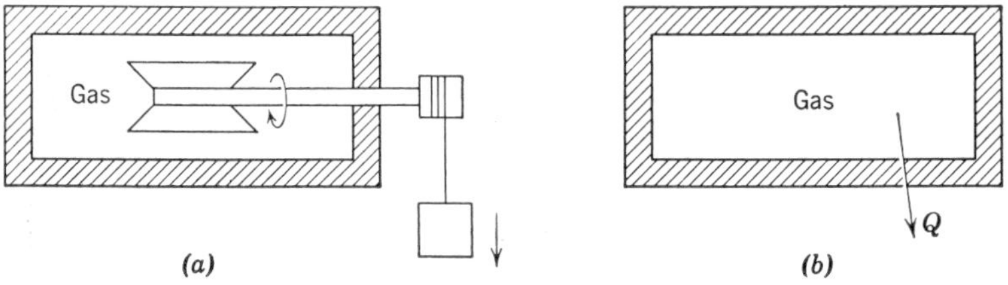

Fig. 5.1 Example of a closed system undergoing a cycle.

Before continuing our discussion, the basis for this law should be pointed out. The basis of every law of nature we set forth is experimental evidence, and this is true also of the first law of thermodynamics. Every experiment that has been conducted thus far has verified the first law either directly or indirectly. The basis of this law is, therefore, experimental evidence; it has never been disproved, and many, many experiments have verified it.

Thus far we have used the unit foot-pounds for work and Btu for heat. We might also use other units, such as joules for work and calories for heat. The first law states that there is a proportionality between the units for heat and work. The magnitude of the proportionality constant J will depend, of course, on the units used for heat and work. Joule in Great Britain and Meyer in Germany did the first accurate work in the 1840's on measurement of the proportionality factor J. The most recent work has been done by Osborne, Stimson, and Ginnings of the National Bureau of Standards. The generally accepted value for J, based on the 60-degree Btu and the foot-pound force, is that $J = 778.2$ ft-lbf/Btu.

However, at the International Steam Table Conference in 1929, the Btu was, in effect, defined in terms of the foot-pound force by the following relation:

$$1 \text{ Btu} = 778.26 \text{ ft-lbf}$$

This definition fixes the magnitude of the Btu in terms of previously defined units, and is called the International British thermal unit. Throughout the rest of the text this will be the British thermal unit with which we are concerned whenever the term is used. Thus, when using the International Btu, the question is not to accurately determine the value of J for the 60 F Btu, but rather how many International Btu's are required to raise 1 lbm of water from 59.5 F to 60.5 F.

For much engineering work, however, the accuracy of other data does not warrant more accuracy than the relation 778 ft-lbf = 1 Btu, and, in general, this relation will be used in the examples given in this text.

It is also evident from this discussion that heat and work can be expressed in the same units. Thus, it is perfectly correct to speak of Btu of work and ft-lbf of heat, or calories of work and joules of heat. Therefore we can write Eq. 5.1 without the proportionality factor J.

$$\oint \delta W = \oint \delta Q \tag{5.2}$$

The implication of the equation thus written is that heat and work are expressed in the same units. In this text, therefore, the proportionality factor J will not be written into equations, but rather the student should realize that each equation must have consistent units throughout.

We also note that

$$1 \text{ hp} = 33{,}000 \text{ ft-lbf/min} = 42.4 \text{ Btu/min} = 2545 \text{ Btu/hr}$$

$$1 \text{ kw} = 44{,}200 \text{ ft-lbf/min} = 56.9 \text{ Btu/min} = 3412 \text{ Btu/hr}$$

5.2 The First Law of Thermodynamics for a Change in State of a Closed System

Equation 5.2 states the first law of thermodynamics for a closed system during a cycle. Many times, however, we are concerned with a process rather than a cycle, and we now consider the first law of thermodynamics for a closed system that undergoes a change of state. This can be done by introducing a new property, which is given the symbol E.

Consider a system that undergoes a cycle, changing from state 1 to state 2 by process A, and returning from state 2 to state 1 by process B. This cycle is shown in Fig. 5.2 on a pressure (or other intensive property)–volume (or other extensive property) diagram.

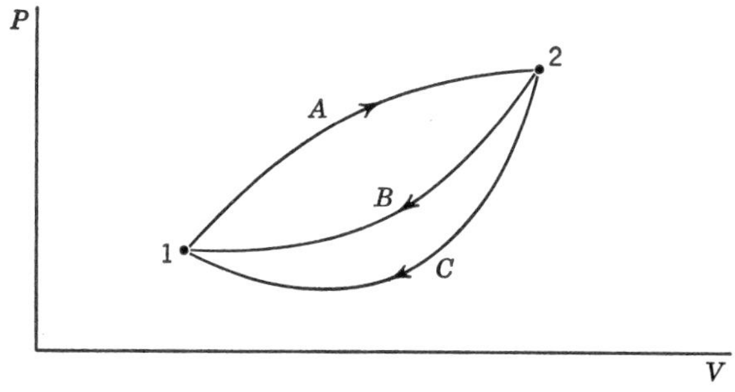

Fig. 5.2 Demonstration of the existence of the thermodynamic property *E*.

From the first law of thermodynamics, Eq. 5.2

$$\oint \delta Q = \oint \delta W$$

Considering the two separate processes we have

$$\int_{1A}^{2A} \delta Q + \int_{2B}^{1B} \delta Q = \int_{1A}^{2A} \delta W + \int_{2B}^{1B} \delta W$$

Now consider another cycle, the system changing from state 1 to state 2 by process A, as before, and returning to state 1 by process C. For this cycle we can write

$$\int_{1A}^{2A} \delta Q + \int_{2C}^{1C} \delta Q = \int_{1A}^{2A} \delta W + \int_{2C}^{1C} \delta W$$

Subtracting the second of these equations from the first, we have

$$\int_{2B}^{1B} \delta Q - \int_{2C}^{1C} \delta Q = \int_{2B}^{1B} \delta W - \int_{2C}^{1C} \delta W$$

or, by rearranging

$$\int_{2B}^{1B} (\delta Q - \delta W) = \int_{2C}^{1C} (\delta Q - \delta W) \tag{5.3}$$

Since B and C represent arbitrary processes between states 1 and 2, we conclude that the quantity $(\delta Q - \delta W)$ is the same for all processes between state 1 and state 2. Therefore, $(\delta Q - \delta W)$ depends only on the initial and final states and not on the path followed between the two states. We conclude that this is a point function, and therefore is the differential of a property of the system. This property is given the symbol E. Thus, we can write

$$\delta Q - \delta W = dE$$

or

$$\delta Q = dE + \delta W \tag{5.4}$$

Note that since E is a property, its derivative is written dE. When Eq. 5.4 is integrated from an initial state 1 to the final state 2, we have

$${}_1Q_2 = E_2 - E_1 + {}_1W_2 \tag{5.5}$$

Where ${}_1Q_2$ is heat transferred to the system during the process from state 1 to state 2.
E_1 and E_2 are the initial and final values of the property E of the system.
${}_1W_2$ is work done by the system during the process.

The physical significance of the property E is that it represents all the energy of the system in the given state. This energy might be present in a variety of forms, such as the kinetic or potential energy of the system as a whole, energy associated with the movement and position of the molecules, chemical energy such as is present in a storage battery, energy present in a charged condenser, or a host of other forms.

In the study of thermodynamics it is most convenient to consider the kinetic and potential energy separately and then to consider all the other energy of the system in a single property which we call the *internal energy*, and to which we give the symbol U. Thus, we would write

$$E = \text{Internal energy} + \text{Kinetic energy} + \text{Potential energy}$$

or

$$E = U + KE + PE$$

The reason for doing this is that the kinetic and potential energy can be specified by the macroscopic parameters of mass, velocity, and elevation. The internal energy U includes all other forms of energy of the system. Since each of these are thermodynamic properties we can write

$$dE = dU + d\,(KE) + d\,(PE) \tag{5.6}$$

The first law of thermodynamics for a change of state of a closed system may therefore be written

$$\delta Q = dU + d\,(KE) + d\,(PE) + \delta W \tag{5.7}$$

In words this equation states that as a system undergoes a change of state, energy may cross the boundary as either heat or work, and each may be positive or negative. The net change in the energy of the system will be exactly equal to the net energy that crossed the boundary of the system. The energy of the system may change in any of three ways, namely, by a change in internal energy, kinetic energy, or potential energy. Since the mass of a closed system is fixed, we can also say that a quantity of matter may have energy in three forms, internal energy, kinetic energy, or potential energy.

This section will be concluded by deriving an expression for the kinetic and potential energy of a closed system. Consider first a closed system that is initially at rest (relative to the earth), and let this system be acted upon by a force F which moves the system a distance dx in the direction of the force. Let there be no heat transfer and no change in internal energy or potential energy. Then, from the first law, Eq. 5.7

$$\delta W = -F\,dx = -d\,KE$$

But

$$F = \frac{ma}{g_c} = \frac{m\,d\overline{V}\,dx}{g_c\,d\tau\,dx} = \frac{m\,dx\,d\overline{V}}{g_c\,d\tau\,dx} = \frac{m}{g_c}\overline{V}\frac{d\overline{V}}{dx}$$

Then

$$F\,dx = \frac{m}{g_c}\overline{V}\,d\overline{V} = d\,KE$$

Integrating

$$\int_{\overline{V}=0}^{\overline{V}} \frac{m}{g_c}\overline{V}\,d\overline{V} = \int_{KE=0}^{KE} d\,KE$$

$$KE = \frac{1}{2}\frac{m\overline{V}^2}{g_c} \qquad d\,KE = \frac{m}{g_c}\overline{V}\,d\overline{V} \tag{5.8}$$

An expression for potential energy can be found in a similar manner. Consider a closed system that is initially at rest and at the elevation of some reference level. Let this system be acted upon by a vertical force F which is of such magnitude that it slowly raises (in elevation) the system an amount dZ. Let the acceleration due to gravity at this point be g. From the first law, Eq. 5.7

$$\delta W = -F\,dZ = -d\,PE$$

But

$$F = \frac{ma}{g_c} = \frac{mg}{g_c}$$

Then

$$d\,PE = \frac{mg}{g_c}\,dZ$$

Integrating

$$\int_{PE=0}^{PE} d\,PE = m\int_{Z=0}^{Z} \frac{g}{g_c}\,dZ$$

Assuming that g does not vary with Z (which is a very reasonable assumption for moderate changes in elevation)

$$PE = \frac{mg}{g_c}Z \tag{5.9}$$

Substituting these expressions for kinetic and potential energy into Eq. 5.7 we have

$$\delta Q = dU + \frac{d(m\overline{V}^2)}{2g_c} + d\left(\frac{mg}{g_c}Z\right) + \delta W \tag{5.10}$$

In the integrated form this equation is

$${}_1Q_2 = U_2 - U_1 + \frac{m(\overline{V}_2^{\,2} - \overline{V}_1^{\,2})}{2g_c} + \frac{mg}{g_c}(Z_2 - Z_1) + {}_1W_2 \tag{5.11}$$

Three observations should be made regarding this equation. The first is that in this section the property E (to which we gave no name) was found to exist, and we were able to write the first law for a change of state using Eq. 5.5. However, rather than deal with this property E, we find it more convenient to consider internal energy, kinetic energy, and potential energy. In general, this will be the procedure followed in the rest of this book.

The second observation is that Eq. 5.11 is in effect a statement of the conservation of energy. The net change of the energy of the closed system is always equal to the net transfer of energy across the system boundary as heat and work. This is somewhat analogous to a joint checking account which a man might have with his wife. There are two ways in which deposits and withdrawals can be made, either by the man or by his wife, and the balance will always reflect the net amount of the transaction. Similarly, there are two ways in which energy can cross the boundary of a closed system, either as heat or work, and the energy of the system will change by the exact amount of the net energy crossing the system boundary.

The third observation is that Eq. 5.11 can give only changes in internal energy, kinetic energy, and potential energy. We can learn nothing about absolute values of these quantities from Eq. 5.11. If we wish to assign values to internal energy, kinetic energy, and potential energy we must assume reference states and assign a value to the quantity in this reference state. The kinetic energy of a body with zero velocity relative to the earth is assumed to be zero. Similarly, the value of the potential energy is assumed to be zero when the body is at some reference elevation. With internal energy, therefore, we must also have a reference state if we wish to assign values of this property. This matter is considered in the following section.

5.3 Internal Energy—A Thermodynamic Property

Internal energy is an extensive property, since if we divide a homogeneous closed system into two equal parts, each part will have one-half of the initial internal energy. Similarly, the kinetic and potential energies are extensive properties.

The symbol U designates the internal energy of a given mass of a substance. Following the convention used with other extensive properties, the symbol u designates the internal energy per unit mass. Therefore

$$U = mu$$

In Chapter 3 it was noted that in the absence of motion, gravity, surface effects, electricity, and magnetism, the state of a pure substance is specified by two independent properties. It is very significant that with these restrictions, one of the independent properties of a pure substance is internal energy. This means, for example, that if we specify the pressure and temperature of superheated steam, the internal energy with reference to some arbitrary base is also specified.

Thus, in a table of thermodynamic properties such as the steam tables, the value of internal energy can be tabulated along with other thermodynamic properties. Table 2 of the steam tables by Keenan and Keyes lists the internal energy for saturated states. Included is the internal energy of saturated liquid u_f, the internal energy of saturated vapor u_g, and the difference between the internal energy of saturated liquid and saturated vapor u_{fg}. The values are given in relation to an arbitrarily assumed reference state. This reference state will be discussed later. The internal energy of saturated steam of a given quality is calculated in the same way specific volume is calculated. The relations are

$$u = (1 - x)u_f + xu_g$$
$$u = u_f + xu_{fg} \tag{5.12}$$
$$u = u_g - (1 - x)u_{fg}$$

The same value is obtained from each equation, but, as was pointed out earlier in the discussion on specific volume, when using a slide rule, for high quality the last equation gives more accuracy, and for low quality the second equation gives more accuracy. As an example, let us determine the internal energy of saturated steam having a pressure of 80 lbf/in.^{2} and a quality of 95 per cent.

$$u = u_g - (1 - x)u_{fg}$$

$$u = 1102.1 - 0.05(820.3) = 1061.1 \text{ Btu/lbm}$$

Example 5.1 A tank containing a fluid is stirred by a paddle wheel. The power input to the paddle wheel is 2 hp. Heat is transferred from the tank at the rate of 1500 Btu/hr. Considering the tank and the fluid as the system determine the change in the internal energy of the system during 1 hr.

$${}_1W_2 = -2(2545) = -5090 \text{ Btu/hr}$$

Since there is no change in kinetic and potential energy, Eq. 5.11 reduces to

$${}_1Q_2 = U_2 - U_1 + {}_1W_2$$

$$-1500 = U_2 - U_1 - 5090$$

$$U_2 - U_1 = 3590 \text{ Btu/hr}$$

Example 5.2 Consider a system composed of a stone having a mass of 10 lbm and a bucket containing 100 lbm of water. Initially the stone is 77.8 ft above the water and the stone and water are at the same temperature. The stone then falls into the water.

Determine ΔU, ΔKE, ΔPE, Q and W for the following cases.

(*a*) At the instant the stone is about to enter the water: Assuming no heat transferred to or from the stone as it falls, we conclude that during the change from the initial state to the state that exists at the moment the stone enters the water,

$${}_1Q_2 = 0 \qquad {}_1W_2 = 0 \qquad \Delta U = 0$$

Therefore Eq. 5.11 reduces to

$$\Delta PE = -\Delta KE = \frac{mg}{g_c}(Z_2 - Z_1)$$

$$= \frac{10 \text{ lbm} \times 32.17 \text{ ft/sec}^2}{32.17 \text{ lbm-ft/lbf-sec}^2} \times (-77.8) \text{ ft} = -778 \text{ ft-lbf}$$

$$= -1 \text{ Btu}$$

That is, $\Delta KE = 1$ Btu and $\Delta PE = -1$ Btu.

(*b*) Just after the stone comes to rest in the bucket:

$$_1Q_2 = 0 \qquad _1W_2 = 0 \qquad \Delta KE = 0$$

Then

$$\Delta PE = -\Delta U = \frac{mg}{g_c}(Z_2 - Z_1) = -1 \text{ Btu}$$

$$\Delta U = 1 \text{ Btu} \qquad \Delta PE = -1 \text{ Btu}$$

(*c*) After enough heat has been transferred so that the stone and water are at the same temperature they were initially: We conclude that $\Delta U = 0$. Therefore in this case

$$\Delta U = 0 \qquad \Delta KE = 0 \qquad _1W_2 = 0$$

$$_1Q_2 = \Delta PE = \frac{mg}{g_c}(Z_2 - Z_1) = -1 \text{ Btu}$$

Example 5.3 A vessel having a volume of 100 ft³ contains 1 ft³ of saturated liquid water and 99 ft³ of saturated water vapor at 14.7 lbf/in.² Heat is transferred until the vessel is filled with saturated vapor. Determine the heat transfer for this process.

Since this is a closed system we can use Eq. 5.11, and since changes in kinetic and potential energy are not involved, this reduces to

$$_1Q_2 = U_2 - U_1 + {_1W_2}$$

Further, the work for this process is zero, and therefore

$$_1Q_2 = U_2 - U_1$$

The thermodynamic properties can be found in Table 2 of the steam tables. The initial internal energy U_1 is the sum of the initial internal energy of the liquid and the vapor.

$$U_1 = m_{1\text{ liq}}\,u_{1\text{ liq}} + m_{1\text{ vap}}\,u_{1\text{ vap}}$$

$$m_{1\text{ liq}} = \frac{V_{\text{liq}}}{v_f} = \frac{1}{0.01672} = 59.81 \text{ lbm}$$

$$m_{1\text{ vap}} = \frac{V_{\text{vap}}}{v_g} = \frac{99}{26.80} = 3.69 \text{ lbm}$$

$$U_1 = 59.8(180.0) + 3.69(1077.5) = 14{,}740 \text{ Btu}$$

To determine u_2 we need to know two thermodynamic properties, since this determines the final state. The properties we know are the quality, $x = 100\%$, and v_2, the final specific volume.

$$m = m_{1\ \text{liq}} + m_{1\ \text{vap}} = 59.81 + 3.69 = 63.50 \text{ lbm}$$

$$v_2 = \frac{V}{m} = \frac{100}{63.50} = 1.575 \text{ ft}^3/\text{lbm}$$

In Table 2 of the steam tables we find that at a pressure of 294 lbf/in.^{2} $v_g = 1.575$, which is the final pressure of the steam. Therefore,

$$u_2 = 1117.0 \text{ Btu/lbm}$$

$$U_2 = mu_2 = 1117.0(63.50) = 70{,}930 \text{ Btu}$$

$$_1Q_2 = U_2 - U_1 = 70{,}930 - 14{,}740 = 56{,}190 \text{ Btu}$$

5.4 The First Law of Thermodynamics for the Open System

In the previous sections of this chapter we have considered the first law of thermodynamics for the closed system, and noted that energy can cross the boundary of the closed system as heat and as work. Many thermodynamic analyses involve an open system, however, and it is essential therefore that we consider the first law for the open system. In the open system, matter crosses the boundary, and in doing so carries a certain amount of energy into the system. Thus, the energy of the system is influenced by heat, work, and the matter crossing the system boundary.

In extending the principle of the conservation of energy to the open system we must ask ourselves the question, how much is the energy of the system increased as a result of matter crossing the system boundary? When we have answered this question we will be able to state the first law of thermodynamics for the open system.

The energy that crosses the boundary includes the following.

(*a*) Internal energy. Each pound of fluid has the internal energy u, and as it crosses the boundary of the system the energy of the system is increased by this amount.

(*b*) Kinetic energy. Since the fluid stream has a certain velocity $\overline{V}$, each pound of fluid carries the kinetic energy $\overline{V}^2/2g_c$ into the system.

(*c*) Potential energy. As pointed out earlier, potential energy is always measured relative to some base. Where Z is the elevation with reference to this base, each pound of fluid entering the system has the potential energy Zg/g_c.

(*d*) Flow energy. This energy is not associated directly with the matter crossing the boundary, but is associated with the fact that somewhere a pumping process occurs which forces the fluid across the system boundary. Sometimes this quantity is also called flow work, for a reason we will consider later.

Let us develop this concept with the aid of Fig. 5.3, which is in essence a hydraulic lift. A pump supplies fluid to the tank and in turn some fluid leaves the tank and enters cylinder *B*, and raises a weight. Assume that the movement of piston *B* is so small that there is a negligible change in pressure due to the increased head at *B*.

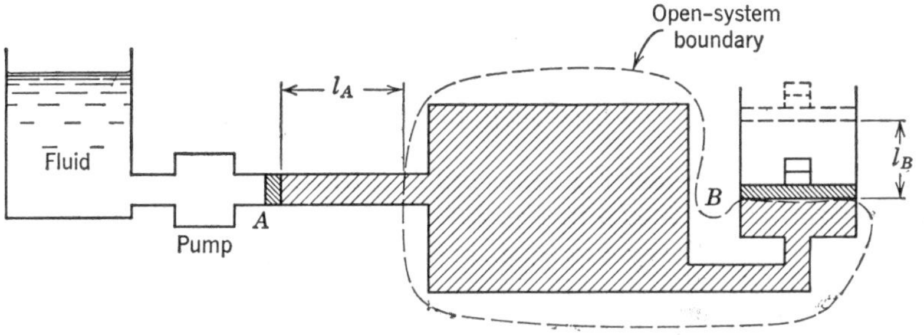

Fig. 5.3 An illustration showing the flow energy associated with the flow of matter across the boundary of an open system.

First of all, we will analyze this as a closed system. Let all of the fluid covered by the cross-hatching constitute the system. At point *A*, the system is separated from the other fluid by a piston *A*. Let us analyze the process that occurs when 1 lbm of fluid leaves the pump, and in turn, 1 lbm of fluid enters the cylinder, thus raising the weight, assuming the process to be adiabatic. We make the following observations regarding this process:

(*a*) The following events happen during this process:

1. Piston *A* moves a distance l_A, which, when multiplied by the cross-sectional area of the pipe, A_A, corresponds to the volume occupied by 1 lbm of the fluid.
2. One pound of fluid enters the tank.
3. One pound of fluid leaves the tank and enters the cylinder *B*.
4. The piston *B* moves a distance l_B, which, when multiplied by the cross-sectional area of the cylinder, A_B, corresponds to the volume occupied by 1 lbm of fluid.

(*b*) Work is done at the two points where the boundary of the system moves, namely, at A and B. The work done at A is

$$W = -P \times A_A \times l_A = -Pv_A$$

The minus sign indicates this work is done on the system. The work done at B is

$$W = P \times A_B \times l_B = Pv_B$$

v_A and v_B are the specific volume at A and B, respectively.

(*c*) There is no change in the state of the fluid, only a change in its position. Therefore, there is no change in the internal energy of the system, and the specific volume at A and B are the same, i.e., $v_A = v_B$.

(*d*) Since $v_A = v_B$, the net work done by the system is zero. This would also follow from the first law for the closed system, since there is no heat transfer and no change in internal energy.

Let us now consider the same problem for the open system, whose boundaries are as shown in Fig. 5.3. One pound of fluid now crosses the boundary of the system at A, and work is done by the system only at B, where the boundary of the system moves as the piston is raised.

We inquire now as to the source of the energy that accomplished this work. First of all, we conclude that it was done by the energy carried across the system boundary by the fluid entering at A, since there is no change in the state of the fluid in the tank. In what form was this energy carried into the control volume by the entering fluid? We might consider internal energy, but we concluded from our closed-system analysis that this work was not associated with a change in internal energy. Kinetic energy is not the answer either, since a numerical check would indicate that the kinetic energy that a low-velocity stream has is usually much less than the amount of work done in raising the weight. Potential energy must also be ruled out, since there is no significant change in elevation. Rather, this work is done by what we call *flow energy*, which is the energy carried into or transmitted across the system boundary as a result of the fact that a pumping process, which occurs somewhere outside the system, causes the fluid to enter the system. We conclude that since the work done at B is Pv, the magnitude of the flow energy per pound mass transmitted across the system boundary is Pv, the product of the pressure and specific volume. This corresponds to the work done on the system by piston A in the closed-system analysis. Thus,

$$\text{Flow energy/pound mass} = Pv$$

As noted previously this quantity is also called flow work. This is a reasonable term since the Pv product represents the work necessary to force the pound mass of fluid into the system. However, as stated in Chapter 4, our definition of work assumes that work is not associated with mass crossing the boundary of the system, and therefore, we prefer to use the term flow energy.

The student is reminded that the units for flow energy must be consistent with the other energy quantities involved in a given equation, as demonstrated in the examples that follow. When pressure is in pounds force per square foot and specific volume is in cubic feet per pound mass, the units of flow energy are foot-pound force per pound mass.

Having determined the various forms of energy associated with a fluid as it crosses the boundary of an open system, we will proceed with our consideration of the first law of thermodynamics for an open system. This is most easily done by extending the principles of the conservation of energy to the open system. Consider the open system shown in Fig. 5.4. A mass of fluid δm_i enters the open system; a mass

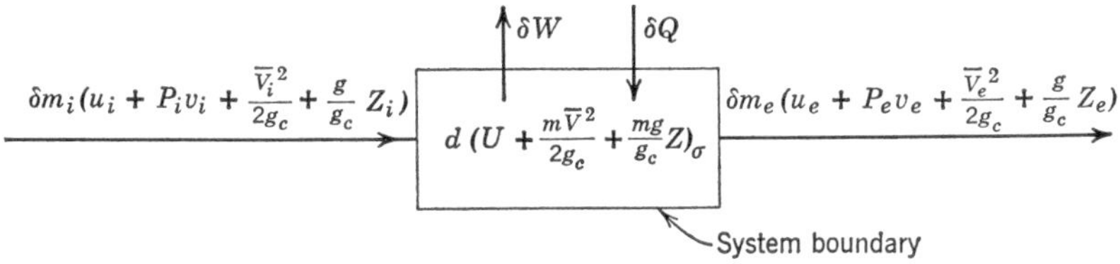

Fig. 5.4 The general case for an open system.

of fluid δm_e leaves the open system; an amount of heat δQ is transferred to the system, an amount of work δW is done by the system; and the total energy of the system changes by the amount

$$d\left(U + \frac{m\overline{V}^2}{2g_c} + \frac{mg}{g_c} Z\right)_\sigma$$

The subscript σ (suggesting the word system) refers to conditions within the boundary of the system.

As the mass δm_i crosses the boundary of the system the energy of the system increases by the amount

$$\delta m_i \left(u_i + P_i v_i + \frac{\overline{V}_i^2}{2g_c} + \frac{g}{g_c} Z_i\right)$$

where the subscripts i (suggesting the word in) refer to the state of the fluid as it enters the system.

Similarly, as the mass of fluid δm_e crosses the boundary of the system the energy of the system decreases by the amount

$$\delta m_e \left(u_e + P_e v_e + \frac{\bar{V}_e^2}{2g_c} + \frac{g}{g_c} Z_e \right)$$

where the subscripts e (suggesting the word exit) refer to the state of the fluid leaving the system.

The total energy within the boundary of the system may change either because of a change in the mass of the system or because of a change in state of the fluid within the system boundary. The increase in the total energy within the boundary is therefore given by

$$d\left(U + \frac{m\bar{V}^2}{2g_c} + \frac{mg}{g_c} Z \right)_\sigma$$

or

$$d\left[(m)_\sigma \left(u + \frac{\bar{V}^2}{2g_c} + \frac{g}{g_c} Z \right)_\sigma \right]$$

From the principle of the conservation of energy we can therefore write,

$$\delta Q + \delta m_i \left(u_i + P_i v_i + \frac{\bar{V}_i^2}{2g_c} + \frac{g}{g_c} Z_i \right)$$
$$= \delta W + \delta m_e \left(u_e + P_e v_e + \frac{\bar{V}_e^2}{2g_c} + \frac{g}{g_c} Z_e \right)$$
$$+ d\left(U + \frac{m\bar{V}^2}{2g_c} + \frac{mg}{g_c} Z \right)_\sigma \quad (5.13)$$

This equation is the first law of thermodynamics for the open system. It is important to note however, that this equation applies equally well to both a closed and open system, because for a closed system $\delta m_i = 0$ and $\delta m_e = 0$, and Eq. 5.13 reduces to Eq. 5.11, which applies to a closed system. Therefore, it is recommended that the student gain a thorough mastery of Eq. 5.13, and apply it in its entirety to a number of problems. In so doing the terms that are not significant for a given problem may be dropped out, and a familiarity with this important equation will thereby be gained.

The first law may also be written as a rate equation by considering a period of time $d\tau$ during which the mass δm_i enters the system, the mass δm_e leaves the system, the heat δQ is transferred to the system, and the work δW is done by the system. Then, as $d\tau \to 0$, we have the following rates:

$$\text{Heat-transfer rate} = \frac{\delta Q}{d\tau} = \dot{Q}$$

$$\text{Power (time rate of work)} = \frac{\delta W}{d\tau} = \dot{W}$$

$$\text{Flow rate into the system} = \frac{\delta m_i}{d\tau} = \dot{m}_i$$

$$\text{Flow rate out of the system} = \frac{\delta m_e}{d\tau} = \dot{m}_e$$

Introducing these into the first law equation, we have

$$\frac{\delta Q}{d\tau} + \frac{\delta m_i}{d\tau}\left(u_i + P_i v_i + \frac{\overline{V}_i^2}{2g_c} + \frac{g}{g_c} Z_i\right) = \frac{\delta W}{d\tau} + \frac{\delta m_e}{d\tau}\left(u_e + P_e v_e + \frac{\overline{V}_e^2}{2g_c} + \frac{g}{g_c} Z_e\right) + \frac{d}{d\tau}\left(U + \frac{m\overline{V}^2}{2g_c} + \frac{mg}{g_c} Z\right)_\sigma$$

$$\dot{Q} + \dot{m}_i\left(u_i + P_i v_i + \frac{\overline{V}_i^2}{2g_c} + \frac{g}{g_c} Z_i\right) = \dot{W} + \dot{m}_e\left(u_e + P_e v_e + \frac{\overline{V}_e^2}{2g_c} + \frac{g}{g_c} Z_e\right) + \frac{d}{d\tau}\left(U + \frac{m\overline{V}^2}{2g_c} + \frac{mg}{g_c} Z\right)_\sigma \quad (5.14)$$

5.5 The Thermodynamic Property Enthalpy

At this point we will consider another thermodynamic property, enthalpy (accent on second syllable). Enthalpy is an extensive property and is given the symbol H; the enthalpy per unit mass is given the symbol h. *Enthalpy* is defined as

$$H = U + PV$$
$$h = u + Pv \quad (5.15)$$

In words, enthalpy per unit mass is the sum of the internal energy u

and the product of the pressure and specific volume. Since each of the quantities on the right side is a thermodynamic property, enthalpy itself is a thermodynamic property. Obviously, in order to add the internal energy and pressure-volume product, the units for both must be the same. In this text the usual unit for enthalpy and internal energy is Btu/lbm. When pressure is in lbf/ft^2 and specific volume in ft^3/lbm, the factor $J = 778$ ft-lbf/Btu must be introduced in the denominator (i.e., Pv/J) in order to express the Pv term in Btu/lbm. However, the factor J will not be carried along in the equations because the essential thing is that consistent units be used, and those units might be Btu, ft-lbf, calories, joules, or any other energy unit.

Having introduced and defined the property enthalpy, we recognize that the sum of the internal energy u and the flow energy Pv in Eq. 5.13 is $(u + Pv)$ and is therefore equal to the enthalpy. The main reason for introducing the property enthalpy is to effect this simplification in problems involving open systems. One important point should be made, however. Enthalpy is always defined as a property equal to the sum of the internal energy and pressure-volume product ($H = U + PV$). Only in the case of a fluid entering or leaving an open system, however, does the PV term represent energy. The air in the room has a certain pressure and a certain volume and a certain internal energy U. The PV product can be calculated and added to the internal energy U to determine the enthalpy of the air. However, the PV product is not an energy quantity (though it has the units of energy, even as torque has the units of energy but is not an energy quantity), and therefore, the enthalpy of the air in the room is not an energy quantity. The only energy of the air in the room is its internal energy.

When we have a fluid crossing the boundary of an open system, the PV product represents flow energy, and the enthalpy of this fluid represents the sum of the internal energy and the flow energy. Therefore, the enthalpy of a substance represents an energy quantity only when this substance crosses the boundary of an open system.

Tables of thermodynamic properties, such as the steam tables, give values of enthalpy. The values given are all relative to a reference state in which the enthalpy is given an arbitrary value, usually zero. In the steam tables, the enthalpy of saturated liquid at 32 F is the reference state and is given a value of zero. In refrigerants, such as ammonia and Freon-12, the reference state is saturated liquid at -40 F, the enthalpy in this reference state being zero. Thus, it is possible to have negative values of enthalpy, as is the case for saturated solid in Table 5 of Keenan and Keyes' steam tables. It should be pointed out that when enthalpy and internal energy are given values relative to the same reference state, as is the case in essentially all thermodynamic tables, the internal energy u is equal to $-Pv$ at the reference state. For example, at 32 F the enthalpy of saturated liquid water is zero. Therefore, in this state the internal energy u is found as follows:

$$u = h - Pv$$

$$u = 0 - \frac{0.08854 \times 144 \times 0.01602}{778} = -0.000263 \text{ Btu/lbm}$$

This is negligible as far as the significant figures of the tables are concerned, but the principle should be kept in mind, as in certain cases it is significant.

In the superheat region of most thermodynamic tables, no value of specific internal energy u is given. However, this is readily calculated from the relation $u = h - Pv$, though it is important to keep the units in mind. As an example, let us calculate the specific internal energy u of superheated steam at 100 lbf/in.^{2}, 500 F.

$$u = h - Pv$$

$$u = 1279.1 - \frac{100 \times 144 \times 5.589}{778} = 1175.6 \text{ Btu/lbm}$$

The enthalpy of a substance in a saturation state is found by using the quality in the same way that the specific volume and internal energy were found. The enthalpy of saturated liquid has the symbol h_f, saturated vapor h_g, and the increase in enthalpy during vaporization, h_{fg}. For a saturation state, the enthalpy can be calculated by one of the following relations:

$$h = (1 - x)h_f + xh_g$$

$$h = h_f + xh_{fg} \tag{5.16}$$

$$h = h_g - (1 - x)h_{fg}$$

The enthalpy of compressed water may be found by using the corrections from Table 4 of the steam tables, in the same manner that the specific volume was found in Chapter 3.

When the property enthalpy is introduced into Eq. 5.13 the following relation is obtained.

$$\delta Q + \delta m_i\left(h_i + \frac{\overline{V}_i^2}{2g_c} + \frac{g}{g_c} Z_i\right) = \delta W + \delta m_e\left(h_e + \frac{\overline{V}_e^2}{2g_c} + \frac{g}{g_c} Z_e\right)$$

$$+ d\left(U + \frac{m\overline{V}^2}{2g_c} + m\frac{g}{g_c} Z\right)_\sigma \tag{5.17}$$

Similarly, when the definition of enthalpy is introduced into Eq. 5.14 we have

$$\dot{Q} + \dot{m}_i\left(h_i + \frac{\overline{V}_i^2}{2g_c} + \frac{g}{g_c}Z_i\right) = \dot{W} + \dot{m}_e\left(h_e + \frac{\overline{V}_e^2}{2g_c} + \frac{g}{g_c}Z_e\right) + \frac{d}{d\tau}\left(mu + \frac{m\overline{V}^2}{2g_c} + \frac{mg}{g_c}Z\right)_\sigma \quad (5.18)$$

Equations 5.17 and 5.18 are the form in which we will usually use the first law, since in most cases it is more convenient to use enthalpy than its equivalent, $u + Pv$.

The following example involves a closed system, but shows how the internal energy of a superheated vapor can be found from the enthalpy, pressure, and specific volume as tabulated.

Example 5.4 A cylinder fitted with a piston has an initial volume of 6 in.^{3}, and is filled with Freon-12 at 20 lbf/in.^{2}, 40 F. The Freon is then compressed until the pressure is 125 lbf/in.^{2} The final temperature is 160 F. During the compression process the heat transfer from the Freon-12 is 0.01 Btu. Determine the work of compression.

For a closed system the first law is

$$_1Q_2 = m(u_2 - u_1) + {_1W_2}$$

The values of u_1 and u_2 can be found from the Freon-12 tables, Table A.3 in the Appendix.

$$u_1 = 83.289 - \frac{20 \times 144 \times 2.137}{778} = 75.379 \text{ Btu/lbm}$$

$$u_2 = 98.023 - \frac{125 \times 144 \times 0.3901}{778} = 87.997 \text{ Btu/lbm}$$

$$m = \frac{V_1}{v_1} = \frac{6}{1728 \times 2.137} = 0.00162 \text{ lbm}$$

$$_1W_2 = -0.01 - 0.00162(87.997 - 75.379) = -0.0304 \text{ Btu}$$

5.6 Conservation of Mass and the Continuity Equation

In section 5.2, the first law of thermodynamics was developed for the closed system that undergoes a change of state. By definition a closed system has a fixed mass, and it is always possible for the closed system to undergo a change of state in which the energy of the system changes.

We know from modern physics that mass and energy are related by the well-known equation

$$E = mc^2 \tag{5.19}$$

where c = velocity of light
E = energy

We might ask the question, does the mass of a system change when its energy increases as a result, say, of heat transfer? The conclusion one reaches from Eq. 5.19 is that it does change. Let us calculate the magnitude of this change of mass for a typical problem, and determine whether this change in mass is significant.

Consider as a system a rigid vessel that contains a 1-lbm mixture of a hydrocarbon fuel (such as gasoline) and air. From our knowledge of combustion, we know that after combustion takes place it will be necessary to transfer about 1250 Btu from the system in order to restore the system to its initial temperature. From the first law,

$$_1Q_2 = U_2 - U_1 + {}_1W_2$$

we conclude, since $_1W_2 = 0$ and $_1Q_2 = -1250$ Btu, that the internal energy of the system decreases by 1250 Btu during the heat-transfer process. Let us now calculate the decrease in mass during this process using Eq. 5.19.

The velocity of light, c, is 9.83×10^8 ft/sec. Therefore:

$$1250 \text{ (Btu)} = m \text{ (lbm)} \times 9.83^2 \times 10^8 \times 10^8 \text{ ft}^2/\text{sec}^2$$

$$\times \frac{\frac{1}{778} \text{ Btu/ft-lbf}}{32.17 \text{ lbm-ft/lbf-sec}^2}$$

$$m = 3.24 \times 10^{-11} \text{ lbm}$$

Therefore, when the energy of the closed system decreases by 1250 Btu, the decrease in mass is 3.24×10^{-11} lbm.

A change in mass of this magnitude cannot be detected by even our most accurate chemical balance. And, certainly, a change in mass of this magnitude is beyond the accuracy of any engineering calculations. Therefore, if we use the laws of conservation of mass and conservation of energy as separate laws, we introduce no significant error in any thermodynamic problem (except for those involving nuclear reactions).

In the open system, the continuity equation is of significance. This equation simply states that the increase in the mass of the open system is equal to the net mass of fluid crossing the boundary of the system. Consider Fig. 5.4, and during the period of time $d\tau$ let mass δm_i enter the open system, the mass δm_e leave the open system, and the change of the mass of the system be dm_σ. Then the conservation of mass requires that

$$dm_\sigma = \delta m_i - \delta m_e$$

or (5.20)

$$(m_2 - m_1)_\sigma = m_i - m_e$$

This can also be written as a rate equation by introducing the time $d\tau$

$$\frac{dm_\sigma}{d\tau} = \frac{\delta m_i}{d\tau} - \frac{\delta m_e}{d\tau} = \dot{m}_i - \dot{m}_e \tag{5.21}$$

It is sometimes convenient to have a general expression for conservation of mass of an open system in terms of the density, velocity, area, and volume. Consider a small element of the surface of the boundary, dA, Fig. 5.5, and let fluid flow out through this surface with a velocity $\overline{V}_e$ at an angle α to the normal through the surface. Let the mass δm_e flow through this element in the time $d\tau$. It follows that

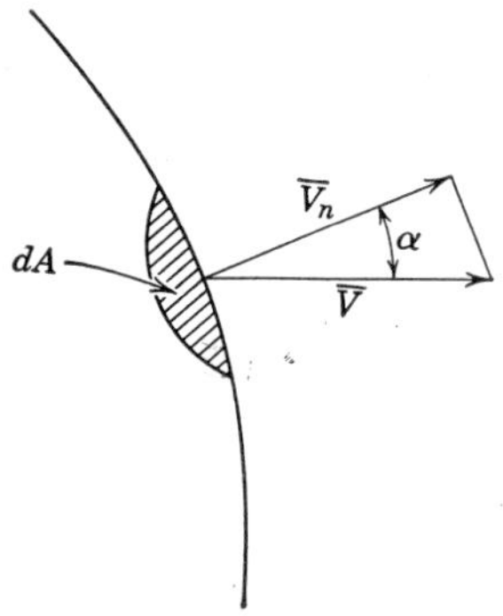

Fig. 5.5 Flow across the boundary of an open system.

$$\delta m_e = d\tau(\rho_e \overline{V}_e \cos \alpha \, dA)$$

The net rate of flow outward across the boundary of the system, $\dot{m}_e$, is found by integrating over the entire area of the boundary.

$$\dot{m}_e = \frac{\delta m_e}{d\tau} = \int_A \rho_e \overline{V}_e \cos \alpha \, dA \tag{5.22}$$

A similar expression can be written for flow into the open system.

$$\dot{m}_i = \frac{\delta m_i}{d\tau} = \int_A \rho_i \overline{V}_i \cos \alpha \, dA \tag{5.23}$$

The mass of the system can be found by considering an element of volume dV. The density of the mass occupying this volume is ρ, and therefore at any instant of time the mass of the system is

$$m_\sigma = \int_V \rho \, dV$$

where $\int_V$ indicates an integration over the entire volume occupied by the system. The rate of change of the mass of the system is therefore

$$\frac{dm_\sigma}{d\tau} = \frac{d}{d\tau}\int_V \rho \, dV = \int_V \frac{d\rho}{d\tau} dV \tag{5.24}$$

Therefore, Eqs. 5.22, 5.23, and 5.24 can be substituted into Eq. 5.21 to give the relation

$$\int_V \frac{d\rho}{d\tau}\,dV = \int_A \rho_i \overline{V}_i \cos\alpha\, dA - \int_A \rho_e \overline{V}_e \cos\alpha\, dA \tag{5.25}$$

From Eq. 5.22 or 5.23 it follows that if we consider the flow in a pipe or duct, and choose the cross-sectional area in such a way that the flow is normal to it (i.e., $\cos\alpha = 1$), and if the velocity is uniform across the section, then the mass rate of flow is given by the relation

$$\dot{m} = \rho \overline{V} A$$

or (5.26)

$$\dot{m}v = \overline{V} A$$

Example 5.5 The flow of steam through a 4-in. diameter pipe is 10,000 lbm/hr. The pressure of the steam is 100 lbf/in.^{2} and the temperature is 500 F. Determine the velocity of the steam.

$$v = 5.589 \text{ ft}^3/\text{lbm}$$

$$A = \frac{\pi(4)^2}{4 \times 144} = 0.0872 \text{ ft}^2$$

$$\dot{m} = \frac{10{,}000}{3{,}600} = 2.78 \text{ lbm/sec}$$

$$\overline{V} = \frac{\dot{m}v}{A} = \frac{2.78 \text{ lbm/sec} \times 5.589 \text{ ft}^3/\text{lbm}}{0.0872 \text{ ft}^2} = 178 \text{ ft/sec}$$

$$\overline{V} = 10{,}680 \text{ ft/min}$$

5.7 The Steady-Flow Process

The most frequently encountered process for the open system is the steady-flow process. A steady-flow process is one that meets the following requirements:

(*a*) The mass rate of flow entering the system does not vary with time and is equal to the mass rate of flow leaving the system. That is, $\dot{m}_i = \dot{m}_e =$ constant. In this case we designate the mass rate of flow as $\dot{m}$. That is, $\dot{m} = \dot{m}_i = \dot{m}_e$.

(*b*) The state, velocity, and elevation of the fluid entering and leaving do not vary with time.

$$(h_i, \overline{V}_i, Z_i, h_e, \overline{V}_e, Z_e) \neq f(\tau)$$

(*c*) The state, velocity, and elevation of the fluid at each point in the system do not vary with time. (The fluid might pass through a relatively rapid intermittent process, as in a reciprocating air compressor, and still be considered to have fulfilled this requirement. Steady flow includes such processes provided that each of these intermittent processes is identical and an integral number of cycles are considered.) There is no change in the total energy of the system and therefore

$$\frac{d}{d\tau}\left(U + \frac{m\overline{V}^2}{2g_c} + \frac{mg}{g_c} Z\right)_\sigma = 0$$

(*d*) The rates at which heat and work cross the system boundary do not vary with time.

$$\dot{Q} \neq f(\tau) \qquad \dot{W} \neq f(\tau)$$

A typical example of a steady-flow process is a steam turbine that operates at constant load, a constant rate of steam flow, and constant rate of heat transfer. It is evident that when the steam turbine is considered as an open system, the requirements for a steady-flow process as given above are all met.

For a steady-flow process, therefore, Eq. 5.18 reduces to

$$\dot{Q} + \dot{m}\left(h_i + \frac{\overline{V}_i^2}{2g_c} + \frac{g}{g_c} Z_i\right) = \dot{m}\left(h_e + \frac{\overline{V}_e^2}{2g_c} + \frac{g}{g_c} Z_e\right) + \dot{W} \qquad (5.27)$$

If we let $\dot{Q}/\dot{m} = q =$ heat transfer per pound of fluid flow, and $\dot{W}/\dot{m} = w =$ work per pound of fluid flow, we have the steady-flow energy equation expressed on a mass basis. That is,

$$q + h_i + \frac{\overline{V}_i^2}{2g_c} + \frac{g}{g_c} Z_i = h_e + \frac{\overline{V}_e^2}{2g_c} + \frac{g}{g_c} Z_e + w \qquad (5.28)$$

This is one of the most important equations used in the application of thermodynamics to engineering problems. Many problems can be solved on either a mass basis (Eq. 5.19) or time basis (Eq. 5.18), the advantage of one or the other depending on the particular problem. Example 5.6 is worked out in both ways.

In later chapters we will have occasion to use the differential-equation form of the steady-flow energy equation. In effect this involves an open system across which the change in the fluid properties is infinitesimal. Thus

$$h_e = h_i + dh$$

$$\frac{\overline{V}_e^2}{2g_c} = \frac{\overline{V}_i^2}{2g_c} + \frac{d\,(\overline{V}^2)}{2g_c}$$

$$\frac{g}{g_c} Z_e = \frac{g}{g_c} Z_i + \frac{g}{g_c} dZ$$

The heat transfer during this steady-flow process is δq and the work done is δw.

The steady-flow energy equation for this process reduces to

$$dh + \frac{d\overline{V}^2}{2g_c} + \frac{g}{g_c} dZ + \delta w - \delta q = 0 \qquad (5.29)$$

Example 5.6 The mass rate of flow into a steam turbine is 10,000 lbm/hr, and the heat transfer from the turbine is 30,000 Btu/hr. The following data are known for the steam entering and leaving the turbine:

	Inlet Conditions	Exit Conditions
Pressure	300 lbf/in.^{2}	16 lbf/in.^{2}
Temperature	700 F	
Quality		100%
Velocity	200 ft/sec	600 ft/sec
Elevation above reference plane	16 feet	10 feet
$g = 32.17$ ft/sec^2		

Determine the power output of the turbine.

From the steam tables: $h_i = 1368.3 \qquad h_e = 1152.0$

Using Eq. 5.18 first we have

$$\dot{Q} + \dot{m}\left(h_i + \frac{\overline{V}_i^2}{2g_c} + \frac{g}{g_c} Z_i\right) = \dot{m}\left(h_e + \frac{\overline{V}_e^2}{2g_c} + \frac{g}{g_c} Z_e\right) + \dot{W}$$

$$-30{,}000 \text{ Btu/hr} + 10{,}000 \text{ lbm/hr}\left(1368.3 \text{ Btu/lbm}\right.$$

$$+ \frac{200 \times 200 \text{ ft}^2/\text{sec}^2}{2 \times 32.17 \text{ lbm-ft/lbf-sec}^2 \times 778 \text{ ft-lbf/Btu}}$$

$$\left. + \frac{16 \text{ ft} \times 32.17 \text{ ft/sec}^2}{32.17 \text{ lbm-ft/lbf-sec}^2 \times 778 \text{ ft-lbf/Btu}}\right)$$

$$= 10{,}000 \frac{\text{lbm}}{\text{hr}}\left(1152.0 \frac{\text{Btu}}{\text{lbm}}\right.$$

$$+ \frac{600 \times 600 \text{ ft}^2/\text{sec}^2}{2 \times 32.17 \text{ lbm-ft/lbf-sec}^2 \times 778 \text{ ft-lbf/Btu}}$$

$$+ \frac{10 \text{ ft} \times 32.17 \text{ ft/sec}^2}{32.17 \text{ lbm-ft/lbf-sec}^2 \times 778 \text{ ft-lbf/Btu}}\Bigg) + \dot{W} \text{ Btu/hr}$$

$$- 30{,}000 + 10{,}000(1368.3 + 0.799 + 0.026) = 10{,}000(1152.0 + 7.2 + 0.0128) + \dot{W}$$

$$\dot{W} = -30{,}000 + 13{,}691{,}000 - 11{,}592{,}000 = 2{,}069{,}000 \text{ Btu/hr}$$

$$\dot{W} = \frac{2{,}069{,}000 \text{ Btu/hr}}{2545 \text{ Btu/hp-hr}} = 814 \text{ hp}$$

If Eq. 5.28 is used, the work per pound mass of fluid flowing is found first.

$$q + h_i + \frac{\overline{V}_i^{\,2}}{2g_c} + \frac{g}{g_c} Z_i = h_e + \frac{\overline{V}_e^{\,2}}{2g_c} + \frac{g}{g_c} Z_e + w$$

$$q = \dot{Q}/\dot{m} = \frac{-30{,}000}{10{,}000} = -3 \text{ Btu/lbm}$$

$$- 3 \text{ Btu/lbm} + 1368.3 \text{ Btu/lbm} + \frac{200 \times 200 \text{ ft}^2/\text{sec}^2}{2 \times 32.17 \text{ lbm-ft/lbf-sec}^2 \times 778 \text{ Btu/lbm}} + \frac{16 \text{ ft} \times 32.17 \text{ ft/sec}}{32.17 \text{ lbm-ft/lbf-sec}^2 \times 778 \text{ ft-lbf/Btu}}$$

$$= 1152 \text{ Btu/lbm} + \frac{600 \times 600 \text{ ft}^2/\text{sec}^2}{2 \times 32.17 \text{ lbm-ft/lbf-sec}^2 \times 778 \text{ ft-lbf/Btu}} + \frac{10 \text{ ft} \times 32.17 \text{ ft/sec}}{32.17 \text{ lbm-ft/lbf-sec}^2 \times 778 \text{ ft-lbf/Btu}} + w \text{ Btu/lbm}$$

$$- 3 + 1368.3 + 0.799 + 0.026 = 1152.0 + 7.2 + 0.0128 + w$$

$$w = 206.9 \text{ Btu/lbm}$$

$$\dot{W} = \frac{206.9 \text{ Btu/lbm} \times 10{,}000 \text{ lbm/hr}}{2545 \text{ Btu/hp-hr}} = 814 \text{ hp}$$

Two further observations can be made by referring to this example. First, in many engineering problems potential energy changes are insignificant when compared with the other energy quantities. In the above example the potential energy change did not change any of the significant figures. In most problems where the change in elevation is small the potential energy terms may be neglected.

Second, if velocities are small, say, under 100 ft/sec, in many cases the kinetic energy is insignificant compared to other energy quantities. Further, when the velocities entering and leaving the system are essentially the same, the change in kinetic energy is small. Since it is the change in kinetic energy that is important in the steady-flow energy equation, the kinetic-energy terms can usually be neglected when there is no significant change in velocity on entering and leaving the system.

Example 5.7 Steam at 100 lbf/in.^{2}, 400 F, enters an insulated nozzle with a velocity of 200 ft/sec. It leaves at a pressure of 20 lbf/in.^{2} and a velocity of 2000 ft/sec. Determine the final temperature or quality of the steam. (If superheated, temperature is the significant property, whereas quality is significant if saturated.)

Considering the nozzle as an open system it is evident that there is no work crossing the boundary of the system. Since the nozzle is insulated, we will assume that there is no heat transfer. Also, since there is no significant change in elevation between the inlet and exit of the nozzle, the change in potential energy is negligible. Therefore, for this process Eq. 5.28 reduces to

$$h_i + \frac{\overline{V}_i^2}{2g_c} = h_e + \frac{\overline{V}_e^2}{2g_c}$$

$$h_e = 1227.6 + \frac{(200)^2}{2 \times 32.2 \times 778} - \frac{(2000)^2}{2 \times 32.2 \times 778}$$

$$h_e = 1227.6 - 79.2 = 1148.4 \text{ Btu/lbm}$$

The two properties of the fluid leaving which we now know are pressure and enthalpy, and therefore the state of this fluid is determined. Since h_e is less than h_g at 20 lbf/in.^{2}, the quality is calculated.

$$h = h_g - (1 - x)h_{fg}$$

$$1148.4 = 1156.3 - (1 - x)960.1$$

$$(1 - x) = \frac{7.9}{960.1} = 0.8\%$$

$$x = 99.2\%$$

Example 5.8 In a refrigeration system, in which Freon-12 is the refrigerant, the Freon enters the compressor at 30 lbf/in.^{2}, 20 F, and leaves the compressor at 160 lbf/in.^{2}, 200 F. The mass rate of flow is 125 lbm/hr, and the power input to the compressor is 1 kw.

On leaving the compressor the Freon enters a water-cooled condenser at 160 lbf/in.^{2}, 180 F, and leaves as a liquid at 100 F at 150 lbf/in.^{2} Water enters the condenser at 55 F and leaves at 75 F. Determine:

(*a*) The heat transfer from the compressor per hour.

(*b*) The rate at which cooling water flows through the condenser.

Consider first the compressor as the system (an open system).

$$\text{work per pound} = w = -\frac{3412}{125} = -27.3 \text{ Btu/lbm}$$

From the Freon-12 tables

$$h_i = 79.76 \text{ Btu/lbm} \qquad h_e = 103.91 \text{ Btu/lbm}$$

Since the velocity entering a refrigeration compressor is low and not greatly different from the velocity leaving, kinetic energy changes, as well as potential energy changes, can be neglected. Using Eq. 5.28

$$q + h_i = w + h_e$$

$$q = -27.30 + 103.91 - 79.76 = -3.15 \text{ Btu/lbm}$$

$$\dot{Q} = 125 \text{ lbm/hr} \times (-3.15) \text{ Btu/lbm} = -394 \text{ Btu/hr}$$

Next we consider the condenser as the open system. The schematic diagram for this condenser is shown in Fig. 5.6.

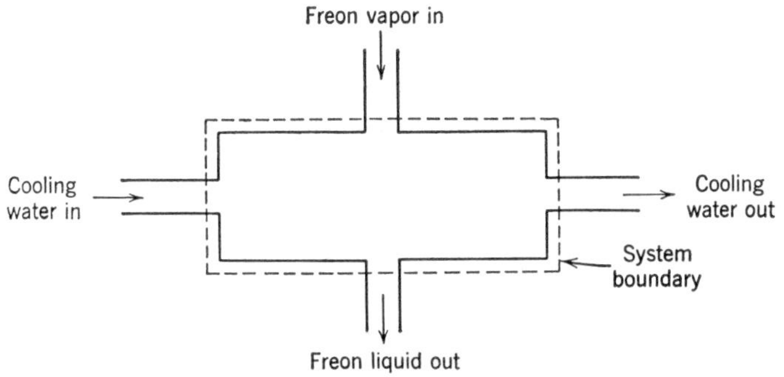

Fig. 5.6 Schematic diagram of a Freon condenser.

With this system we have two fluid streams, the Freon and the water, entering and leaving the system. It is reasonable to assume that the kinetic and potential energy changes are negligible. We note that the work is zero, and we make the other reasonable assumption that there is no heat transfer across the boundary of the system. Therefore Eq. 5.27 reduces to

$$\dot{m}_i h_i = \dot{m}_e h_e$$

However, in this case we have two streams entering and two streams leaving. Using the subscript r for refrigerant and w for water,

$$\dot{m}_r h_{ir} + \dot{m}_w h_{iw} = \dot{m}_r h_{er} + \dot{m}_w h_{ew}$$

From the Freon-12 and steam tables we have

$$h_{ir} = 100.34 \text{ Btu/lbm} \qquad h_{iw} = 23.07 \text{ Btu/lbm}$$

$$h_{er} = 31.10 \qquad h_{ew} = 43.03$$

Solving the above equation for m_w, the rate of flow of water, we have

$$\dot{m}_w = \dot{m}_r \frac{(h_i - h_e)_r}{(h_e - h_i)_w} = 125 \text{ lbm/hr} \frac{(100.34 - 31.10) \text{ Btu/lbm}}{(43.03 - 23.07) \text{ Btu/lbm}}$$

$$= 433 \text{ lbm/hr}$$

This problem can also be solved by considering only one fluid. This is done by having the boundary of the open system include the tubes that separate the Freon-12 from the water. In this case we have heat transfer across the boundary of the system, since the energy crosses the tubes due to the temperature difference between the Freon-12 and the water. This is shown schematically in Fig. 5.7, and the solution follows.

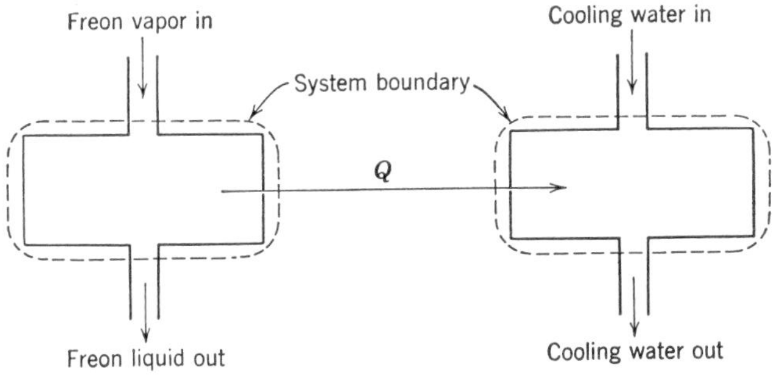

Fig. 5.7 Schematic diagram of Freon condenser considering two open systems.

The heat transfer for the open system involving Freon is calculated first.

$$\dot{Q} = \dot{m}_r(h_e - h_i)$$

$$\dot{Q} = 125 \text{ lbm/hr } (31.10 - 100.34) \text{ Btu/lbm} = -8655 \text{ Btu/hr}$$

This is also the heat transfer to the other open system, but for this system $\dot{Q} = +8655$ Btu/hr.

$$\dot{Q} = \dot{m}_w(h_e - h_i)$$

$$\dot{m}_w = \frac{8655 \text{ Btu/hr}}{(43.03 - 23.07) \text{ Btu/lbm}} = 433 \text{ lbm/hr}$$

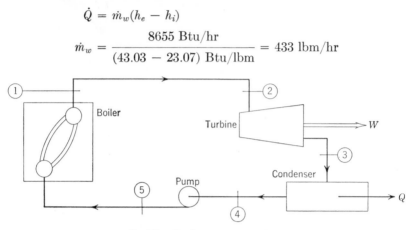

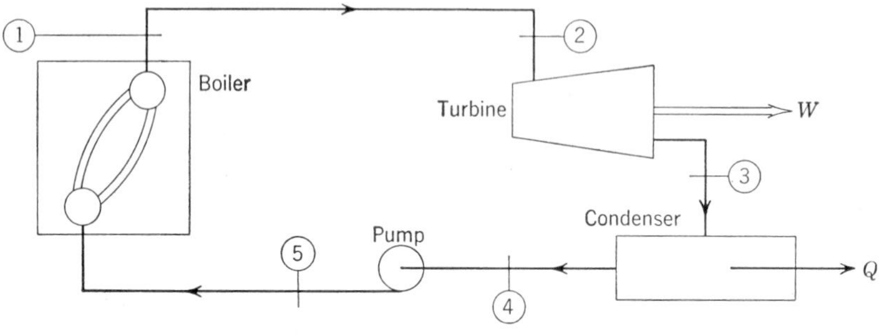

Fig. 5.8 Simple steam power plant.

Example 5.9 Consider the simple steam power plant, as described in Chapter 1 and shown in Fig. 5.8. The following data are for such a power plant.

Location	Pressure	Temperature or Quality
Leaving boiler	400 lbf/in.^{2}	600 F
Entering turbine	380 lbf/in.^{2}	560 F
Leaving turbine, entering condenser	2 lbf/in.^{2}	93%
Leaving condenser, entering pump	1.9 lbf/in.^{2}	115 F
Pump work = 3 Btu/lbm		

Determine the following quantities per pound mass:
(*a*) Heat transfer in line between boiler and turbine.
(*b*) Turbine work.
(*c*) Heat transfer in condenser.
(*d*) Heat transfer in boiler.

In working with such a cycle, one recognizes that there is a certain advantage in assigning a number to various points in the cycle. For this reason the subscripts "*i*" and "*e*" in the steady-flow energy equation are often replaced by appropriate numbers. Thus, for the turbine

$$q_{\text{turb}} + h_2 + \frac{\overline{V}_2^2}{2g_c} + \frac{g}{g_c} Z_2 = w_{\text{turb}} + h_3 + \frac{\overline{V}_3^2}{2g_c} + \frac{g}{g_c} Z_3$$

For the boiler (no work is involved)

$$q_{\text{boiler}} + h_5 + \frac{\overline{V}_5^2}{2g_c} + \frac{g}{g_c} Z_5 = h_1 + \frac{\overline{V}_1^2}{2g_c} + \frac{g}{g_c} Z_1$$

We might also use the alternate but equally acceptable notation

$$q_{\text{turb}} = {}_2q_3 \qquad q_{\text{boiler}} = {}_5q_1$$

In this problem we will neglect changes in kinetic energy and potential energy.

The following property values are obtained from the steam tables, where the subscripts refer to Fig. 5.8:

$$h_1 = 1306.9 \text{ Btu/lbm}$$

$$h_2 = 1285.2 \text{ Btu/lbm}$$

$$h_3 = 1116.2 - 0.07(1022.2) = 1044.7 \text{ Btu/lbm}$$

$$h_4 = 82.9 \text{ Btu/lbm}$$

(*a*) Considering the pipe line between boiler and turbine,

$${}_1q_2 + h_1 = h_2$$

$${}_1q_2 = h_2 - h_1 = 1285.2 - 1306.9 = -21.7 \text{ Btu/lbm}$$

(*b*) A turbine is essentially an adiabatic machine, and therefore we can write

$$h_2 = h_3 + {}_2w_3$$

$${}_2w_3 = 1285.3 - 1044.7 = 240.6 \text{ Btu/lbm}$$

(*c*) For the condenser the work is zero.

$${}_3q_4 + h_3 = h_4$$

$${}_3q_4 = 82.9 - 1044.7 = -961.8 \text{ Btu/lbm}$$

(*d*) The enthalpy at point 5 may be found by considering the steady-flow energy equation across the pump

$${}_4w_5 + h_4 = h_5$$

$$h_5 = 3.0 + 82.9 = 85.9 \text{ Btu/lbm}$$

(*e*) For the boiler we can write

$${}_5q_1 + h_5 = h_1$$

$${}_5q_1 = 1306.9 - 85.9 = 1221.0 \text{ Btu/lbm}$$

5.8 Flow into and out of a Vessel

In the last section we considered the steady-flow process for an open system. In this section we consider the process in which a fluid flows into or out of a vessel. The solution to this problem simply involves the use of Eq. 5.17. However, care must be exercised in the selection of a system, and this can be best demonstrated by an example.

Example 5.10 Steam at a pressure of 200 lbf/in.^{2}, 600 F, is flowing in a pipe, Fig. 5.9. Connected to this pipe through a valve is an evacu-

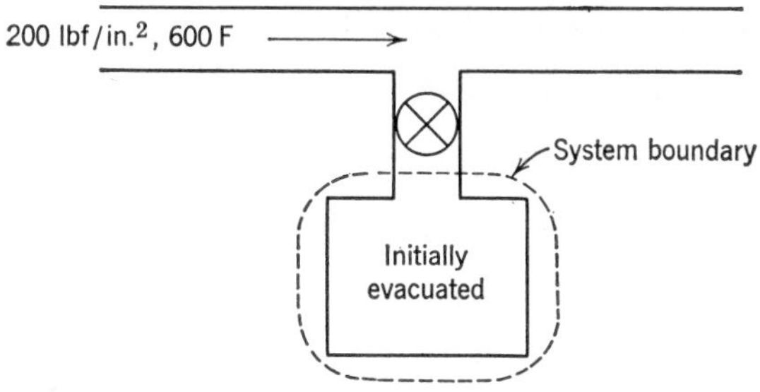

Fig. 5.9 Flow into an evacuated vessel—open-system analysis.

ated tank. The valve is opened and the tank fills with steam, until the pressure is 200 lbf/in.^{2}, and then the valve is closed. The process takes place adiabatically and kinetic energies and potential energies are negligible. Determine the final temperature of the steam.

Consider first the open system shown in Fig. 5.9. Considering Eq. 5.17 we note that

$$\delta Q = 0 \qquad \delta m_e = 0 \qquad \delta W = 0$$

Therefore,

$$\delta m_i h_i = dU$$

Integrating, and noting that h_i is constant, we have

$$h_i \int_0^{m_i} \delta m = \int_0^{U_2} dU$$

$$m_i h_i = U_2 = m_2 u_2$$

But

$$m_i = m_2$$

and therefore

$$h_i = u_2$$

Therefore, the final internal energy of the steam in the tank is equal to the enthalpy of the steam entering the tank.

From the steam tables

$$h_i = u_2 = 1322.1 \text{ Btu/lbm}$$

Since the final pressure is given as 200 lbf/in.^{2}, we know two properties of the final state and therefore the final state is determined. However, the steam table does not list the internal energies of superheated steam, and therefore one must calculate a few internal energies before an interpolation can be made for the final temperature. The temperature corresponding to a pressure of 200 lbf/in.^{2} and an internal energy of 1322.1 Btu/lbm is found to be 885 F.

This problem can also be solved using a closed-system analysis. In Fig. 5.10 let the system be composed of the mass of steam which will

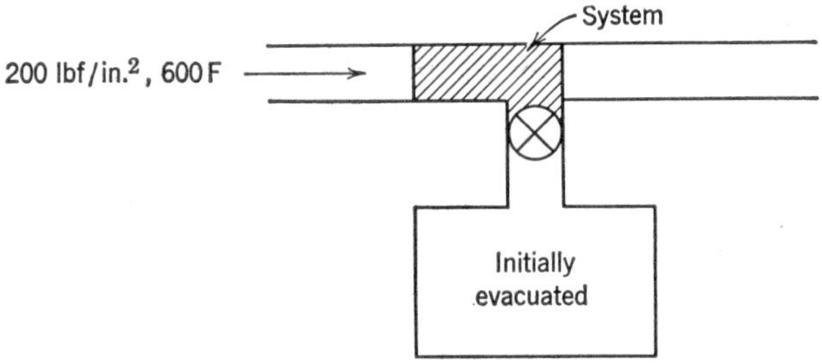

Fig. 5.10 Flow into an evacuated vessel—closed-system analysis.

enter the tank, plus the evacuated tank. The process is adiabatic, but we must examine the boundaries for work. The boundaries of the system certainly move, and we realize that the steam in the pipe does work on the steam that comprises the system, the amount of work being

$$-W = P_1V_1 = mP_1v_1$$

Writing the first law for the closed system, Eq. 5.11, and noting that kinetic and potential energies are being neglected, we have

$${}_1Q_2 = U_2 - U_1 + {}_1W_2$$

$$0 = U_2 - U_1 - P_1V_1$$

$$0 = mu_2 - mu_1 - mP_1v_1 = mu_2 - mh_1$$

Therefore,

$$u_2 = h_1,$$

which is the same conclusion which was reached using the open system.

Example 5.11 Let the tank of the previous problem have a volume of 10 ft^3 and initially contain saturated steam at 50 lbf/in.^{2} The valve is then opened and steam from the line at 200 lbf/in.^{2}, 600 F, flows into the tank until the pressure is 200 lbf/in.^{2}

Calculate the mass of steam that flows into the tank. Using Eq. 5.17 we note that

$$\delta Q = 0 \qquad \delta m_e = 0 \qquad \delta W = 0$$

Again neglecting kinetic and potential energies, we have

$$\delta m_i h_i = dU$$

Integrating

$$h_i \int_0^{m_i} \delta m_i = \int_{U_1}^{U_2} dU$$

$$m_i h_i = U_2 - U_1 = m_2 u_2 - m_1 u_1$$

But, from the continuity equation,

$$m_i = m_2 - m_1$$

Therefore,

$$(m_2 - m_1)h_i = m_2 u_2 - m_1 u_1$$

$$m_2(h_i - u_2) = m_1(h_i - u_1)$$

There are two unknowns in this equation; namely, m_2 and u_2. However, we have one additional equation, namely,

$$m_2 v_2 = V_T = 10 \text{ ft}^3$$

These two equations can be solved simultaneously by trial and error. The correct solution is shown below:

$$v_1 = 8.515 \text{ ft}^3/\text{lbm} \qquad m_1 = \frac{10}{8.515} = 1.173 \text{ lbm}$$

$$h_i = 1322.1 \text{ Btu/lbm} \qquad u_1 = 1095.3 \text{ Btu/lbm}$$

Assume

$$m_2 = 3.05 \text{ lbm}$$

Then

$$v_2 = \frac{10}{3.05} = 3.28 \text{ ft}^3/\text{lbm}$$

At

$$P_2 = 200 \text{ lbf/in.}^2 \qquad \text{and} \qquad v_2 = 3.28 \text{ ft}^3$$

$$u_2 = 1234.8 \text{ Btu/lbm}$$

We can now solve for m_2 from the other equation, and check on the correctness of our assumption.

$$m_2 = m_1 \frac{(h_1 - u_1)}{(h_i - u_2)} = \frac{1.173(1322.1 - 1095.3)}{1322.1 - 1234.8} = \frac{265}{87.3} = 3.05 \text{ lbm}$$

This checks our original assumption.

The mass of steam that flows into the tank is

$$m_2 - m_1 = 3.05 - 1.17 = 1.88 \text{ lbm}$$

A plot of the assumed m_2 vs. calculated m_2 is of help in arriving quickly at a solution to such a problem.

Example 5.12 A tank of 50 ft^3 volume contains saturated ammonia at a pressure of 200 lbf/in.^{2} Initially the tank contains 50 per cent liquid and 50 per cent vapor by volume. Vapor is withdrawn from the top of the tank until the pressure is 100 lbf/in.^{2} Assuming that only vapor (i.e., no liquid) leaves and that the process is adiabatic, calculate the mass of ammonia that is withdrawn.

Considering Eq. 5.17, we note that

$$\delta Q = 0 \qquad \delta m_i = 0 \qquad \delta W = 0$$

Kinetic and potential energies can also be neglected. Therefore:

$$dU + \delta m_e h_e = 0$$

The vapor leaving will always be saturated, and since the pressure is changing, h_e is not constant. However, at 200 lbf/in.^{2}, $h_g = 632.7$ Btu/lbm and at 100 lbf/in.^{2}, $h_g = 626.5$ Btu/lbm. Since the change in h_g during this process is small, we may assume with good accuracy that h_e is the average of the two values given above. Therefore

$$h_e = 629.6 \text{ Btu/lbm}$$

On integrating, we have

$$m_2 u_2 - m_1 u_1 + m_e h_e = 0$$

But

$$m_e = m_1 - m_2$$

$$m_2 u_2 - m_1 u_1 = m_2 h_e - m_1 h_e$$

$$m_2(h_e - u_2) = m_1 h_e - m_1 u_1$$

The following values are from the ammonia tables:

$$v_{f1} = 0.02732 \text{ ft}^3/\text{lbm} \qquad v_{g1} = 1.502 \text{ ft}^3/\text{lbm}$$

$$v_{f2} = 0.02584 \qquad v_{g2} = 2.952$$

$$u_{f1} = 150.9 - \frac{200 \times 144 \times 0.0273}{778} = 149.9 \text{ Btu/lbm}$$

$$u_{f2} = 104.7 - \frac{100 \times 144 \times 0.0258}{778} = 104.2$$

$$u_{g1} = 632.7 - \frac{200 \times 144 \times 1.502}{778} = 577.1$$

$$u_{g2} = 626.5 - \frac{100 \times 144 \times 2.952}{778} = 571.9$$

$$u_{fg2} = 571.9 - 104.2 = 467.7$$

Calculating first the initial mass, m_1, in the tank: The mass of the liquid initially present, m_{f1}, is

$$m_{f1} = \frac{25}{0.02732} = 915 \text{ lbm}$$

Similarly, the initial mass of vapor, m_{g1}, is

$$m_{g1} = \frac{25}{1.502} = 16.6 \text{ lbm}$$

$$m_1 = m_{f1} + m_{g1} = 915 + 17 = 932 \text{ lbm}$$

$$m_1 h_e = 932 \times 629.6 = 586{,}800 \text{ Btu}$$

$$m_1 u_1 = m_{f1} u_{f1} + m_{g1} u_{g1} = 915 \times 149.9 + 16.6 \times 577.0 = 146{,}700 \text{ Btu}$$

Substituting these into the energy equation

$$m_2(h_e - u_2) = m_1 h_e - m_1 u_1$$

$$= 586{,}800 - 146{,}700 = 442{,}100 \text{ Btu}$$

Again we have a trial-and-error solution.

Assume $m_2 = 850$ lbm. Then

$$v_2 = \tfrac{50}{850} = 0.0588 \text{ ft}^3/\text{lbm}$$

The final quality and internal energy can now be calculated.

$$v_2 = 0.0588 = 0.0258 + x_2(2.952 - 0.0258)$$

$$x_2 = \frac{0.0334}{2.926} = 0.01127$$

$$u_2 = 104.2 + 0.01127(467.7) = 109.5 \text{ Btu/lbm}$$

This assumed value of m_2 is now checked by the energy equation

$$m_2 = \frac{m_1 h_e - m_1 u_1}{h_e - u_2} = \frac{442{,}100}{629.6 - 109.5} = 850 \text{ lbm}$$

The mass of ammonia that was withdrawn, m_e, is

$$m_e = m_1 - m_2 = 932 - 850 = 82 \text{ lbm}$$

5.9 The Constant-Volume and Constant-Pressure Specific Heats and the Joule-Thomson Coefficient

The constant-volume specific heat, the constant-pressure specific heat, and the Joule-Thomson coefficient are each defined in terms of properties that have already been considered, and therefore the definitions are given here. The full significance of these quantities will be evident in later chapters.

The *constant-volume specific heat* C_v is defined by the relation

$$C_v \equiv \left(\frac{\partial u}{\partial t}\right)_v \tag{5.30}$$

The *constant-pressure specific heat* C_p is defined by the relation

$$C_p \equiv \left(\frac{\partial h}{\partial t}\right)_P \tag{5.31}$$

The *Joule-Thomson coefficient* μ_J is defined by the relation

$$\mu_J \equiv \left(\frac{\partial t}{\partial P}\right)_h \tag{5.32}$$

Note that each of these quantities is defined in terms of properties, and therefore the constant-volume and constant-pressure specific heats and the Joule-Thomson coefficient are thermodynamic properties of a substance. These definitions assume constant composition, and also that there are no surface, electrical, or magnetic effects.

The term *heat capacity* is often used instead of specific heat. However, neither term adequately conveys the nature of the thermodynamic property involved. For example, consider the two identical closed systems of Fig. 5.11. In the first, 100 Btu of heat is transferred to the system, and in the second, 100 Btu of work is done on the system. Thus, the change of internal energy is the same in each, and therefore the final state and the final temperature are the same in each. Therefore, in accordance with Eq. 5.30, exactly the same value for the average constant-volume specific heat would be found for the substance for the

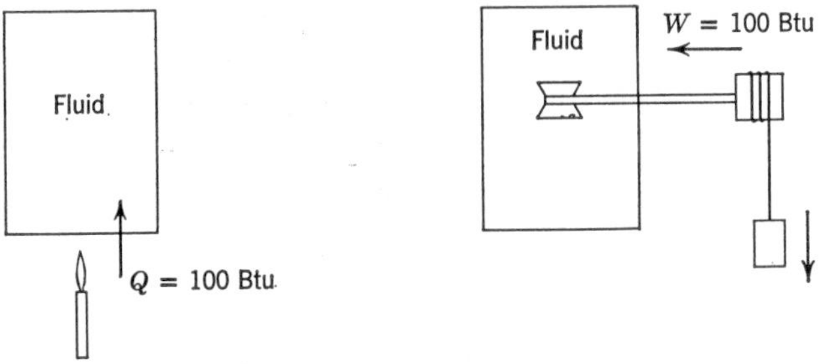

Fig. 5.11 Sketch showing two ways in which a given ΔU may be achieved.

two processes. However, if the specific heat is defined in terms of the heat transfer, different values will be obtained for the specific heat, since in the first case the heat transfer is finite and in the second case it is zero. Thus, although the terms specific heat and heat capacity are not the most appropriate, it is always important to keep in mind that these are thermodynamic properties, and are defined by Eqs. 5.30 and 5.31. When other effects are involved, additional specific heats can be defined, such as specific heat at constant magnetization for a system involving magnetic effects.

Example 5.13 Estimate the constant-pressure specific heat of steam at 100 lbf/in.^{2}, 500 F.

If we consider a change of state at constant pressure, Eq. 5.31 may be written

$$C_p \approx \left(\frac{\Delta h}{\Delta t}\right)_P$$

From the steam tables,

$$\text{At } 100 \text{ lbf/in.}^2, 480 \text{ F}, h = 1269.0$$

$$\text{At } 100 \text{ lbf/in.}^2, 520 \text{ F}, h = 1289.2$$

Since we are interested in C_p at 100 lbf/in.^{2}, 500 F,

$$C_p \approx \frac{20.2}{40} = 0.505 \text{ Btu/lbm F}$$

5.10 The Constant-Pressure, Quasistatic Process for a Closed System

The constant-pressure process for a closed system is of particular interest if it is quasistatic. Consider Fig. 5.12 in which heat is transferred

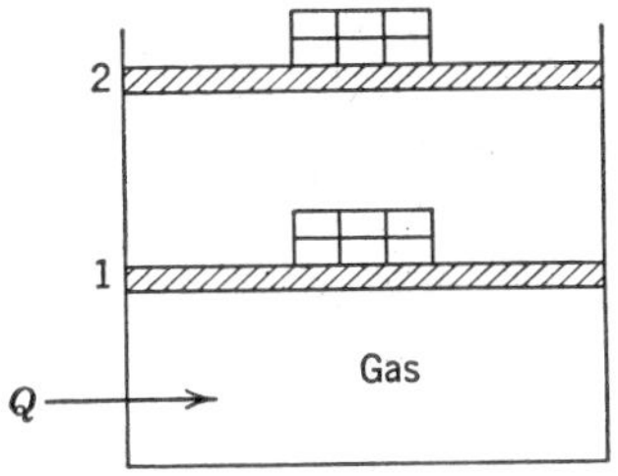

Fig. 5.12 The constant-pressure quasistatic process.

to the gas while the pressure remains constant. Since this is a closed system, we can apply the first law and write

$$_1Q_2 = U_2 - U_1 + {_1W_2}$$

The work can be calculated from the relation

$$_1W_2 = \int_1^2 P\,dV$$

Since the pressure is constant

$$_1W_2 = P\int_1^2 dV = P(V_2 - V_1) = P_2V_2 - P_1V_1$$

Therefore

$$_1Q_2 = U_2 - U_1 + P_2V_2 - P_1V_1 = H_2 - H_1$$

On a unit-mass basis we have

$$_1q_2 = h_2 - h_1$$

Thus, in a constant-pressure quasistatic process, the heat transfer is equal to the change in enthalpy. Earlier in this chapter we stressed the fact that enthalpy had significance as an energy quantity only in a flow process. This is one exception to that statement, and the only reason for it is that in a constant-pressure process the work done during the process is equal to the difference in the PV product. If pressure had not been constant this would not be true. The student is urged to solve all problems from basic relations and is cautioned against extending this simplified relation for constant-pressure processes to other processes where it does not apply.

Example 5.14 A cylinder fitted with a piston has a volume of 2 ft^3 and contains steam at 50 lbf/in.^{2}, 300 F. Heat is transferred to the steam until the temperature is 500 F, while the pressure remains constant.

Determine the heat transfer, the work, and the change in internal energy for this process.

This is a closed system, and changes in kinetic and potential energy are not significant. Therefore

$$_1Q_2 = m(u_2 - u_1) + {_1W_2}$$

$$_1W_2 = \int_1^2 P\,dV = P\int_1^2 dV = P(V_2 - V_1) = m(P_2v_2 - P_1v_1)$$

Therefore

$$_1Q_2 = m(u_2 - u_1) + m(P_2v_2 - P_1v_1) = m(h_2 - h_1)$$

$$m = \frac{V_1}{v_1} = \frac{2}{8.773} = 0.228 \text{ lbm}$$

$$h_1 = 1184.3 \qquad h_2 = 1283.9$$

$$_1Q_2 = 0.228(1283.9 - 1184.3) = 22.7 \text{ Btu}$$

$$_1W_2 = mP(v_2 - v_1) = \frac{0.228 \times 50 \times 144}{778}(11.309 - 8.773)$$

$$= 5.34 \text{ Btu}$$

Therefore

$$U_2 - U_1 = {_1Q_2} - {_1W_2} = 22.7 - 5.3 = 17.4 \text{ Btu}$$

This can be checked by finding u_1 and u_2

$$u_1 = h_1 - P_1v_1 = 1184.3 - \frac{50 \times 144 \times 8.773}{778}$$

$$= 1103.0 \text{ Btu/lbm}$$

$$u_2 = h_2 - P_2v_2 = 1283.9 - \frac{50 \times 144 \times 11.309}{778}$$

$$= 1179.2 \text{ Btu/lbm}$$

$$U_2 - U_1 = m(u_2 - u_1) = 0.228(1179.2 - 1103.0) = 17.4 \text{ Btu}$$

5.11 The Throttling Process

A throttling process is a steady-flow process across a restriction, with a resulting drop in pressure. A typical example is the flow through a partially opened valve or a restriction in the line. In most cases this occurs so rapidly and in such a small space, that there is neither sufficient time nor a large enough area for much heat transfer. Therefore, we usually may assume such processes to be adiabatic.

If we consider an open system, shown in Fig. 5.13, we can write the steady-flow energy equation for this process. There is no work, no

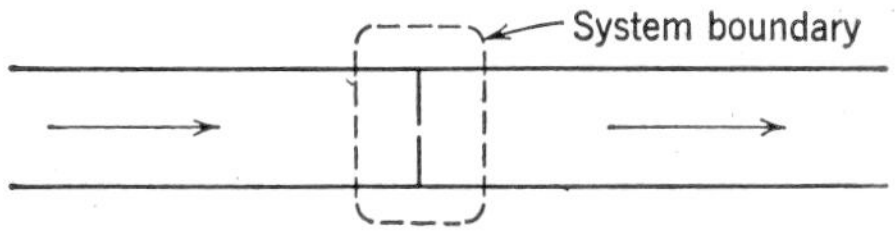

Fig. 5.13 The throttling process.

change in potential energy, and we make the reasonable assumption that there is no heat transfer. The steady-flow energy equation, Eq. 5.28, reduces to

$$h_i + \frac{\overline{V}_i^2}{2g_c} = h_e + \frac{\overline{V}_e^2}{2g_c}$$

The specific volume always increases in such a process, and, therefore, if the pipe is of constant diameter, the kinetic energy of the fluid increases. In many cases, however, this increase in kinetic energy is small (or perhaps the diameter of the exit pipe is larger than that of the inlet pipe) and we can say with a high degree of accuracy that in this process the final and initial enthalpies are equal.

It is for such a process that the Joule-Thomson coefficient, μ_J, is significant. It was defined previously as

$$\mu_J = \left(\frac{\partial t}{\partial P}\right)_h$$

A positive Joule-Thomson coefficient means that the temperature drops during throttling, and when the Joule-Thomson coefficient is negative the temperature rises during throttling.

Example 5.15 Steam at 100 lbf/in.^{2}, 500 F is throttled to 20 lbf/in.^{2} Changes in kinetic energy are negligible for this process. Determine the final temperature and specific volume of the steam, and the average Joule-Thomson coefficient.

For this process,

$$h_i = h_e = 1279.1 \text{ Btu/lbm}$$

$$P_e = 20 \text{ lbf/in.}^2$$

These two properties determine the final state. From the superheat table for steam

$$t_e = 484.2 \text{ F}$$

$$\mu_{J\,(\text{av})} = \left(\frac{\Delta t}{\Delta P}\right)_h = \frac{-15.8 \text{ F}}{-80 \text{ lbf/in.}^2} = 0.198 \text{ in.}^2\text{-F/lbf}$$

Frequently a throttling process involves a change in the phase of the fluid. Examples of this are the throttling calorimeter and the expansion valve in a vapor-compression refrigeration system. The following examples illustrate a thermodynamic analysis of these devices.

Example 5.16 The throttling calorimeter is a device by which the enthalpy of saturated steam flowing in a line and having a small amount of entrained liquid may be determined. A throttling calorimeter is shown in Fig. 5.14. The easiest measurements to make in order

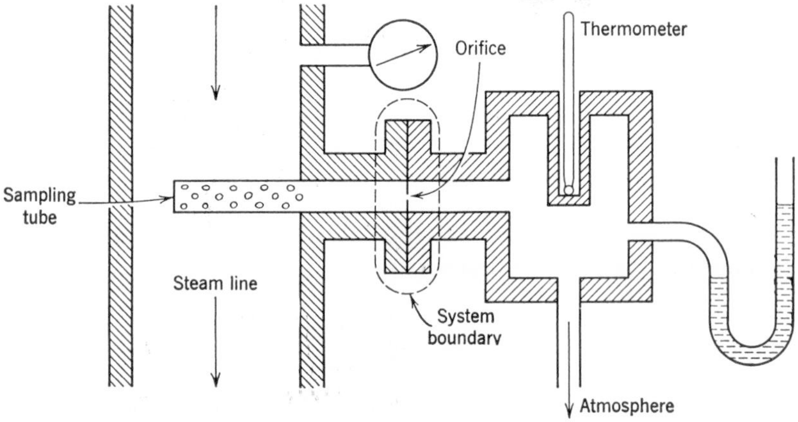

Fig. 5.14 The throttling calorimeter.

to determine the state of the steam are pressure and temperature. However, since the steam flowing in the line is saturated, pressure and temperature measurements will not fix the state. Consider now the process in which some of the steam from the pipe enters the sampling tube, flows through the orifice and into the calorimeter, and then to the atmosphere. This is a process for which the initial and final enthalpies are equal, as may be demonstrated by considering the open system shown. Further, the properties of steam (and many other substances) are such that for certain initial pressures and qualities, the steam in the calorimeter will be superheated. This is illustrated in the following calculation.

The following data are given:

Pressure in steam line:	125 lbf/in.^{2}
Pressure in calorimeter:	16 lbf/in.^{2}
Temperature in calorimeter:	240 F

Determine the quality of the steam in the pipe.

The enthalpy of the steam in the calorimeter, as found from the steam tables, is 1163.7 Btu/lbm, which is also the enthalpy of the steam in the line. We also know the pressure in the line, and, therefore, the state is determined and the quality may be found.

$$h_{\text{line}} = 1163.7 = h_{g(\text{line})} - (1 - x)_{\text{line}}\, h_{fg(\text{line})}$$

$$1163.7 = 1191.1 - (1 - x)_{\text{line}}\, 875.4$$

$$(1 - x)_{\text{line}} = \frac{27.4}{875.4} = 0.0313$$

$$x_{\text{line}} = 0.9687 \text{ or } 96.87\%$$

The lowest quality that can be measured at any given pressure is determined by the requirement that the steam in the calorimeter be superheated. As a precautionary measure, it is usually required that the steam in the calorimeter be superheated at least 10 F.

Example 5.17 The throttling process across the expansion valve or through the capillary tube in a vapor compression refrigeration cycle was illustrated in Chapter 1. In this process the pressure of the refrigerant drops from the high pressure in the condenser to the low pressure in the evaporator. Also during this process some of the liquid flashes into vapor. If we consider this process adiabatic, the quality of the refrigerant entering the evaporator can be calculated.

Consider the following process, in which ammonia is the refrigerant. The ammonia enters the expansion valve at a pressure of 225 lbf/in.^{2} and a temperature of 90 F. Its pressure on leaving the expansion valve is 38.5 lbf/in.^{2} Calculate the quality of the ammonia leaving the expansion valve.

In this case the enthalpy entering and leaving are equal.

$$h_i = h_e$$

From the ammonia tables

$$h_i = 143.6 \text{ Btu/lbm}$$

(The enthalpy of a slightly compressed liquid is essentially a function of the temperature.)

$$h_e = 143.6 = 53.8 + x_e(561.1)$$

$$x_e = \frac{89.7}{561.1} = 0.160 = 16.0\%$$

PROBLEMS

5.1 At a toy fair there are four types of power automobiles. A is battery operated; B has an electric motor that utilizes a-c current; C has a spring that can be wound with a key; and D has a charged capsule of CO_2 gas and operates like a rocket. Examine each of the automobiles as it operates for heat, work, and changes in internal energy. In each case consider the entire car as the system.

5.2 A certain hydraulic transmission has an efficiency of 94%. This means that the power output is 94% of the power input.

Assuming steady-state conditions (i.e., no change in the internal energy or kinetic energy of the system), determine the heat transfer per hour for a power input of 200 hp.

5.3 The average heat transfer from a person to the surroundings when he is not actively working is about 400 Btu/hr. Suppose that in an auditorium containing 1000 people the ventilation system fails.

(*a*) How much does the internal energy of the air in the auditorium increase during the first 15 minutes after the ventilation system fails?

(*b*) Considering the auditorium and all the people as a system, and assuming no heat transfer to the surroundings, how much does the internal energy of the system change? How do you account for the fact that the temperature of the air increases?

5.4 A sealed "bomb" containing certain chemicals is placed in a tank of water which is open to the atmosphere. When the chemicals react, heat is transferred from the bomb to the water, causing the temperature of the water to rise. A stirring device is used to circulate the water, and the power input to the rod driving the stirrer is 0.04 hp.

In a 15-min period the heat transfer from the bomb is 1000 Btu and the heat transfer from the tank to the surrounding air is 50 Btu. Assuming no evaporation of water, determine the increase in the internal energy of the water.

5.5 On a warm summer day a housewife decides to beat the heat by closing the windows and doors in the kitchen, and opening the refrigerator door. At first she feels cool and refreshed, but after a while the effect begins to wear off.

Evaluate this situation as it relates to the first law, considering the following systems: the refrigerator, the room excluding the refrigerator, and the room including the refrigerator.

5.6 A radiator of a steam heating system has a volume of 0.7 ft^3. When the radiator is filled with dry saturated steam at a pressure of 17 lbf/in.^{2} all valves to the radiator are closed. How much heat will have been transferred to the room when the pressure of the steam is 10 lbf/in^2.?

5.7 A pressure vessel having a volume of 5 ft^3 contains steam at the critical point. Heat is removed until the pressure is 400 lbf/in.^{2} Determine the heat transferred from the steam.

5.8 A rigid vessel having a volume of 20 ft^3 is filled with steam at 100 lbf/in.^{2}, 500 F. Heat is transferred from the steam until it exists as saturated vapor. Calculate the heat transferred during this process.

5.9 A tank of 10 ft^3 volume contains saturated ammonia vapor at 120 F. Calculate the heat transfer when the temperature of the ammonia is decreased to 40 F.

5.10 A sealed tube has a volume of 2 in.^{3} and contains a certain fraction of liquid and vapor H_2O in equilibrium at 14.7 lbf/in.^{2} The fraction of liquid and vapor is such that when heated the steam will pass through the critical point. Calculate the heat transfer when the steam is heated from the initial state at 14.7 lbf/in.^{2} to the critical state.

5.11 A steam boiler has a total volume of 80 ft^3. The boiler initially contains 60 ft^3 of liquid water and 20 ft^3 of vapor in equilibrium at 14.7 lbf/in.^{2} The boiler is fired up and heat is transferred to the water and steam in the boiler. Somehow, the valves on the inlet and discharge of the boiler are both left closed. The relief valve lifts when the pressure reaches 800 lbf/in.^{2} How much heat was transferred to the water and steam in the boiler before the relief valve lifted?

5.12 A tank of Freon-12 contains half liquid and half vapor by volume, and has a total volume of 4 ft^3. The initial temperature is

120 F. Calculate the heat transfer when the contents are cooled to 80 F.

5.13 An insulated and evacuated vessel of 1 ft^3 volume contains a capsule of water at 100 lbf/in.^{2}, 300 F. The volume of the capsule is 0.1 ft^3. The capsule breaks and the contents fill the entire volume. What is the final pressure?

5.14 The tank *A* (Fig. 5.15) contains 1 lbm saturated Freon-12 vapor at a temperature of 80 F. The valve is then opened slightly and Freon flows slowly into the cylinder until the pressure is 20

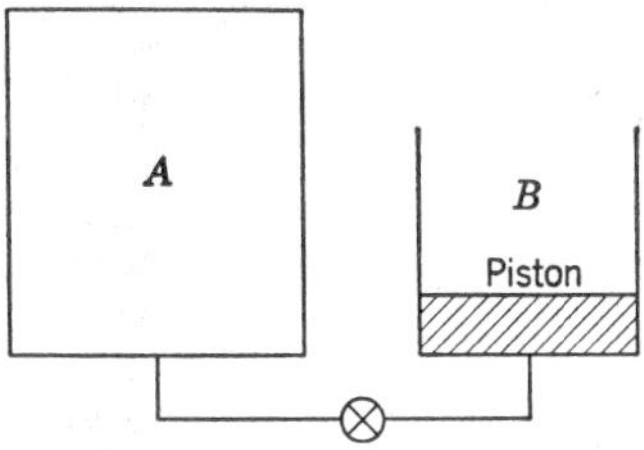

Fig. 5.15 Sketch for Problem 5.14.

lbf/in.^{2} During this process, heat is transferred to the Freon-12 so that the temperature remains constant at 80 F. Calculate the heat transfer during this process.

5.15 A cylinder fitted with a piston contains 5 lbm of H_2O at a pressure of 200 lbf/in.^{2} and a quality of 80%. The piston is restrained by a spring which is so arranged that for zero volume in the cylinder the spring is fully extended. The spring force is proportional to the spring displacement. The weight of the piston may be neglected so that the force on the spring is exactly balanced by the pressure forces on the H_2O.

Heat is transferred to the H_2O until its volume is 150% of the initial volume.

(*a*) What is the final pressure?

(*b*) What is the quality (if saturated) or temperature (if superheated) in the final state?

(*c*) Draw a *P-V* diagram and determine the work.

(*d*) Determine the heat transfer.

5.16 Freon-12 vapor enters a compressor at 25 lbf/in.^{2}, 40 F, and the mass rate of flow is 5 lbm/min. What is the smallest diameter tubing that can be used if the velocity of refrigerant must not exceed 20 ft/sec?

5.17 Steam which is contained in a cylinder expands against a piston. Following are the conditions before and after expansion.

Before Expansion		After Expansion	
Pressure:	150 lbf/in.^{2}	Pressure:	16 lbf/in.^{2}
Temperature:	500 F	Volume:	0.6 ft^3
Volume:	0.1 ft^3		

The heat transfer during expansion = −0.8 Btu. Calculate the work done during this process.

5.18 One lbm of steam is confined inside a spherical elastic membrane or balloon which supports an internal pressure proportional to its diameter. The initial condition of the steam is 10 lbf/in.^{2}, 400 F. Heat is transferred to the steam until the pressure reaches 12 lbf/in.^{2} Determine:

(*a*) The final temperature.
(*b*) The heat transfer.

5.19 The following data are for a simple steam power plant of the type shown in Fig. 5.8.

Rate of steam flow = 200,000 lbm/hr
Power to pump = 400 hp
Pipe diameters:
  Steam generator to turbine: 8 in.
  Condenser to steam generator: 3 in.
Pressures and temperatures at various points in the cycle:

Location	Pressure	Quality or Temperature	Velocity
Entering turbine	800 lbf/in.^{2}	900 F	
Leaving turbine, entering condenser	1.6 lbf/in.^{2}	92%	600 ft/sec
Leaving condenser, entering pump	1.5 lbf/in.^{2}	110 F	
Leaving pump	900 lbf/in.^{2}		
Entering economizer	890 lbf/in.^{2}	115 F	
Leaving economizer, entering steam generator	860 lbf/in.^{2}	350 F	
Leaving steam generator	830 lbf/in.^{2}	920 F	

Calculate:

(*a*) Power output of turbine.
(*b*) Heat transfer per hour in condenser, economizer, and steam generator.
(*c*) Diameter of pipe connecting the turbine to the condenser.

(*d*) Gallons of cooling water per minute through the condenser if the temperature of the cooling water increases from 55 F to 75 F in the condenser.

5.20 A certain process used in making fresh water from salt water involves the use of a compressor to compress the vapor. Saturated vapor enters the compressor at 14.7 lbf/in.^{2}, the vapor leaves the compressor at 19 lbf/in.^{2}, 240 F. The velocities entering and leaving the compressor are essentially the same. The rate of flow is 100 lbm/hr and the heat transferred from the compressor is 750 Btu/hr. Calculate the power required to drive the compressor.

5.21 Steam enters the nozzle of a turbine with a low velocity at a pressure of 400 lbf/in.^{2}, 600 F, and leaves the nozzle at 250 lbf/in.^{2} at a velocity of 1540 ft/sec. The rate of flow of steam is 3000 lbm/hr. Calculate the quality or temperature of the steam leaving the nozzle and the exit area of the nozzle.

5.22 In a certain process industry it is necessary to compress 500 lbm/hr of steam from an initial state of 20 lbf/in.^{2}, 230 F to a pressure of 70 lbf/in.^{2} The temperature after compression is 460 F. Heat transfer from the compressor is 10,000 Btu per hour. Kinetic and potential energy changes may be neglected. What horsepower is required to drive the compressor?

5.23 Recently construction of a one-mile-high skyscraper was proposed. Suppose that in such a skyscraper heating steam is to be supplied to the top floor via a vertical pipe. Steam enters the pipe at ground level as dry saturated vapor at a pressure of 30 lbf/in.^{2} At the top of the pipe the pressure is 15 lbf/in.^{2}, and the heat transfer from the steam as it flows up the pipe is 50 Btu/lbm. What is the quality of the steam at the top of the pipe?

5.24 Water vapor is compressed in a centrifugal compressor. Dry saturated vapor at 100 F enters the compressor and the vapor leaves at 4 lbf/in.^{2}, 340 F. Heat is transferred from the vapor during the compression process at the rate of 2000 Btu/hr. The rate of flow of vapor is 300 lbm/hr. Calculate the horsepower required to drive the compressor.

5.25 Consider the steam turbine-driven water pump, which is shown schematically in Fig. 5.16. The enthalpy of the water is increased 4 Btu/lbm during the pumping process. The initial state of the water and the initial and final states of the steam are shown. Consider both the pumping process and expansion process adiabatic.

(*a*) What is the rate of steam flow required to drive the turbine?

(*b*) If the turbine is replaced with an electric motor, what capacity (hp) motor is required?

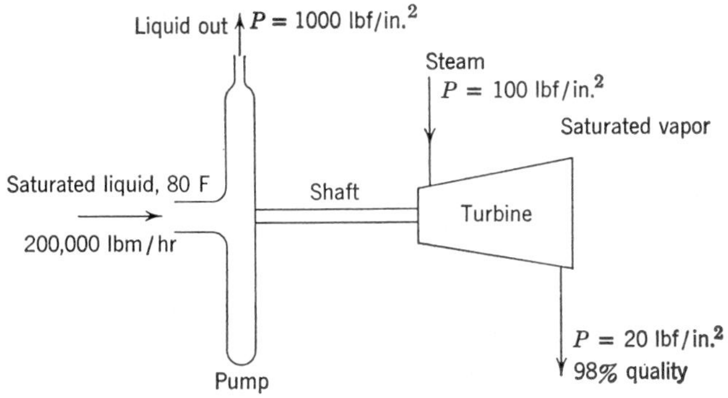

Fig. 5.16 Sketch for Problem 5.25.

5.26 The following data are for a refrigeration cycle which has a hermetically sealed compressor and uses Freon-12 as a refrigerant.

	Pressure	Temperature
Leaving compressor	180 lbf/in.^{2}	240 F
Entering condenser	178 lbf/in.^{2}	220 F
Leaving condenser, entering expansion valve	175 lbf/in.^{2}	100 F
Leaving expansion valve, entering evaporator	29 lbf/in.^{2}	
Leaving evaporator	27 lbf/in.^{2}	20 F
Entering compressor	25 lbf/in.^{2}	40 F

Rate of flow of Freon = 200 lbm/hr
Power input to compressor = 2.5 hp

Calculate:
(*a*) The heat transfer per hour from the compressor.
(*b*) The heat transfer per hour from the Freon in the condenser.
(*c*) The heat transfer per hour to the Freon in the evaporator.

5.27 The following data are from the test of a large refrigeration unit utilizing ammonia as the refrigerant.

	Pressure	Temperature
Leaving compressor	260 lbf/in.^{2}	?
Entering condenser	250 lbf/in.^{2}	200 F
Leaving condenser, entering expansion valve	240 lbf/in.^{2}	105 F
Leaving expansion valve, entering the evaporator	35 lbf/in.^{2}	
Leaving the evaporator, entering compressor	30 lbf/in.^{2}	20 F

Assume no heat transfer from the compressor. The power input to the compressor is 140 hp. The capacity of the plant is 1,000,000 Btu/hr (i.e., the heat transfer to the refrigerant in the evaporator).

(*a*) Determine the rate of flow of the ammonia.

(*b*) What is the temperature of the ammonia leaving the compressor?

(*c*) What is the rate of heat transfer from the ammonia in the condenser?

5.28 A schematic diagram of a hydroelectric plant is shown in Fig. 5.17. Water enters the inlet conduit at the level of the lake, and work is done by the water in the hydraulic turbine. Assume no change in

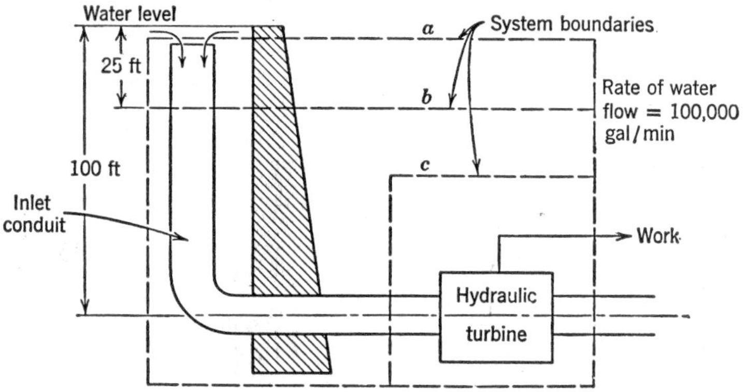

Fig. 5.17 Sketch for Problem 5.28.

kinetic energy or internal energy u in the inlet conduit, and that the kinetic energy and internal energy of the water leaving the hydraulic turbine is the same as that in the inlet conduit. Calculate the power developed by the hydraulic turbine considering each of the three systems shown.

5.29 In certain situations, when only superheated steam is available, a need for saturated steam may arise for a specific purpose. This can be accomplished in a desuperheater, in which case water is sprayed into the superheated steam in such amounts that the steam leaving the superheater is dry and saturated. The following data apply to such a desuperheater, which operates as a steady-flow process. Superheated steam at the rate of 2000 lbm/hr, at 400 lbf/in.^{2}, 600 F, enter the desuperheater. Water at 420 lbf/in.^{2}, 100 F, also enters the desuperheater. The dry saturated vapor leaves at 380 lbf/in.^{2} Calculate the rate of flow of water.

5.30 Liquid ammonia at a temperature of 60 F and a pressure of 200 lbf/in.^{2} is mixed in a steady-flow process with saturated ammonia vapor at a pressure of 200 lbf/in.^{2} The mass rate of flow of liquid and vapor are equal, and after mixing the pressure is 180 lbf/in.^{2} and the quality is 85%. Determine the heat transfer per lbm of mixture.

5.31 Three hundred lbm of steam per hr at 100 lbf/in.^{2}, 600 F, are mixed with water at 100 lbf/in.^{2}, 80 F in a steady-flow adiabatic process. Calculate the rate of flow of the water if the mixture leaves as dry saturated vapor at 90 lbf/in.^{2}

5.32 Liquid water at a pressure of 100 lbf/in.^{2} and a temperature of 80 F enters a 1-in. diameter tube at the rate of 0.8 ft^3/min. Heat is transferred to the water so that it leaves as saturated vapor at 90 lbf/in.^{2} Determine the heat transfer per min.

5.33 Ammonia vapor flows through a pipe at a pressure of 140 lbf/in.^{2} and a temperature of 100 F. Attached to the pipeline is an evacuated vessel having a volume of 1 ft^3. The valve in the line to this evacuated vessel is opened, and the pressure in the vessel comes to 140 lbf/in.^{2}, at which time the valve is closed. If this process occurred adiabatically, how much ammonia flowed into the vessel?

5.34 A tank having a volume of 200 ft^3 contains saturated vapor (steam) at a pressure of 20 lbf/in.^{2} Attached to this tank is a line in which vapor at 100 lbf/in.^{2}, 400 F, flows. Steam from this line enters the vessel until the pressure is 100 lbf/in.^{2} If there is no heat transfer from the tank and the heat capacity of the tank is neglected, calculate the mass of steam that enters the tank.

5.35 An insulated tank having a volume of 20 ft^3 contains dry saturated steam at 800 lbf/in.^{2} Steam is withdrawn through a line from the top of the tank until the pressure is 400 lbf/in.^{2} Calculate the mass of steam that is withdrawn, assuming, that at any moment the tank contains a homogenous mixture of liquid and vapor, and that this homogeneous mixture is withdrawn from the tank.

5.36 Occasionally a steam accumulator is used to drive a steam locomotive or for other purposes. Consider an accumulator that contains steam at 400 lbf/in.^{2} Assume that the steam in the tank is dry and saturated and that only saturated vapor is withdrawn. The volume of the accumulator is 80 ft^3. Steam is withdrawn from the accumulator until the pressure is reduced to 50 lbf/in.^{2}

What mass of vapor is withdrawn from the tank? (A step-by-step solution is suggested for this problem.)

5.37 A pressure vessel having a volume of 30 ft^3 contains saturated steam at 500 F. The vessel initially contains 50% vapor and 50%

liquid by volume. Liquid is withdrawn slowly from the bottom of the tank, and heat is transferred to the tank in order to maintain constant temperature. Determine the heat transfer to the tank when half of the contents of the tank has been removed.

5.38 A popular demonstration involves making ice by pumping a vacuum over liquid water until the pressure is reduced to less than the triple-point pressure. Such an apparatus is shown schematically in Fig. 5.18. In this case the tank has a total volume of 1 ft^3. Initially

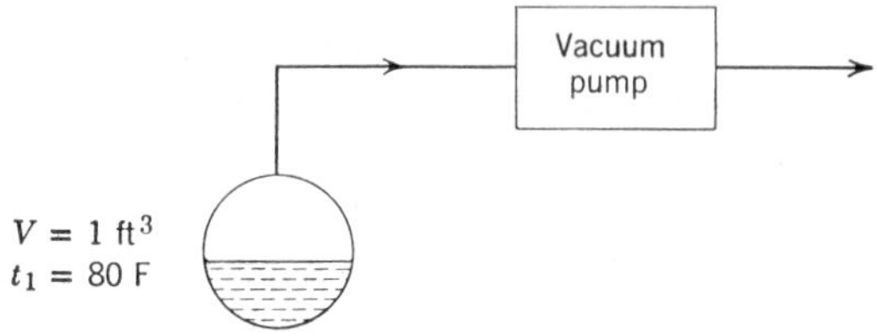

Fig. 5.18. Sketch for Problem 5.38.

the tank contains 0.9 ft^3 of saturated vapor (H_2O) and 0.1 ft^3 of saturated liquid. The initial temperature is 80 F. Assume no heat transfer during this process.

(*a*) Determine what fraction of the initial mass will be pumped off when the liquid and vapor first reach 32 F.

(*b*) Determine what fraction of the initial mass can be solidified.

5.39 Consider the arrangement shown in Fig. 5.19. The throttle valve is opened until the pressure in the cylinder reaches 75 lbf/in.^{2}

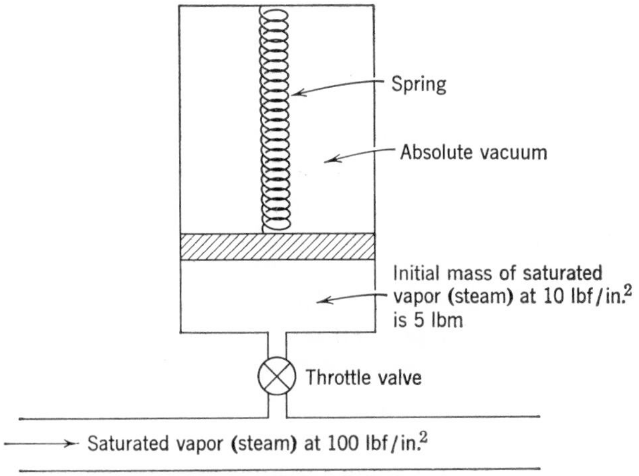

Fig. 5.19 Sketch for Problem 5.39.

During the period the valve is open, heat transfer to the cylinder contents is 1000 Btu, the work done on the spring is 2000 Btu and 10 lbm of steam flow into the cylinder. Determine the final quality (if saturated) or temperature (if superheated) of the steam.

5.40 In this problem we compare three different systems that involve flow into a vessel from a line in which steam at 100 lbf/in.², 500 F is flowing. In each case the process is adiabatic. Determine the final temperature of the steam in each of the following:

(*a*) The vessel is initially evacuated, and steam flows in until the pressure is 100 lbf/in.²

(*b*) Steam flows into a cylinder fitted with a piston which is restrained by a spring. Assume no pressure on the back side of the piston, and that the pressure of the steam in the cylinder is proportional to the displacement. Steam flows into the cylinder until the pressure is 100 lbf/in.² (Fig. 5.20.)

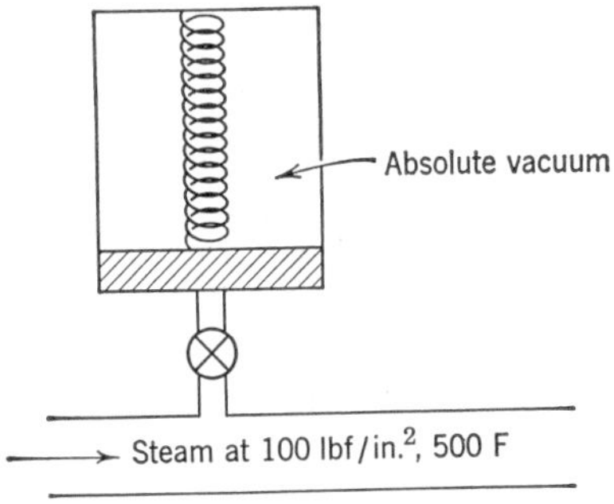

Fig. 5.20 Sketch for Problem 5.40b.

(*c*) Steam flows into a cylinder with a piston which is so weighted that a constant pressure of 100 lbf/in.² is maintained in the cylinder. Steam flows into the cylinder causing the piston to rise. (Fig. 5.21.)

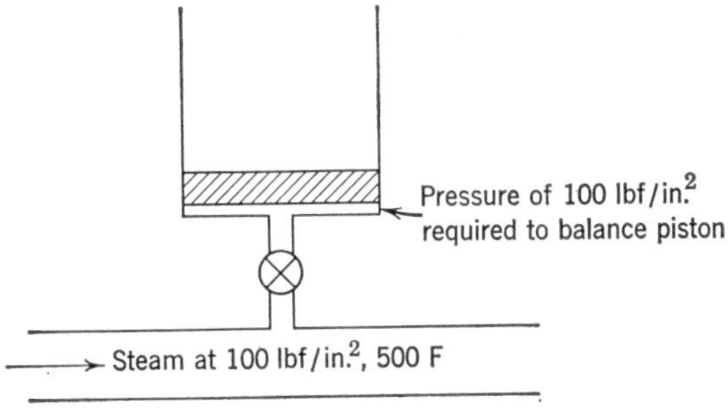

Fig. 5.21 Sketch for Problem 5.40c.

5.41 A throttling calorimeter is used to measure the quality of steam in a pipe in which the absolute pressure is 180 lbf/in.^{2} A mercury manometer is used to measure the pressure in the calorimeter, and shows a pressure in the calorimeter of 2.5 in. Hg above atmospheric pressure. If a minimum of 10 degrees of superheat is required in the calorimeter, what is the minimum quality of steam that can be determined? The barometer reads 29.43 in. Hg.

5.42 Water at a pressure of 1500 lbf/in.^{2}, 350 F, is throttled to a pressure of 20 lbf/in.^{2} in an adiabatic process. What is the quality after throttling?

5.43 Freon-12 at 180 lbf/in.^{2}, 100 F flows through the expansion valve in a vapor-compression refrigeration system. The pressure leaving the valve is 20 lbf/in.^{2} Determine the quality of the Freon leaving the expansion valve. State clearly all assumptions you made in solving this problem.

5.44 Consider the "do-it-yourself" power plant to be operated on the kitchen stove, shown in Fig. 5.22. A steam kettle having a volume

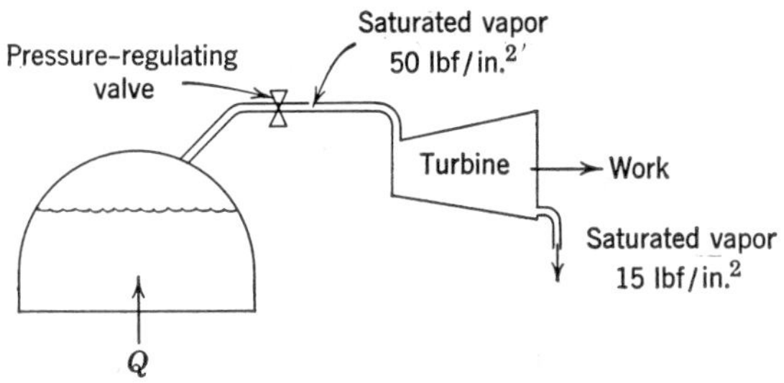

Fig. 5.22. Sketch for Problem 5.44.

of 1 ft^3 contains 90% liquid and 10% vapor (by volume) at 14.7 lbf/in.^{2} The pressure-regulating valve is then adjusted to 50 lbf/in.^{2}, and heat is transferred to the water. When the pressure reaches 50 lbf/in.^{2}, saturated vapor flows from the kettle to the turbine. The steam leaves the turbine as saturated vapor at 15 lbf/in.^{2} This process is continued until the kettle is filled with 10% liquid and 90% vapor by volume (at 50 lbf/in.^{2}). There is no heat transfer from the turbine.

(*a*) Determine the total work done by the turbine during this process.

(*b*) Determine the total heat transfer during this process.

5.45 The capacity and performance of small refrigeration compressors are frequently measured by an apparatus known as a calorimeter, which is shown schematically in Fig. 5.23. In this apparatus a water-cooled condenser is utilized, and the desired compressor discharge pressure is maintained by regulating the flow of cooling water. The evaporator is surrounded by a secondary fluid which is maintained at the ambient temperature by means of an electric heater. Thus, it may be assumed that there is no heat transfer from the secondary refrigerant to the surroundings, and that the heat transfer from the secondary refrigerant to the main refrigerant is equal to the electric input to the heater.

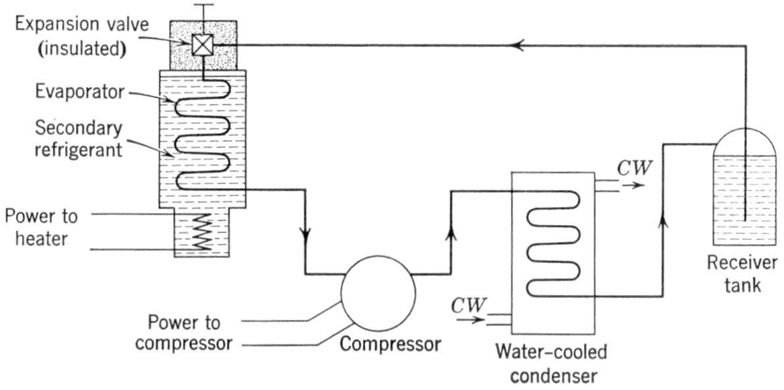

Fig. 5.23 Sketch for Problem 5.45.

The test essentially involves measuring the power input to the compressor and to the heater while maintaining the prescribed pressures and temperatures throughout the cycle.

The following data were obtained from the test of a compressor rated at $\frac{1}{9}$ horsepower which utilizes Freon-12 as a refrigerant. Assume standard atmospheric pressure.

Pressure of Freon-12 in evaporator	39.04 in. Hg gage
Pressure of Freon-12 entering expansion valve	180 lbf/in.^{2} gage
Temperature of Freon-12 entering expansion valve	104 F
Temperature of Freon-12 leaving evaporator	90 F
Ambient temperature	90 F
Power input to calorimeter	105 watts
Power input to compressor motor	125 watts

(*a*) How many lbm of Freon-12 does the compressor pump per hr?

(*b*) What is the ratio of refrigerating effect to work input? We shall note in the next chapter that this parameter is called the coefficient of performance.

(*c*) If the compressor pumped the same rate as determined in (*a*), but the liquid entered the expansion valve at a temperature of 90 F, what would the coefficient of performance be?

5.46 A spherical drop of liquid is suspended in an infinite atmosphere as shown in Fig. 5.24. Owing to differences in temperature it is

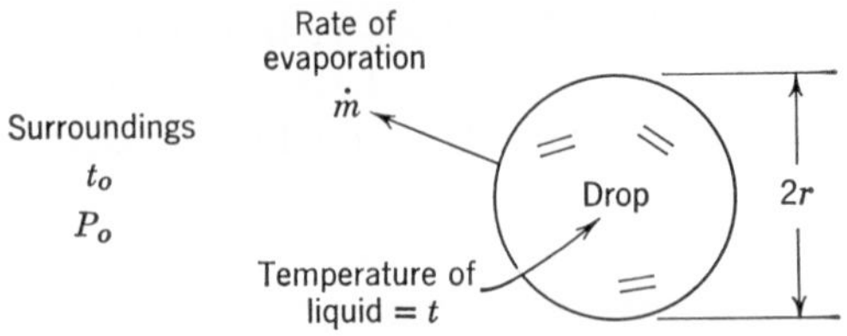

Fig. 5.24 **Sketch for Problem 5.46.**

exchanging heat with its surroundings. It also is losing mass by evaporation to the surroundings.

The instantaneous rate of heat transfer $\dot{q}$ is expressed by

$$\dot{q} = KA(t_o - t)$$

where K is a constant and A is the surface area. The instantaneous rate of mass transfer is given by the symbol $\dot{m}$. It may be assumed that at a given instant of time the sphere is uniform in temperature and always at a pressure equal to that of the surroundings, P_o. The vapor of the evaporating liquid in the region adjacent to the drop is saturated.

Derive an expression for the instantaneous time rate of change of the drop temperature, $\partial t/\partial \tau$, in terms of the significant physical quantities.

CHAPTER 6

THE SECOND LAW OF THERMODYNAMICS

The first law of thermodynamics states that during any cycle a closed system undergoes, the cyclic integral of the heat is equal to the cyclic integral of the work. The first law, however, places no restrictions on the direction of the heat and the work. A cycle in which a given amount of heat is transferred from the system and an equal amount of work is done on the system satisfies the first law just as well as does a cycle in which the directions of the heat and work are reversed. However, we know from our experience that this does not necessarily work, and this kind of experimental evidence is the basis of the second law of thermodynamics.

In its broader significance the second law involves the fact that processes proceed in a certain direction, and not in the opposite direction. A hot cup of coffee cools by virtue of heat transfer to the surroundings, but heat will not flow from the surroundings to the hotter cup of coffee. Gasoline is used as a car drives up a hill, but on coasting down the hill, the fuel level in the gasoline tank cannot be restored to its original level. Such familiar observations as these, and a host of others, are evidence of the validity of the second law of thermodynamics.

We will first consider this second law for a closed system and in the next chapter we will extend the principles to an open system.

6.1 Heat Engines and Heat Pumps

Consider the closed system and the surroundings that were cited in connection with the first law and are shown in Fig. 6.1. Let the gas

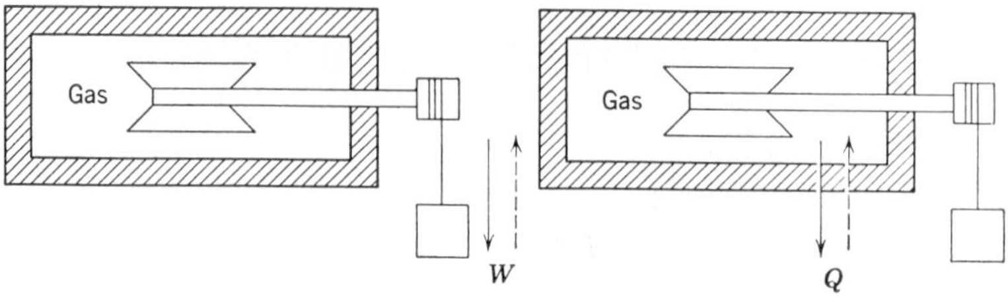

Fig. 6.1 A closed system that undergoes a cycle involving work and heat.

constitute the system and, as in our discussion of the first law, let this system undergo a cycle in which work is first done on the system by the paddle wheel as the weight is lowered. Then let the cycle be completed by transferring heat to the surroundings.

We know from our experience, however, that we cannot reverse this cycle. That is, if we transfer heat to the gas, as shown by the dotted arrow, the temperature of the gas will increase, but the paddle wheel will not turn and raise the weight. With the given surroundings (the container, the paddle wheel, and the weight) this system can operate in a cycle in which the heat transfer and work are both negative, but it cannot operate in a cycle in which both the heat transfer and work are positive, even though this would not violate the first law.

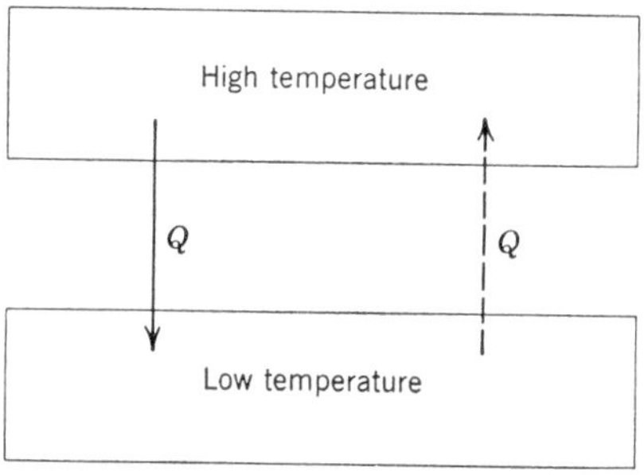

Fig. 6.2 An example showing the impossibility of completing a cycle by transferring heat from a low-temperature body to a high-temperature body.

Consider another cycle that we know from our experience to be impossible. Let two systems, one at a high temperature and the other at a low temperature, undergo a process in which a quantity of heat is transferred from the high-temperature system to the low-temperature system. We know that this process can take place. We also know that the reverse process, in which heat is transferred from the low-temperature system to the high-temperature system, does not occur, and that it is impossible to complete the cycle by heat transfer only. This is illustrated in Fig. 6.2.

These two illustrations lead us to the consideration of the heat engine and heat pump (i.e., refrigerator). With the heat engine we can have a system that operates in a cycle and has a net positive work and a net positive heat transfer. With the heat pump we can have a system that operates in a cycle and has heat transferred to it from a low-temperature body and heat transferred from it to a high-temperature body, though work is required to do this. Two simple heat engines and a simple heat pump will be considered.

The first heat engine is shown in Fig. 6.3, and consists of a cylinder fitted with appropriate stops and a piston. Let the gas in the cylinder constitute the system. Initially the piston rests on the lower stops,

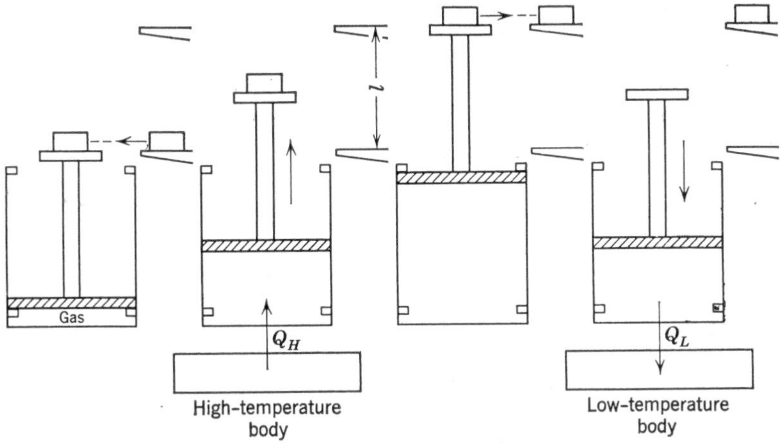

Fig. 6.3 A simple heat engine.

with a weight on the platform. Let the system now undergo a process in which heat is transferred from some high-temperature body to the gas, causing it to expand and raise the piston to the upper stops. At this point the weight is removed. Now let the system be restored to its initial state by transferring heat from the gas to a low-temperature body, thus completing the cycle. Since the weight was raised during the cycle, it is evident that work was done by the gas during the cycle. From the first law we conclude that the net heat transfer was positive and equal to the work done during the cycle.

Such a device is called a heat engine, and the substance to which and from which heat is transferred is called the working substance or working fluid. A heat engine may be defined as a device that operates in a thermodynamic cycle and does a certain amount of net positive work as a result of heat transfer from a high-temperature body and to a low-temperature body. Often the term heat engine is used in a broader sense to include all devices that produce work, either through heat transfer or combustion, even though the device does not operate in a thermodynamic cycle. The internal-combustion engine and the gas turbine are examples of such devices, and calling these heat engines is an acceptable use of the term. In this chapter, however, we are concerned with the more restricted form of heat engine, as defined above, which operates on a thermodynamic cycle.

A simple steam power plant is an example of a heat engine in this restricted sense. Each component in this plant may be analyzed by a steady-flow process, but considered as a whole it may be considered a heat engine (Fig. 6.4) in which water (steam) is the working fluid.

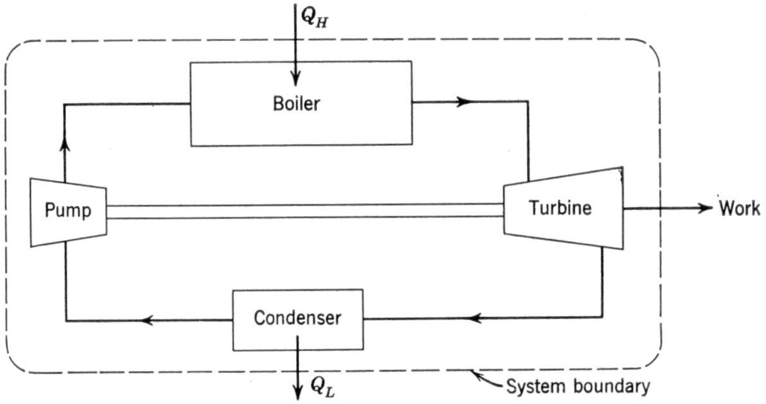

Fig. 6.4 A heat engine involving steady-flow processes.

An amount of heat, Q_H, is transferred from a high-temperature body, which may be the products of combustion in a furnace, a reactor, or a secondary fluid which in turn has been heated in a reactor. In Fig. 6.4 the turbine is shown schematically as driving the pump, indicating that what is significant is the net work which is delivered during the cycle. The quantity of heat, Q_L, is rejected to a low-temperature body, which is usually the cooling water in a condenser. Thus, the simple steam power plant is a heat engine in the restricted sense, for it has a working fluid, to which and from which heat is transferred, and which does a certain amount of work as it undergoes a cycle. By use of a heat engine we are able to have a system operate in a cycle and have the net work and net heat transfer both positive, which we were not able to do with the system and surroundings of Fig. 6.1.

One should note that in using the symbols Q_H and Q_L we have departed from our sign connotation for heat, because for a heat engine

Q_L is negative when the working fluid is considered as the system. In this chapter it will be advantageous to use the symbol Q_H to represent the heat transfer to or from the high-temperature body, and Q_L the heat transfer to or from the low-temperature body. The direction of the heat transfer will be evident in each case from the context.

At this point it is appropriate to introduce the concept of thermal efficiency of a heat engine. In general we say that efficiency is the ratio of output (the energy sought) to input (the energy that costs), but these must be clearly defined. At the risk of oversimplification we may say that in a heat engine the energy sought is the work, and the energy that costs money is the heat from the high-temperature source (indirectly, the cost of the fuel). *Thermal efficiency* is defined as:

$$\eta_{\text{thermal}} = \frac{W \text{ (energy sought for)}}{Q_H \text{ (energy that costs)}} = \frac{Q_H - Q_L}{Q_H} = 1 - \frac{Q_L}{Q_H} \tag{6.1}$$

The second cycle we were not able to complete was the one that involved the impossibility of transferring heat directly from a low-temperature body to a high-temperature body. This can of course be done with a refrigerator, or, as it is frequently called, a heat pump. To illustrate this, let us consider a vapor-compression refrigerator cycle, as shown in Fig. 6.5. The working fluid is the refrigerant, such

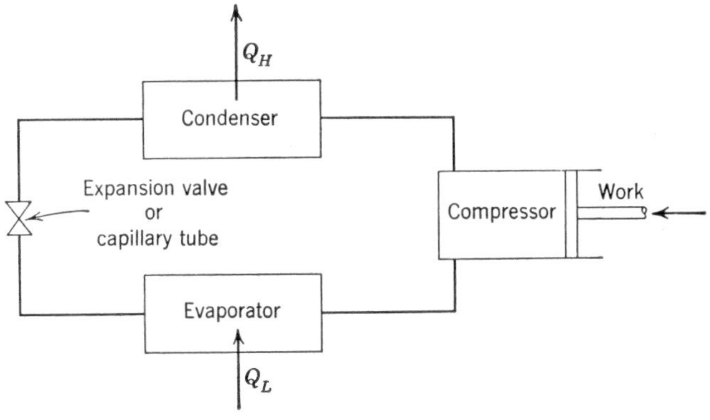

Fig. 6.5 A simple refrigeration cycle.

as a Freon or ammonia, which goes through a thermodynamic cycle. Heat is transferred to the refrigerant in the evaporator, where its pressure and temperature are low. Work is done on the refrigerant in the compressor and heat is transferred from it in the condenser, where its pressure and temperature are high. The pressure drop occurs as the refrigerant flows through the throttle valve.

Thus, in a heat pump we have a device which operates in a cycle, which requires work, and which accomplishes the objective of transferring heat from a low-temperature body to a high-temperature body.

The "efficiency" of a refrigerator is expressed in terms of the *coefficient of performance*, which we designate with the symbol β. In the case of a refrigerator the objective (i.e., energy sought) is Q_L, the heat transferred from the refrigerated space, and the energy that costs is the work W. Thus the coefficient of performance, β,* is

$$\beta = \frac{Q_L \text{ (energy sought for)}}{W \text{ (energy that costs)}} = \frac{Q_L}{Q_H - Q_L} = \frac{1}{Q_H/Q_L - 1} \tag{6.2}$$

Before stating the second law, the concept of a thermal reservoir should be introduced. A thermal reservoir is a body to which and from which heat can be transferred indefinitely without change in the temperature of the reservoir. Thus, a thermal reservoir always remains at constant temperature. The ocean and the atmosphere approach this definition very closely. Frequently it will be useful to designate a high-temperature reservoir and a low-temperature reservoir. Sometimes a reservoir from which heat is transferred is called a source, and a reservoir to which heat is transferred is called a sink.

6.2 Second Law of Thermodynamics

On the basis of the matter considered in the previous section we are now ready to state the *second law of thermodynamics*. There are two classical statements of the second law, known as the Kelvin-Planck statement and the Clausius statement.

* It should be noted that a heat pump can be used with either of two objectives in mind. It can be used as a refrigerator, in which case the primary objective is Q_L, the heat transferred to the refrigerant from the refrigerated space. It can also be used as a heating system (usually referred to as a heat pump), in which case the objective is Q_H, the heat transferred from the refrigerant to the high-temperature body, which is the space to be heated. In this case Q_L is transferred to the refrigerant from the ground, the atmospheric air, or well water. The coefficient of performance in this case, β', is

$$\beta' = \frac{Q_H \text{ (energy sought for)}}{W \text{ (energy that costs)}} = \frac{Q_H}{Q_H - Q_L} = \frac{1}{1 - Q_L/Q_H}$$

It also follows that for a given cycle,

$$\beta' - \beta = 1$$

Unless otherwise specified, the term coefficient of performance will always refer to a refrigerator as defined by Eq. 6.2.

Kelvin-Planck statement It is impossible to construct a device which will operate in a cycle and produce no effect other than the raising of a weight and the exchange of heat with a single reservoir.

This statement ties in with our discussion of the heat engine, and in effect it says that it is impossible to construct a heat engine which operates in a cycle and receives a given amount of heat from a high-temperature body and does an equal amount of work. The only alternative is that some heat must be transferred from the working fluid at a lower temperature to a low-temperature body. Thus, work can be done by the transfer of heat only if there are two temperature levels involved, and heat is transferred from the high-temperature body to the heat engine and also from the heat engine to the low-temperature body. This implies that it is impossible to build a heat engine having a thermal efficiency of 100 per cent.

Clausius statement It is impossible to construct a device which operates in a cycle and produces no effect other than the transfer of heat from a cooler body to a hotter body.

This statement is related to the heat pump, and in effect says that it is impossible to construct a heat pump that operates without an input of work. This also implies that the coefficient of performance is always less than infinity.

Discussion of the two statements Regarding these two statements, three observations should be made. The first is that both of these are negative statements. It is of course impossible to "prove" a negative statement. However, we can say that the second law of thermodynamics (like every other law of nature) rests on experimental evidence. Every relevant experiment that has been conducted has either directly or indirectly verified the second law, and no experiment has ever been conducted that contradicts the second law. The basis of the second law is therefore experimental evidence.

A second observation is that these two statements of the second law are equivalent. Two statements are equivalent if the truth of each statement implies the truth of the other, or if the violation of each statement implies the violation of the other. That a violation of the Clausius statement implies a violation of the Kelvin-Planck statement may be shown as follows. The device at the left in Fig. 6.6 is a heat pump that requires no work, and thus violates the Clausius statement. Let an amount of heat Q_L be transferred from the low-temperature reservoir to this heat pump, and let the same amount of heat Q_L be transferred to the high-temperature reservoir. Let an amount of heat Q_H, which is greater than Q_L, be transferred from the high-temperature reservoir to the heat engine, and let the engine reject the amount of heat Q_L as it does an amount of work W (which equals $Q_H - Q_L$). Since there is no net heat transfer to the low-temperature reservoir, the low-temperature reservoir, the heat engine, and the heat pump can together be considered as a device that operates in a cycle and produces no effect other than the raising of a weight (work) and the exchange of heat with a single reservoir. Thus, a violation of the Clausius statement implies a violation of the Kelvin-Planck statement. The complete equivalence of these two statements is established when it is

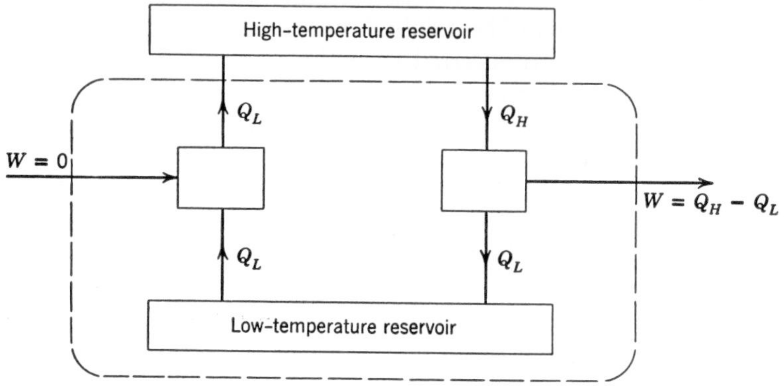

Fig. 6.6 Demonstration of the equivalence of the two statements of the second law.

also shown that a violation of the Kelvin-Planck statement implies a violation of the Clausius statement. This is left as an exercise for the student.

The third observation is that frequently the second law of thermodynamics has been stated as the impossibility of constructing a perpetual-motion machine of the second kind. A perpetual-motion machine of the first kind would create work from nothing or create mass-energy, thus violating the first law. A perpetual-motion machine of the second kind would violate the second law, and a perpetual-motion machine of the third kind would have no friction, and thus run indefinitely but would produce no work.

A heat engine that violated the second law could be made into a perpetual-motion machine of the second kind as follows. Consider Fig. 6.7, which might be the power plant of a ship. An amount of heat Q_L is transferred from the ocean to a high-temperature body by means of a heat pump. The work required is W', and the heat transferred to the high-temperature body is Q_H; let the same amount of heat be transferred to a heat engine, which violates the Kelvin-Planck statement of the second law, and does an amount of work $W = Q_H$.

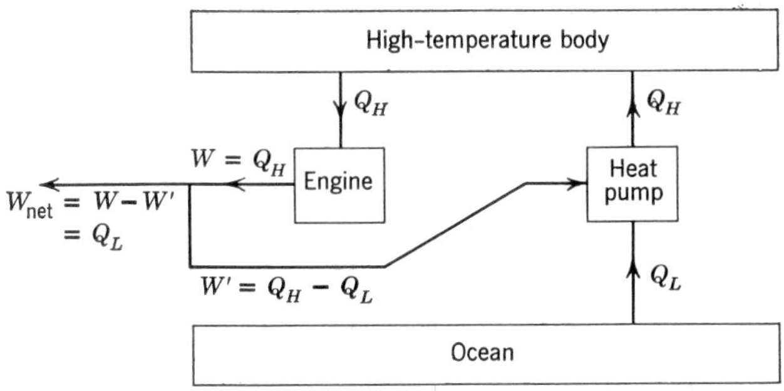

Fig. 6.7 A perpetual-motion machine of the second kind.

Of this work an amount of work $Q_H - Q_L$ is required to drive the heat pump, leaving the net work ($W_{net} = Q_L$) available for driving the ship. Thus, we have a perpetual-motion machine in the sense that work is done by utilizing freely available sources of energy such as the ocean or atmosphere.

6.3 The Reversible Process

The question that now logically arises is this. If it is impossible to have a heat engine of 100 per cent efficiency, what is the maximum efficiency one can have? The first step in the answer to this question is to define an ideal process, which is called a reversible process.

A *reversible process* for a system is defined as a process, which once having taken place, can be reversed and leaves no change in either the system or surroundings.

Let us illustrate the significance of this definition for a gas contained in a cylinder which is fitted with a piston. Consider first Fig. 6.8, in which a gas (which we define as the system) at high pressure is restrained by a piston which is secured by a pin. When the pin is removed, the piston is raised and forced abruptly against the stops. Some work is done by the system, since the piston has been raised a certain amount. Suppose we wish to restore the system to its initial state. One way of doing this would be to exert a force on the piston, thus compressing the gas until the pin could again be inserted in the piston. Since the work done on the gas in this reverse process is greater than the work done by the gas in the initial process, an amount of heat must be transferred from the gas in order that the system have the same internal energy it had originally. Thus, the system is restored to its initial state, but the surroundings have changed by virtue of the fact that work was required to force the piston down and heat was transferred to the surroundings. Thus, the initial process is an irreversible one because it could not be reversed without leaving a change in the surroundings.

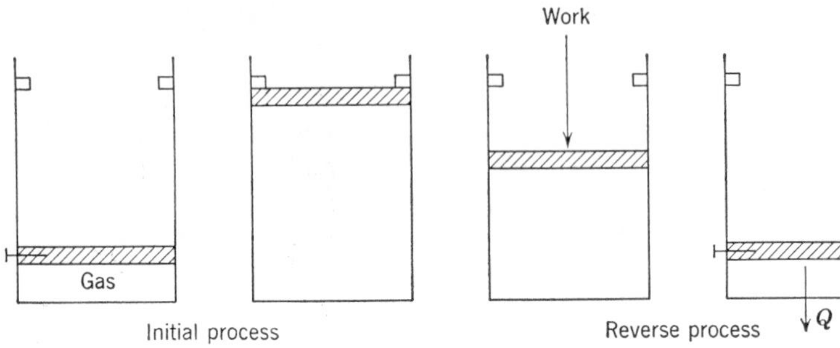

Fig. 6.8 An example of an irreversible process.

In Fig. 6.9 let the gas in the cylinder comprise the system and let the piston be loaded with a number of weights. Let the weights be slid off horizontally one at a time, allowing the gas to expand and do work in raising the weights which remain on the piston. As the size of the weights is made smaller and their number is increased, we approach a process that can be reversed, for at each level of the piston during the reverse process there will be a small weight which is exactly at the level of the platform and thus can be placed on the platform

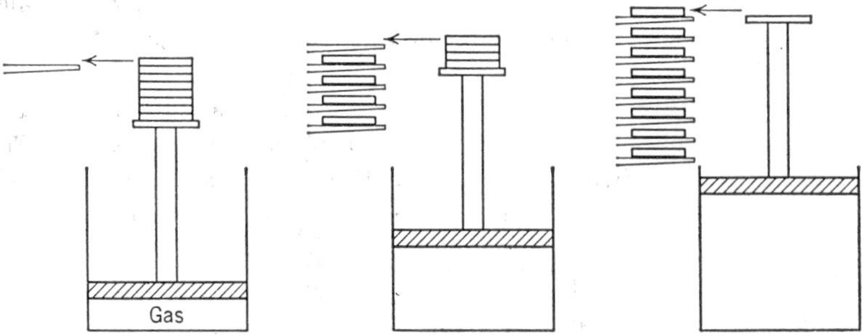

Fig. 6.9 An example of a process that approaches being reversible.

without requiring work. In the limit, therefore, as the weights become very small, the reverse process can be accomplished in such a manner that both the system and surroundings are in exactly the same state they were initially. Such a process is a reversible process.

6.4 Factors That Render Processes Irreversible

There are many factors that make processes irreversible, four of which are considered in this section.

Friction It is readily evident that friction makes a process irreversible, but a brief illustration may clarify the point. Let a block and an inclined plane comprise a system, Fig. 6.10, and let the block be pulled up the inclined plane by weights which are lowered. A certain amount

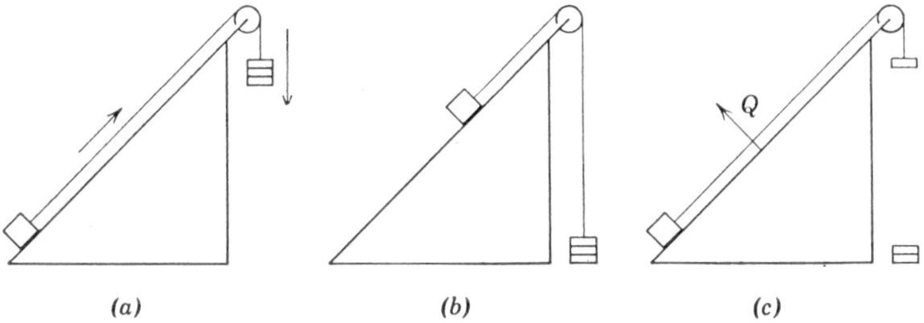

Fig. 6.10 Demonstration of the fact that friction makes processes irreversible.

of work is required to do this. Some of this work is required to overcome the friction between the block and the plane, and some is required to increase the potential energy of the block. The block can be restored to its initial position by removing some of the weights, thus allowing the block to slide down the plane. Some heat transfer from the system to the surroundings will no doubt be required to restore the block to its initial temperature. Since the surroundings are not restored to their initial state at the conclusion of the reverse process, we conclude that friction has rendered the process irreversible. Another type of frictional effect is that associated with the flow of viscous fluids in pipes and passages and in the movement of bodies through viscous fluids.

Unrestrained expansion The classic example of an unrestrained expansion is shown in Fig. 6.11, in which a gas is separated from a vacuum by a membrane. Consider the process that occurs when the membrane breaks and the gas fills the entire vessel. It can be shown that this is an irreversible process by considering the process necessary to restore the system to its original state. This would involve compressing the gas and transferring heat from the gas until its initial state was reached. Since the work and heat transfer involve a change in the surroundings, the surroundings are not restored to their initial state, indicating that the unrestrained expansion was an irreversible process. The process described in Fig. 6.8 was also an example of an unrestrained expansion.

In the reversible expansion of a gas there must be only an infinitesimal difference between the force exerted by the gas and the restraining force, so the rate at which the boundary moves will be infini-

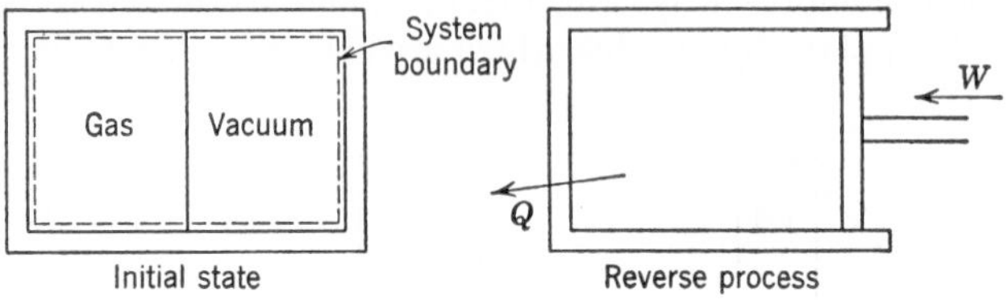

Fig. 6.11 **Demonstration of the fact that unrestrained expansion makes processes irreversible.**

tesimal. However, actual cases involve a finite difference in forces which gives rise to a finite rate of movement of the boundary, and thus are irreversible in some degree.

Heat transfer through a finite temperature difference Consider as a system a high-temperature body and a low-temperature body, and let heat be transferred from the high-temperature body to the low-temperature body. The only way in which the system can be restored to its initial state is to provide refrigeration, which requires work from the surroundings, and some heat transfer to the surroundings will also be necessary. Because of the heat transfer and the work, the surroundings are not restored to their original state, indicating that the process was irreversible.

An interesting question now arises. Heat is defined as energy which is transferred due to a temperature difference. We have just shown that heat transfer through a temperature difference is an irreversible process. Therefore, how can we have a reversible heat-transfer process? A heat-transfer process approaches a reversible process as the temperature difference between the two bodies approaches zero. Therefore, we define a reversible heat-transfer process as one in which the heat is transferred through an infinitesimal temperature difference. We realize of course that to transfer a finite amount of heat through an infinitesimal temperature difference would require an infinite amount of time, or infinite area. Therefore, all actual heat-transfer processes are through a finite temperature difference and are therefore irreversible, and the greater the temperature difference the greater the irreversibility. We will find however, that the concept of reversible heat transfer is very useful in describing ideal processes.

Mixing of two different substances This process is illustrated in Fig. 6.12, in which two different gases are separated by a membrane. Let the membrane break and a homogeneous mixture of oxygen and nitrogen fill the entire volume. This process will be considered in some detail in Chapter 9. We can say here that this may be considered as a special case of an unrestrained expansion, for each gas undergoes an

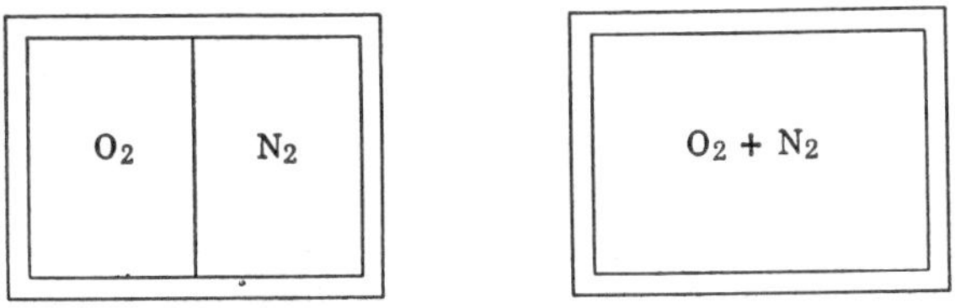

Fig. 6.12 Demonstration of the fact that the mixing of two different substances is an irreversible process.

unrestrained expansion as it fills the entire volume. A certain amount of work is necessary to separate these gases, as is shown by a typical air-separation plant, in which air is separated into its various components.

Other factors There are a number of other factors that make processes irreversible which will not be considered in detail here. Among these are electrical-resistance and hysteresis effects. A combustion process as it ordinarily takes place is also an irreversible process.

It is frequently advantageous to distinguish between internal and external irreversibility. Figure 6.13 shows two identical systems to

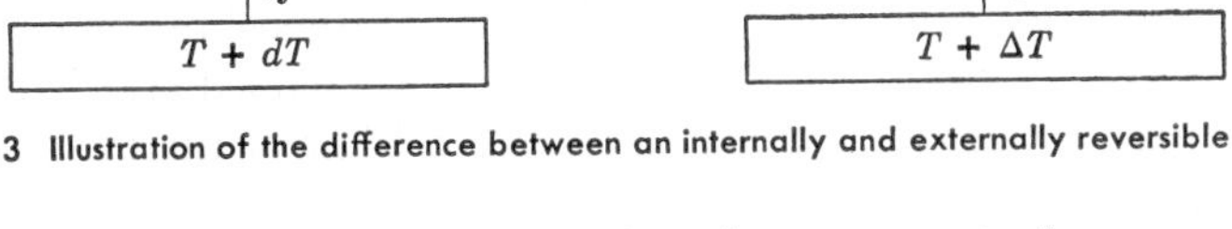

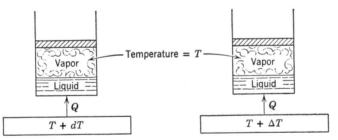

Fig. 6.13 Illustration of the difference between an internally and externally reversible process.

which heat is transferred. Assuming the systems to be a pure substance, the temperature remains constant during the heat-transfer process. In one the heat is transferred from a reservoir at a temperature $T + dT$, and in the other the reservoir is at a much higher temperature, $T + \Delta T$, than the system. The first is a reversible heat-transfer process and the second is an irreversible heat-transfer process. However, as far as the system itself is concerned, it passes through exactly the same states in both processes, which we assume are reversible. Thus, we can say in the second case that the process is internally reversible but externally irreversible because the irreversibility occurs outside the system.

One should also note the general interrelation of reversibility, equilibrium, and time. In a reversible process the deviation from equilibrium is infinitesimal, and therefore it occurs at an infinitesimal rate. Since it is desirable that actual processes proceed at a finite rate, the deviation from equilibrium must be finite, and therefore the actual process is irreversible in some degree, and the greater the deviation from equilibrium, the greater the irreversibility, and the more rapidly the process will occur. It should also be noted that the quasistatic process, which was described in Chapter 2, is a reversible process, and hereafter the term reversible process will be used.

6.5 The Carnot Cycle

Having defined the reversible process and considered some factors that make processes irreversible, let us again pose the question raised in Section 6.3, namely, if the efficiency of all heat engines is less than 100 per cent, what is the most efficient cycle we can have? Let us answer this question for a heat engine that receives its heat from a high-temperature reservoir and rejects heat to a low-temperature reservoir. Since we are dealing with reservoirs we recognize that both the high temperature and the low temperature are constant and remain constant regardless of the amount of heat transferred.

Let us assume that this heat engine, which operates between the given high-temperature and low-temperature reservoirs, operates on a cycle in which every process is reversible. If every process is reversible, the cycle is also reversible, and if the cycle is reversed, the heat engine becomes a heat pump. In the next section we will show that this is the most efficient cycle that can operate between two constant-temperature reservoirs. It is called the Carnot cycle, and is named after a French engineer, Sadi Carnot, who stated the second law of thermodynamics in 1824.

We now turn our attention to a consideration of the Carnot cycle. Figure 6.14 shows a power plant which is similar in many respects to a simple steam power plant and which we assume operates on the Carnot cycle. Assume the working fluid to be a pure substance, such as steam. Heat is transferred from the high-temperature reservoir to the water (steam) in the boiler. For this to be a reversible heat transfer, the temperature of the water (steam) must be only infinitesimally lower than the temperature of the reservoir. This also implies, since the temperature of the reservoir remains constant, that the temperature of the water must remain constant. Therefore, the first process in the Carnot cycle is a reversible isothermal process in which heat is transferred from

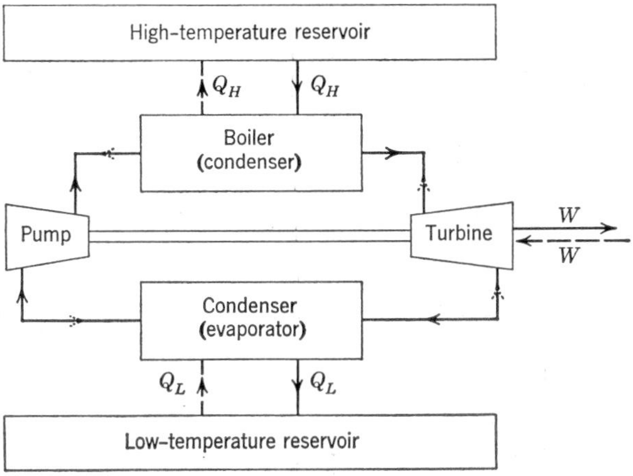

Fig. 6.14 Example of a heat engine that operates on a Carnot cycle.

the high-temperature reservoir to the working fluid. A change of phase from liquid to vapor at constant pressure is of course an isothermal process for a pure substance.

The next process occurs in the turbine. It occurs without heat transfer and is therefore adiabatic. Since all processes in the Carnot cycle are reversible, this must be a reversible adiabatic process, during which the temperature of the working fluid decreases from the temperature of the high-temperature reservoir to the temperature of the low-temperature reservoir.

In the next process heat is rejected from the working fluid to the low-temperature reservoir. This must be a reversible isothermal process in which the temperature of the working fluid is infinitesimally higher than that of low-temperature reservoir.

The final process, which completes the cycle, is a reversible adiabatic process in which the temperature of the working fluid increases from the low temperature to the high temperature. If this were to be done with water (steam) as the working fluid, it would involve taking a mixture of liquid and vapor from the condenser and compressing it. This would be very inconvenient in practice and therefore in all power plants the working fluid is completely condensed in the condenser, and the pump handles only the liquid phase.

Since the Carnot cycle is reversible, every process could be reversed, in which case it would be a heat pump. The heat pump is shown by the dotted line and parentheses in Fig. 6.14. The temperature of the working fluid in the evaporator would be infinitesimally less than the temperature of the low-temperature reservoir, and in the condenser it is infinitesimally higher than that of the high-temperature reservoir.

It should be emphasized that the Carnot cycle can be executed in many different ways. Many different working substances can be used, such as a gas or a paramagnetic substance in a magnetic field. There are also various possible arrangements of machinery. For example, a Carnot cycle can be devised that takes place entirely within a cylinder, using a gas as a working substance, as shown in Fig. 6.15.

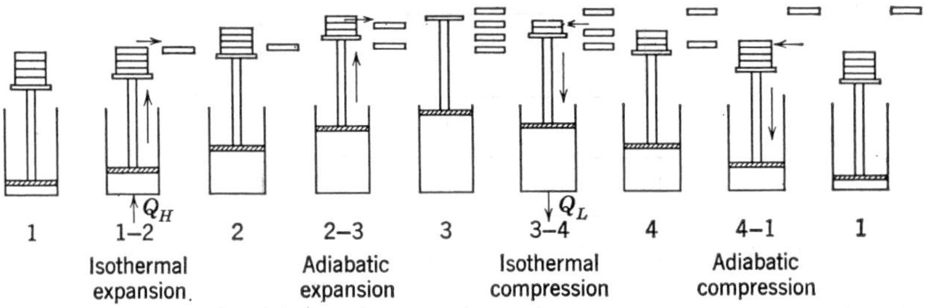

Fig. 6.15 Example of a closed system operating on a Carnot cycle.

The important point to be made here is that the Carnot cycle, regardless of what the working substance may be, always has the same four basic processes. These are:

(*a*) A reversible isothermal process in which heat is transferred to or from the high-temperature reservoir.

(*b*) A reversible adiabatic process in which the temperature of the working fluid decreases from the high temperature to the low temperature.

(*c*) A reversible isothermal process in which heat is transferred to or from the low-temperature reservoir.

(*d*) A reversible adiabatic process in which the temperature of the working fluid increases from the low temperature to the high temperature.

6.6 Two Propositions Regarding the Efficiency of a Carnot Cycle

There are two important propositions regarding the efficiency of a Carnot cycle:

First proposition It is impossible to construct an engine that operates between two given reservoirs that is more efficient than a reversible engine operating between the same two reservoirs.

The proof of this statement involves a "thought experiment." An initial assumption is made, and it is then shown that this assumption leads to impossible conclusions. The only possible conclusion is that the initial assumption was incorrect.

Let us assume that there is an irreversible engine operating between two given reservoirs that has a greater efficiency than a reversible engine operating between the same two reservoirs. Let the heat transfer to the irreversible engine be Q_H, the heat rejected be Q_L', and the work be $W_{\text{I.E.}}$ (which equals $Q_H - Q_L'$) as shown in Fig. 6.16. Let the

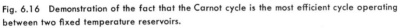

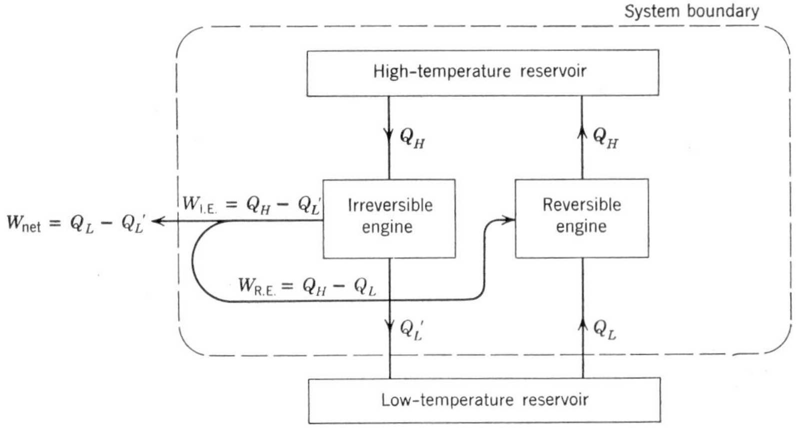

Fig. 6.16 Demonstration of the fact that the Carnot cycle is the most efficient cycle operating between two fixed temperature reservoirs.

reversible engine operate as a heat pump (since it is reversible this is possible) and let the heat transfer with the low-temperature reservoir be Q_L, the heat transfer with the high-temperature reservoir be Q_H, and the work required be $W_{\text{R.E.}}$ (which equals $Q_H - Q_L$).

Since the initial assumption was that the irreversible engine is more efficient, it follows (because Q_H is the same for both engines) that $Q_L' < Q_L$ and $W_{\text{I.E.}} > W_{\text{R.E.}}$. Now the irreversible engine can drive the reversible engine and still deliver the net work W_{net} (which equals $W_{\text{I.E.}} - W_{\text{R.E.}} = Q_L - Q_L'$). However, if we consider the two engines and the high-temperature reservoir as a system, as indicated in Fig. 6.16, we have a system that operates in a cycle, exchanges heat with a single reservoir, and does a certain amount of work. However, this would constitute a violation of the second law and we conclude that our initial assumption (that the irreversible engine is more efficient than the reversible engine) is incorrect, and we conclude that we cannot have an irreversible engine which is more efficient than a reversible engine operating between the same two reservoirs.

Second proposition All engines that operate on the Carnot cycle between two given constant-temperature reservoirs have the same efficiency. The proof of this proposition is similar to that outlined above, and involves the assumption that there is one Carnot cycle which is more efficient than another Carnot cycle operating between the same temperature reservoirs. Let the Carnot cycle with the higher efficiency replace the irreversible cycle of the previous argument, and the Carnot cycle with the lower efficiency operate as the heat pump. The proof proceeds with the same line of reasoning as in the first proposition. The details are left as an exercise for the student.

6.7 The Thermodynamic Temperature Scale

In discussing the matter of temperature in Chapter 2 it was pointed out that the zeroth law of thermodynamics provides a basis for temperature measurement, but that a temperature scale must be defined in terms of a particular thermometer substance and device. A temperature scale that is independent of any particular substance, which might be called an absolute temperature scale, would be most desirable. In the last paragraph we noted that the efficiency of a Carnot cycle is independent of the working substance and depends only on the temperature. This would provide a suitable basis for such an absolute temperature scale, which we will call the thermodynamic temperature scale.

The concept of this temperature scale may be developed with the aid of Fig. 6.17. The efficiency of the Carnot cycle is given by Eq. 6.1, and since this is a function only of the temperatures of the two reservoirs we may write

$$\eta_{\text{thermal}} = 1 - \frac{Q_L}{Q_H} = f(t_L, t_H) = f(T_L, T_H) \tag{6.3}$$

where T designates absolute or thermodynamic temperature.

There are many functional relations possible to relate Q_L and Q_H to T_L and T_H, which will serve to define the absolute scale. The relation that has been selected for the thermodynamic scale of temperature is

$$\frac{T_H}{T_L} = \frac{Q_H}{Q_L} \tag{6.4}$$

With absolute temperatures so defined the efficiency of a Carnot cycle may be expressed in terms of the absolute temperatures.

$$\eta_{\text{thermal}} = 1 - \frac{Q_L}{Q_H} = 1 - \frac{T_L}{T_H} \tag{6.5}$$

This means that if the thermal efficiency of a Carnot cycle operating between two given constant-temperature reservoirs is known, the ratio of the two absolute temperatures is also known.

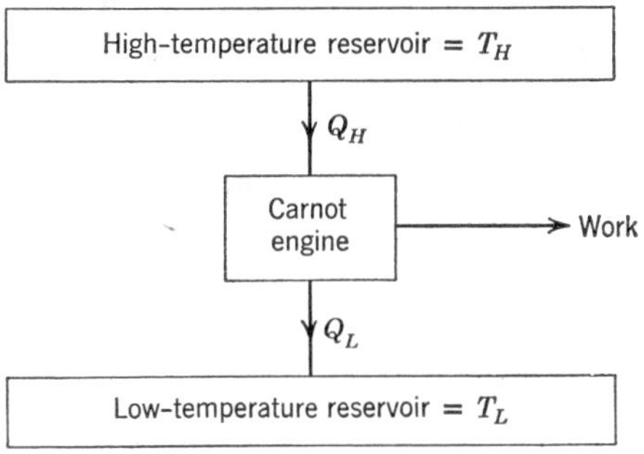

Fig. 6.17 The Carnot cycle as a basis of the thermodynamic scale of temperature.

In order to assign values of absolute temperatures, however, one other relation between T_L and T_H must be known. On the absolute Fahrenheit scale, called the Rankine scale, this second relation is the fact that there are 180 degrees between the ice point and steam point, just as on the Fahrenheit scale. This fact, along with Eq. 6.5, determines the relation between the Fahrenheit and the Rankine scales.

To illustrate this point, suppose we had a heat engine operating on the Carnot cycle that received heat at the temperature of the steam point and rejected heat at the temperature of the ice point. If the efficiency of such an engine could be measured, it would be found to be 26.80 per cent. Therefore, from Eq. 6.5

$$\eta_{\text{th}} = 1 - \frac{T_L}{T_H} = 1 - \frac{T_{\text{ice point}}}{T_{\text{steam point}}} = 0.2680$$

$$\frac{T_{\text{ice point}}}{T_{\text{steam point}}} = 0.7320$$

We also have the relation

$$T_{\text{steam point}} - T_{\text{ice point}} = 180$$

Solving these two equations simultaneously we find

$$T_{\text{steam point}} = 671.69 \text{ R} \qquad T_{\text{ice point}} = 491.69 \text{ R}$$

Temperatures on the Rankine scale are designated by R.

It follows that temperatures on the Fahrenheit and Rankine scales are related as follows:

$$t(°\text{F}) + 459.69 = T(°\text{R})$$

The absolute scale related to the Centigrade scale is the Kelvin scale, designated by K. On both these scales there are 100 degrees between the ice point and the steam point. Therefore, if we wished to use our engine operating on the Carnot cycle between the steam point and the ice point we would have the relations:

$$T_{\text{steam point}} - T_{\text{ice point}} = 100$$

$$\frac{T_{\text{ice point}}}{T_{\text{steam point}}} = 0.7320$$

Solving these two equations simultaneously we find

$$T_{\text{steam point}} = 373.16 \text{ K} \qquad T_{\text{ice point}} = 273.16 \text{ K}$$

It follows that

$$t(°\text{C}) + 273.16 = T(°\text{K})$$

The International temperature scale which was described in Chapter 2 was so defined that it conforms very closely to the thermodynamic temperature scale as defined above.

It is particularly significant that the zero point on the absolute temperature scale is defined by the fact that at absolute zero no heat is rejected by an engine operating on the Carnot cycle. Thus, if a reversible engine could reject heat at absolute zero, it would have a thermal efficiency of 100 per cent, because the thermal efficiency of a Carnot cycle is given by the relation

$$\eta_{\text{th}} = 1 - \frac{T_L}{T_H}$$

It also follows from Eq. 6.5 that the efficiency of a Carnot cycle may be increased by increasing T_H or by decreasing T_L. This is a very important principle for heat engines in general, even though they do not operate on the Carnot cycle. For example, the lower the temperature at which heat can be rejected in a steam power plant the higher the efficiency. However, the lowest temperature at which heat can be rejected is the temperature of the atmosphere or the ocean or a river, and this always becomes a very definite lower limit for T_L. On the other hand, efficiency can be increased by increasing the temperature at which heat is transferred to the working fluid. The limiting factor in this case is the materials available to withstand the high temperatures involved. The steady increase in the maximum temperature and pressure in steam generators over the years is evidence of the progress that has been made in the effort to increase efficiency by increasing T_H.

For a refrigerator the coefficient of performance may be written in terms of the absolute temperatures:

$$\beta = \frac{1}{T_H/T_L - 1} \tag{6.6}$$

From this it follows that the coefficient of performance decreases as T_L decreases and as T_H increases. It also follows that as T_L approaches zero the coefficient of performance for a refrigerator also approaches zero.

PROBLEMS

6.1 Determine the thermal efficiency of the power plant described in Problem 5.19.

6.2 Calculate the coefficient of performance of the refrigeration cycle described in Problem 5.26.

6.3 Prove that a device that violates the Kelvin-Planck statement of the second law also violates the Clausius statement of the second law.

6.4 Suggest a number of factors that would make the cycle described in Problem 5.19 an irreversible cycle.

6.5 One thousand Btu of heat are transferred from a reservoir at 600 F to an engine that operates on the Carnot cycle. The engine rejects heat to a reservoir at 80 F. Determine the thermal efficiency of the cycle and the work done by the engine.

6.6 A refrigerator that operates on a Carnot cycle is required to transfer 10,000 Btu/min from a reservoir at −20 F to the atmosphere at 80 F. What is the power required?

6.7 The maximum allowable temperature of the working fluid is usually determined by metallurgical considerations. In a certain power plant this temperature is 1300 F. Nearby is a river in which the water has a temperature of 48 F. What is the maximum possible efficiency for this power plant?

6.8 An inventor claims to have developed a refrigeration unit which maintains the refrigerated space at 20 F while operating in a room where the temperature is 80 F, and which has a coefficient of performance of 8.5. How do you evaluate his claim? How would you evaluate his claim of a coefficient of performance of 8.0?

6.9 It is proposed to heat a house using a heat pump. The heat transfer from the house is 50,000 Btu/hr. The house is to be maintained at 75 F while the outside air is at a temperature of 20 F. What is the minimum power required to drive the heat pump?

CHAPTER 7

ENTROPY

In our consideration of the second law of thermodynamics in the last chapter, we dealt only with thermodynamic cycles, which is a very important and useful approach. However, in many cases we are concerned with processes rather than cycles. Thus, we might be interested in the second-law analysis of processes we encounter daily, such as the combustion process in an automobile, the cooling of a cup of coffee, or the chemical processes that take place in our bodies.

It would be most desirable to be able to deal with the second law quantitatively as well as qualitatively. In dealing with the first law we stated the law in terms of a cycle, but then found a property, the internal energy, which enabled us to use the first law quantitatively for processes. Similarly we have stated the second law for a cycle, and in this chapter we will find that the second law leads to another property, entropy, which enables us to treat the second law quantitatively. Energy and entropy are both abstract concepts that man has devised to aid in describing certain observations. Thermodynamics has been described as the science of energy and entropy, and as our study of thermodynamics proceeds the student will realize the truth of this statement.

7.1 Some General Observations

In Chapter 4 the work done at the moving boundary of a system as it undergoes a quasistatic process was found to be given by the relation

$$\delta W = P\,dV$$

Since a reversible process is a quasistatic process this relation also gives the work done at the moving boundary of a system during a reversible process.

Consider a reversible cycle in which the only work involved is done at the moving boundary of the system. For such a cycle we can write

$$\oint \delta Q = \oint \delta W = \oint P\,dV \tag{7.1}$$

We note that the work is given by the cyclic integral of the product of the intensive property P and the change in the extensive property V. The question that now arises is whether, for this reversible cycle, the heat might also be given by the cyclic integral of the product of an intensive property and the change in an extensive property. The intensive property most closely associated with heat is temperature. The extensive property necessary for an affirmative answer to the above question is entropy, as will be shown in this chapter. Entropy is given the symbol S, capital S denoting the total entropy for a system, and the lower-case s the specific entropy (entropy per unit mass). We now proceed to a consideration of the definition of entropy.

7.2 The Inequality of Clausius

The first step in our consideration of the property we call entropy is to establish the *inequality of Clausius*, which is

$$\oint \frac{\delta Q}{T} \leq 0$$

The inequality of Clausius is a corollary or consequence of the second law of thermodynamics, and may be developed by consideration of Fig. 7.1, which shows a constant-temperature reservoir at temperature T_o, a reversible engine that operates in a cycle, and a second device (engine) that operates in a cycle. Let us assume that the reversible engine makes some integral number of cycles for each cycle of the second engine.

As shown in Fig. 7.1, δQ_o is the heat transferred to the reversible engine from the reservoir, and the heat rejected by the reversible engine at temperature T is δQ. This temperature T may vary from cycle to cycle. The work done by the second engine is δW_E, and this is equal to the heat transfer δQ.

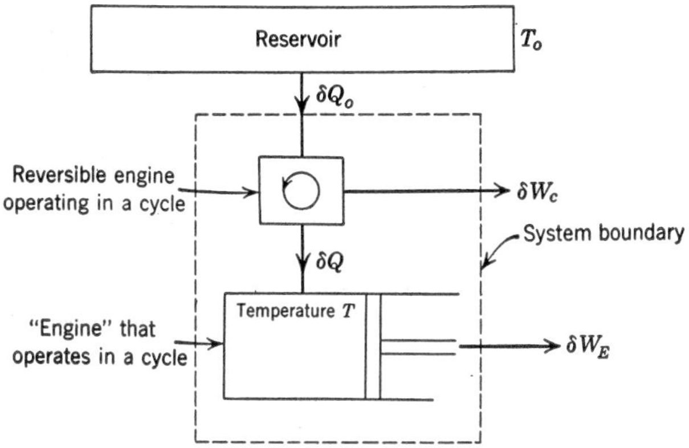

Fig. 7.1 Sketch for demonstrating the inequality of Clausius.

If we consider the two engines together as a system that operates in a cycle we can write

$$\oint \delta W_c = \oint \delta Q \frac{T_o}{T} - \oint \delta Q = \oint \delta Q \left(\frac{T_o}{T} - 1\right)$$

$$\oint \delta W_E = \oint \delta Q$$

$$\oint \delta W = \oint \delta W_c + \oint \delta W_E = \oint \delta Q \left(\frac{T_o}{T} - 1\right) + \oint \delta Q = \oint \frac{T_o}{T} \delta Q$$

Now we immediately recognize that if we consider the two engines together as a system that operates in a cycle, the net work cannot be positive, because we have a system that operates in a cycle, exchanges heat with a single reservoir, and does a net amount of work. This would be a violation of the second law.

Therefore, the work cannot be positive as shown in Fig. 7.1. Rather, the only way these two engines together can operate in a cycle is as shown in Fig. 7.2. In this case the reversible engine operates as a heat pump, and the second "engine" operates in a cycle in which the work is negative and an amount of heat equal to this work is transferred

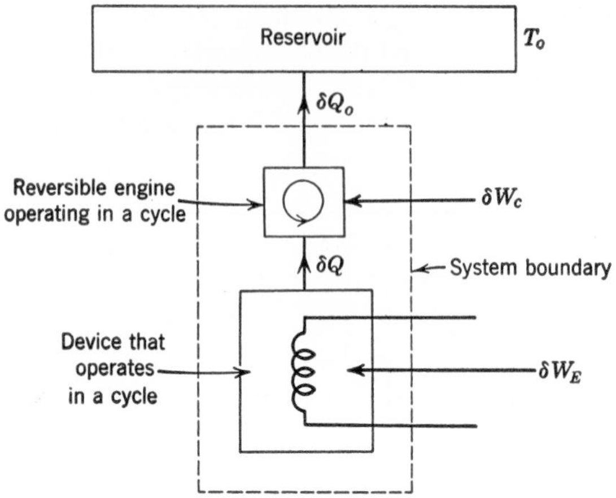

Fig. 7.2 Sketch for demonstrating the inequality of Clausius.

from the system. It is obvious that we do not have a second engine at all, but rather a device such as the electric heater shown in Fig. 7.2. Therefore we conclude that

$$\oint \delta W \leq 0$$

Substituting in Eq. 7.1 it also follows that

$$\oint \frac{T_o}{T}\,\delta Q \leq 0$$

Since T_o is always positive it follows that

$$\oint \frac{\delta Q}{T} \leq 0 \tag{7.2}$$

This is the inequality of Clausius.

It may be shown that the equality sign in Eq. 7.2 holds for a reversible cycle. If the inequality holds for a reversible cycle, it would necessarily mean that for the reverse cycle we would have

$$\oint W_{\text{net}} > 0$$

This however would violate the second law, and the only possibility is that the equality holds for a reversible cycle. Therefore

$$\oint \left(\frac{\delta Q}{T}\right)_{\text{rev}} = 0 \tag{7.3}$$

The significance of the inequality of Clausius may be illustrated by considering the simple steam power plant cycle shown in Fig. 7.3. This cycle is slightly different from the usual cycle for steam power plants in that the pump handles a mixture of liquid and vapor in such proportions that saturated liquid leaves the pump and enters the boiler. Suppose that someone reports that the pressures and quality at various points in the cycle are as given in Fig. 7.3. Does this cycle satisfy the inequality of Clausius?

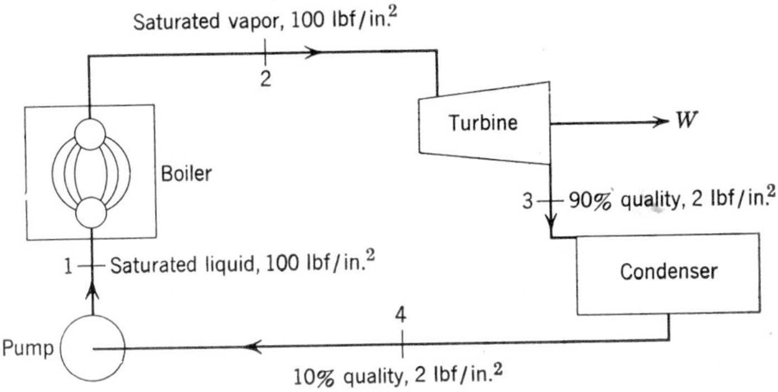

Fig. 7.3 A simple steam power plant that demonstrates the inequality of Clausius.

Heat is transferred in two places, the boiler and the condenser. Therefore

$$\oint \frac{\delta Q}{T} = \int \left(\frac{\delta Q}{T}\right)_{\text{boiler}} + \int \left(\frac{\delta Q}{T}\right)_{\text{condenser}}$$

Since the temperature remains constant in both the boiler and condenser this may be integrated as follows:

$$\oint \frac{\delta Q}{T} = \frac{1}{T_1}\int_1^2 \delta Q + \frac{1}{T_3}\int_3^4 \delta Q = \frac{{}_1Q_2}{T_1} + \frac{{}_3Q_4}{T_3}$$

Let us consider a 1-lb mass as the working fluid.

$${}_1q_2 = h_2 - h_1 = 888.8 \text{ Btu/lbm} \qquad t_1 = 327.8 \text{ F}$$

$${}_3q_4 = h_4 - h_3 = 196.3 - 1014.0 = -818.8 \text{ Btu/lbm} \qquad t_3 = 126.1 \text{ F}$$

Therefore

$$\oint \frac{\delta Q}{T} = \frac{888.8}{327.8 + 459.7} - \frac{818.8}{126.1 + 459.7} = -0.269 \text{ Btu/lbm R}$$

Thus this cycle satisfies the inequality of Clausius, which is equivalent to saying that it does not violate the second law of thermodynamics.

7.3 Entropy—A Property of a System

By the use of Eq. 7.3 and Fig. 7.4 it may be shown that the second law of thermodynamics leads to a property of a system—which we call entropy. Let a closed system undergo a reversible process from state 1 to state 2 along path A, and let the cycle be completed along path B, which is also reversible.

Since this is a reversible cycle we can write

$$\oint \frac{\delta Q}{T} = 0 = \int_{1A}^{2A} \frac{\delta Q}{T} + \int_{2B}^{1B} \frac{\delta Q}{T}$$

Now consider another reversible cycle, which has the same initial process, but let the cycle be completed along path C. For this cycle we can write

$$\oint \frac{\delta Q}{T} = 0 = \int_{1A}^{2A} \frac{\delta Q}{T} + \int_{2C}^{1C} \frac{\delta Q}{T}$$

Subtracting the second equation from the first we have

$$\int_{2B}^{1B} \frac{\delta Q}{T} = \int_{2C}^{1C} \frac{\delta Q}{T}$$

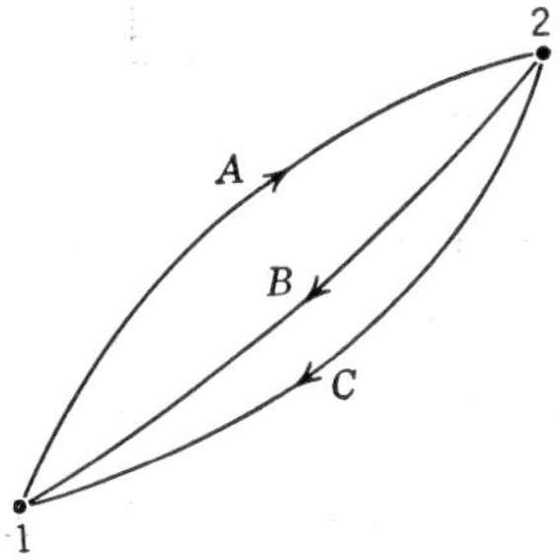

Fig. 7.4 Two reversible cycles demonstrating the fact that entropy is a property of a substance.

Since the $\int \delta Q/T$ is the same for all reversible paths between state 2 and state 1, we conclude that this quantity is independent of the path and is a function of the end states only, and is therefore a property. This is the property called entropy. *Entropy* is a thermodynamic property that is defined by the relation

$$dS = \left(\frac{\delta Q}{T}\right)_{\text{rev}} \tag{7.4}$$

It is important to note that entropy is here defined in terms of a reversible process.

The change in the entropy of a system as it undergoes a change of state may be found by integrating Eq. 7.4. Thus

$$S_2 - S_1 = \int_1^2 \left(\frac{\delta Q}{T}\right)_{\text{rev}} \tag{7.5}$$

In order to integrate this, the relation between T and Q must be known, and illustrations will be given subsequently. The important point to note here is that since entropy is a property, the change in the entropy of a substance in going from one state to another is the same for all processes, both reversible and irreversible, between these two states. Equation 7.5 enables us to find the change in entropy only along a reversible path. However, once it has been evaluated, this is the magnitude of the entropy change for all processes between these two states, because entropy is a property of a substance.

Equation 7.5 enables us to determine changes of entropy, but tells us nothing about absolute values of entropy. However, the third law of thermodynamics, which is discussed in Chapter 15, in essence states that the entropy of a pure substance approaches zero at the absolute zero of temperature. This gives rise to absolute values of entropy and is particularly important when chemical reactions are involved.

However, when no change of composition is involved, it is quite adequate to give values of entropy relative to some arbitrarily selected reference state. This is the procedure followed in most tables of thermodynamic properties, such as the steam tables and ammonia tables. Therefore, until absolute entropy is introduced in Chapter 15, we will give values of entropy relative to some reference state.

A word should be added here regarding the role of T as an integrating factor. We noted in Chapter 4 that Q is a path function, and therefore δQ is an inexact differential. However, an inexact differential may be converted to an exact differential by introduction of an integrating factor, and in this case the integrating factor is $1/T$. Thus, δQ is an inexact differential but $(\delta Q/T)_{\text{rev}}$ is an exact differential.

7.4 The Entropy of a Pure Substance

Since entropy is an extensive property of a system, values of specific entropy (entropy per unit mass) are tabulated in tables of thermodynamic properties in the same manner as specific volume and specific enthalpy. The units of specific entropy in the steam tables, Freon-12 tables, and ammonia tables are Btu/lbm R, and the values are given relative to an arbitrary reference state. In the steam tables the entropy of saturated liquid at 32 F is given the value of zero. For most refrigerants, such as Freon-12 and ammonia, the entropy of saturated liquid at −40 F is assigned the value of zero.

In the saturation region the specific entropy may be calculated using the quality, the relations being similar to those for specific volume and specific enthalpy.

$$s = (1 - x)s_f + xs_g$$

$$s = s_f + xs_{fg} \tag{7.6}$$

$$s = s_g - (1 - x)s_{fg}$$

The entropy of a compressed liquid is found in the same manner as the other properties. These properties are primarily a function of the temperature, and are not greatly different from those for saturated

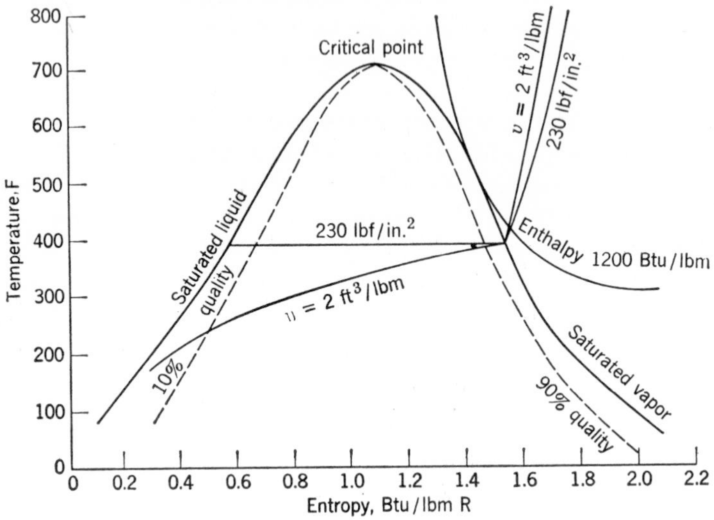

Fig. 7.5 Temperature-entropy diagram for steam.

liquid at the same temperature. Table 4 of Keenan and Keyes' steam tables gives the correction for compressed liquid water as has been discussed previously.

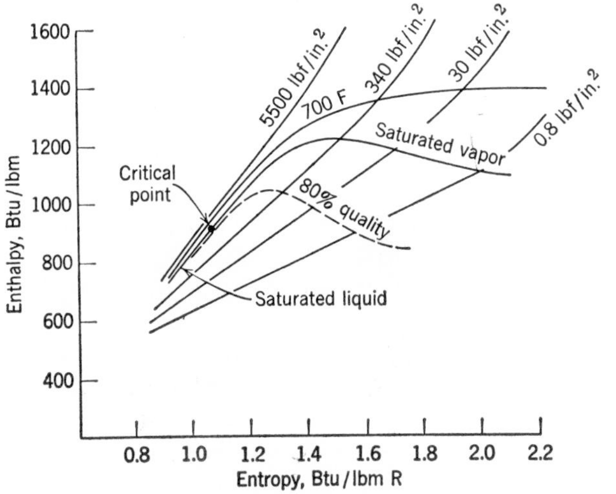

Fig. 7.6 Enthalpy-entropy diagram for steam. (Not to scale.)

The thermodynamic properties of a substance are often shown on a temperature-entropy diagram, and an enthalpy-entropy diagram, which is also called a Mollier diagram. Figures 7.5 and 7.6 show the essential elements of temperature-entropy and enthalpy-entropy diagrams for steam. The general features are the same for all substances. A more complete temperature-entropy diagram for steam is shown on Fig. 9 of the steam tables.

These diagrams are valuable both as a means of presenting thermodynamic data and also because they enable one to visualize the changes of state that occur in various processes, and as our study progresses the student should acquire facility in visualizing thermodynamic processes on these diagrams. The temperature-entropy diagram is particularly important for this.

One further observation should be made here concerning the compressed-liquid lines on the temperature-entropy diagram for water.

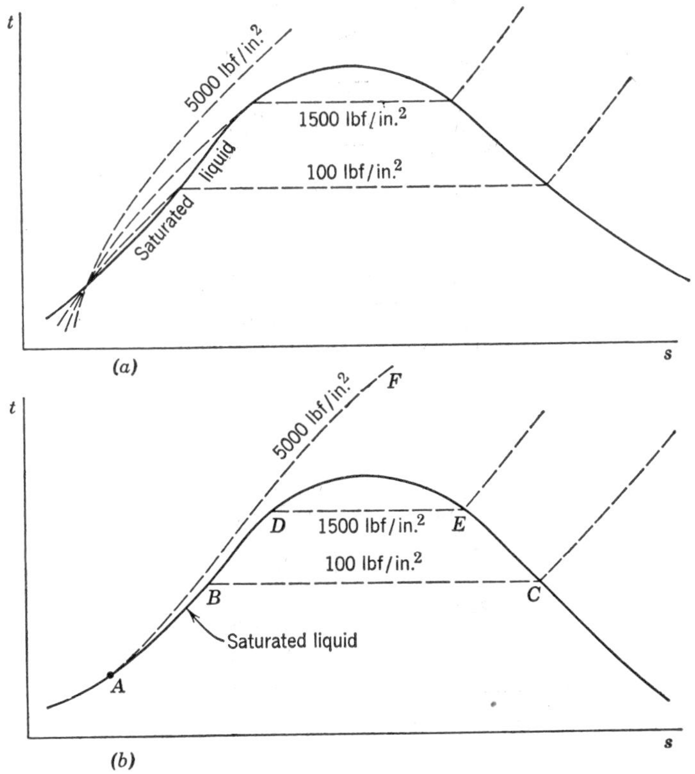

Fig. 7.7 Temperature-entropy diagram to show properties of a compressed liquid.

Reference to Table 4 of the steam tables indicates that the entropy correction is negative at all the listed temperatures except at 32 F. If these corrections are considerably magnified they appear as shown in Fig. 7.7*a*. It can be shown that each constant-pressure line crosses the saturated-liquid line at the point of maximum density, which is about 39 F for water (the exact temperature at which maximum density occurs varies slightly with pressure). For lower temperatures the entropy correction is positive. It is important to understand the general shape of these lines when showing the pumping process in steam power-plant cycles.

Having made this observation for water, it should be stated that for most substances the magnitude of the entropy correction is so small that usually a process in which liquid is heated at constant pressure is shown as coinciding with the saturated-liquid line until the saturation temperature is reached (Fig. 7.7*b*). Thus, if water at 1500 lbf/in.^{2} is heated from 32 F saturation temperature, it would be shown by line ABD, which coincides with the saturated-liquid line.

7.5 Entropy Change in Reversible Processes

Having established the fact that entropy is a thermodynamic property of a system, its significance in various processes will now be considered. In this section we will limit ourselves to closed systems and reversible processes, and consider the Carnot cycle, reversible heat-transfer processes, and reversible adiabatic processes.

Carnot cycle Let the working fluid of a heat engine operating on the Carnot cycle comprise the system. The first process is the isothermal transfer of heat to the working fluid from the high-temperature reservoir. For this we can write

$$S_2 - S_1 = \int_1^2 \left(\frac{\delta Q}{T}\right)_{\text{rev}}$$

Since this is a reversible process in which the temperature of the working fluid remains constant, it can be integrated to give

$$S_2 - S_1 = \frac{1}{T_H}\int_1^2 \delta Q = \frac{{}_1Q_2}{T_H}$$

This process is shown in Fig. 7.8*a*, and the area under line 1–2, area 1–2–*b*–*a*–1, represents the heat transferred to the working fluid during the process.

The second process of a Carnot cycle is a reversible adiabatic one. From the definition of entropy

$$dS = \left(\frac{\delta Q}{T}\right)_{\text{rev}}$$

it is evident that the entropy remains constant in a reversible adiabatic process. A constant-entropy process is called an isentropic process. Line 2–3 represents this process, and this process is concluded at state 3 when the temperature of the working fluid reaches T_L.

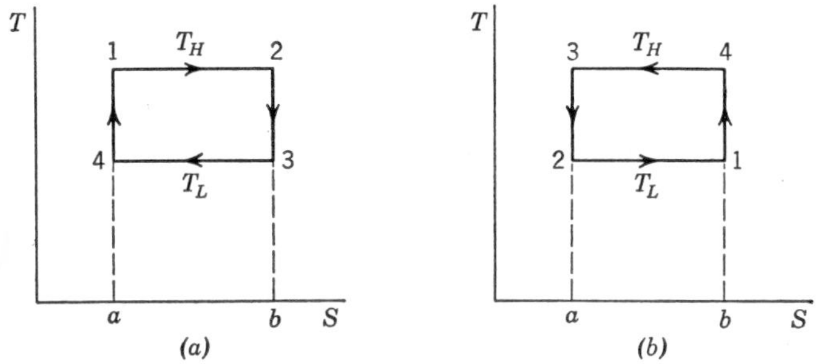

Fig. 7.8 The Carnot cycle on the temperature-entropy diagram.

The third process is the reversible isothermal process in which heat is transferred from the working fluid to the low-temperature reservoir. For this we can write

$$S_4 - S_3 = \int_3^4 \frac{\delta Q}{T_L} = \frac{1}{T_L}\int_3^4 \delta Q = \frac{{}_3Q_4}{T_L}$$

Since during this process the heat transfer is negative (as regards the working fluid), the entropy of the working fluid decreases during this process. Also, since the final process 4–1, which completes the cycle, is a reversible adiabatic process (and therefore isentropic), it is evident that the entropy decrease in process 3–4 must exactly equal the entropy increase in process 1–2. The area under line 3–4, area 3–4–a–b–3, represents the heat transferred from the working fluid to the low-temperature reservoir.

Since the net work of the cycle is equal to the net heat transfer, it is evident that area 1–2–3–4–1 represents the net work of the cycle. The efficiency of the cycle may also be expressed in terms of areas.

$$\eta_{\text{th}} = \frac{W_{\text{net}}}{Q_H} = \frac{\text{area } 1\text{–}2\text{–}3\text{–}4\text{–}1}{\text{area } 1\text{–}2\text{–}b\text{–}a\text{–}1}$$

Some statements made about efficiencies in the last chapter may now be understood graphically. For example, increasing T_H while T_L remains constant increases the efficiency. Decreasing T_L as T_H remains constant increases the efficiency. It is also evident that the efficiency approaches 100 per cent as the absolute temperature at which heat is rejected approaches zero.

If the cycle is reversed, we have a heat pump, and the Carnot cycle for a heat pump is shown in Fig. 7.8*b*. Notice in this case that the entropy increases at T_L, since heat is transferred to the working fluid at T_L. The entropy decreases at T_H due to heat transfer from the working fluid.

Reversible heat-transfer processes Actually we are concerned here with processes that are internally reversible, i.e., processes where there are no irreversibilities within the boundary of the system. For such processes the heat transfer to or from a system can be shown as an area on a temperature-entropy diagram. For example, consider the change of state from saturated liquid to saturated vapor at constant pressure. This would correspond to the process 1–2 on the T-s diagram of Fig. 7.9 (note that absolute temperature is required here), and the area

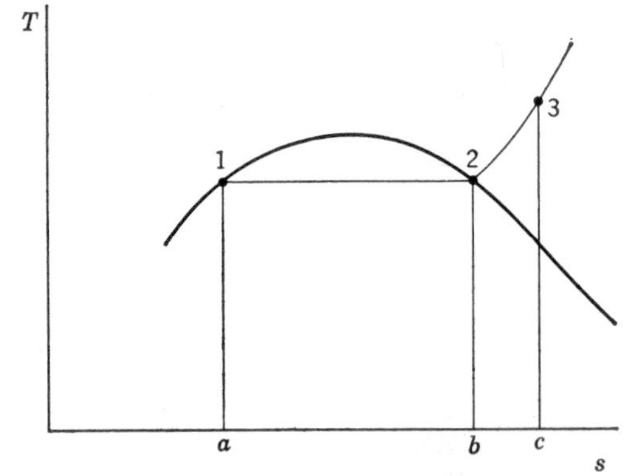

Fig. 7.9 A temperature-entropy diagram to show areas which represent heat transfer for an internally reversible process.

1–2–*b*–*a*–1 represents the heat transfer. Since this is a constant-pressure process, the heat transfer per unit mass is equal to h_{fg}. Thus, we can write

$$s_2 - s_1 = s_{fg} = \int_1^2 \left(\frac{\delta q}{T}\right)_{\text{rev}} = \frac{1}{T}\int_1^2 \delta q = \frac{{}_1q_2}{T} = \frac{h_{fg}}{T}$$

Let us verify this relation for steam at 100 lbf/in.² From the steam tables we have

$$h_{fg} = 888.8 \text{ Btu/lbm}$$

$$s_{fg} = 1.1286 \text{ Btu/lbm R}$$

$$T = 327.81 + 459.69 = 787.50$$

Checking: $$s_{fg} = \frac{h_{fg}}{T} = \frac{888.8}{787.50} = 1.1286 \text{ Btu/lbm R}$$

If heat is transferred to the saturated vapor at constant pressure, the steam is superheated along line 2–3. For this process we can write

$$_2q_3 = \int_2^3 \delta q = \int_2^3 T\,ds$$

Since T is not constant this cannot be integrated unless we know a relation between temperature and entropy. However, we do realize that the area under line 2–3, area 2–3–c–b–2, represents $\int_2^3 T\,ds$, and therefore represents the heat transferred during this process.

The important conclusion to draw here is that for processes that are internally reversible, the area underneath the process line on a temperature-entropy diagram represents the quantity of heat transferred. This is not true for irreversible processes, as will be demonstrated later.

Reversible adiabatic process There are many situations in which essentially adiabatic processes take place. We have already noted that in such cases the ideal process, which is a reversible adiabatic process, is isentropic. We will consider here an example of a reversible adiabatic process for a closed system, and consider in a later section the reversible adiabatic process for an open system. We will also note in a later section that by comparing an actual process with the ideal or isentropic process we have a basis for defining the efficiency of a machine.

Example 7.1 Consider a cylinder fitted with a piston which contains saturated Freon-12 vapor at 20 F. Let this vapor be compressed in a reversible adiabatic process until the pressure is 150 lbf/in.^{2} Determine the work per pound of Freon-12 for this process.

From the first law we conclude that

$$_1q_2 = u_2 - u_1 + {_1w_2} = 0$$

$$_1w_2 = u_1 - u_2$$

State 1 is specified by the statement of the problem and therefore the initial entropy is known. We also conclude, since the process is reversible and adiabatic, that $s_1 = s_2$. Therefore we know entropy and pressure in the final state, which is sufficient to specify the final state, since we are dealing with a pure substance.

From the Freon-12 tables

$$u_1 = h_1 - P_1v_1 = 79.385 - \frac{35.736 \times 144 \times 1.0988}{778} = 72.11 \text{ Btu/lbm}$$

$$s_1 = s_2 = 0.16719 \text{ Btu/lbm R}$$

$$P_2 = 150 \text{ lbf/in.}^2$$

Therefore, from the superheat tables for Freon-12

$$t_2 = 122.9 \text{ F} \qquad h_2 = 90.330 \qquad v_2 = 0.28270$$

$$u_2 = 90.330 - \frac{150 \times 144 \times 0.28270}{778} = 82.48 \text{ Btu/lbm}$$

$${}_1w_2 = u_1 - u_2 = 72.11 - 82.49 = -10.37 \text{ Btu/lbm}$$

7.6 Entropy Changes of a Closed System During an Irreversible Process

Consider the cycles shown in Fig. 7.10. The cycle made up of proc-

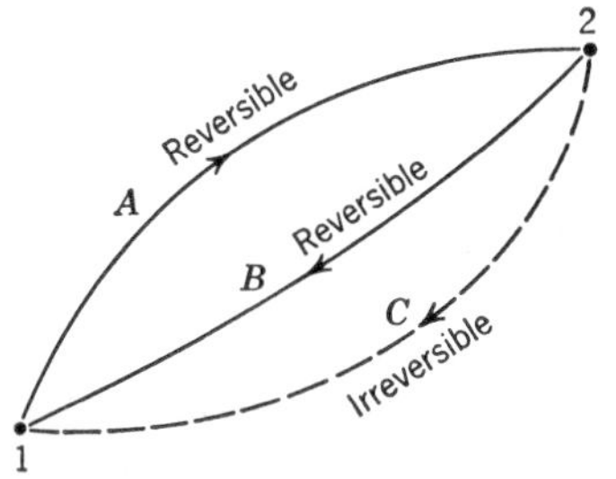

Fig. 7.10 Entropy change for irreversible processes of a closed system.

esses A and B is a reversible cycle. Therefore we can write

$$\oint \frac{\delta Q}{T} = 0 = \int_{1A}^{2A} \frac{\delta Q}{T} + \int_{2B}^{1B} \frac{\delta Q}{T}$$

The cycle made up of processes A and C is an irreversible cycle. Therefore, for this cycle the inequality of Clausius may be applied, giving the result

$$\oint \frac{\delta Q}{T} = \int_{1A}^{2A} \frac{\delta Q}{T} + \int_{2C}^{1C} \frac{\delta Q}{T} \le 0$$

Subtracting the second equation from the first we have

$$\int_{2B}^{1B} \frac{\delta Q}{T} - \int_{2C}^{1C} \frac{\delta Q}{T} \geq 0$$

Since path B is reversible, and since entropy is a property

$$\int_{2B}^{1B} \frac{\delta Q}{T} = \int_{2B}^{1B} dS = \int_{2C}^{1C} dS$$

Therefore

$$dS \geq \frac{\delta Q}{T} \tag{7.7}$$

or

$$S_2 - S_1 \geq \int_1^2 \frac{\delta Q}{T}$$

The equality sign holds for reversible processes and the inequality sign holds for irreversible processes. The implication of this equation is that the effect of irreversibility is always to increase the entropy of a system.

The change of entropy in an irreversible process for a closed system can be given in terms of an equality by introducing the concept of lost work, to which we give the symbol LW.

This concept of lost work can be described with the aid of Fig. 7.11, in which a gas is separated from a vacuum by a membrane. Let the

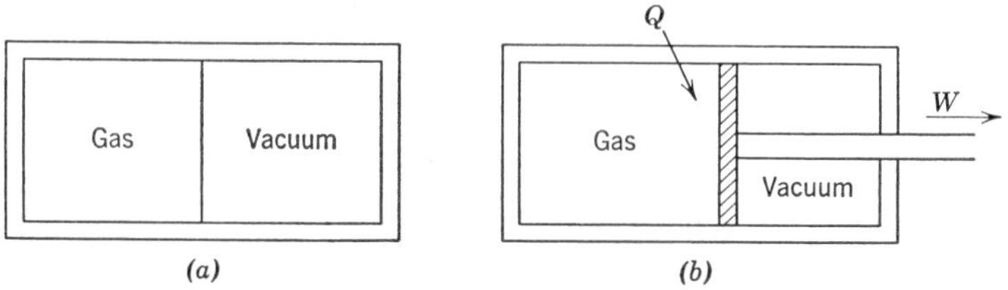

Fig. 7.11 A process that demonstrates lost work.

membrane have a small rupture so that the gas slowly fills the entire volume. At this point let us assume that the nature of the gas is such that its final temperature is the same as the initial temperature. As we have noted previously, this is an irreversible process, and as such we would like to compare it to a reversible process that has the same final state.

Such a reversible process might be achieved as shown in Fig. 7.11*b*, in which the gas expands against a frictionless piston, thus doing work. If the temperature is to remain constant during this reversible process, as it did in the irreversible process, heat must be transferred reversibly from a reservoir of suitable temperature.

The major difference between these two processes is that in the irreversible process the work is zero, whereas in the reversible process the greatest possible work is done. Thus, we might speak of the lost work of an irreversible process. Every irreversible process has associated with it a certain amount of lost work. In Fig. 7.11 we have shown the two extreme cases, a reversible process and one in which the work was zero. Between these two extremes are processes that have various degrees of irreversibility, i.e., the lost work may vary from zero to some maximum value. In the case of an expanding fluid we therefore write

$$P\,dV = \delta W + \delta LW \tag{7.8}$$

The entropy change in an irreversible process can be written in terms of the lost work as follows:

$$dS = \frac{\delta Q + \delta LW}{T} \tag{7.9}$$

Per unit mass the equation is written

$$ds = \frac{\delta q + \delta lw}{T}$$

In a reversible process the lost work is zero, and Eq. 7.9 reduces to the definition of entropy given by Eq. 7.4.

$$dS = \left(\frac{\delta Q}{T}\right)_{\text{rev}}$$

Some important conclusions can now be drawn from Eq. 7.7 or Eq. 7.9. First of all, there are two ways in which the entropy of a system can be increased, namely, by transferring heat to it and by having it undergo an irreversible process. Since the lost work cannot be less than zero, there is only one way in which the entropy of a system can be decreased, and that is to transfer heat from the system.

Second, the change in entropy of a system can be separated into the change due to heat transfer and the change due to irreversibilities. Frequently the increase in entropy due to irreversibilities is called the irreversible production of entropy.

Finally, as we have already noted, for an adiabatic process $\delta Q = 0$, and in this case the increase in entropy is always associated with lost work, i.e., the irreversibilities.

One other point regarding the showing of irreversible processes on P-v and T-s diagrams should be made. The work for an irreversible process is not equal to $\int P\,dv$ and the heat transfer is not equal to $\int T\,ds$. Therefore, the area underneath the path does not represent work and heat on the P-v and T-s diagrams respectively. In fact, in many cases we are not certain of the exact state through which a system passes when it undergoes an irreversible process. For this reason it is advantageous to show irreversible processes as dotted lines and reversible processes as solid lines. Thus, the area underneath a dotted line will never represent work or heat. For example, the processes of Fig. 7.11 would be shown on P-v and T-s diagrams as in Fig. 7.12.

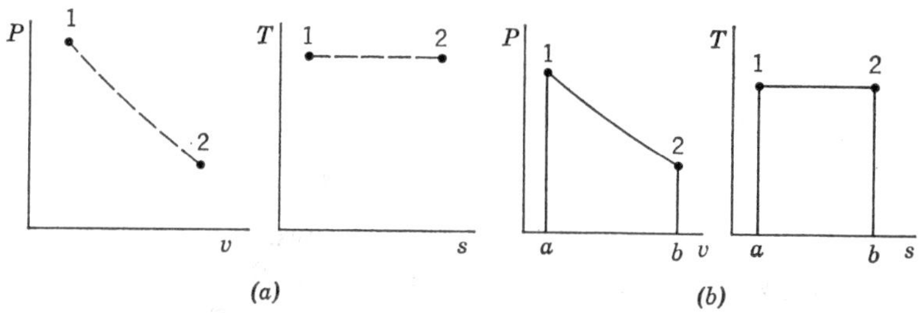

Fig. 7.12 Reversible and irreversible processes on pressure-volume and temperature-entropy diagrams.

Figure 7.12*a* shows an irreversible process, and since the heat transfer and work for this process is zero, the area underneath the path line has no significance. Figure 7.12*b* shows the reversible process, and area 1–2–*b*–*a*–1 represents the work on the P-v diagram and the heat transfer on the T-s diagram.

7.7 Entropy Changes for an Open System

In an open system the entropy is increased because the mass that crosses the boundary of the system has entropy. Thus, as the mass δm_i enters the system, the entropy is increased by the amount $\delta m_i s_i$. Similarly, as the mass δm_e leaves the system the entropy decreases by the amount $\delta m_e s_e$.

Thus, for an open system we can write

$$dS_\sigma \geq \frac{\delta Q}{T} + \delta m_i s_i - \delta m_e s_e \tag{7.10}$$

where dS_σ designates the change in the entropy of the system.

Or, if we wish to use the concept of lost work,

$$dS_\sigma = \frac{\delta Q + \delta LW}{T} + \delta m_i s_i - \delta m_e s_e \tag{7.11}$$

In steady-flow processes there is no change in the entropy of the system (i.e., $dS_\sigma = 0$), and $\delta m_i = \delta m_e = \delta m$. Therefore, for a steady-flow process we can write

$$\delta m(s_e - s_i) \geq \frac{\delta Q}{T} \tag{7.12}$$

or

$$\delta m(s_e - s_i) = \frac{\delta Q + \delta LW}{T} \tag{7.13}$$

For a steady-flow adiabatic process we conclude from Eq. 7.12 that

$$s_e \geq s_i \tag{7.14}$$

That is, in a steady-flow adiabatic process the entropy of the fluid leaving must be equal to or greater than the entropy of the fluid coming in. Since the equality holds for a reversible process, we conclude that for a reversible, steady-flow, adiabatic process

$$s_e = s_i \tag{7.15}$$

The following examples deal with a second-law analysis of a variety of processes.

Example 7.2 Steam enters a steam turbine at a pressure of 100 lbf/in.^{2}, a temperature of 500 F, and a velocity of 200 ft/sec. The steam leaves the turbine at a pressure of 20 lbf/in.^{2} and a velocity of 600 ft/sec. Determine the work per pound of steam if the process is reversible and adiabatic.

The schematic diagram and the T-s diagram for this process are shown in Fig. 7.13.

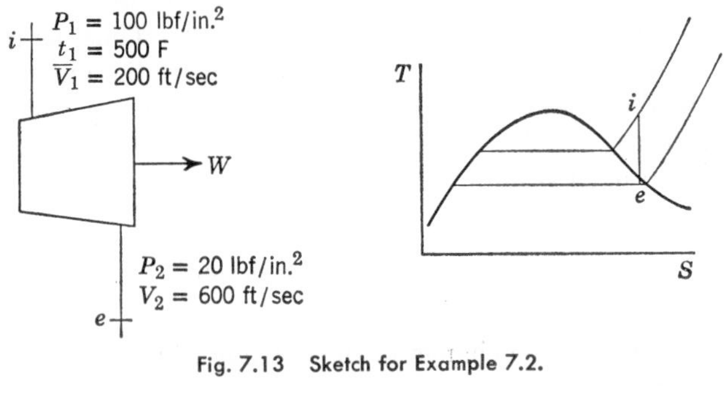

Fig. 7.13 Sketch for Example 7.2.

From the steady-flow energy equation we conclude for this process that

$$h_i + \frac{\overline{V}_i^2}{2g_c} = h_e + \frac{\overline{V}_e^2}{2g_c} + w$$

Since this is a steady-flow, reversible, adiabatic process, $s_e = s_i$, and therefore the two properties known in the final state are entropy and pressure. From the steam tables

$$h_i = 1279.1 \text{ Btu/lbm} \qquad s_i = 1.7085 \text{ Btu/lbm R}$$

The two properties known in the final state are pressure and entropy.

$$P_e = 20 \text{ lbf/in.}^2 \qquad s_e = s_i = 1.7085 \text{ Btu/lbm R}$$

Therefore the quality and enthalpy leaving the turbine can be determined.

$$s_e = 1.7085 = s_g - (1-x)_e s_{fg} = 1.7319 - (1-x)_e 1.3962$$

$$(1-x)_e = \frac{0.0234}{1.3962} = 0.01676$$

$$h_e = h_g - (1-x)_e h_{fg} = 1156.3 - 0.01676(960.1) = 1140.2 \text{ Btu/lbm}$$

Therefore the work per pound of steam for this isentropic process may be found using the steady-flow energy equation

$$w = 1279.1 - 1140.2 + \frac{(200)^2 - (600)^2}{2 \times 32.17 \times 778} = 132.5 \text{ Btu/lbm}$$

Example 7.3 Consider the reversible adiabatic flow of steam through a nozzle. Steam enters the nozzle at 100 lbf/in.^{2}, 500 F, with a velocity of 100 ft/sec. The pressure leaving the nozzle is 40 lbf/in.^{2} Determine the exit velocity from the nozzle, assuming reversible, adiabatic, steady flow.

This process is shown in Fig. 7.14. Frequently numbers are used instead of letters to indicate the state of the fluid entering and leaving, and this procedure is followed in this example.

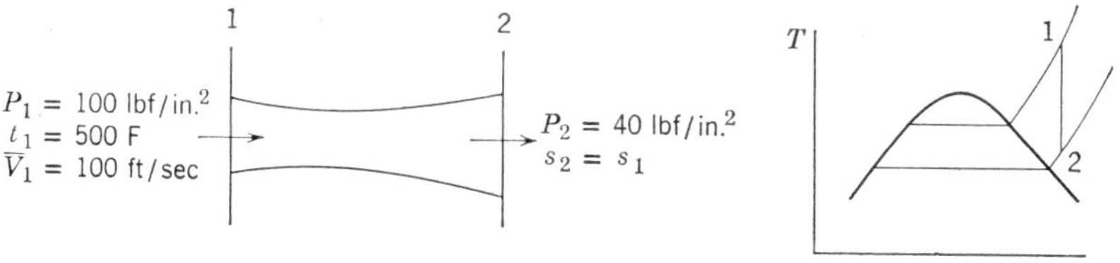

Fig. 7.14 Sketch for Example 7.3.

Since the heat transfer, work, and changes in potential energy are zero, the steady-flow energy equation reduces to

$$h_1 + \frac{\overline{V}_1^2}{2g_c} = h_2 + \frac{\overline{V}_2^2}{2g_c}$$

The two properties known in the final state are entropy and pressure.

$$s_2 = s_1 = 1.7085 \text{ Btu/lbm R} \qquad P_2 = 40 \text{ lbf/in.}^2$$

Therefore, from the superheat tables

$$h_1 = 1279.1 \text{ Btu/lbm} \qquad h_2 = 1193.8 \text{ Btu/lbm}$$

$$\frac{\overline{V}_2^2}{2g_c} = h_1 - h_2 + \frac{\overline{V}_1^2}{2g_c} = 1279.1 - 1193.8 + \frac{100 \times 100}{2 \times 32.17 \times 778}$$

$$= 85.5 \text{ Btu/lbm}$$

$$\overline{V}_2 = \sqrt{2 \times 32.17 \times 778 \times 85.5} = 2070 \text{ ft/sec}$$

Example 7.4 An inventor reports that he has a refrigeration compressor that receives saturated Freon-12 vapor at 0 F and delivers the vapor at 150 lbf/in.^{2}, 120 F. The compression process is adiabatic. Does the process described violate the second law?

For a steady-flow, adiabatic process the second law requires that $s_e \geq s_i$. However, for the compression process described, $s_i = 0.16888$ Btu/lbm R; $s_e = 0.16629$ Btu/lbm R. Therefore, the process described does violate the second law because $s_e < s_i$, whereas for this process the second law requires that $s_e \geq s_i$.

7.8 Two Important Relations for a Pure Substance Involving Entropy

The first law of thermodynamics for a closed system is given by the relation

$$\delta Q = dU + \frac{m d\overline{V}^2}{2g_c} + mdZ \frac{g}{g_c} + \delta W$$

In the absence of changes in kinetic and potential energy this reduces to

$$\delta Q = dU + \delta W \tag{7.16}$$

For a reversible process

$$\delta Q = T\,dS$$

$$\delta W = P\,dV - \mathfrak{J}\,dl - \mathcal{S}\,dA - \mathcal{E}\,dZ$$

If the only work involved in a reversible process is the work done at the moving boundary of the system, the equation for work is

$$\delta W = P\,dV$$

Therefore, for a reversible process in which the only work involved is the work done at the moving boundary of the system

$$T\,dS = dU + P\,dV \tag{7.17}$$

However, once we have written this equation we realize that it involves only changes in properties, and involves no path functions. We conclude therefore, that this equation is valid for all processes, both reversible and irreversible, and that it applies to the substance undergoing a change of state as the result of flow across the boundary of the open system as well as to the substance comprising a closed system.

The fact that Eq. 7.17 applies to both reversible and irreversible processes may be further explained by comparing the first law with Eq. 7.16.

$$\delta Q = dU + \delta W$$

$$T\,dS = dU + P\,dV$$

For an irreversible process $T\,dS$ is different from δQ, by the same amount that $P\,dV$ is different from δW.

This also ties in with our concept of lost work, for

$$T\,dS = \delta Q + \delta LW$$

$$P\,dV = \delta W + \delta LW$$

Substituting these relations into Eq. 7.16 we have

$$T\,dS = dU + P\,dV \tag{7.17}$$

Thus, we can say that $T\,dS$ differs from δQ by the lost work δLW, and $P\,dV$ differs from δW by the lost work δLW.

From our definition of enthalpy we have the relation

$$dH = dU + P\,dV + V\,dP$$

Substituting this relation into Eq. 7.17 we obtain another important relation, namely,

$$T\,dS = dH - V\,dP \tag{7.18}$$

Frequently Eqs. 7.17 and 7.18 are written per unit mass.

$$T\,ds = du + P\,dv \tag{7.19}$$

$$T\,ds = dh - v\,dP \tag{7.20}$$

Equations 7.17 and 7.18 are two very important relations which will be referred to on a number of occasions. It is important to remember that these equations apply to a pure substance for both reversible and irreversible processes in the absence of such effects as surface tension and electricity.

7.9 Some Relations Involving Steady-Flow Processes

Several important equations can be obtained by introducing entropy into the steady-flow energy equation. This equation in terms of differentials, Eq. 5.29, is as follows:

$$dh + \frac{d\overline{V}^2}{2g_c} + \frac{g}{g_c}\,dZ + \delta w - \delta q = 0 \tag{7.21}$$

If we substitute Eqs. 7.9 and 7.20 into the above equation, we have

$$T\,ds + v\,dP + \frac{d\overline{V}^2}{2g_c} + \frac{g}{g_c}\,dZ + \delta w - T\,ds + \delta lw = 0$$

This reduces to

$$v\,dP + \frac{d\overline{V}^2}{2g_c} + \frac{g}{g_c}\,dZ + \delta w + \delta lw = 0 \tag{7.22}$$

For a reversible process the lost work is zero and Eq. 7.22 reduces to

$$v\,dP + \frac{d(\overline{V})^2}{2g_c} + \frac{g}{g_c}\,dZ + \delta w = 0 \tag{7.23}$$

This is an important and useful equation when dealing with reversible processes and two points will be made regarding it. The first is that when Eq. 7.23 is integrated with constant specific volume we have

$$v(P_2 - P_1) + \frac{\overline{V}_2^{\,2} - \overline{V}_1^{\,2}}{2g_c} + \frac{g}{g_c}(Z_2 - Z_1) + w = 0$$

When the work is zero this equation reduces to the well-known Bernoulli equation, which is usually introduced in fluid mechanics courses,

$$v(P_2 - P_1) + \frac{\overline{V}_2^{\,2} - \overline{V}_1^{\,2}}{2g_c} + \frac{g}{g_c}(Z_2 - Z_1) = 0$$

Thus the steady-flow energy equation reduces to the Bernoulli equation for a reversible process in which the work is zero. However, because we cannot have reversible heat transfer to a flowing fluid, the statement is usually made that the Bernoulli equation applies to a reversible adiabatic steady-flow process in which the work is zero.

The second point is that Eq. 7.23 is a very useful equation in finding the work in a steady-flow reversible adiabatic process. Let us consider the significance of this equation for a process in which the changes in kinetic and potential energy are negligible. Where 1 and 2 represent the states entering and leaving the system we can write

$$w = \int_1^2 \delta w = -\int_1^2 v\, dP \qquad (7.25)$$

On the P-v diagram shown in Fig. 7.15, area 1–2–a–b–1 represents

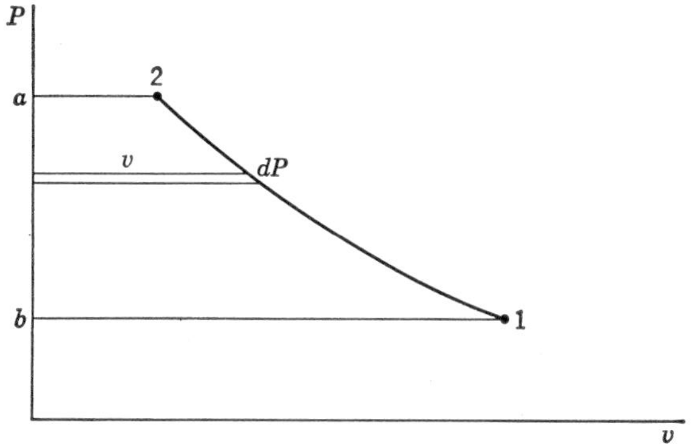

Fig. 7.15 The significance of the $\int v\, dP$ on the pressure-volume diagram.

$\int v\, dP$, and therefore represents the negative work for the process. Since this is a compression process the work would be negative.

It is important to understand the situation in which work per unit mass is equal to $\int P\, dv$ and when it is equal to $-\int v\, dP$. In Chapter 4 we found that the work done per unit mass at the moving boundary of a closed system in a reversible process was equal to $\int P\, dv$. We have now found that work done per unit mass in a reversible steady-flow process with negligible changes in kinetic and potential energy for an open system is $-\int v\, dP$. The relation between these two may be clarified by reference to Fig. 7.16, which refers to an air compressor that has no clearance. (The argument that follows holds equally well for an air compressor with clearance, but is more complicated.) Let the piston be initially at top dead center. As the air flows through the inlet pipe it has a certain amount of flow energy, and as it enters the cylinder it does work on the piston as it moves the piston to the right. If the pressure in the inlet pipe is exactly equal to the pressure in the cylinder, the flow energy is exactly equal to the work done on the piston, and both can be represented by area 1–*b*–*c*–*d*–*e*. If we assume that 1 lbm of fluid enters the cylinder, this area is equal to P_1v_1.

Next the inlet valve closes and the air is compressed until the discharge pressure is reached at state 2. During this process the pound mass of air could be treated as a closed system, and the work done on the air would be equal to $\int P\,dv$, represented by area 1–2–*d*–*e*–1.

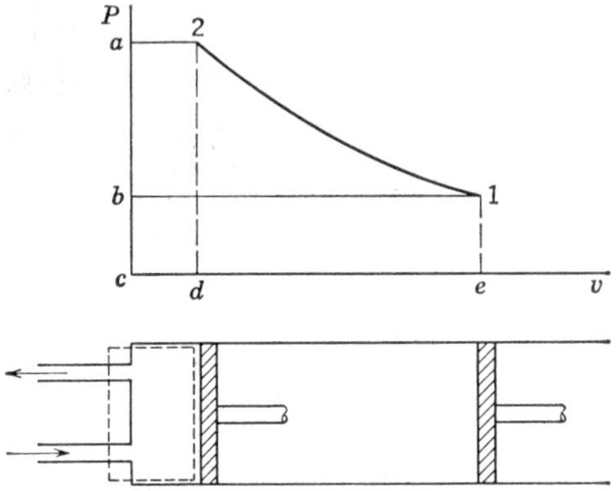

Fig. 7.16 A sketch to show the relation of $\int P\,dv$ and $\int v\,dP$.

When the discharge pressure is reached the exhaust valve opens, and work is done on the air by the piston as it forces the air into the discharge line. As the air flows into the discharge line it has an amount of flow energy exactly equal to the work done on it in forcing it from the cylinder, provided the pressure in the discharge line equals the pressure in the cylinder. This is represented by area 2–*a*–*c*–*d*–2, and for the 1 lbm is equal to P_2v_2.

The net work is equal to area 1–2–*a*–*b*–1, which is equal to $-\int_1^2 v\,dP$. It is also helpful here to recall a familiar relation from calculus.

$$\int_1^2 P\,dv = P_2v_2 - P_1v_1 - \int_1^2 v\,dP \tag{7.26}$$

The first term represents the work done by the piston during the process from state 1 to 2 and the next two terms represent the work done by the piston during the discharge and inlet strokes respectively. The last term represents the net cycle work.

One very useful application of Eq. 7.25 is to calculate the work for the reversible adiabatic (and therefore isentropic) pumping process of a liquid in a steady-flow process. Since the fluid is essentially incompressible, Eq. 7.25 can be integrated as follows,

$$-w = \int_1^2 v\,dP = v\int_1^2 dP = v(P_2 - P_1)$$

If there is a significant change in specific volume the average specific volume should be used.

Example 7.5 Calculate the work per pound to pump water isentropically from 100 lbf/in.^{2}, 80 F to 1000 lbf/in.^{2} From the steam tables $v_1 = 0.01608$ ft^3/lbm, $s_1 = 0.0932$ Btu/lbm R. Assuming the specific volume to remain constant and using Eq. 7.25

$$-w = \int_1^2 v\,dP = v(P_2 - P_1) = 0.01608(1000 - 100) \times \frac{144}{778}$$

$$= 2.68 \text{ Btu/lbm}$$

Figure 3 of Keenan and Keyes' steam tables enables one to determine the work for a steady-flow process directly. The ordinate is $(h - h_f)_s$, which gives the difference between enthalpy in a given state and the enthalpy of saturated liquid that has the same entropy. Thus, the pump work for the example given above is found by subtracting $(h - h_f)_s$ at 100 lbf/in.^{2} and 0.0932 Btu/lbm R from $(h - h_f)_s$ at 1000 lbf/in.^{2} and the same entropy. Thus, we have

$$-w = (h - h_f)_s \text{ at } 1000 \text{ lbf/in.}^2 - (h - h_f)_s \text{ at } 100 \text{ lbf/in.}^2$$

$$-w = 3.0 - 0.3 = 2.7 \text{ Btu/lbm R}$$

It should be noted that Table 4 of the steam tables gives the change of enthalpy with pressure in a constant-temperature process, whereas Figure 3 gives the change of enthalpy with pressure in an isentropic process.

7.10 Principle of the Increase of Entropy

In this section we consider the change in the entropy of a system and its surroundings.

Consider the process shown in Fig. 7.17, in which a quantity of heat δQ, is transferred from a system at temperature T to the surroundings at temperature T_o, and let the work of this process be δW (either posi-

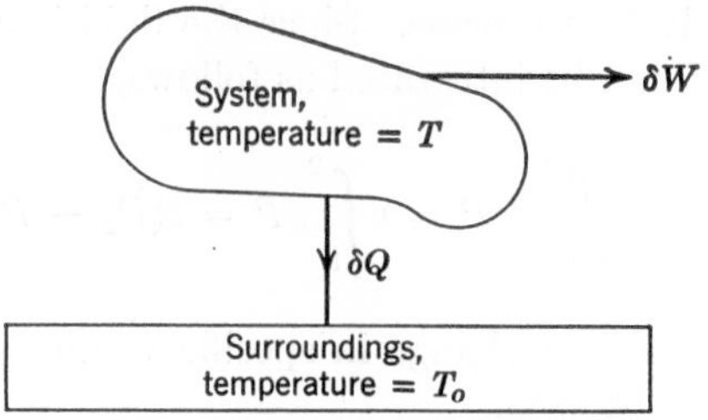

Fig. 7.17 Entropy change for the system plus surroundings.

tive or negative). For this process we can apply Eq. 7.7 to the system and write (δQ is negative for the system)

$$dS_{\text{system}} \geq \frac{-\delta Q}{T}$$

For the surroundings we can write

$$dS_{\text{surroundings}} = \frac{\delta Q}{T_o}$$

The total change of entropy is therefore

$$dS_{\text{system}} + dS_{\text{surroundings}} \geq \frac{-\delta Q}{T} + \frac{\delta Q}{T_o}$$

$$\geq \delta Q\left(\frac{-1}{T} + \frac{1}{T_o}\right)$$

The same conclusion would be reached in the case of an open system, because the change in the entropy of the system would be

$$dS_\sigma \geq -\frac{\delta Q}{T} + \delta m_i s_i - \delta m_e s_e$$

The change in the entropy of the surroundings would be

$$dS_{\text{surroundings}} = \frac{\delta Q}{T_o} - \delta m_i s_i + \delta m_e s_e$$

Therefore,

$$dS_{\text{system}} + dS_{\text{surroundings}} \geq \frac{-\delta Q}{T} + \frac{\delta Q}{T_o} \geq \delta Q\left(-\frac{1}{T} + \frac{1}{T_0}\right)$$

Since $T > T_o$ it follows that

$$-\frac{1}{T} + \frac{1}{T_o} > 0$$

and therefore

$$dS_{\text{system}} + dS_{\text{surroundings}} \geq 0 \tag{7.27}$$

This means that processes involving an interaction of a system and its surroundings will take place only if the net entropy change is greater than zero, or in the limit, remains constant. The constant-entropy process is reversible when both the system and surroundings are considered.

Equation 7.27 is known as the principle of the increase of entropy. All processes that take place are irreversible, and therefore the entropy is continually increasing. Only those processes that have associated with them an increase in entropy of the system plus surroundings will take place. This applies to the combustion of fuel in our automobile engines, to the cooling of our coffee, and to processes that take place in our bodies.

A simple illustration may help to clarify this point.

Example 7.6 Suppose 1 lbm of saturated water vapor at 212 F is condensed to saturated liquid at 212 F in a constant-pressure process by heat transfer to the surrounding air, which is at 80 F. What is the net increase in entropy of the system plus surroundings?

For the system, from the steam tables

$$\Delta S_{\text{system}} = -s_{fg} = -1.4446 \text{ Btu/lbm R}$$

Considering the surroundings

$$Q_{\text{to surroundings}} = h_{fg} = 970.3 \text{ Btu}$$

$$\Delta S_{\text{surroundings}} = \frac{Q}{T_o} = \frac{970.3}{540} = 1.7968 \text{ Btu/R}$$

$$\Delta S_{\text{system}} + \Delta S_{\text{surroundings}} = -1.4446 + 1.7968 = 0.3522 \text{ Btu/R}$$

This increase in entropy is in accordance with the principle of the increase of entropy, and tells, as does our experience, that this process will take place.

It is interesting to note how this heat transfer from the water to the surroundings might have taken place reversibly. Suppose that an engine operating on the Carnot cycle received heat from the water and rejected heat to the surroundings, as shown in Fig. 7.18. The decrease in the entropy of the water is equal to the increase in the entropy of the surroundings.

$$\Delta S_{\text{system}} = -1.4446 \text{ Btu/lbm R}$$

$$\Delta S_{\text{surroundings}} = 1.4446 \text{ Btu/lbm R}$$

$$Q_{\text{to surroundings}} = T\,\Delta S = 540(1.4446) = 780.1 \text{ Btu/lbm}$$

$$W = Q_H - Q_L = 970.3 - 780.1 = 190.2 \text{ Btu/lbm}$$

Since this is a reversible cycle, the engine could be reversed and operated as a heat pump. For this cycle the work input to the heat pump would be 190.2 Btu/lbm.

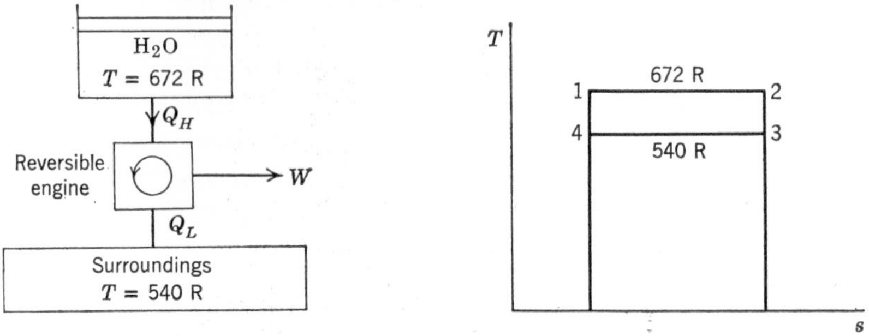

Fig. 7.18 Reversible heat transfer with the surroundings.

This principle of the increase of entropy is the essential thrust of the second law of thermodynamics, and will be developed further in Chapter 10 where the concepts of availability and irreversibility are developed.

7.11 Some General Remarks Regarding Entropy

Very few students of thermodynamics have failed to either ask or be asked the question, usually in a jesting mood, "What is entropy?" We have spent several pages now in discussing the matter, and it might very well be that the student has obtained a good grasp of the material covered, and yet has a very vague concept of entropy. A few remarks might be in order here.

First of all, it would be just as difficult to answer the question "What is energy?" Yet because the term energy is more familiar to us we use it every day, and do not worry about the fact that we really can't explain it. It has meaning to us because we associate it with facts we observe every day, which are actually the facts expressed by the first law. Entropy is much like this, since entropy is a property that enables us to quantitatively express the second law. However, entropy is not as much a part of our everyday vocabulary as energy is. If it were we would use the word daily, since we have many occasions each day to observe the second law of thermodynamics.

A second point to be made regarding entropy is that frequently entropy is associated with probability. From this point of view an irreversible increase in entropy would be associated with a change of state from a less probable state to a more probable state. For example, to use a previous example, one is more likely to find gas on both sides of the ruptured membrane of Fig. 6.11 than to find a gas on one side and a vacuum on the other. The direction of the process is therefore from less probable states to more probable states, and associated with this is an increase in entropy.

A final point to be made is that the second law of thermodynamics and the principle of increase in entropy have great philosophical implications. The question that arises is how did the universe get into the state of reduced entropy in the first place, since all natural processes known to us tend to increase entropy? Are there processes unknown to us, such as a "continual creation," which tend to decrease entropy, and thus offset the increase in entropy associated with the natural processes known to us? On the other end of the scale the question that arises is what is the future of the universe? Will it come to a uniform temperature and maximum entropy, at which time life will be impossible? Quite obviously we cannot give conclusive answers to these questions on the basis of the second law only, but they are certainly topics that illustrate its philosophical implications. The author has found that the second law tends to increase his conviction that there is a Creator who has the answer for the future destiny of man and the universe.

PROBLEMS

7.1 A heat engine operates on the Carnot cycle and receives 500 Btu from a reservoir at 1000 F, and rejects heat at 80 F.

(*a*) Show the cycle on a *T*-*s* diagram, considering the working fluid as the system.

(*b*) Calculate the work and efficiency of the cycle.

(*c*) Calculate the change in entropy of the high-temperature and low-temperature reservoirs.

(*d*) Suppose the engine operates on the cycle indicated above, but the temperature of the high-temperature reservoir is increased to 1500 F. (Heat is then transferred from the reservoir at 1500 F to the working fluid at 1000 F.) The low-temperature reservoir remains at 80 F. Determine the entropy change for each reservoir.

7.2 A Carnot cycle utilizes steam as the working fluid, and has an efficiency of 20%. Heat is transferred to the working fluid at 400 F and during this process the working fluid changes from saturated liquid to saturated vapor.

(*a*) Show this cycle on a *T-s* diagram that includes the saturated-liquid and saturated-vapor lines.

(*b*) Calculate the quality at the beginning and end of the heat-rejection process.

(*c*) Calculate the work per lbm of steam.

7.3 A Carnot cycle heat pump (refrigerator) has ammonia as the working fluid. Heat is transferred from the ammonia at 100 F, and during this process the ammonia changes from saturated vapor to saturated liquid. Heat is transferred to the working fluid at 0 F.

(*a*) Show this cycle on a *T-s* diagram.

(*b*) What is the quality at the beginning and end of the isothermal process at 0 F?

(*c*) What is the coefficient of performance of this cycle as a refrigerator?

7.4 A pressure vessel contains steam at 100 lbf/in.^{2}, 600 F. A valve at the top of the pressure vessel is opened, allowing steam to escape. Assume that at any instant the steam that remains in the pressure vessel has undergone a reversible adiabatic process. Determine the fraction of steam that has escaped when the steam remaining in the pressure vessel is saturated vapor.

7.5 A cylinder fitted with a piston is filled with steam at 100 lbf/in.^{2}, 500 F. The steam expands in a reversible adiabatic process until the pressure is 20 lbf/in.^{2} The initial volume of the steam in the cylinder is 1 ft^3. Determine the work done during this process.

7.6 Ammonia at 50 lbf/in.^{2}, 40 F is compressed in a cylinder by a piston to a pressure of 240 lbf/in.^{2} in a reversible adiabatic process. Determine the work of compression per lbm of ammonia.

7.7 Plot to scale a temperature-entropy diagram for Freon-12, showing the following lines:

(*a*) Saturated liquid and saturated vapor.

(*b*) 20 lbf/in.^{2} and 200 lbf/in.^{2} constant-pressure lines.

(*c*) 90 Btu/lbm and 100 Btu/lbm constant-enthalpy lines.

(*d*) 1 ft^3/lbm constant specific-volume line.

7.8 Freon-12 at 50 lbf/in.^{2}, 200 F is compressed in a cylinder in a reversible isothermal process to 400 lbf/in.^{2} Determine the heat transfer and work per lbm. Show the process on a *T-s* diagram.

7.9 Saturated water vapor at 100 lbf/in.^{2} undergoes a reversible isothermal process in a cylinder until the pressure reaches 20 lbf/in.^{2} Calculate the heat transfer and work per lbm for this process. Show the process on a *T-s* diagram.

7.10 A refrigeration compressor receives ammonia at 30 lbf/in.^{2},

20 F and delivers it at 230 lbf/in.^{2} Determine the work of compression if the process is reversible and adiabatic.

7.11 A centrifugal compressor receives dry saturated water vapor at 50 F and compresses it to 4 lbf/in.^{2} The volume rate of flow into the compressor is 1000 ft^3/min. Assuming the compression process to be reversible and adiabatic, determine the power required to drive the compressor.

7.12 A steam turbine receives steam at 100 lbf/in.^{2}, 500 F. The steam expands in a reversible adiabatic process and leaves the turbine as saturated vapor. The power output of the turbine is 20,000 hp. What is the rate of steam flow in lbm/hr to the turbine?

7.13 A refrigerator utilizes Freon-12 as the refrigerant and handles 200 lbm/hr. The Freon enters the compressor at 30 lbf/in.^{2}, 20 F, and leaves at 175 lbf/in.^{2} What hp motor will be required to drive the compressor if the compression process is reversible and adiabatic?

7.14 A design for a turbine has been proposed involving the reversible, isothermal, steady flow of steam through the turbine. Saturated vapor at 100 lbf/in.^{2} enters the turbine and the steam leaves at 20 lbf/in.^{2} Determine the work per lbm of steam flowing through the turbine.

7.15 A diffuser is a device in which a fluid flowing at high velocity is decelerated in such a way that the pressure increases during the process. Steam at 20 lbf/in.^{2}, 300 F enters the diffuser with a velocity of 2000 ft/sec and leaves with a velocity of 200 ft/sec. If the process is reversible and occurs without heat transfer, what is the final pressure and temperature of the steam? A Mollier diagram may be helpful in solving this problem.

7.16 Steam at 400 lbf/in.^{2}, 600 F expands through a nozzle to 300 lbf/in.^{2} at the rate of 20,000 lbm/hr. If the process occurs isentropically and the initial velocity is low, calculate:

(*a*) The velocity leaving the nozzle.

(*b*) The exit area of the nozzle.

7.17 A contact feedwater heater for heating the water going into a boiler operates on the principle of mixing steam and water. For the change of state shown in Fig. 7.19, calculate the increase in entropy per hr, assuming a steady-flow adiabatic process.

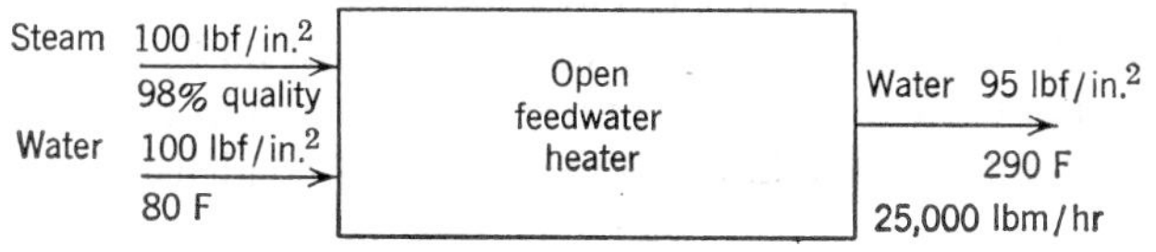

Fig. 7.19 Sketch for Problem 7.17.

7.18 Steam at 100 lbf/in.^{2}, 500 F flows in a pipe. Attached to the pipe is an evacuated vessel. The valve connecting the evacuated vessel to the pipe is opened, and steam flows into the vessel until the pressure reaches 100 lbf/in.^{2} Determine the increase in entropy per lbm of steam, assuming the process to be adiabatic.

7.19 In a refrigeration plant liquid ammonia enters the expansion valve at 200 lbf/in.^{2}, 80 F. The pressure leaving the expansion valve is 40 lbf/in.^{2}, and changes in kinetic energy are negligible. What is the increase in entropy per lbm? Show this process on a temperature-entropy diagram.

7.20 Freon-12 flows through a capillary tube. The entering state is 150 lbf/in.^{2}, 80 F, the pressure leaving is 30 lbf/in.^{2}, and the process is adiabatic. What is the increase in entropy per lbm of Freon-12 flowing through the capillary tube? Show the initial and final states on a T-s diagram.

7.21 A salesman reports that he has a steam turbine available that delivers 3800 hp. The steam enters the turbine at 100 lbf/in.^{2}, 500 F and leaves the turbine at a pressure of 2 lbf/in.^{2}, and the required rate of steam flow is 30,000 lbm/hr.

(*a*) How do you evaluate his claim?

(*b*) Suppose he changed his claim and said the required steam flow was 34,000 lbm/hr.

7.22 Show that the following relation holds for a system that involves a stretched wire.

$$T\,dS = dU - \mathfrak{I}\,dL$$

7.23 In a steam power plant water enters the pump at 20 lbf/in.^{2} and leaves at 1400 lbf/in.^{2} The initial temperature of the water is 120 F. Calculate the work per lbm for this process if it occurs isentropically. Check your answer using Figure 3 of Keenan and Keyes' steam tables.

7.24 A liquid ammonia pump handles 5000 lbm/hr. The liquid enters the pump at 0 F, 35 lbf/in.^{2} and leaves at 200 lbf/in.^{2} What hp motor will be required to drive the pump if the flow is reversible and adiabatic?

7.25 Steam enters a condenser at 2 lbf/in.^{2}, 90% quality, and leaves as saturated liquid. The cooling water enters at 50 F, and such large quantities are used that there is no significant rise in its temperature. What is the net increase in entropy per lbm of condensate leaving the condenser?

7.26 A refrigerator requires 2 hp and has a coefficient of performance of 4.0. The temperature of the refrigerated space is 0 F and the temperature of the surroundings to which heat is rejected is 90 F. Determine the net increase in entropy per hr.

7.27 Tank A initially contains 10 lbm of steam at 100 lbf/in.^{2}, 600 F, and is connected through a valve to a cylinder fitted with a

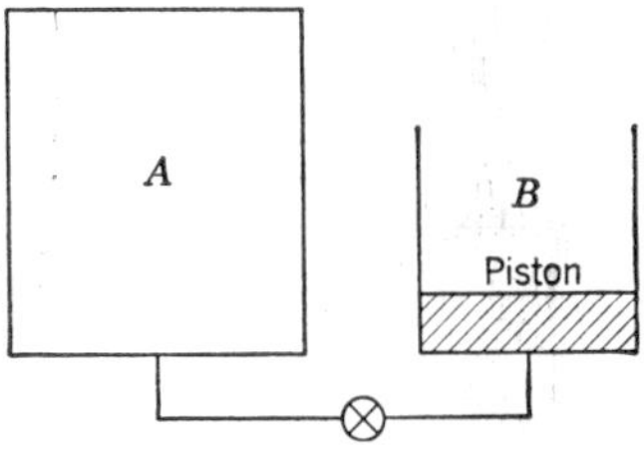

Fig. 7.20 Sketch for Problem 7.27.

frictionless piston, as shown in Fig. 7.20. A pressure of 20 lbf/in.^{2} is required to balance the weight of the piston.

The connecting valve is opened until the pressure in A equals 20 lbf/in.^{2} Assume the entire process to be adiabatic, and that the steam that finally remains in A has undergone a reversible adiabatic process.

Determine the work done against the piston and the final temperature of the steam in cylinder B.

7.28 Repeat Problem 7.27, but let the piston initially rest on stops so that the initial volume of the cylinder B is 10 ft^3, as shown in Fig. 7.21. Assume that this volume is initially evacuated.

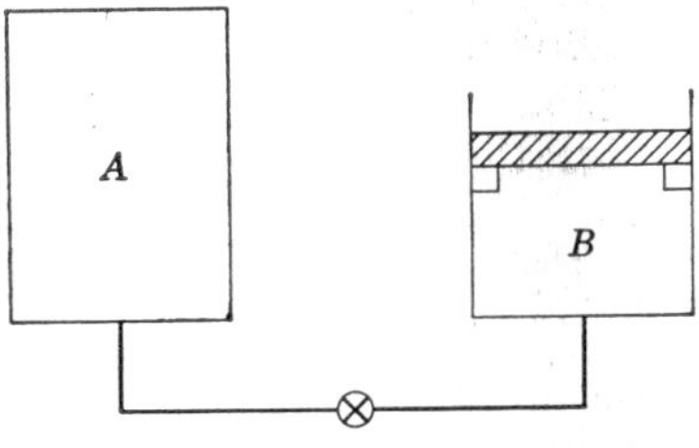

Fig. 7.21 Sketch for Problem 7.28.

7.29 A reversible steam turbine drives a reversible water pump (see Fig. 5.16). Both the pump and turbine are adiabatic. Saturated water at 80 F enters the pump, and the water leaves at 1000 lbf/in.^{2} Saturated steam at 100 lbf/in.^{2} enters the turbine. The exhaust pressure on the turbine is 20 lbf/in.^{2} The rate of water flow is 200,000 lbm/hr. What is the required rate of steam flow?

7.30 The equipment shown in Fig. 7.22 is used to fill bottles of Freon-12 for shipping. The shipping bottles are initially evacuated. Each bottle contains 100 lbm of Freon-12 when it is filled, and the volume of liquid Freon-12 is 85% of the total, and the volume of vapor is 15% of the total. Heat is transferred from the bottle during the

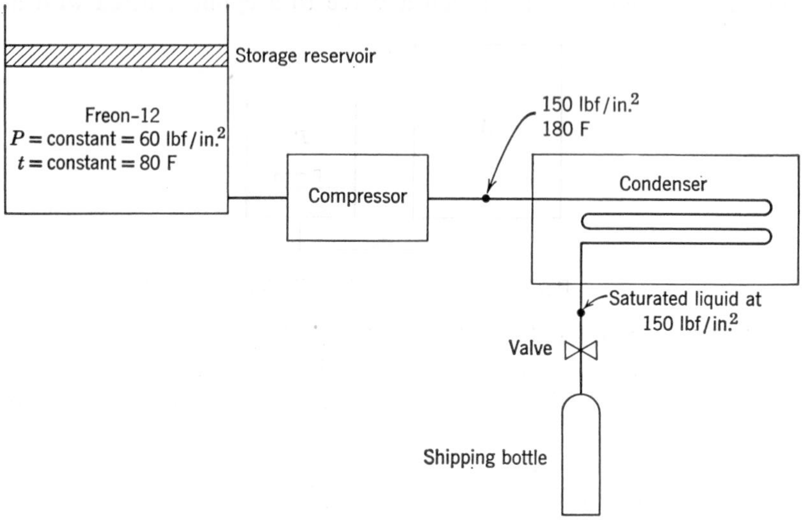

Fig. 7.22 Sketch for Problem 7.30.

filling process so that the temperature of the Freon-12 when the filling process has been completed is 80 F. The compression process is adiabatic.

(*a*) Determine the volume of a shipping bottle.

(*b*) Determine the work of compression for filling one bottle.

(*c*) Determine the heat transfer in the condenser for filling one bottle.

(*d*) Determine the heat transfer from the bottle for filling one bottle.

(*e*) If all the heat transfer from the condenser and bottle is to the surroundings at a temperature of 80 F, determine the net increase in entropy of the system and the surroundings for the filling of one bottle.

(*f*) Prove that the compression process indicated does not violate the second law of thermodynamics.

CHAPTER 8

IDEAL GASES

In our study of thermodynamics to this point we have used tables of thermodynamic properties, such as the steam tables and ammonia tables, as our source of information on the properties of substances. In this chapter we turn our attention to ideal gases, also called perfect gases. One of the unique things about ideal gases is that the relations between the various thermodynamic properties are given by simple mathematical equations, and therefore less thermodynamic data are necessary to solve problems involving ideal gases.

In this chapter we will consider the definition of an ideal gas and its relation to real gases, the relations between the thermodynamic properties of an ideal gas, and some problems involving ideal gases and the first and second laws of thermodynamics.

8.1 Definition of an Ideal Gas

An *ideal gas* may be defined as a substance that has the equation of state

$$Pv = RT \tag{8.1}$$

R is a constant having a particular value for each substance, and is called the gas constant. The gas constant has units which depend on the units used for pressure, specific volume, and temperature. In general the units for R used in this text will be ft-lbf/lbm R. These units result from expressing the pressure in lbf/ft^2, specific volume in ft^3/lbm, and temperature in °R. The value of R for a number of substances is given in Table A.6 of the Appendix.

If both sides of Eq. 8.1 are multiplied by the mass m, the equation of state becomes

$$PV = mRT \tag{8.2}$$

It follows from Eq. 8.2 that for any process involving an ideal gas, the initial and final states are related by the equation

$$\frac{P_1V_1}{T_1} = \frac{P_2V_2}{T_2} \tag{8.3}$$

Thus an ideal gas is a gas which follows Boyle's and Charles' laws.

Frequently it is advantageous to work with ideal gases on a mole basis. The mass of a substance is equal to the product of the number of moles n and the molecular weight M.

$$m = nM$$

Therefore, on substituting this into Eq. 8.2

$$PV = nMRT \tag{8.4}$$

The molal specific volume, $\bar{v}$, was defined in Chapter 2 as the volume per mole V/n). Therefore

$$P\bar{v} = MRT \tag{8.5}$$

Avogadro's law states that equal volumes of different ideal gases at the same pressure and temperature contain the same number of molecules. Since the molecular weight gives the relative mass of individual molecules, it follows that equal volumes of different ideal gases at the same pressure and temperature contain the same number of moles (the mass of one mole is numerically equal to the molecular weight of the substance).

Consider equal volumes of ideal gases A and B at the same pressure and temperature. For these two gases Eq. 8.4 is

$$P_AV_A = n_AM_AR_AT_A \qquad P_BV_B = n_BM_BR_BT_B$$

Since equal volumes of different ideal gases contain the same number of moles it follows that

$$\frac{P_AV_A}{n_AT_A} = \frac{P_BV_B}{n_BT_B} = M_AR_A = M_BR_B$$

Therefore the product MR is a constant for all ideal gases, and is called the universal gas constant, designated by $\bar{R}$. That is,

$$MR = \bar{R} \tag{8.6}$$

The value of the universal gas constant depends on the units used. For the most frequently used units in this text the value of $\bar{R}$ is as follows:

$$\bar{R} = 1545 \text{ ft-lbf/lb-mole R}$$

$$\bar{R} = 1.986 \text{ Btu/lb-mole R}$$

$$\bar{R} = 1.986 \text{ cal/gm-mole K}$$

Example 8.1 What is the mass of air contained in a room 20 ft × 30 ft × 12 ft if the pressure is 14.7 lbf/in.^{2}, and the temperature is 80 F? Assume air to be an ideal gas.

Using Eq. 8.2, and the value of R from Table A.6,

$$m = \frac{PV}{RT} = \frac{14.7 \times 144 \text{ lbf/ft}^2 \times 7200 \text{ ft}^3}{53.34 \text{ ft-lbf/lbm R} \times 540 \text{ R}} = 529 \text{ lbm}$$

Example 8.2 A tank has a volume of 15 ft^3 and contains 20 lbm of an ideal gas having a molecular weight of 24. The temperature is 80 F. What is the pressure?

The gas constant is first determined.

$$R = \frac{\bar{R}}{M} = \frac{1545 \text{ ft-lbf/lb-mole R}}{24 \text{ lbm/lb-mole}} = 64.4 \text{ ft-lbf/lbm R}$$

We now solve for P.

$$P = \frac{mRT}{V} = \frac{20 \text{ lbm} \times 64.4 \text{ ft-lbf/lbm R} \times 540 \text{ R}}{144 \text{ in.}^2/\text{ft}^2 \times 15 \text{ ft}^3} = 321 \text{ lbf/in.}^2$$

8.2 The Relation of Ideal Gases to Real Gases

The question considered in this section is the conditions under which an actual gas may be considered an ideal gas. The equation of state for an ideal gas, $Pv = RT$, holds for all gases as the pressure approaches zero. As the pressure increases above the limiting value of zero, real gases deviate from the ideal-gas equation of state, the deviation being the greatest near the critical point. In general, for a given pressure, the deviation increases as the temperature decreases, being maximum for the saturated-vapor state. This is illustrated in Fig. 8.1, which is a temperature-entropy diagram for steam. For all points to the right of the line marked 1 per cent deviation, the deviation from the ideal-gas equation of state is less than 1 per cent and for points to the left of this line the deviation is greater than 1 per cent. Thus, at 1 lbf/in.^{2} the ideal-gas equation of state holds very well even for saturated vapor. As the pressure increases a greater amount of superheat is necessary for a deviation of less than 1 per cent.

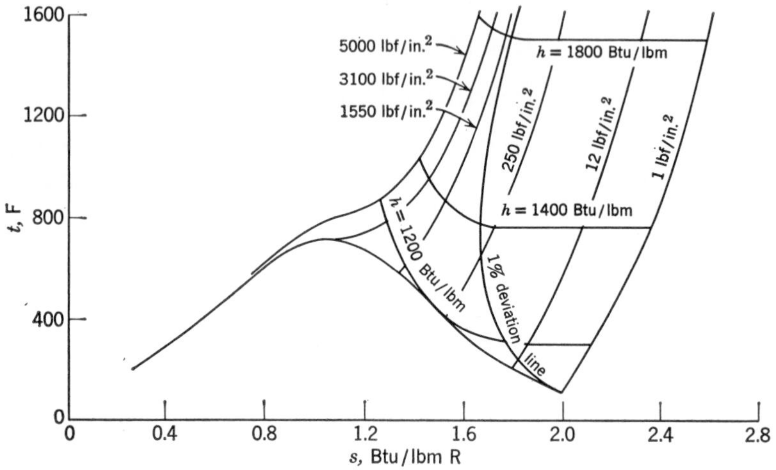

Fig. 8.1. Temperature-entropy diagram for steam showing the line where the deviation from the ideal-gas equation of state is 1 per cent.

Although the diagram of Fig. 8.1 is for steam, it is a typical diagram for all substances. The critical temperature of air is 239 R. Thus, air at room temperature is considerably above the critical temperature and the ideal-gas equation of state is quite accurate to pressures of several hundred pounds per square inch. However, when dealing with the liquefaction of air, considerable error might result if the vapor were treated as an ideal gas.

In general the term vapor is applied to a substance in the gaseous phase if it is saturated or slightly superheated, whereas the term gas is applied to the same substance when highly superheated. There is no sharp line of demarcation between a vapor and a gas.

8.3 The Internal Energy and Enthalpy of an Ideal Gas

It can be shown that the specific internal energy u of an ideal gas is a function of temperature only. This is a consequence of the fact that the equation of state is $Pv = RT$, and will be demonstrated conclusively in Chapter 14. Thus, we can write

$$u = f(T)$$

This means that an ideal gas at a given temperature has a certain definite specific internal energy u, regardless of the pressure.

In 1843 Joule demonstrated this fact when he conducted the following experiment, which is one of the classical experiments in thermodynamics (Fig. 8.2). Two pressure vessels, connected by a pipe and valve,

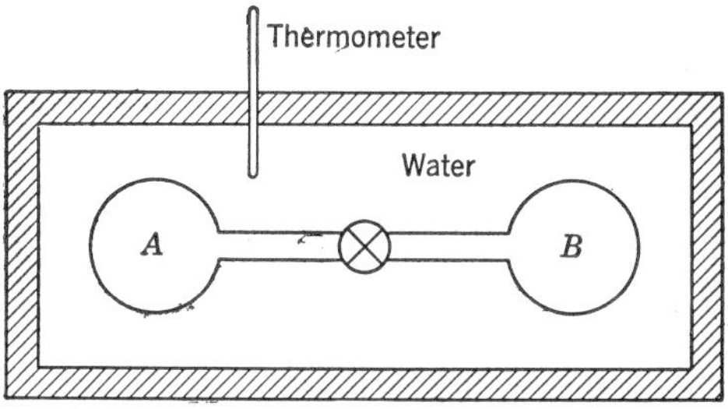

Fig. 8.2 Apparatus for conducting Joule's experiment.

were immersed in a bath of water. Initially vessel A contained air at 22 atm pressure and vessel B was highly evacuated. When thermal equilibrium was attained the valve was opened, allowing the pressures in A and B to equalize. No change in the temperature of the bath was detected during or after this process. Because there was no change in the temperature of the bath, Joule concluded that there had been no heat transferred to the air. Since the work was also zero, he concluded from the first law of thermodynamics that there was no change in the internal energy of the gas. Since the pressure and volume changed during this process, one concludes that internal energy is not a function of pressure and volume. When very careful experiments are made with real gases a very small change in temperature may be detected. This is because real gases do not conform exactly to the definition of an ideal gas.

The relation between the specific internal energy u and the temperature can be established by using the definition of constant-volume specific heat given by Eq. 5.30

$$C_v = \left(\frac{\partial u}{\partial t}\right)_v$$

Since the internal energy of an ideal gas is not a function of volume, for an ideal gas we can write

$$C_{v\infty} = \frac{du}{dt}$$

$$du = C_{v\infty}\, dt \tag{8.7}$$

For a given mass m

$$dU = mC_{v\infty}\, dt \tag{8.8}$$

From the definition of enthalpy and the equation of state of an ideal gas, it follows that

$$h = u + Pv = u + RT \tag{8.9}$$

Since R is a constant and u is a function of temperature only, it follows that the specific enthalpy of an ideal gas is also a function of temperature only. That is,

$$h = f(t)$$

The relation between enthalpy and temperature is found from the constant-pressure specific heat as defined in Eq. 5.31

$$C_p = \left(\frac{\partial h}{\partial t}\right)_P$$

Since the specific enthalpy of an ideal gas is a function of the temperature only, and is independent of the pressure, it follows that

$$C_{po} = \frac{dh}{dt}$$

$$dh = C_{po}\, dt \tag{8.10}$$

For a given mass m

$$dH = mC_{po}\, dt \tag{8.11}$$

The consequences of Eqs. 8.7 and 8.10 are demonstrated by Fig. 8.3, which shows two lines of constant temperature. Since internal energy

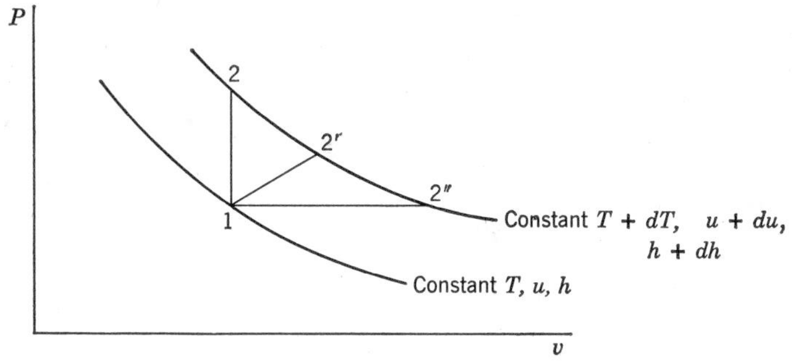

Fig. 8.3 Pressure-volume diagram for an ideal gas.

and enthalpy are functions of temperature only, these lines of constant temperature are also lines of constant internal energy and constant enthalpy. From state 1 the high temperature can be reached by a variety of paths, and in each case the final state is different. However, regardless of the path, the change in internal energy and enthalpy are the same, for lines of constant temperature are also lines of constant u and constant h.

This is pertinent in regard to the temperature-entropy diagram of an ideal gas, shown in Fig. 8.4. (In understanding such a diagram it is

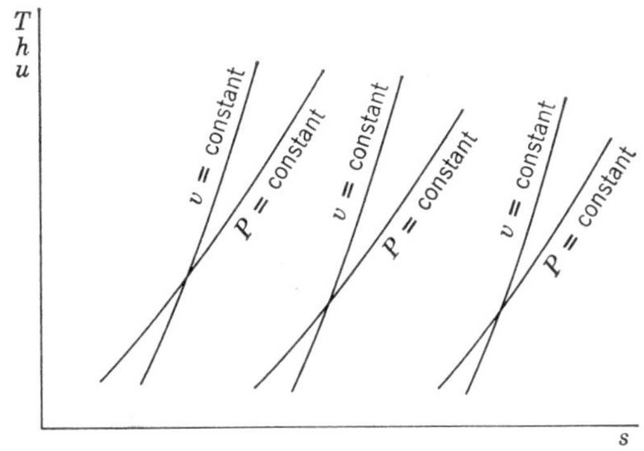

Fig. 8.4 Temperature-entropy, internal-energy-entropy, enthalpy-entropy diagram for an ideal gas.

helpful to recall that this diagram represents the superheated-vapor region of a more complete diagram such as is shown in Fig. 8.1.) Since lines of constant temperature are also lines of constant enthalpy and constant internal energy, Fig. 8.4 can also be considered as an h-s or a u-s diagram, which are often very useful.

This matter is also demonstrated by the constant-enthalpy lines of Fig. 8.1. It will be noted that to the right of the 1 per cent deviation line the constant-enthalpy lines are essentially horizontal, which is true for an ideal gas. The constant-enthalpy lines are very steep near the critical point, indicating a large deviation from an ideal gas.

8.4 Specific Heat of Ideal Gases

Since the internal energy and enthalpy of an ideal gas are a function of temperature only, it follows that the constant-volume and constant-pressure specific heats are also a function of temperature only. That is,

$$C_{v\infty} = f(T) \qquad C_{po} = f(T)$$

Since all gases approach the ideal gas as the pressure approaches zero, the ideal-gas specific heat for a given substance is often called the zero-pressure specific heat, and the zero-pressure constant-pressure specific

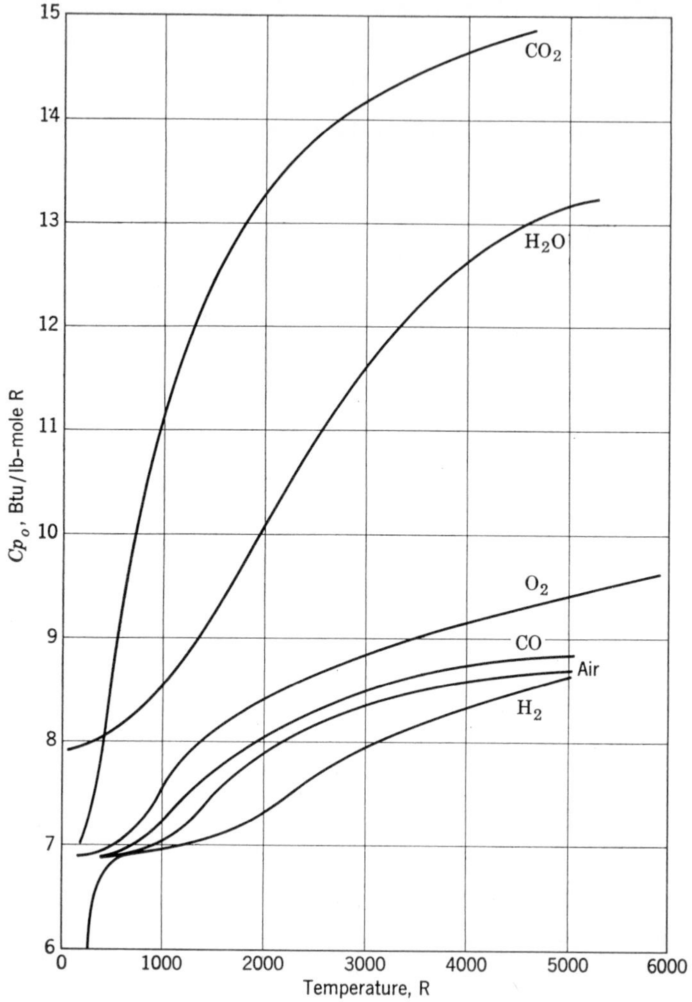

Fig. 8.5 **Constant-pressure specific heats for a number of gases at zero pressure.**

heat is given the symbol C_{po}. The zero-pressure constant-volume specific heat is given the symbol $C_{v\infty}$, since the specific volume becomes very large as the pressure approaches zero. C_{po} as a function of temperature for a number of different substances is shown in Fig. 8.5. Empirical equations expressing C_{po} as a function of temperature have been developed for many substances, and Table A.7 in the Appendix gives a number of these.

A very important relation between the constant-pressure and constant-volume specific heats of an ideal gas may be developed from the definition of enthalpy.

$$h = u + Pv = u + RT$$

Differentiating, and substituting Eqs. 8.7 and 8.10,

$$dh = du + R\,dT$$

$$C_{po}\,dT = C_{vo}\,dT + R\,dT$$

Therefore,

$$C_{po} - C_{vo} = R \tag{8.12}$$

On a mole basis this equation would be written

$$\bar{C}_{po} - \bar{C}_{vo} = \bar{R} \tag{8.13}$$

This tells us that the difference between the constant-pressure and constant-volume specific heats of an ideal gas is always constant, though both are a function of temperature. Thus, using the value of $\bar{C}_{po}$ from Table A.7, the constant-volume specific heat of oxygen, $\bar{C}_{vo}$, would be written

$$\bar{C}_{vo} = \bar{C}_{po} - \bar{R}$$

$$\bar{C}_{vo} = 11.515 - \frac{172}{\sqrt{T}} + \frac{1530}{T} - 1.986 \text{ Btu/lb-mole R}$$

$$= 9.529 - \frac{172}{\sqrt{T}} + \frac{1530}{T} \text{ Btu/lb-mole R}$$

The equations from Table A.7 can be substituted into Eqs. 8.8 and 8.11, and the change in internal energy and enthalpy can be found by integration.

Example 8.3 Calculate the change of enthalpy as 1 lbm of oxygen is heated from 500 R to 2000 R.

Using the specific-heat equation from Table A.7, and integrating Eq. 8.11, we have

$$\bar{h}_2 - \bar{h}_1 = \int_{T_1}^{T_2} \bar{C}_{po}\,dT = \int_{T_1}^{T_2} 11.515\,dT - \frac{172\,dT}{T^{1/2}} + 1530\frac{dT}{T}$$

$$\bar{h}_2 - \bar{h}_1 = 11.515(T_2 - T_1) - 172 \times 2(T_2^{1/2} - T_1^{1/2}) + 1530 \ln \frac{T_2}{T_1}$$

$$\bar{h}_{2000} - \bar{h}_{500} = 17{,}280 - 7690 + 2120 = 11{,}710 \text{ Btu/lb-mole}$$

$$h_{2000} - h_{500} = \frac{\bar{h}_{2000} - \bar{h}_{500}}{M} = \frac{11{,}710}{32} = 366 \text{ Btu/lbm}$$

The average specific heat for any process is defined by the relation

$$C_{p(\text{av})} = \frac{\int_{T_1}^{T_2} C_p \, dT}{T_2 - T_1} \tag{8.14}$$

Thus, the average specific heat for Example 8.3 is

$$C_{p(\text{av})} = \frac{366 \text{ Btu/lbm}}{(2000 - 500) \text{ R}} = 0.244 \text{ Btu/lbm R}$$

The integration of the specific-heat equation, as shown above, is a very time consuming procedure, and for this reason the *Gas Tables*, by Keenan and Kaye were developed. In these tables the enthalpy and internal energy of a number of different gases, including air, are given. A summary of the tables for air is given in Table A.8 of the Appendix. In effect, the values of internal energy and enthalpy given in these tables were obtained by integrating the zero-pressure specific-heat equations from absolute zero to the given temperature. The internal energy and enthalpy of the gases are assumed to be zero at absolute zero.

8.5 Entropy Change of Ideal Gases

Two very useful equations for computing entropy change of an ideal gas can be developed from Eqs. 7.19 and 7.20 by substituting Eqs. 8.7 and 8.10, as follows:

$$T\,ds = du + P\,dv$$

For an ideal gas

$$du = C_{v\infty}\,dT \qquad \text{and} \qquad \frac{P}{T} = \frac{R}{v}$$

Therefore,

$$ds = C_{v\infty}\frac{dT}{T} + \frac{R\,dv}{v} \tag{8.15}$$

$$s_2 - s_1 = \int_1^2 C_{v\infty}\frac{dT}{T} + R \ln \frac{v_2}{v_1} \tag{8.16}$$

Similarly

$$T\,ds = dh - v\,dP$$

For an ideal gas

$$dh = C_{po}\,dT \qquad \text{and} \qquad \frac{v}{T} = \frac{R}{P}$$

Therefore,

$$ds = C_{po}\frac{dT}{T} - R\frac{dP}{P} \tag{8.17}$$

$$s_2 - s_1 = \int_1^2 C_{po}\frac{dT}{T} - R\ln\frac{P_2}{P_1} \tag{8.18}$$

In order to integrate Eqs. 8.16 and 8.18, the relation between specific heat and temperature must be known. The empirical equations of Table A.7 can be used, as is illustrated in the example that follows.

Example 8.4 Consider the same example as cited previously, in which oxygen is heated from 500 R to 2000 R. Assume that during this process the pressure dropped from 30 lbf/in.^{2} to 20 lbf/in.^{2}

Calculate the change in entropy per lbm.

Using Eq. 8.18 and the appropriate equation from Table A.7,

$$\bar{s}_2 - \bar{s}_1 = \int_{T_1}^{T_2} 11.515\frac{dT}{T} - \frac{172\,dT}{T^{3/2}} + \frac{1530\,dT}{T^2} - \bar{R}\int_{P_1}^{P_2}\frac{dP}{P}$$

$$= \Big[115.15\ln T + 2\times 172\,T^{-1/2} - 1530\,T^{-1}\Big]_{T_1}^{T_2} - \bar{R}\ln\frac{P_2}{P_1}$$

$$= 11.515\ln\frac{2000}{500} + 2\times 172\left[\frac{1}{\sqrt{2000}} - \frac{1}{\sqrt{500}}\right]$$

$$-1530\left[\frac{1}{2000} - \frac{1}{500}\right] - 1.986\ln\frac{20}{30}$$

$$= 15.98 - 7.70 + 2.29 + 0.80 = 11.37 \text{ Btu/lb-mole R}$$

$$s_2 - s_1 = \frac{\bar{s}_2 - \bar{s}_1}{M} = \frac{11.37}{32} = 0.356 \text{ Btu/lbm R}$$

Changes in entropy can also be found from the *Gas Tables.* The method used is developed from Eq. 8.17. The entropy in an arbitrary reference state, where the temperature is T_o and the pressure is 1 atm is assumed to be zero. Therefore, at a given temperature T and pressure P the entropy s is found from Eq. 8.17 to be

$$s = \int_{T_o}^{T} C_{po}\frac{dT}{T} - R\ln P$$

where P = pressure in atmospheres.

In the *Gas Tables* the quantity ϕ is defined as

$$\phi = \int_{T_o}^{T} C_{po} \frac{dT}{T}$$

Then,

$$s = \phi - R \ln P$$

The change of entropy between states 1 and 2 is

$$s_2 - s_1 = \phi_2 - \phi_1 - R \ln \frac{P_2}{P_1} \tag{8.19}$$

This equation, as well as Eqs. 8.16 and 8.18, holds for all processes, both reversible and irreversible.

In Chapter 15 the matter of absolute entropy will be introduced, and it will be shown that the value of ϕ in the *Gas Tables* is actually the absolute entropy at 1 atm pressure and the given temperature.

Example 8.5 Calculate the change in entropy per pound of air as air is heated from 540 R to 1200 R while the pressure drops from 50 lbf/in.^{2} to 40 lbf/in.^{2}

From the *Gas Tables*

$$\phi_1 = 0.6008 \text{ Btu/lbm R} \qquad \phi_2 = 0.7963 \text{ Btu/lbm R}$$

Using Eq. 8.19

$$s_2 - s_1 = 0.7963 - 0.6008 - \frac{53.34}{778} \ln \frac{40}{50}$$

$$s_2 - s_1 = 0.2108 \text{ Btu/lbm R}$$

8.6 Reversible Adiabatic Processes Using the Gas Tables

The *Gas Tables* can be used for reversible adiabatic processes by employing the relative pressure P_r and relative specific volume v_r. The definition of these terms and the derivation follow.

For the reversible adiabatic process

$$T\,ds = dh - v\,dP = 0$$

Therefore,

$$dh = C_{po}\,dT = v\,dP = RT \frac{dP}{P}$$

$$\frac{dP}{P} = \frac{C_{po}}{R} \frac{dT}{T}$$

Let this equation be integrated between a reference state having a temperature T_o and a pressure P_o, and a given arbitrary state having a temperature T and a pressure P. Then

$$\ln \frac{P}{P_o} = \frac{1}{R} \int_{T_o}^{T} C_{po} \frac{dT}{T}$$

The right side of this equation is a function of temperature only. The relative pressure P_r is defined as

$$\ln P_r \equiv \ln \frac{P}{P_o} = \frac{1}{R} \int_{T_o}^{T} C_{po} \frac{dT}{T} = \frac{\phi}{R} \tag{8.20}$$

Thus, a value of P_r can be tabulated as a function of temperature.

If we consider two states, 1 and 2, along a constant-entropy line it follows from Eq. 8.20 that

$$\frac{P_1}{P_2} = \left(\frac{P_{r1}}{P_{r2}}\right)_{s=\text{constant}} \tag{8.21}$$

This equation states that the ratio of the relative pressures for two states having the same entropy is equal to the ratio of the absolute pressures.

The development of the relative specific volume is similar, and the ratio of the relative specific volumes v_r in an isentropic process is equal to the ratio of the specific volumes. That is,

$$\frac{v_1}{v_2} = \left(\frac{v_{r1}}{v_{r2}}\right)_{s=\text{constant}} \tag{8.22}$$

Example 8.6 Air expands from an initial pressure of 50 lbf/in.^{2}, 600 F to 20 lbf/in.^{2} in a reversible adiabatic steady-flow process. The changes in kinetic energy are negligible. Calculate the work per lbm.

The work can be found from the steady-flow energy equation, which for this process reduces to

$$w = h_1 - h_2$$

From the *Gas Tables*

$$T_1 = 1060 \text{ R} \qquad h_1 = 255.96 \text{ Btu/lbm} \qquad P_{r1} = 15.203$$

From Eq. 8.21

$$P_{r2} = P_{r1} \times \frac{P_2}{P_1} = 15.203 \times \frac{20}{50} = 6.081$$

From the *Gas Tables*

$$T_2 = 822 \text{ R} \qquad h_2 = 197.18$$

$$w = h_1 - h_2 = 255.96 - 197.18 = 58.8 \text{ Btu/lbm}$$

8.7 Simplified Relations Assuming Constant Specific Heat

Examination of Fig. 8.5 indicates that at the ambient temperature and below, the zero-pressure specific heat of many substances is essentially constant, and at temperatures above the ambient temperature the variation with temperatures is fairly small. Therefore, in many cases only a small error will result if the specific heat is assumed to be constant. This is particularly true if the mean specific heat over the temperature range involved is used.

If the specific heat is assumed to be constant, the relations between the various thermodynamic properties are greatly simplified, and equations for various processes can be developed. In this section we will consider a few of the relations that result from the assumption of constant specific heat.

First of all, when the specific heats are constant, Eqs. 8.8 and 8.11 are readily integrated.

$$U_2 - U_1 = \int_1^2 dU = mC_{v\infty} \int_1^2 dT = mC_{v\infty}(T_2 - T_1) \tag{8.23}$$

$$H_2 - H_1 = \int_1^2 dH = mC_{po} \int_1^2 dT = mC_{po}(T_2 - T_1) \tag{8.24}$$

Similarly, from Eqs. 8.16 and 8.18 the change in entropy becomes

$$S_2 - S_1 = mC_{v\infty} \ln \frac{T_2}{T_1} + mR \ln \frac{V_2}{V_1} \tag{8.25}$$

$$S_2 - S_1 = mC_{po} \ln \frac{T_2}{T_1} - mR \ln \frac{P_2}{P_1} \tag{8.26}$$

It is advantageous to introduce the specific heat ratio k, which is defined as the ratio of constant-pressure to the constant-volume specific heat at zero pressure.

$$k = \frac{C_{po}}{C_{v\infty}} \tag{8.27}$$

Since the difference between C_{po} and $C_{v\infty}$ is a constant (Eq. 8.13) and since C_{po} and $C_{v\infty}$ are functions of temperature, it follows that k is also a function of temperature. However, in this section we are considering the specific heat constant, and with this assumption k is also constant.

From the definition of k and Eq. 8.12 it follows that

$$C_{v\infty} = \frac{R}{k-1} \qquad C_{po} = \frac{kR}{k-1} \tag{8.28}$$

Some very useful and simple relations for the reversible adiabatic process can be developed when the specific heats are assumed to be constant:

For the reversible adiabatic process, $ds = 0$. Therefore,

$$T\,ds = du + P\,dv = C_{v\infty}\,dT + P\,dv = 0$$

From the equation of state for an ideal gas

$$dT = \frac{1}{R}(P\,dv + v\,dP)$$

Therefore,

$$\frac{C_{v\infty}}{R}(P\,dv + v\,dP) + P\,dv = 0$$

Equation 8.28 may be now introduced.

$$\frac{1}{k-1}(P\,dv + v\,dP) + P\,dv = 0$$

$$v\,dP + kP\,dv = 0$$

$$\frac{dP}{P} + k\frac{dv}{v} = 0$$

Since k is assumed to be constant,

$$Pv^k = \text{constant} \tag{8.29}$$

This equation holds for all reversible adiabatic processes involving an ideal gas with constant specific heat. It is usually advantageous to express this constant in terms of the initial and final states.

$$Pv^k = P_1 v_1{}^k = P_2 v_2{}^k = \text{constant} \tag{8.30}$$

From this equation the following expressions relating the initial and final states of an isentropic process can be derived.

$$\frac{P_2}{P_1} = \left(\frac{v_1}{v_2}\right)^k = \left(\frac{V_1}{V_2}\right)^k \tag{8.31}$$

$$\frac{T_2}{T_1} = \left(\frac{P_2}{P_1}\right)^{(k-1)/k} = \left(\frac{v_1}{v_2}\right)^{k-1} \tag{8.32}$$

With the assumption of constant specific heat some convenient equations can be derived for the work done during an adiabatic process. Consider first a closed system consisting of an ideal gas that undergoes a process in which work is done only at the moving boundary.

$$Q = m(u_2 - u_1) + W = 0$$

$$W = -m(u_2 - u_1) = -mC_{v\infty}(T_2 - T_1)$$

$$= \frac{mR}{1-k}(T_2 - T_1) = \frac{P_2V_2 - P_1V_1}{1-k} \tag{8.33}$$

Consider an ideal gas that undergoes a steady-flow adiabatic process in which changes of kinetic and potential energy are negligible. In this case

$$w = h_1 - h_2 = C_{po}(T_1 - T_2)$$

$$= \frac{kR}{k-1}(T_1 - T_2) = \frac{k}{k-1}(P_1v_1 - P_2v_2) \tag{8.34}$$

It should be noted that Eqs. 8.33 and 8.34 apply to adiabatic processes only. Since no assumption was made regarding reversibility, they apply to both reversible and irreversible processes. Frequently these equations are derived for reversible processes by starting with the relation $w = \int P\,dV$ for the closed system and $w = -\int v\,dP$ for the open system. This derivation is left as an exercise for the student.

8.8 The Reversible Polytropic Process

When a gas undergoes a reversible process in which there is heat transfer, the process frequently takes place in such a manner that a plot of log P vs. log V is a straight line. For such a process

$$PV^n = \text{constant}$$

This is called a *polytropic process.*

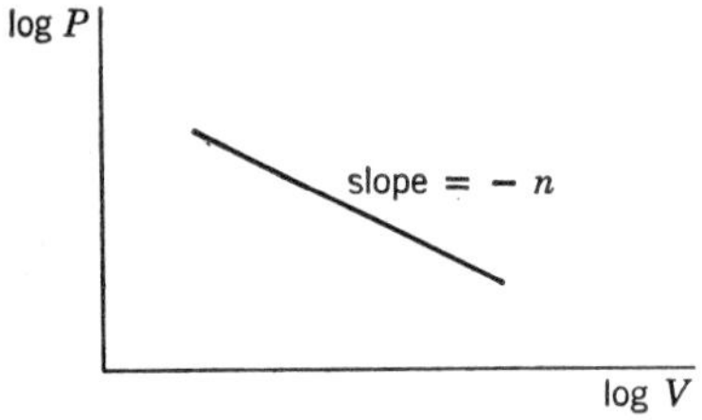

Fig 8.6 Example of a polytropic process.

An example is the expansion of the combustion gases in the cylinder of a water-cooled reciprocating engine. If the pressure and volume during a polytropic process are measured, as might be done with an engine indicator, and the logarithms of the pressure and volume are plotted, the result would be similar to Fig. 8.6. From this figure it follows that

$$\frac{d \ln P}{d \ln V} = -n$$

$$d \ln P + n \ln V = 0$$

If n is a constant (which implies a straight line on the log P vs. log V plot), this can be integrated to give the following relation:

$$PV^n = \text{constant} = P_1V_1{}^n = P_2V_2{}^n \tag{8.35}$$

From this it is evident that the following relations can be written for a polytropic process.

$$\frac{P_2}{P_1} = \left(\frac{V_1}{V_2}\right)^n$$

$$\frac{T_2}{T_1} = \left(\frac{P_2}{P_1}\right)^{(n-1)/n} = \left(\frac{V_1}{V_2}\right)^{n-1} \tag{8.36}$$

For the reversible polytropic process an expression for the work done by a closed system can be derived from the relations $W = \int P\,dV$ and $PV^n = $ constant.

$$W = \int_1^2 P\,dV = \text{constant} \int_1^2 \frac{dV}{V^n}$$

$$= \frac{P_2V_2 - P_1V_1}{1-n} = \frac{mR(T_2 - T_1)}{1-n} \tag{8.37}$$

In a similar manner an expression can be derived for the work done in a steady-flow, reversible process with negligible changes in kinetic and potential energies, from the relations $w = -\int v\,dP$ and $Pv^n =$ constant.

$$w = -\int_1^2 v\,dP = -\text{constant} \int_1^2 \frac{dP}{P^{1/n}}$$

$$= -\frac{n}{n-1}(P_2v_2 - P_1v_1) = -\frac{nR}{n-1}(T_2 - T_1) \tag{8.38}$$

The polytropic processes for various values of n are shown in Fig. 8.7 on a P-v and T-s diagram. The values of n for some familiar processes are also given.

Isobaric process	$n = 0$
Isothermal process	$n = 1$
Isentropic process	$n = k$
Isovolumic process	$n = \infty$

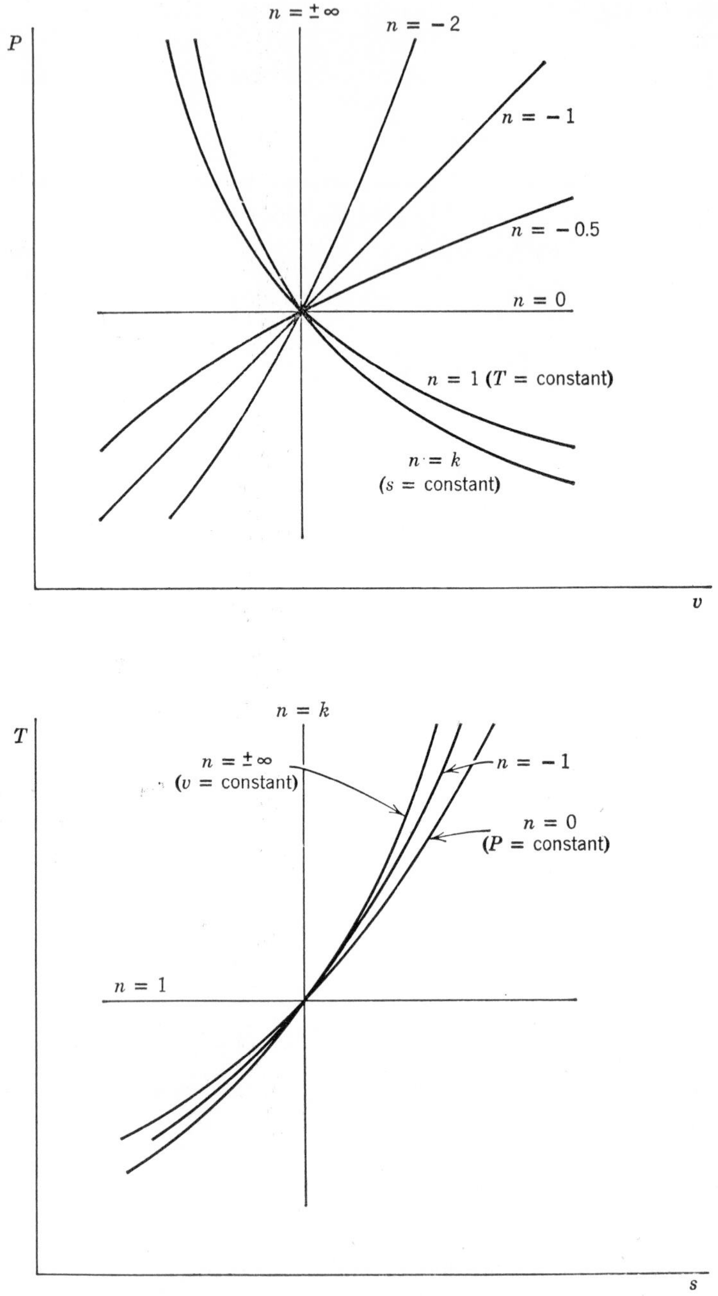

Fig. 8.7 Polytropic processes on *P-v* and *T-s* diagrams.

Example 8.7 Nitrogen is compressed in a reversible process in a cylinder from 14.7 lbf/in.^{2}, 60 F to 60 lbf/in.^{2} During the compression process the relation between pressure and volume is $PV^{1.3}$ = constant. Calculate the work and heat transfer per pound, and show this process on a P-v and T-s diagram.

$$P_1 = 14.7 \text{ lbf/in.}^2 \qquad P_2 = 60 \text{ lbf/in.}^2$$

$$T_1 = 520 \text{ R}$$

From Eq. 8.36 T_2 can be calculated.

$$\frac{T_2}{T_1} = \left(\frac{P_2}{P_1}\right)^{(n-1)/n} = \left(\frac{60}{14.7}\right)^{(1.3-1)/1.3} = 1.383$$

$$T_2 = 520 \times 1.383 = 719 \text{ R}$$

Since this is a closed system the work can be found using Eq. 8.37.

$${}_1w_2 = \frac{R(T_2 - T_1)}{1 - n} = \frac{55.15(719 - 520)}{(1 - 1.3)778} = -47.0 \text{ Btu/lbm}$$

The heat transfer can be calculated using the first law. $C_{v\infty}$ may be assumed to be constant over this range in temperature.

$${}_1q_2 = u_2 - u_1 + {}_1w_2 = C_{v\infty}(t_2 - t_1) + {}_1w_2$$

$$= 0.177(719 - 520) - 47.0 = -11.8 \text{ Btu/lbm}$$

This process is shown on the P-v and T-s diagrams of Fig. 8.8.

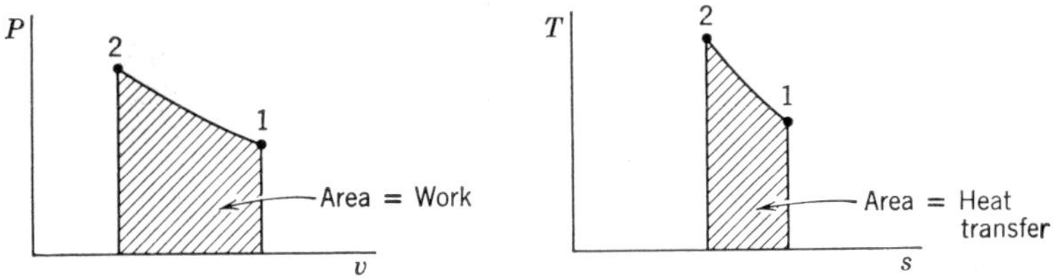

(a) (b)

Fig. 8.8 Diagram for Example 8.7.

8.9 The Reversible Isothermal Process

The reversible isothermal process is of particular interest. In this case

$$PV = \text{constant} = P_1V_1 = P_2V_2 \tag{8.39}$$

The work done in a nonflow isothermal process can be found by integrating the equation

$$_1W_2 = \int_1^2 P\,dV$$

The integration is as follows:

$$_1W_2 = \int_1^2 P\,dV = \text{constant} \int_1^2 \frac{dV}{V} = P_1V_1 \ln \frac{V_2}{V_1} = P_1V_1 \ln \frac{P_1}{P_2}$$

$$_1W_2 = mRT \ln \frac{V_2}{V_1} = mRT \ln \frac{P_1}{P_2} \tag{8.40}$$

For the reversible isothermal steady-flow process with the negligible changes in kinetic and potential energy the work is calculated from Eq. 7.25.

$$w = -\int_1^2 v\,dP = -\text{ constant} \int_1^2 \frac{dP}{P}$$

$$= -P_1v_1 \ln \frac{P_2}{P_1} = P_1v_1 \ln \frac{P_1}{P_2} \tag{8.41}$$

Note that the work is the same for the closed system as for the steady-flow process.

Since there is no change in internal energy or enthalpy in an isothermal process, the heat transfer is equal to the work (neglecting changes in kinetic and potential energy). Therefore, we could have derived Eqs. 8.40 and 8.41 by calculating the heat transfer.

For example, using Eq. 7.19

$$\int_1^2 T\,ds = {}_1q_2 = \int_1^2 du + \int_1^2 P\,dv$$

But $du = 0$ and therefore,

$$_1q_2 = \int_1^2 P\,dv = P_1v_1 \ln \frac{v_2}{v_1}$$

which is identical to Eq. 8.40.

8.10 The Ideal-Gas Temperature Scale

Reference was made in Section 2.10 to the ideal-gas temperature scale. Both constant-pressure and constant-volume gas thermometers have been used. Consider first how an ideal gas might be used to measure temperature in a constant-volume gas thermometer, which is shown schematically in Fig. 8.9. Let the gas bulb be placed in the

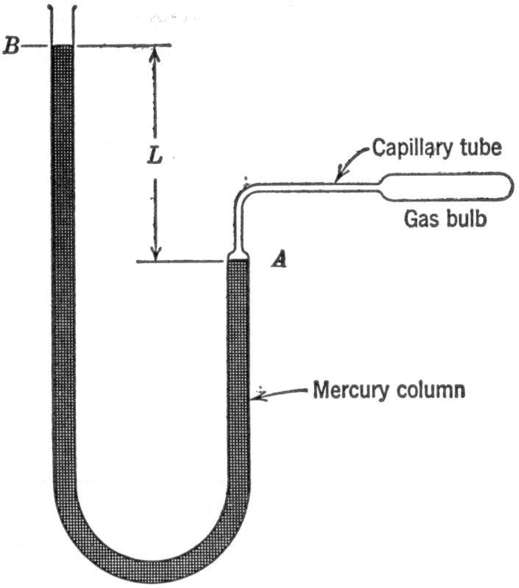

Fig. 8.9 Schematic diagram of a constant-volume gas thermometer.

location where the temperature is to be measured, and let the mercury column be so adjusted that the level of mercury stands at the reference mark A. Thus the volume of the gas remains constant. Assume that the gas in the capillary tube is at the same temperature as the gas in the bulb. Then the pressure of the gas, which is indicated by the height L of the mercury column, is an indication of the temperature.

Let the pressure which is associated with the temperature of the triple point of water (273.16 K) be first measured, and let us designate this pressure $P_{\text{t.p.}}$. Then, from the definition of an ideal gas, any other temperature T could be determined from a pressure measurement P by the relation

$$T = 273.16\left(\frac{P}{P_{\text{t.p.}}}\right)$$

The temperature so measured is referred to as the ideal-gas temperature.

From a practical point of view we have the problem that no gas behaves exactly like an ideal gas. However, we do know that as the pressure approaches zero, the behavior of all gases approaches that of an ideal gas. Suppose then, that a series of measurements is made with varying amounts of gas in the gas bulb. This means that the pressure measured at the triple point, and also the pressure at any other temperature, will vary. If the indicated temperature T_i (obtained by assuming that the gas is ideal) is plotted against the pressure of gas with the bulb at the triple point of water, a curve such as the one shown in Fig. 8.10 is obtained. When this curve is extrapolated to zero pressure,

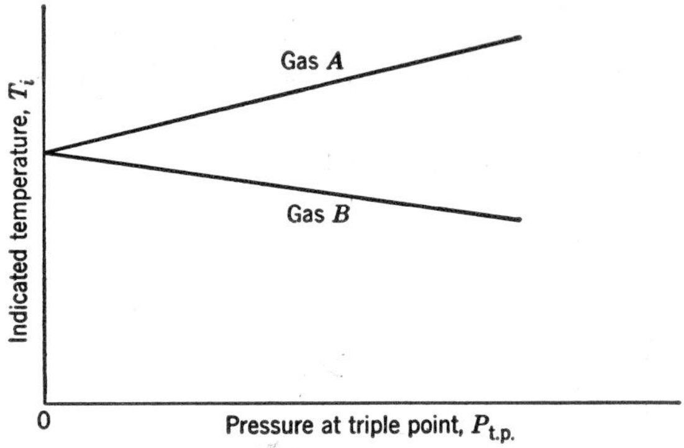

Fig. 8.10 Sketch showing how the ideal-gas temperature is determined.

the correct ideal-gas temperature is obtained. Different curves might result from different gases, but they should all indicate the same temperature at zero pressure.

There are many sources of error which must be taken into account in the actual experimental work, necessitating careful design of the apparatus and accurate calculation of errors.

PROBLEMS

8.1 A spherical balloon has a radius of 20 ft. The atmospheric pressure is 14.7 lbf/in.^{2} and the temperature is 60 F.

(*a*) Calculate the mass and the number of moles of air this balloon displaces.

(*b*) If the balloon is filled with helium at 14.7 lbf/in.^{2}, 60 F, what is the mass and the number of moles of helium?

8.2 Nitrogen is heated in a steady-flow process. The initial pressure and temperature are 100 lbf/in.^{2} and 100 F respectively. The final pressure is 80 lbf/in.^{2} and the final temperature is 2000 F. Determine the heat transfer and the change of entropy during this process using the following data:

(*a*) The specific heat data given by the appropriate equation in Table A.7.

(*b*) Assume the specific heat to be constant at the value given in Table A.6.

8.3 Ten lbm of air are heated from 14.7 lbf/in.^{2}, 100 F to 500 F in a constant-volume process. Calculate the change in internal energy, the change in enthalpy, the heat transfer, and the work.

8.4 Repeat Problem 8.3 for a constant-pressure process.

8.5 Air contained in a cylinder fitted with a piston is initially at 200 lbf/in.^{2}, 1800 F, and expands to 20 lbf/in.^{2} in a reversible adiabatic process. By means of data given in the Air Tables (Table A.8 in the Appendix) determine the following:

(*a*) The final temperature.
(*b*) The final specific volume.
(*c*) The change in internal energy per lbm.
(*d*) The change in enthalpy per lbm.
(*e*) The work done per lbm by the air during this expansion.

8.6 Air undergoes a steady-flow reversible adiabatic process. The initial state is 200 lbf/in.^{2}, 1800 F and the final pressure is 20 lbf/in.^{2} Changes in kinetic and potential energy are negligible. By means of data given in the Air Tables (Table A.8 in the Appendix) determine the following:

(*a*) The final temperature.
(*b*) The final specific volume.
(*c*) The change in internal energy per lbm.
(*d*) The change in enthalpy per lbm.
(*e*) The work per lbm.

8.7 Repeat Problems 8.5 and 8.6 assuming constant specific heats at the value given in Table A.6.

8.8 Derive the relations (Eq. 8.28)

$$\bar{C}_{vo} = \frac{\bar{R}}{k-1} \qquad \bar{C}_{po} = \frac{k\bar{R}}{k-1}$$

8.9 Starting with the relation $w = -\int_1^2 v\,dP$, show that the work done per lbm of fluid flow in a steady-flow, reversible adiabatic process, involving an ideal gas with constant specific heat, and with no changes in kinetic or potential energy, is given by the relation

$$w = \frac{kRT_1}{k-1}\left[1 - \left(\frac{P_2}{P_1}\right)^{(k-1)/k}\right]$$

8.10 Show that the change of entropy of an ideal gas with constant specific heats is given by the relation

$$S_2 - S_1 = mC_{po} \ln \frac{V_2}{V_1} + mC_{vo} \ln \frac{P_2}{P_1}$$

8.11 An ideal gas with constant specific heat enters a nozzle with a velocity V_i and leaves with a velocity V_e after undergoing a reversible adiabatic expansion. Show that the velocity leaving is given by the relation

$$\overline{V}_e = \sqrt{\overline{V}_i^2 + \frac{2g_ckRT}{k-1}\left[1 - \left(\frac{P_2}{P_1}\right)^{(k-1)/k}\right]}$$

8.12 Derive Eqs. 8.37 and 8.38.

8.13 Twenty ft^3 of air at a pressure of 58 lbf/in.^{2} and a temperature of 520 R expand reversibly in a cylinder to a pressure of 14.7 lbf/in.^{2} The final volume is 78.9 ft^3. Assuming constant specific heat for this process calculate:

(*a*) The heat transfer during the expansion.
(*b*) The change of entropy during this process.

8.14 Air enters the compressor of a gas turbine at 14.0 lbf/in.^{2}, 60 F at the rate of 4000 ft^3/min. Air leaves at 60 lbf/in.^{2} Calculate the hp required to drive this compressor if the process is reversible and adiabatic. Changes in kinetic and potential energy are negligible.

8.15 The efficiency of an actual compressor with the same operating conditions as in Problem 8.14 is 78%. (The efficiency of such a compressor is defined as the ratio of the isentropic work to the actual work.) What is the hp required to drive such a compressor and the temperature of the air leaving?

8.16 Air is compressed in a reversible isothermal steady-flow process from 15 lbf/in.^{2}, 100 F to 100 lbf/in.^{2} Calculate the work of compression per lbm, the change of entropy per lbm, and the heat transfer per lbm.

8.17 Air is compressed in a reversible steady-flow polytropic process from 15 lbf/in.^{2}, 100 F to 100 lbf/in.^{2}, and during this process the relation $PV^{1.25}$ = constant applies. Calculate the work, heat transfer, and change of entropy per lbm.

8.18 Air is compressed in a reversible adiabatic steady-flow process from 15 lbf/in.^{2}, 100 F to 100 lbf/in.^{2} Calculate the work and heat transfer per lbm. Show the processes of Problems 8.16, 8.17, and 8.18 on the same P-v and T-s diagram. Why are most reciprocating compressors water jacketed?

8.19 Air is throttled from 100 lbf/in.^{2}, 100 F to 15 lbf/in.^{2} in a steady-flow adiabatic process. Changes in kinetic and potential energy are small. Calculate the change of entropy per lbm of air and show the initial and final states on P-v and T-s diagrams.

8.20 A Carnot engine has 1 lbm of air as the working fluid. Heat is received at 1200 R and rejected at 500 R. At the beginning of the heat addition process the pressure is 100 lbf/in.^{2}, and during this process the volume triples. Calculate the net cycle work per lbm of air.

8.21 Nitrogen, at the rate of 1000 ft^3/min is compressed adiabatically in an axial flow compressor (steady flow) from 14.0 lbf/in.^{2} to 50 lbf/in.^{2} The temperature entering the compressor is 70 F and leaving the compressor it is 340 F. The velocity of the nitrogen entering the compressor is 500 ft/sec and leaving it is 20 ft/sec. Calculate the shaft hp of the compressor.

8.22 Helium which is contained in a cylinder fitted with a piston expands reversibly according to the relation $PV^{1.5}$ = constant. The initial volume of the helium is 2 ft^3, the initial pressure is 70 lbf/in.^{2}, and the initial temperature is 400 R. After expansion the pressure is 30 lbf/in.^{2} Calculate the work done and heat transfer during the expansion.

8.23 Nitrogen expands in a nozzle from a pressure of 100 lbf/in.^{2} to 60 lbf/in.^{2} The initial temperature is 200 F and the velocity entering the nozzle is 300 ft/sec. The rate of flow of nitrogen is 2000 lbm/hr. Calculate the exit area of the nozzle if the flow is reversible and adiabatic.

8.24 Air flows in a pipeline at a pressure of 100 lbf/in.^{2} and a temperature of 80 F. Connected to this tank is an evacuated vessel. When the valve on this tank is opened, air flows into the tank until the pressure is 100 lbf/in.^{2} If this process occurs adiabatically, what is the final temperature of the air?

8.25 A tank of 20 ft^3 volume contains air at 100 lbf/in.^{2}, 100 F. A valve on the tank is opened and the pressure in the tank drops quickly to 20 lbf/in.^{2} If the air that remains in the tank has undergone a reversible adiabatic process, calculate the final mass of air in the tank.

8.26 At one time a certain scientist felt that the equivalent of the following process violated the second law of thermodynamics. (See Fig. 8.11.) Initially compartments A and B, which are separated by a

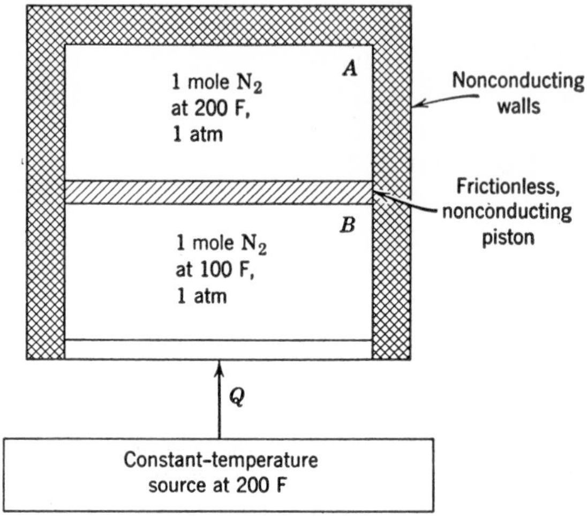

Fig. 8.11 Sketch for Problem 8.26.

frictionless nonconducting piston of negligible mass, each contain 1 mole of nitrogen at 1 atm pressure. The initial temperature in compartment A is 200 F and in compartment B it is 100 F. Heat is transferred from a constant-temperature reservoir at 200 F to the nitrogen in compartment B until it reaches a temperature of 200 F. During the heat transfer process the piston will move as necessary to maintain the same pressure in both compartments.

(*a*) Determine the final temperature in compartment A.

(*b*) Prove that even though the final temperature in compartment A is greater than 200 F that this process does not violate the second law.

8.27 Water at 70 F is pumped from a lake to an elevated storage tank. The average elevation of the water in the tank is 100 ft above the surface of the lake, and the volume of the tank is 10,000 gal. Initially the tank contains air at 14.7 lbf/in.^{2}, 70 F, and the tank is closed so that the air is compressed as the water enters the bottom of the tank. The pump is operated until the tank is three-quarters full. The temperature of the air and water remain constant at 70 F. Determine the work input to the pump.

8.28 An ideal gas with constant specific heat flows in a pipe at a pressure P and absolute temperature T. Attached to this pipe is an evacuated vessel. The valve connecting this evacuated vessel to the pipe is opened, and gas rushes into the evacuated vessel until the pressure in the vessel equals the line pressure.

Show that if this process occurs adiabatically, and if constant specific heat is assumed, the final temperature of the gas in the tank is kT. ($k = C_{po}/C_{vo}$.)

8.29 A large tank having a volume of 20 ft^3 is connected to a small tank having a volume of 4 ft^3. The large tank contains air initially at 100 lbf/in.^{2}, 80 F and the small tank is initially evacuated. A valve in the connecting pipe between the two tanks is suddenly opened and is closed when the tanks come to pressure equilibrium. The air in the large tank may be assumed to have undergone a reversible process and the entire process is adiabatic. What is the final mass of air in the small tank? What is the final temperature of the air in the small tank?

8.30 In 1819 Desormes and Clement measured the value of the specific heat ratio k for air in the following manner. The air in a large pressure vessel was maintained at a pressure P_1, slightly above atmospheric pressure, and both the air and the vessel were at atmospheric temperature. The valve in a line connecting the vessel to the atmosphere was quickly opened until the pressure in the tank equaled the atmospheric pressure P_o, at which moment the valve was closed. After the vessel and the air had again come to thermal equilibrium with the atmosphere (the atmospheric temperature remains constant), the final pressure in the vessel, P_2, was noted. Find an expression for k in terms of P_1, P_2, and P_o. It may be assumed that there is no heat transfer to the air during the time when air is escaping from the vessel.

8.31 A frictionless, thermally conducting piston separates the air and water in the cylinder shown in Fig. 8.12. The initial volume of

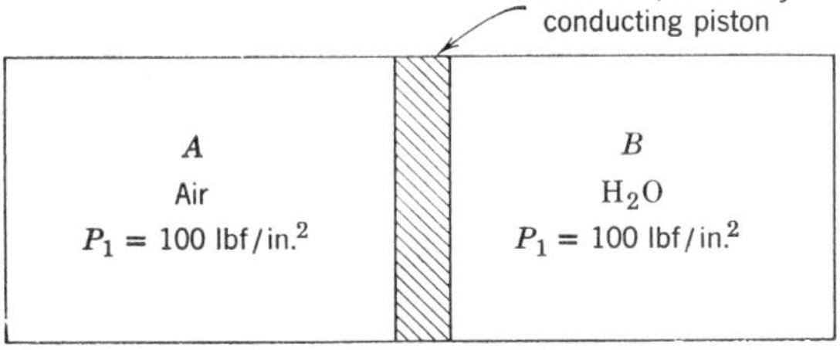

Fig. 8.12 Sketch for Problem 8.31.

A and B are equal, the volume of each being 10 ft^3. The initial pressure in both A and B is 100 lbf/in.^{2} The volume of the liquid in B is 2% of the total volume of B. Heat is transferred to both A and B until all the liquid in B evaporates.

(*a*) Determine the total heat transfer during this process.

(*b*) Determine the work done by the piston on the air and the heat transfer to the air.

CHAPTER 9

MIXTURES INVOLVING IDEAL GASES

Many thermodynamic problems involve a homogeneous mixture of different gases. The most familiar example is air, which consists of a mixture of nitrogen, oxygen, a small amount of argon, and traces of other gases. The principles involved in determining the thermodynamic properties of a mixture of ideal gases from the properties of the individual constituents are considered in the first part of this chapter. The word inert means there are no chemical reactions between the component gases.

Frequently a mixture involves a substance, such as water, which may condense or evaporate during a given process. The heating and cooling of air in air conditioning frequently involves such processes. The last part of this chapter is concerned with definitions and principles involved in using a mixture of inert ideal gases and a low-pressure vapor which may undergo a change of phase.

9.1 Some Definitions for Gaseous Mixtures

Consider a mixture of three gases, a, b, and c, at a pressure P and a temperature T, and having a volume V. The total mass m_T, of the mixture is the sum of the mass of the individual constituents.

$$m_T = m_a + m_b + m_c \tag{9.1}$$

The *mass fraction* of any component is the ratio of the mass of that component to the total mass. The mass fraction is given the symbol mf. Thus

$$mf_a = \frac{m_a}{m_T} \qquad mf_b = \frac{m_b}{m_T} \qquad mf_c = \frac{m_c}{m_T} \tag{9.2}$$

From Eqs. 9.1 and 9.2 one concludes that

$$mf_a + mf_b + mf_c = 1$$

The total number of moles of mixture n_T is equal to the sum of the moles of the individual gases.

$$n_T = n_a + n_b + n_c \tag{9.3}$$

The *mole fraction* of any component is the ratio of the number of moles of that component to the total number of moles. The mole fraction is given the symbol x (since quality and mole fraction are rarely involved in the same problem in this text, no confusion should arise in using the symbol x for both). Thus

$$x_a = \frac{n_a}{n_T} \qquad x_b = \frac{n_b}{n_T} \qquad x_c = \frac{n_c}{n_T} \tag{9.4}$$

It also follows that

$$x_a + x_b + x_c = 1$$

The concept of volume fraction is best defined with the aid of Fig. 9.1. Let the gases be separated into individual containers at the pressure and temperature of the mixture and let the volumes of the separated gases be designated V_a, V_b, V_c. The *volume fraction* of any component is defined as the ratio of the volume of the separated gas to the volume of the mixture. The volume fraction is designated by the symbol vf. Thus

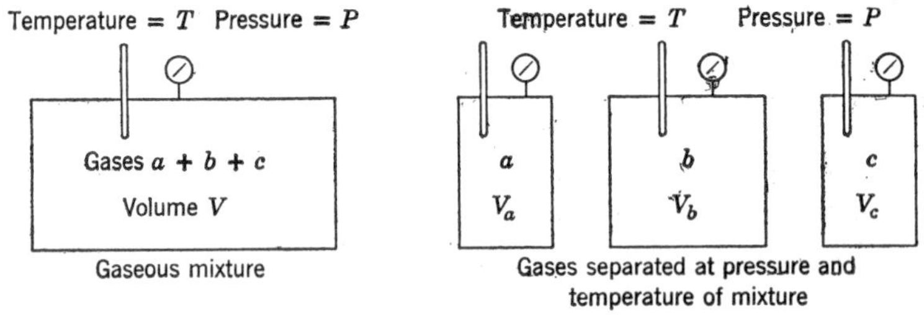

Fig. 9.1 Separation of a gaseous mixture into components at the pressure and temperature of the mixture.

$$vf_a = \frac{V_a}{V} \qquad vf_b = \frac{V_b}{V} \qquad vf_c = \frac{V_c}{V} \tag{9.5}$$

The *partial pressure* of any component in a gaseous mixture is defined as the pressure of the component if it alone occupied the entire volume at the temperature of the mixture. This is illustrated in Fig. 9.2.

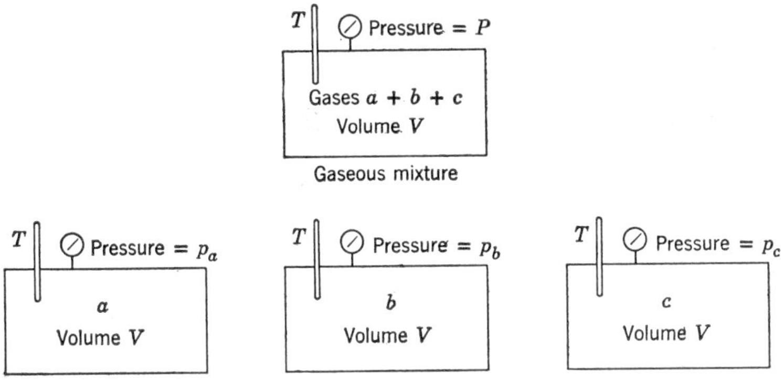

Fig. 9.2. Separation of a gaseous mixture into components at the volume and temperature of the mixture.

Partial pressures are designated by p. Thus, the partial pressure of gas a is designated p_a. The ratio of the partial pressure of any given component to the total pressure is called the partial-pressure ratio, but no particular symbol is assigned to this term.

9.2 The Gibbs-Dalton Law

The properties of a mixture of ideal gases can be found from the properties of the individual gases by applying the Gibbs-Dalton Law. For sake of clarity this law is stated here in three parts. The assumption is made that the mixture can also be considered an ideal gas.

1. The pressure of a mixture of ideal gases is equal to the sum of partial pressures of the individual components. This may also be stated in equation form, and for the example above we can write

$$P = p_a + p_b + p_c \tag{9.6}$$

This statement is also known as Dalton's law of partial pressures. It holds exactly for a mixture of ideal gases, and in this case the mixture is an ideal gas. In practice this law holds with a high degree of accuracy, even when the behavior of individual gases deviates somewhat from that of an ideal gas.

2. The internal energy of a mixture of ideal gases is equal to the sum of the internal energies of the individual components at the pres-

sure and temperature of the mixture. This can be illustrated by reference to Fig. 9.3. Suppose the partitions that separate the three ideal gases are ruptured. Considering all the gases as the system, we recognize that this is a constant internal energy process. Since there is no change in the internal energy of any of the components, there is

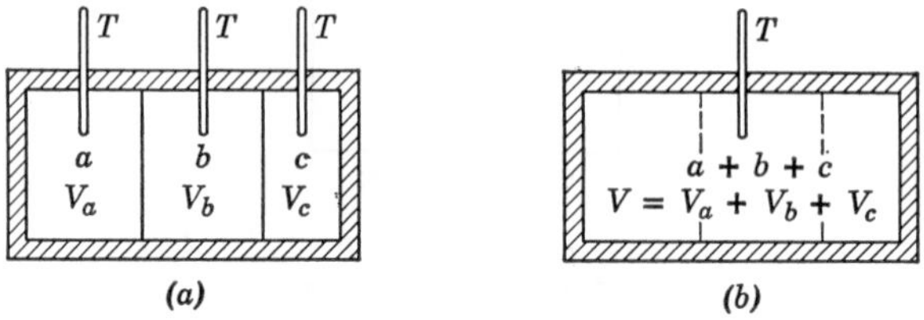

Fig. 9.3 Adiabatic mixing of ideal gases that are initially at the same temperature and pressure.

no change in temperature, and the internal energy of the gaseous mixture is the sum of the internal energies of the component gases. For the mixture we can write

$$U_T = U_a + U_b + U_c \tag{9.7}$$

where U_T is the internal energy of the mixture.

3. The entropy of a mixture of ideal gases is equal to the sum of the entropies of the component gases. In Fig. 9.3*b* the entropy of the mixture is equal to the sum of the entropies of the component gases as they exist in the mixture. Thus, the entropy of each component gas must be found at its partial pressure. In equation form we can write

$$S_T = S_a + S_b + S_c$$

where S_T is the entropy of the mixture. It should be pointed out that there is an increase in entropy of the system when changing from the state of Fig. 9.3*a* to the state of Fig. 9.3*b*. The entropy of the mixture is found by adding the entropies of the component gases as they exist in the mixture.

9.3 Relations Involving Pressure, Volume, and Composition

As a consequence of Dalton's law of partial pressures it can be shown that the volume fraction, mole fraction, and partial-pressure ratio are all equal for a mixture of ideal gases. This may be demonstrated by reference to Figs. 9.1 and 9.2. For the mixture and separated gases of Fig. 9.1 we can write

$$PV = n_T\bar{R}T \qquad PV_a = n_a\bar{R}T \qquad PV_b = n_b\bar{R}T \qquad PV_c = n_c\bar{R}T \tag{9.8}$$

Each of the equations for a component can be divided by the equation for the mixture. For component a

$$\frac{V_a}{V} = vf_a = \frac{n_a}{n_T} = x_a \tag{9.9}$$

Thus, for each component the volume fraction is equal to the mole fraction.

If Eqs. 9.8 are each solved for the number of moles and substituted into Eq. 9.2, we conclude that

$$V = V_a + V_b + V_c \tag{9.10}$$

The concept that the volume of a mixture is equal to the sum of the volumes of the separated components at the pressure and temperature of the mixture is known as Amagat's law. This law (the question arises as to whether this should properly be called a law) holds exactly for ideal gases but mixtures involving real gases may deviate from it somewhat.

Repeating this procedure for Fig. 9.2 we have

$$PV = n_T\bar{R}T \qquad p_aV = n_a\bar{R}T \qquad p_bV = n_b\bar{R}T \qquad p_cV = n_c\bar{R}T \tag{9.11}$$

Again the equation for the component gases can be divided by the equation for the mixture. For component a

$$\frac{p_a}{P} = \frac{n_a}{n_T} = x_a \tag{9.12}$$

Thus, the partial-pressure ratio is equal to the mole fraction. Equating Eqs. 9.9 and 9.12 we have

$$x_a = vf_a = \frac{p_a}{P_T} \tag{9.13}$$

In words this equation states that for each component in a mixture of ideal gases the mole fraction, volume fraction, and partial-pressure ratio are all equal.

This fact is useful in converting an analysis of a gaseous mixture on a volume basis to a mass basis and vice versa, and in determining the molecular weight and gas constant for a gaseous mixture. A number of equations can be derived, as shown in the following paragraphs. However, the calculation can be done in table form with great clarity. The example given below illustrates this approach, and is essentially self-explanatory. In Table 9.1 it is assumed that the analysis is given on a volumetric basis and it is required to find the analysis on a mass basis, and the molecular weight and gas constant for the mixture.

TABLE 9.1

Constituent	Per Cent by Volume	Lb Moles per Mole of Mixture		Molecular Weight		Mass (lbm) per Mole of Mixture	Analysis on Mass Basis, Per Cent
CO_2	12	0.12	×	44.0	=	5.28	$\frac{5.28}{30.08} = 17.55$
O_2	4	0.04	×	32.0	=	1.28	$\frac{1.28}{30.08} = 4.26$
N_2	82	0.82	×	28.0	=	22.96	$\frac{22.96}{30.08} = 76.33$
CO	2	0.02	×	28.0	=	0.56	$\frac{0.56}{30.08} = 1.86$
						30.08	100.00

Molecular weight of mixture = 30.08

$$R \text{ for mixture} = \frac{\bar{R}}{M} = \frac{1545}{30.08} = 51.4 \text{ ft-lbf/lbm R}$$

The key to the solution is that the mole fraction is equal to the volume fraction.

The same example is used in the following table to illustrate the procedure for determining the analysis on a volumetric basis when the analysis on a mass basis is given. Frequently the analysis on a mass basis is called a gravimetric analysis.

TABLE 9.2

Constituent	Mass Fraction		Molecular Weight		Lb Moles per Pound of Mixture	Mole Fraction	Volumetric Analysis, Per Cent
CO_2	0.1755	÷	44.0	=	0.00399	0.120	12.0
O_2	0.0426	÷	32.0	=	0.00133	0.040	4.0
N_2	0.7633	÷	28.0	=	0.02726	0.820	82.0
CO	0.0186	÷	28.0	=	0.00066	0.020	2.0
					0.03324	1.000	100.0

$$M = \frac{1}{\text{moles/lbm mixture}} = \frac{1}{0.03324} = 30.08$$

$$R = \frac{\bar{R}}{M} = \frac{1545}{30.08} = 51.4 \text{ ft-lbf/lbm R}$$

An expression for the gas constant R for a mixture can be derived by writing the equation of state for the mixture and the component gases of Fig. 9.2 in terms of mass.

$$PV = mRT \qquad p_aV = m_aR_aT \qquad p_bV = m_bR_bT \qquad p_cV = m_cR_cT$$

Since

$$P = p_a + p_b + p_c$$

$$\frac{m_TRT}{V} = \frac{m_aR_aT}{V} + \frac{m_bR_bT}{V} + \frac{m_cR_cT}{V}$$

Writing this in terms of the mass fraction mf we have

$$R = mf_aR_a + mf_bR_b + mf_cR_c \tag{9.14}$$

The expression for the molecular weight M of a mixture can be found from Eq. 9.14 by substituting the relations $\overline{R} = MR$.

$$\frac{\overline{R}}{M} = mf_a\frac{\overline{R}}{M_a} + mf_b\frac{\overline{R}}{M_b} + mf_c\frac{\overline{R}}{M_c}$$

$$M = \frac{1}{mf_a/M_a + mf_b/M_b + mf_c/M_c} \tag{9.15}$$

9.4 Internal Energy, Enthalpy, and Specific Heats of Gaseous Mixtures

As stated previously, from the Gibbs-Dalton law we conclude that

$$U_T = U_a + U_b + U_c$$

or

$$m_Tu_T = m_au_a + m_bu_b + m_cu_c$$

Differentiating we have, for a constant mass,

$$m_T\,du_T = m_a\,du_a + m_b\,du_b + m_c\,du_c$$

Since the mixture and component gases are considered ideal gases we can substitute Eq. 8.8 and write

$$m_TC_{v(m)}\,dT = m_aC_{v(a)}\,dT + m_bC_{v(b)}\,dT + m_cC_{v(c)}\,dT \tag{9.16}$$

where $C_{v(m)}$ is the constant-volume specific heat of the mixture. It is understood that all the C_v's are $C_{v\infty}$'s, but the ∞'s are left out to avoid undue complexity in writing the equations. From Eq. 9.16 we can write an expression for the constant-volume specific heat of the mixture.

$$C_{v(m)} = \frac{m_a}{m_T}C_{v(a)} + \frac{m_b}{m_T}C_{v(b)} + \frac{m_c}{m_T}C_{v(c)}$$

$$= mf_aC_{v(a)} + mf_bC_{v(b)} + mf_cC_{v(c)} \tag{9.17}$$

A similar equation can be derived for the molal specific heat at constant volume.

$$U_T = U_a + U_b + U_c$$

$$n_T \bar{u}_m = n_a \bar{u}_a + n_b \bar{u}_b + n_c \bar{u}_c$$

$$n_T \bar{C}_{v(m)}\, dT = n_a \bar{C}_{v(a)}\, dT + n_b \bar{C}_{v(b)}\, dT + n_c \bar{C}_{v(c)}\, dT$$

$$\bar{C}_{v(m)} = \frac{n_a}{n_T} \bar{C}_{v(a)} + \frac{n_b}{n_T} \bar{C}_{v(b)} + \frac{n_c}{n_T} \bar{C}_{v(c)}$$

$$= x_a \bar{C}_{v(a)} + x_b \bar{C}_{v(b)} + x_c \bar{C}_{v(c)} \tag{9.18}$$

From Eqs. 9.7 and 9.11 we conclude that the enthalpy of a mixture of ideal gases is equal to the sum of the enthalpy of the individual components

$$H = H_a + H_b + H_c \tag{9.19}$$

By a derivation parallel to that given above for the constant-volume specific heats, an expression for the constant-pressure specific heat on both a mass basis and molal basis can be found. This derivation is left as an exercise for the student. The equations are as follows, assuming that these are zero-pressure specific heats.

$$C_{p(m)} = mf_a C_{p(a)} + mf_b C_{p(b)} + mf_c C_{p(c)} \tag{9.20}$$

$$\bar{C}_{p(m)} = x_a \bar{C}_{p(a)} + x_b \bar{C}_{p(b)} + x_c \bar{C}_{p(c)}$$

Example 9.1 Calculate the constant-volume and constant-pressure specific heat of air at 80 F from the specific heat of the individual components as given in Table A.6. For this purpose let us consider air as having the following composition on a volumetric basis:

Constituent	Volume Fraction	Mass Fraction
O_2	0.2099	0.232
N_2	0.7803	0.755
A	0.0098	0.0135

Using Eq. 9.17 we can calculate $C_{v(m)}$.

$$C_{v(m)} = mf_{O_2} C_{v(O_2)} + mf_{N_2} C_{v(N_2)} + mf_A C_{v(A)}$$

$$= 0.232(0.157) + 0.755(0.177) + 0.0135(0.0756)$$

$$= 0.0364 + 0.1336 + 0.0010 = 0.171 \text{ Btu/lbm R}$$

$C_{p(m)}$ can be calculated using Eq. 9.20.

$$C_{p(m)} = mf_{O_2}C_{p(O_2)} + mf_{N_2}C_{p(N_2)} + mf_A C_{p(A)}$$
$$= 0.232(0.219) + 0.755(0.248) + 0.0135(0.1253)$$
$$= 0.0508 + 0.1872 + 0.0017 = 0.240 \text{ Btu/lbm R}$$

9.5 Entropy of a Mixture of Ideal Gases

The third part of the Gibbs-Dalton law states that the entropy of a mixture of ideal gases is equal to the sum of the entropies of the component gases as they exist in the mixture. Since entropy values are usually given relative to an arbitrary reference state, it is necessary to define this reference state and the magnitude of the entropy in it.

For fixed composition of a mixture of ideal gases a value is assigned to the entropy of the particular mixture in a specified reference state. In all other states entropy values are given relative to this reference state. Since there is no change in composition the mixture is treated as an ideal gas and the pertinent equations of Chapter 8 apply. This was in fact done in Chapter 8 when dealing with air.

However, it may sometimes be desirable to assign an arbitrary reference state to each of the components of a mixture. The entropy of the mixture is found by summing up the entropies of the components as they exist in the mixture, the entropy of each component having been found relative to its reference state. This procedure can best be demonstrated by an example.

Example 9.2 Calculate the entropy of air at 20 lbf/in.^{2}, 200 F, assuming that the entropy of oxygen and nitrogen are both zero at 14.7 lbf/in.^{2}, 0 F. For this calculation assume that air is 79 per cent nitrogen and 21 per cent oxygen by volume. This is equivalent to an analysis on a mass basis of 76.7 per cent nitrogen and 23.3 per cent oxygen. Assume constant specific heats as given in Table A.6.

The entropy of the nitrogen in the mixture is found from Eq. 8.26,

$$(s - s_o)_{N_2} = C_{p(N_2)} \ln \frac{T}{T_o} - R_{N_2} \ln \frac{p_{N_2}}{p_o}$$

The subscript o refers to the reference state, and p_{N_2} is the partial pressure of the nitrogen in the mixture.

From Eq. 9.12 p_{N_2} can be calculated.

$$\frac{p_{N_2}}{P} = x_{N_2} \qquad p_{N_2} = 0.79(20) = 15.8 \text{ lbf/in.}^2$$

$$(s - s_o)_{N_2} = 0.248 \ln \frac{660}{460} - \frac{55.15}{778} \ln \frac{15.8}{14.7} = 0.0844 \text{ Btu/lbm N}_2 \text{ R}$$

Similarly for the oxygen as it exists in the mixture:

$$\frac{p_{O_2}}{P} = x_{O_2} \qquad p_{O_2} = 0.21(20) = 4.2 \text{ lbf/in.}^2$$

$$(s - s_o)_{O_2} = 0.219 \ln \frac{660}{460} - \frac{48.28}{778} \ln \frac{4.2}{14.7} = 0.1567 \text{ Btu/lbm } O_2 \text{ R}$$

The mass fraction of nitrogen in air, mf_{N_2}, is 0.767, and the mass fraction of oxygen, mf_{O_2}, is 0.233. Further, we have assumed that $s_0 = 0$ for both O_2 and N_2 at 14.7 lbf/in.^{2}, 0 F. Therefore, relative to this basis, the entropy of 1 lbm of air is

$$s_{air}(20 \text{ lbf/in.}^2, 200 \text{ F}) = 0.767(0.0844) + 0.233(0.1567)$$
$$= 0.1012 \text{ Btu/lbm R}$$

Example 9.3 Let n_a moles of gas a at a given pressure and temperature be mixed with n_b moles of gas b at the same pressure and temperature in an adiabatic constant-volume process, shown in Fig. 9.4. Determine the increase in entropy for this process.

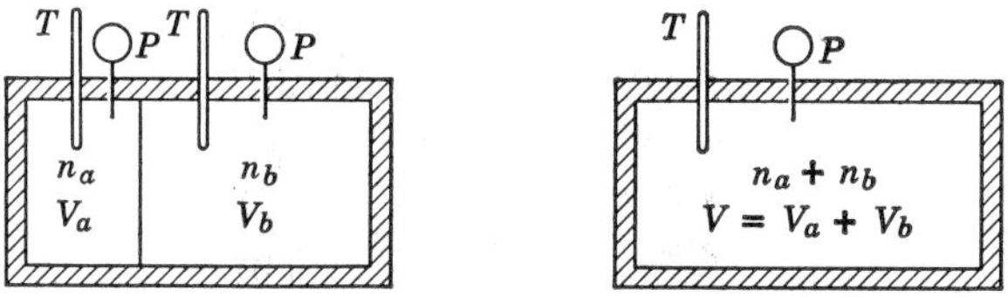

Fig. 9.4 Sketch for Example 9.3.

The final partial pressure of gas a is p_a and for gas b it is p_b. Since there is no change in temperature Eq. 8.26 reduces to

$$(S_2 - S_1)_a = -n_a\bar{R} \ln \frac{p_a}{P} = -n_a\bar{R} \ln x_a$$

$$(S_2 - S_1)_b = -n_b\bar{R} \ln \frac{p_b}{P} = -n_b\bar{R} \ln x_b$$

The total change in entropy is the sum of the entropy changes for gases a and b.

$$S_2 - S_1 = -\bar{R}(n_a \ln x_a + n_b \ln x_b) \tag{9.21}$$

This equation can be written for the general case of mixing any number of components at the same pressure and temperature as follows:

$$S_2 - S_1 = -\bar{R}\Sigma n_k \ln x_k \tag{9.22}$$

The interesting thing to note about this equation is that the increase in entropy depends only on the number of moles of component gases, and is independent of the composition of the gas. For example, when 1 mole of oxygen and 1 mole of nitrogen are mixed the increase in entropy is the same as when 1 mole of hydrogen and 1 mole of nitrogen are mixed. But we also know that if 1 mole of nitrogen is "mixed" with another mole of nitrogen there is no increase in entropy. The question that arises is how dissimilar must the gases be in order to have an increase in entropy? The answer lies in our ability to distinguish between the two gases. The entropy increases whenever we can distinguish between the gases being mixed. When we are not able to distinguish between the gases, there is no increase in entropy.

Example 9.4 A mixture consisting of 2 lbm of helium and 5 lbm of nitrogen at 14.7 lbf/in.^{2}, 80 F is compressed in a reversible adiabatic process to 100 lbf/in.^{2} Assume constant specific heat as given in Table A.6. Calculate:

(*a*) The initial and final partial pressures of each component.
(*b*) The final temperature.
(*c*) The change in internal energy of the mixture during this process.
(*d*) The change in enthalpy of the mixture during this process.
(*e*) The change of entropy of the helium during this process.
(*f*) The change of entropy of the nitrogen during this process.

Solution.

(*a*) To find the partial pressures one must know the mole fractions.

$$n_{\mathrm{He}} = \frac{2.0}{4.00} = 0.500 \qquad n_{\mathrm{N_2}} = \frac{5}{28.0} = 0.1787$$

$$n_T = 0.5000 + 0.1787 = 0.6787$$

$$x_{\mathrm{He}} = \frac{n_{\mathrm{He}}}{n_T} = \frac{0.5}{0.6787} = 0.737 \qquad x_{\mathrm{N_2}} = \frac{n_{\mathrm{N_2}}}{n_T} = \frac{0.1787}{0.6787} = 0.263$$

$$x_{\mathrm{He}} = \frac{p_{\mathrm{He}}}{P_T} = 0.737 \qquad p_{1\mathrm{He}} = 0.737(14.7) = 10.84 \text{ lbf/in.}^2$$

$$p_{2\mathrm{He}} = 0.737(100) = 73.7 \text{ lbf/in.}^2$$

$$x_{\mathrm{N_2}} = \frac{p_{N_2}}{P_T} = 0.263 \qquad p_{1\mathrm{N_2}} = 0.263(14.7) = 3.86 \text{ lbf/in.}^2$$

$$p_{2\mathrm{N_2}} = 0.263(100) = 26.3 \text{ lbf/in.}^2$$

(*b*) The final temperature can be found using Eq. 8.32

$$\frac{T_2}{T_1} = \left(\frac{P_2}{P_1}\right)^{(k-1)/k}$$

We must first know the specific heat ratio k, and this in turn must be found from the constant-volume and constant-pressure specific heats. These can be found from Eqs. 9.17 and 9.20.

$$C_{v(m)} = mf_{\mathrm{He}}C_{v\mathrm{He}} + mf_{\mathrm{N_2}}C_{v(\mathrm{N_2})}$$

$$= \tfrac{2}{7}(0.753) + \tfrac{5}{7}(0.177) = 0.341 \text{ Btu/lbm R}$$

$$C_{p(m)} = mf_{\mathrm{He}}C_{p\mathrm{He}} + mf_{\mathrm{N_2}}C_{p\mathrm{N_2}}$$

$$= \tfrac{2}{7}(1.25) + \tfrac{5}{7}(0.248) = 0.534 \text{ Btu/lbm R}$$

For this mixture, the specific heat ratio k_m is

$$k_m = \frac{C_{p(m)}}{C_{v(m)}} = \frac{0.534}{0.341} = 1.565$$

$$\frac{T_2}{T_1} = \left(\frac{100}{14.7}\right)^{(1.565-1)/1.565} = 2.00$$

$$T_2 = 540(2.00) = 1080$$

(*c*) From Eq. 8.23

$$U_2 - U_1 = mC_{v\infty}(T_2 - T_1) = 7 \times 0.341(1080 - 540) = 1288 \text{ Btu}$$

(*d*) From Eq. 8.24

$$H_2 - H_1 = mC_{po}(T_2 - T_1) = 7 \times 0.534(1080 - 540) = 2018 \text{ Btu}$$

(*e*) The change of entropy of the helium can be found by use of Eq. 8.26. The initial and final partial pressures of the helium must be used.

$$(S_2 - S_1)_{\mathrm{He}} = m\left(C_{po} \ln \frac{T_2}{T_1} - R \ln \frac{p_2}{p_1}\right)_{\mathrm{He}}$$

$$= 2\left(1.25 \ln \frac{1080}{540} - \frac{386.0}{778} \ln \frac{73.7}{10.84}\right) = -0.175 \text{ Btu/R}$$

(*f*) Similarly for the nitrogen

$$(S_2 - S_1)_{\mathrm{N_2}} = m\left(C_{po} \ln \frac{T_2}{T_1} - R \ln \frac{p_2}{p_1}\right)_{\mathrm{N_2}}$$

$$= 5\left(0.248 \ln \frac{1080}{540} - \frac{55.15}{778} \ln \frac{26.3}{3.86}\right) = 0.175 \text{ Btu/R}$$

One observes that the entropy of the helium decreases by the same amount that the entropy of the nitrogen increases. Thus, there is no net change in entropy. This phenomenon is best explained by the use of the T-s diagram of Fig. 8.7. For the expansion of this problem $k = 1.567$, whereas k for helium is 1.66 and k for nitrogen is 1.40. Therefore, when the helium undergoes the expansion governed by the relation $PV^{1.567}$ = constant, the entropy decreases, and when nitrogen undergoes this same expansion the entropy increases. This is shown in Fig. 9.5.

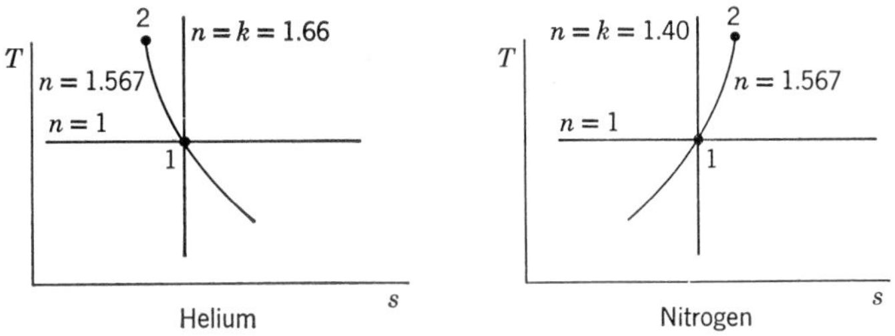

Fig. 9.5 Temperature-entropy diagram for Example 9.4.

9.6 Some Definitions Regarding a Mixture of a Gas and a Vapor

Frequently one of the components of a gaseous mixture is a vapor. Since the pressure of the vapor in the mixture is its partial pressure, the vapor is often at such a low pressure that it can be treated as an ideal gas, even though it is a slightly superheated or saturated vapor. When the temperature of a gas-vapor mixture is lowered sufficiently, some of the vapor may condense or solidify. (It will condense to a liquid if the partial pressure is above the triple-point pressure, and solidify if the pressure is below the triple-point pressure.) The reverse process, in which some solid or liquid may sublime or evaporate, is also possible. Thus, the composition of the gaseous mixture may change as the component that exists as a vapor condenses or evaporates. The remainder of the chapter concerns this problem.

The most familiar example is atmospheric air, and many of the definitions have been first stated for air-water vapor mixtures. It should be emphasized, however, that the principles are basic and apply to all mixtures where both the gases and the vapor can be treated as an ideal gas. Therefore the general term gas-vapor mixture will be used, although most of the applications will be to air-water vapor mixtures.

The *dew point* of a gas-vapor mixture is the temperature at which the vapor condenses or solidifies when it is cooled at constant pressure. This is shown on the T-s diagram for the vapor shown in Fig. 9.6. Suppose that the temperature of the gaseous mixture and the partial pressure of the vapor in the mixture are such that the vapor is initially superheated at state 1. If the mixture is cooled at constant pressure the partial pressure of the vapor remains constant until point 2 is reached, and then condensation will begin. The temperature at state 2 is the dew-point temperature. Line 1–3 has been shown on the diagram to demonstrate that if the mixture is cooled at constant volume the condensation begins at point 3, which is slightly lower than the dew-point temperature.

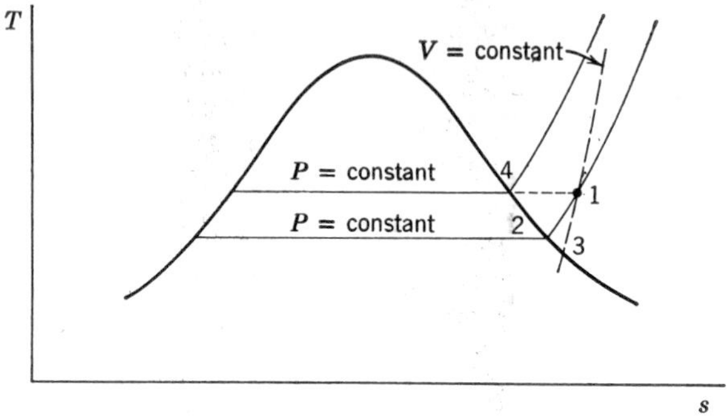

Fig. 9.6 Temperature-entropy diagram to show definition of the dew point.

If the vapor is at the saturation pressure and temperature, the mixture is referred to as a saturated mixture, and for an air-water vapor mixture, the term saturated air is used.

The *relative humidity* ϕ is defined as the ratio of the mole fraction of the vapor in the mixture to the mole fraction of vapor in a saturated mixture at the same temperature and total pressure. If the vapor is considered an ideal gas, the definition reduces to the ratio of the partial pressure of the vapor as it exists in the mixture, p_v, to the saturation pressure of the vapor at the same temperature, p_g.

$$\phi = \frac{p_v}{p_g}$$

In terms of the numbers on the T-s diagram of Fig. 9.6, the relative humidity ϕ would be

$$\phi = \frac{p_1}{p_4}$$

Since we are considering the vapor a perfect gas it is readily shown that the relative humidity can also be defined in terms of specific volume or density.

$$\phi = \frac{p_v}{p_g} = \frac{\rho_v}{\rho_g} = \frac{v_g}{v_v} \tag{9.23}$$

The *humidity ratio* ω of an air-water vapor mixture is defined as the ratio of the mass of water vapor m_v to the mass of dry air m_a. The term dry air is used to emphasize that this refers to air only and not to the water vapor. The term specific humidity is used synonymously with humidity ratio.

$$\omega = \frac{m_v}{m_a} \tag{9.24}$$

The definition would be identical for any other gas-vapor mixture, and the subscript a would refer to the gas exclusive of the vapor. Since we are considering both the vapor and the mixture ideal gases, a very useful expression for humidity ratio in terms of partial pressures can be developed.

$$m_v = \frac{p_v V}{R_v T} = \frac{p_v V M_v}{\bar{R}T} \qquad m_a = \frac{p_a V}{R_a T} = \frac{p_a V M_a}{\bar{R}T}$$

Then

$$\omega = \frac{p_v V/R_v T}{p_a V/R_a T} = \frac{R_a p_v}{R_v p_a} = \frac{M_v p_v}{M_a p_a} \tag{9.25}$$

For an air-water vapor mixture this reduces to

$$\omega = 0.622\frac{p_v}{p_a} \tag{9.26}$$

The *degree of saturation* μ is defined as the ratio of the actual humidity ratio to the humidity ratio of a saturated mixture at the same temperature and total pressure.

An expression for the relation between the relative humidity ϕ and the humidity ratio ω can be found by solving Eqs. 9.23 and 9.26 for p_v and equating them. The resulting relation for an air-water vapor mixture is

$$\phi = \frac{\omega p_a}{0.622 p_g} \tag{9.27}$$

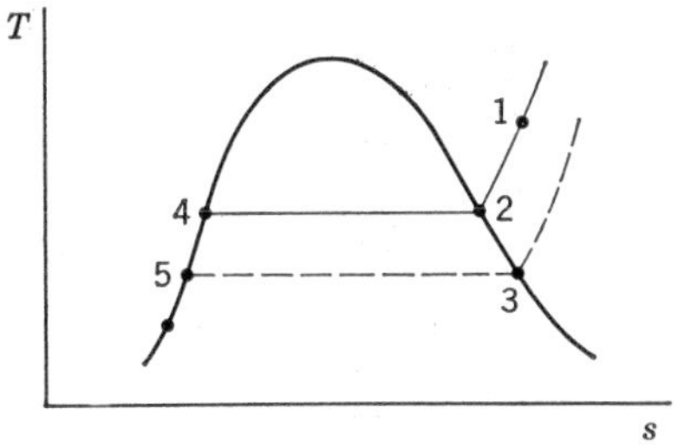

Fig. 9.7 Temperature-entropy diagram to show the cooling of a gas-vapor mixture at a constant pressure.

A few words should also be said about the nature of the process that occurs when a gas-vapor mixture is cooled at constant pressure. Suppose that the vapor is initially superheated at state 1 in Fig. 9.7. As the mixture is cooled at constant pressure the partial pressure of the vapor remains constant until the dew point is reached at point 2, at which point the vapor in the mixture is saturated. The first vapor that condenses is at state 4, and is in equilibrium with the vapor. As the temperature is lowered further, more of the vapor condenses, which lowers the partial pressure of the vapor in the mixture. The vapor that remains in the mixture is always saturated, and the liquid is in equilibrium with it. For example, when the temperature is reduced to T_3, the vapor in the mixture is at state 3, and its partial pressure is the saturation pressure corresponding to T_3. The liquid in equilibrium with it is at state 5.

Example 9.5 Consider 2000 ft^3 of an air-water vapor mixture at 14.70 lbf/in.^{2}, 90 F, 70 per cent relative humidity. Calculate the humidity ratio, dew point, mass of air, and mass of vapor.

From Eq. 9.23 and the steam tables

$$\phi = 0.70 = \frac{p_v}{p_g}$$

$$p_v = 0.70(0.6982) = 0.4887 \text{ lbf/in.}^2$$

The dew point is the saturation temperature corresponding to this pressure, which is 78.9 F.

The partial pressure of the air is

$$p_a = P - p_v = 14.70 - 0.49 = 14.21 \text{ lbf/in.}^2$$

The humidity ratio can be calculated from Eq. 9.26.

$$\omega = 0.622 \times \frac{p_v}{p_a} = 0.622 \times \frac{0.4887}{14.21} = 0.02135$$

The mass of air is

$$m_a = \frac{p_a V}{R_a T} = \frac{14.21 \times 144 \times 2000}{53.34 \times 550} = 139.6 \text{ lbm}$$

The mass of the vapor can be calculated by using the humidity ratio or by using the ideal gas equation of state.

$$m_v = \omega m_a = 0.02135(139.6) = 2.98 \text{ lbm}$$

$$m_v = \frac{0.4887 \times 144 \times 2000}{85.7 \times 550} = 2.98 \text{ lbm}$$

Example 9.6 Calculate the amount of water vapor condensed if the mixture of Example 9.5 is cooled to 40 F in a constant-pressure process.

At 40 F the mixture is saturated, since this is below the dew-point temperature. Therefore

$$p_{v2} = p_{g2} = 0.1217 \text{ lbf/in.}^2$$

$$p_{a2} = 14.7 - 0.12 = 14.58 \text{ lbf/in.}^2$$

$$\omega_2 = 0.622 \times \frac{0.1217}{14.58} = 0.00520$$

The amount of water vapor condensed is equal to the difference between the initial and final mass of water vapor.

$$\text{mass of vapor condensed} = m_a(\omega_1 - \omega_2) = 139.6(0.02135 - 0.0052) = 2.25 \text{ lbm}$$

9.7 The First Law Applied to Gas-Vapor Mixtures

In applying the first law of thermodynamics to gas-vapor mixtures it is helpful to realize that because of our assumption that ideal gases are involved, the various components can be treated separately when calculating changes of internal energy, enthalpy, and entropy. Therefore, in dealing with air-water vapor mixtures the changes in enthalpy of the water vapor can be found from the steam tables and the ideal gas relations can be applied to the air. This is illustrated by the examples that follow.

Example 9.7 An air-conditioning unit is shown in Fig. 9.8, with pressure, temperature, and relative humidity data. Calculate the

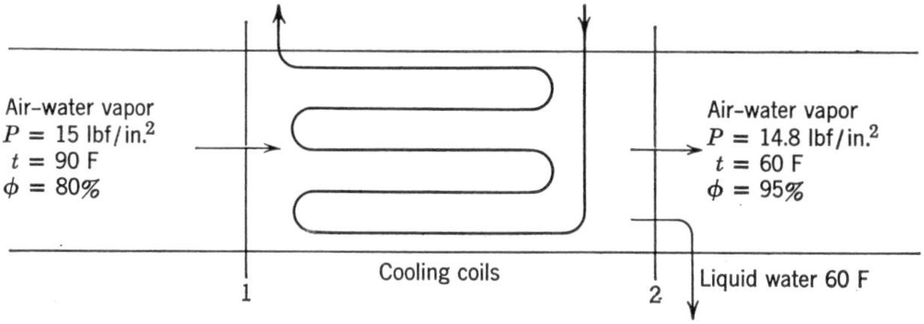

Fig. 9.8 Sketch for Example 9.7.

heat transfer per pound of dry air, assuming that changes in kinetic energy are negligible.

Let us consider a steady-flow process for an open system that excludes the cooling coils. The steady-flow energy equation reduces to

$$Q + H_1 = H_2$$

The enthalpy entering (H_1) and leaving (H_2) includes the enthalpy of both the air and the water. The mass of vapor entering per pound of dry air is ω_1, the mass of vapor leaving is ω_2, and the amount of water condensed per pound of dry air is $\omega_1 - \omega_2$. Therefore, designating by q the heat transfer per pound of dry air, the first law can be written

$$q + h_{a1} + \omega_1 h_{v1} = h_{a2} + \omega_2 h_{v2} + (\omega_1 - \omega_2)h_{l2}$$

$$q = C_{p(a)}(t_2 - t_1) + \omega_2 h_{v2} - \omega_1 h_{v1} + (\omega_1 - \omega_2)h_{l2}$$

$$p_{v1} = \phi p_{g1} = 0.80(0.6982) = 0.5586 \text{ lbf/in.}^2$$

$$\omega_1 = \frac{R_a}{R_v}\frac{p_{v1}}{p_{a1}} = 0.622\frac{0.5586}{15.00 - 0.56} = 0.0241$$

$$p_{v2} = \phi_2 p_{g2} = 0.95(0.2563) = 0.2435$$

$$\omega_2 = \frac{R_a}{R_v} \times \frac{p_{v2}}{p_{a2}} = 0.622 \times \frac{0.2435}{14.80 - 0.24} = 0.0104$$

The value for $C_{p(a)}$ can be found from Table A.6, and values for the enthalpy of water can be taken from the steam tables. (Since the water vapor at these low pressures is being considered an ideal gas, the enthalpy of the water vapor is a function of the temperature only. Therefore, the enthalpy of slightly superheated water vapor is equal to the enthalpy of saturated vapor at the same temperature.)

$$q = 0.240(60 - 90) + 0.0104(1088.0) - 0.0241(1100.9) + (0.0241 - 0.0104)28.06$$

$$= -7.20 + 11.32 - 26.53 + 0.38 = -22.03 \text{ Btu/lbm dry air}$$

Example 9.8 A tank has a volume of 10 ft^3 and contains nitrogen and water vapor. The temperature of the mixture is 120 F and the total pressure is 30 lbf/in.^{2} The partial pressure of the water vapor is 0.8 lbf/in.^{2} Calculate the heat transfer when the contents of the tank are cooled to 50 F.

This is a constant-volume process, and since the work is zero the first law reduces to

$$Q = U_2 - U_1 = m_{N_2}C_{v(N_2)}(t_2 - t_1) + (m_2u_2)_v + (m_2u_2)_l - (m_1u_1)_v$$

As written this equation assumes that some of the vapor condensed. This must be checked, however, as shown below.

The mass of nitrogen and water vapor can be calculated using the ideal-gas equation of state.

$$m_{N_2} = \frac{p_{N_2}V}{R_{N_2}T} = \frac{29.2 \times 144 \times 10}{55.15 \times 580} = 1.314 \text{ lbm}$$

$$m_{v1} = \frac{p_{v1}V}{R_vT} = \frac{0.8 \times 144 \times 10}{85.76 \times 580} = 0.0232 \text{ lbm}$$

If condensation takes place the final state of the vapor will be saturated vapor at 50 F. In this case

$$m_{v2} = \frac{p_{v2}V}{R_vT} = \frac{0.1781 \times 144 \times 10}{85.76 \times 510} = 0.00586 \text{ lbm}$$

Since this is less than the original mass of vapor there must have been condensation.

The mass of liquid that is condensed, m_{l2}, is

$$m_{l2} = m_{v1} - m_{v2} = 0.0232 - 0.0059 = 0.0173 \text{ lbm}$$

The internal energy of the water vapor is equal to the internal energy of saturated water vapor at the same temperature. Therefore,

$$u_{1v} = 1113.7 - \frac{1.6924 \times 144 \times 203.27}{778}$$
$$= 1050.1 \text{ Btu/lbm}$$
$$u_{2v} = 1083.7 - \frac{0.1781 \times 144 \times 1703.2}{778}$$
$$= 1027.6 \text{ Btu/lbm}$$

$$Q = 1.314 \times 0.177(50 - 120) + 0.00586(1027.6)$$
$$+ 0.0173(18.07) - 0.0232(1050.1)$$
$$= -16.31 + 6.02 + 0.31 - 24.26 = -34.24 \text{ Btu}$$

9.8 The Adiabatic Saturation Process

An important process involving an air-water vapor mixture is the adiabatic saturation process, in which an air-vapor mixture comes in contact with a body of water in a well-insulated duct (Fig. 9.9). If

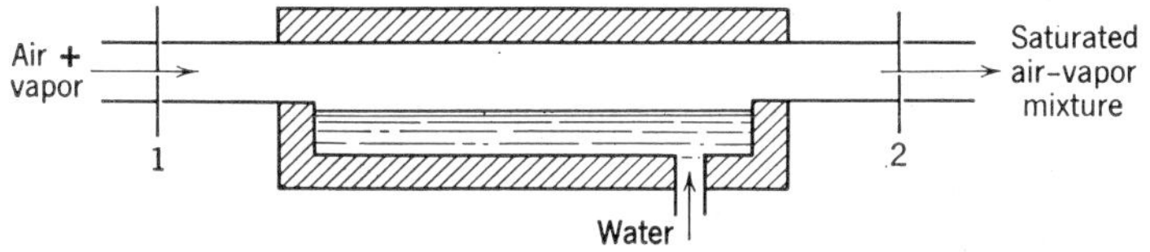

Fig. 9.9 The adiabatic saturation process.

the initial relative humidity is less than 100 per cent some of the water will evaporate and the temperature of the air-vapor mixture will decrease. If the mixture leaving the duct is saturated and if the process is adiabatic, the temperature of the mixture on leaving is known as the adiabatic saturation temperature. In order for this to take place as a steady-flow process, make-up water at the adiabatic saturation temperature is added at the same rate at which water is evaporated. The pressure is assumed to be constant.

Considering the adiabatic saturation process to be a steady-flow process, and neglecting changes in kinetic and potential energy, the first law reduces to

$$h_{a1} + \omega_1 h_{v1} + (\omega_2 - \omega_1)h_{l2} = h_{a2} + \omega_2 h_{v2}$$

$$\omega_1(h_{v1} - h_{l2}) = c_{pa}(t_2 - t_1) + \omega_2(h_{v2} - h_{l2})$$

$$\omega_1(h_{v1} - h_{l2}) = c_{pa}(t_2 - t_1) + \omega_2 h_{fg2} \qquad (9.28)$$

The most significant point to be made regarding the adiabatic saturation process is that the temperature of the mixture on leaving, the *adiabatic saturation temperature,* is a function of the pressure, temperature, and relative humidity of the entering air-vapor mixture and of the exit pressure. Thus, the relative humidity and humidity ratio of the entering air-vapor mixture can be determined from measurements of the pressure and temperature of the air-vapor mixture entering and leaving the adiabatic saturator. Since these measurements are relatively easy to make, this is one means of determining the humidity of an air-vapor mixture.

Example 9.9 The pressure of the mixture entering and leaving the adiabatic saturator is 14.7 lbf/in.^{2}, the entering temperature is 84 F, and the temperature leaving, which is the adiabatic saturation temperature, is 70 F. Calculate the humidity ratio and relative humidity of the air-water vapor mixture entering.

Since the water vapor leaving is saturated, $p_{v2} = p_{g2}$, and ω_2 can be calculated.

$$\omega_2 = 0.622 \times \frac{0.3631}{14.7 - 0.36} = 0.01573$$

ω_1 can be calculated using Eq. 9.28.

$$\omega_1 = \frac{0.24(70 - 84) + 0.01573 \times 1054.3}{1098.4 - 38.0} = \frac{-3.36 + 16.60}{1060.4} = 0.0125$$

$$\omega_1 = 0.622 \times \frac{p_{v1}}{14.7 - p_{v1}} = 0.0125$$

$$p_{v1} = 0.289$$

$$\phi_1 = \frac{p_{v1}}{p_{g1}} = \frac{0.289}{0.577} = 0.501$$

9.9 Wet-Bulb and Dry-Bulb Temperatures

The humidity of an air-water vapor mixture is usually found from dry-bulb and wet-bulb data. These data are obtained by use of a psychrometer, which involves the flow of air past a wet-bulb and dry-bulb thermometer. The bulb of the wet-bulb thermometer is covered with a cotton wick which is saturated with water. The dry-bulb thermometer is used simply to measure the temperature of the air. The flow of air may be maintained by a fan, as in the continuous-flow psychrometer shown in Fig. 9.10, or by moving the thermometer

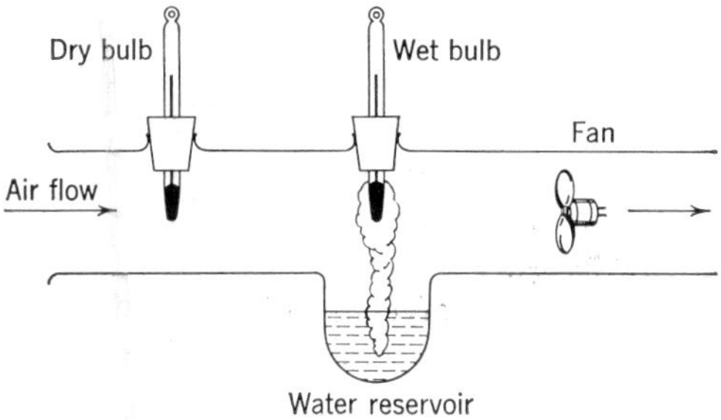

Fig. 9.10 Steady-flow apparatus for measuring wet- and dry-bulb temperature.

through the air as is done in the sling psychrometer, which consists of a wet-bulb and a dry-bulb thermometer so mounted that they can be whirled.

The processes that take place at the wet-bulb thermometer are somewhat involved. First of all, if the air-water vapor mixture is not saturated, some of the water in the wick will evaporate and diffuse into the surrounding air. A drop in the temperature of the water in the wick will be associated with this evaporation. However, as soon as the temperature of the water drops, there will be heat transfer to the water both from the air and from the thermometer. Finally a steady state, which is determined by heat and mass transfer rates, will be reached. In general, air velocities upwards of 700 ft/min are desirable so that the convective heat transfer is large compared to radiant heat transfer.

The psychrometric chart is the most convenient method of determining relative humidity and humidity ratio from wet-bulb and dry-bulb data, although equations that have been developed can also be used. A psychrometric chart is included in the Appendix, Table A.12.

The difference between the wet-bulb temperature and adiabatic saturation temperature should be carefully noted. The wet-bulb temperature is influenced by heat and mass transfer rates, whereas the adiabatic saturation temperature simply involves equilibrium between the entering air-vapor mixture and water at the adiabatic saturation temperature. However, it happens that the wet-bulb temperature and the adiabatic saturation temperature are approximately equal for air-water vapor mixtures at atmospheric temperature and pressure. This is not necessarily true at temperatures and pressures that deviate significantly from ordinary atmospheric conditions, or for other gas-vapor mixtures.

9.10 The Psychrometric Chart

Properties of air-water vapor mixtures are given in graphical form on psychrometric charts. These are available in a number of different forms, and only the main features are considered here.

The basic psychrometric chart consists of a plot of dry-bulb temperature as abscissa and humidity ratio as ordinate. If we fix the total pressure for which the chart is to be constructed (which is usually one standard atmosphere), lines of constant relative humidity and wet-bulb temperature can be drawn on the chart, because for a given dry-bulb temperature, total pressure, and humidity ratio, the relative humidity and wet-bulb temperature are fixed. The partial pressure of the water vapor is fixed by the humidity ratio and total pressure, and therefore a second ordinate scale can be constructed indicating the partial pressure of the water vapor.

Most charts give the enthalpy of an air-vapor mixture per pound of dry air. The values given assume that the enthalpy of the dry air is zero at 0 F, and the enthalpy of the vapor is taken from the steam tables (which are based on the assumption that the enthalpy of saturated liquid is zero at 32 F). This procedure is quite satisfactory because we are usually concerned only with differences in enthalpy. The fact that lines of constant enthalpy are essentially parallel to lines of constant wet-bulb temperature is evident from the fact that the wet-bulb temperature is essentially equal to the adiabatic saturation temperature. Thus, in Fig. 9.9, if we neglect the enthalpy of the liquid entering the adiabatic saturator, the enthalpy of the air-vapor mixture entering is equal to the enthalpy of the saturated air-vapor mixture leaving, and a given adiabatic saturation temperature fixes the enthalpy of the mixture entering.

Some charts are available that give corrections for variation from standard atmospheric pressures. Before using a given chart one should be certain to understand the assumptions made in constructing it, and that it is applicable to the particular problem at hand.

PROBLEMS

9.1 Following is the volumetric analysis of a gaseous mixture:

Constituent	% by Volume
N_2	60
CO_2	22
CO	11
O_2	7

(*a*) Determine the analysis on a mass basis.

(*b*) What is the mass of 1000 ft^3 of this gas when the pressure is 15.0 $lbf/in.^2$ and the temperature 80 F?

9.2 The mixture of Problem 9.1 is heated in a steady-flow process from 100 F to 500 F. Calculate the necessary heat transfer per lbm of mixture.

9.3 Nitrogen and hydrogen are mixed in a steady-flow adiabatic process in the ratio of 3 lbm of hydrogen per lbm of nitrogen. The hydrogen enters at 20 $lbf/in.^2$, 100 F, and the nitrogen at 20 $lbf/in.^2$, 500 F. The pressure after mixing is 18 $lbf/in.^2$ Determine the final temperature of the mixture and the net entropy change per lbm of mixture.

9.4 A mixture of CO_2 and O_2 is expanded in a cylinder in a reversible adiabatic process from 30 $lbf/in.^2$, 200 F to 15 $lbf/in.^2$ The mole fraction of the CO_2 is 0.2. Determine the work done per mole of mixture, the final temperature, and the entropy change of the CO_2 and the O_2.

9.5 A tank having a volume of 10 ft^3 contains oxygen at 50 $lbf/in.^2$, 80 F. Nitrogen at a pressure of 100 $lbf/in.^2$, 300 F flows from a pipe into the tank until the pressure reaches 90 $lbf/in.^2$ The entire process is adiabatic. Determine the final temperature of the mixture and the change in entropy for this process.

9.6 A room of dimensions 10 ft x 20 ft x 8 ft contains an air-water vapor mixture at a total pressure of 14.7 $lbf/in.^2$ and a temperature of 90 F. The partial pressure of the water vapor is 0.3 $lbf/in.^2$ Calculate:

(*a*) The humidity ratio.

(*b*) The dew point.

(*c*) The total mass of water vapor in the room.

9.7 An air-water vapor mixture enters an air-conditioning unit at a pressure of 20 $lbf/in.^2$, a temperature of 90 F, and a relative humidity of 80%. The mass of dry air entering per min is 100 lbm. The air-vapor mixture leaves the air-conditioning unit at 18 $lbf/in.^2$, 55 F,

100% relative humidity. The moisture condensed leaves at 55 F. Determine the heat transfer per min from the air.

9.8 An air-vapor mixture enters a cooler-dehumidifier at the rate of 1000 ft^3/min. The mixture enters at 14.7 lbf/$in.^2$, 100 F, and a relative humidity of 80%. The humidity ratio on leaving is decreased to one-third of that on entering. The relative humidity on leaving is 100%, and the pressure is 14.7 lbf/$in.^2$ Determine the required heat transfer per min.

9.9 A certain apparatus involves a precise knowledge of the amount of water vapor in an electrical conductivity test cell. This is obtained by charging the bomb with a mixture of nitrogen and water vapor at 1000 lbf/$in.^2$ pressure at a temperature of 80 F. The water vapor in the mixture is saturated. What is the humidity ratio of the mixture in the cell?

9.10 Atmospheric air enters a two-stage compressor at 14.2 lbf/$in.^2$, 90 F, and 70% relative humidity. The volume rate of flow into the compressor is 500 ft^3/min. On leaving the first stage the air enters the intercooler. The pressure at the exit of the intercooler is 60 lbf/$in.^2$ The air leaves the second stage at 255 lbf/$in.^2$, 320 F and then enters the aftercooler. The air leaves the aftercooler at 250 lbf/$in.^2$, 100 F. This is shown schematically in Fig. 9.11.

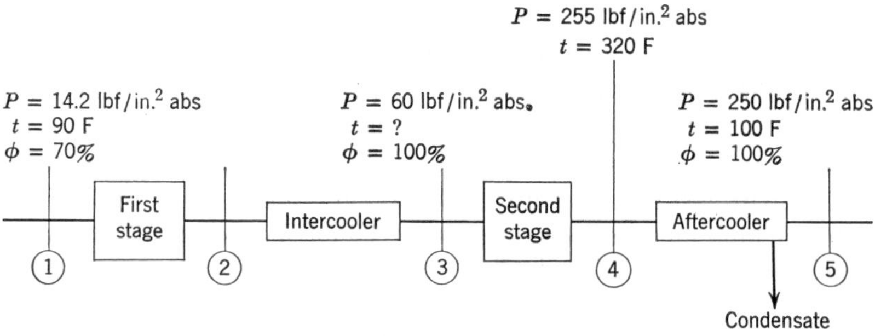

Fig. 9.11 Sketch for Problem 9.10.

(*a*) What will be the temperature of the air leaving the intercooler if the relative humidity on leaving is 100%, but no moisture is condensed in the intercooler?

(*b*) How much moisture is condensed per hr in the aftercooler, assuming no condensation in the intercooler?

9.11 A tank contains an air-vapor mixture at a temperature of 110 F and a relative humidity of 30% and a pressure of 1 atm.

(*a*) What is the dew point of the mixture?

(*b*) At what temperature will condensation begin if the mixture is cooled down?

(*c*) How much heat transfer is required to cool the contents of the tank to 60 F?

9.12 (*a*) Determine (by a first-law analysis) the humidity ratio and relative humidity of an air-water vapor mixture that has a dry-bulb temperature of 85 F, an adiabatic saturation temperature of 76 F, and a pressure of 14.7 lbf/in.^{2}

(*b*) By use of the psychrometric chart determine the humidity ratio and relative humidity of an air-water vapor mixture that has a dry-bulb temperature of 85 F, a wet-bulb temperature of 76 F, and a pressure of 14.7 lbf/in.^{2}

9.13 When limited quantities of cooling water for a condenser are available, a cooling tower is often used (Fig. 9.12). Consider the case

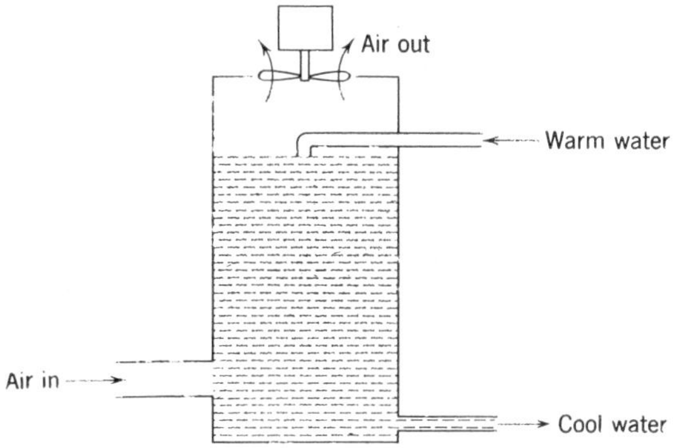

Fig. 9.12 **Sketch for Problem 9.13.**

where 20,000 lbm/hr of water at 105 F enters the top of the cooling tower, and the cool water leaves the bottom at 65 F. The air-water vapor mixture enters the bottom of the cooling tower at 14.7 lbf/in.^{2}, and has a dry-bulb temperature of 72 F and a wet-bulb temperature of 60 F. The air-water vapor mixture leaving the tower has a pressure of 14.3 lbf/in.^{2}, a temperature of 90 F, and a relative humidity of 80%. Determine the lbm of dry air per min that must be used and the fraction of the incoming water that evaporates. Assume the process to be adiabatic.

9.14 In areas where the temperature is high and the humidity is low, some measure of air conditioning can be achieved by evaporative cooling. This involves spraying water into the air, which subsequently evaporates with a resulting decrease in the temperature of the mixture. Such a scheme is shown in Fig. 9.13.

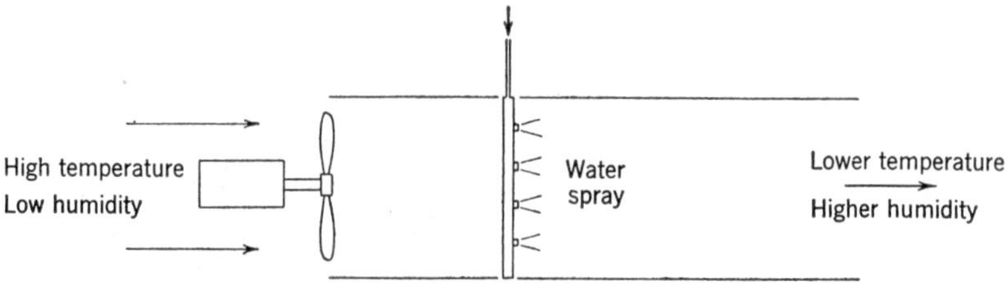

Fig. 9.13 Sketch for Problem 9.14.

Consider the case of atmospheric air at 100 F, 10% relative humidity, and 14.7 lbf/in.^{2} Cooling water at 50 F is sprayed into the air. If the air-water vapor mixture is to be cooled to 80 F, what will be the relative humidity? What are the disadvantages of this approach to air conditioning?

9.15 One method of removing moisture from atmospheric air is to cool the air so that the moisture condenses or freezes out. Suppose an experiment requires a humidity ratio of 0.0001. To what temperature must the air be cooled at a pressure of 1 atm in order to achieve this humidity? To what temperature must it be cooled if the pressure is 100 atm?

CHAPTER 10

AVAILABILITY, IRREVERSIBILITY, AND EFFICIENCY

In effect this chapter is a continuation of Chapter 7 which discussed the second law of thermodynamics. This consideration of the second law was interrupted to consider ideal gases and mixtures of ideal gases, so that now we are able to consider the application of the second law to systems involving a pure substance, ideal gases, and gaseous mixtures. In this chapter attention is focused on the matter of availability. The concept of availability involves the maximum work that can be done by a system in any given state. Closely associated with availability are the concepts of irreversibility and efficiency, both of which are used to indicate the relation of the actual work to the maximum work.

10.1 Introduction to the Concept of Availability

A study of the Carnot cycle affords a convenient starting point for a study of availability. Consider the Carnot cycle shown in Fig. 10.1. The amount of heat Q_H (area 1–2–b–a–1) is transferred from the high-temperature reservoir at temperature T_H to the engine operating on the Carnot cycle. Of this energy transferred to the engine as heat, the amount Q_L (area 4–3–b–a–4) is eventually transferred to the surroundings at temperature T_o. Since this is the lowest possible temperature at which heat can be rejected (at the given location), the maximum amount of work that can be done is W_c (where $W_c = Q_H - Q_L$ = area 1–2–3–4–1), and this maximum amount of work is called the availability of the energy that was transferred from the high-temperature reservoir to the engine. That portion which was rejected, Q_L, is sometimes referred to as the unavailable energy (in the sense that it is not available for doing work).

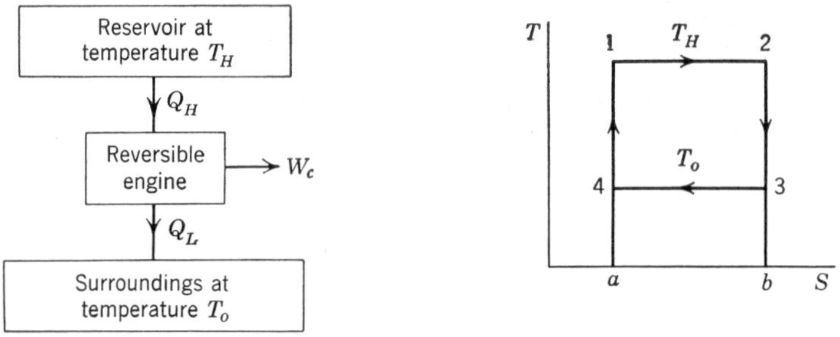

Fig. 10.1 A Carnot cycle.

The concept of irreversibility can also be introduced in relation to the cycle just described. Suppose an actual engine, Fig. 10.2, operates between the same reservoir and surroundings as described in Fig. 10.1, and assume that the same amount of heat, Q_H (area 1–2–b–a–1), is transferred to the actual engine. The actual engine will have irreversibilities associated with it, and therefore, the actual work W will be less than the work W_c of the reversible engine. It follows that the heat rejected by the actual engine, Q_L' (area 4–3′–c–a–4), will be greater than the heat rejected by the reversible engine. The net work of the

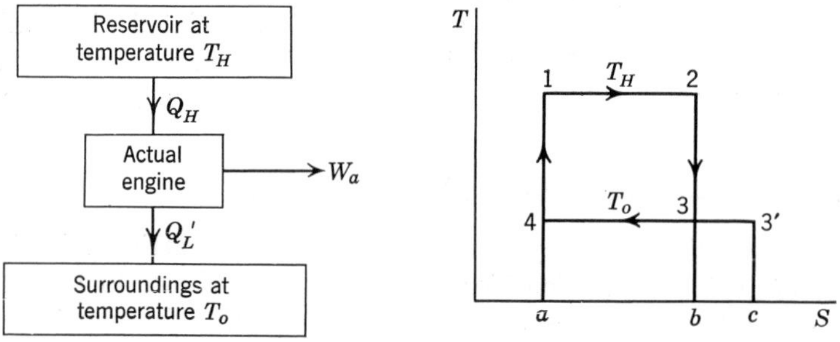

Fig. 10.2 An actual cycle.

actual engine is equal to $Q_H - Q_L'$ (area 1–2–b–a–1 minus area 4–3′–c–a–4). The irreversibility I for the actual engine can be defined as the difference between the work for the reversible engine and the work for the actual engine.

$$I = W_c - W$$

It can be shown that this irreversibility is equal to area 3–3′–c–b–3 or $T_o \Delta S_{net}$. This is readily shown by noting that

$$W_c = Q_H - Q_L = \text{Area } 1\text{–}2\text{–}b\text{–}a\text{–}1 - \text{Area } 4\text{–}3\text{–}b\text{–}a\text{–}4$$

$$W = Q_H - Q_L' = \text{Area } 1\text{–}2\text{–}b\text{–}a\text{–}1 - \text{Area } 4\text{–}3'\text{–}c\text{–}a\text{–}4$$

Therefore,

$$I = W_c - W = -(\text{Area } 4\text{–}3\text{–}b\text{–}a\text{–}4) - (-\text{Area } 4\text{–}3'\text{–}c\text{–}a\text{–}4)$$
$$= \text{Area } 3\text{–}3'\text{–}c\text{–}b\text{–}3 = T_o \Delta S_{net}$$

We have considered the terms availability and irreversibility in relation to a Carnot cycle. Many times, however, we are not concerned with a cycle but rather with a process. For example, if we have a tank of compressed air we might wish to know the maximum amount of work that could be done by this air in this tank. That is, we would like to know the availability of the air in the tank. We might ask the same question regarding a tank of gasoline, a storage battery, or the exhaust gases from an engine. The rest of this chapter is a more general approach to availability and irreversibility. The first step is to define reversible work for the general case.

10.2 Reversible Work

Reversible work, as used in this chapter and throughout the book, is the maximum work that can be done by a given system during a given change of state, assuming heat transfer only with the surroundings. That is, if the initial and final states of a system are specified, the reversible work refers to the maximum work that can be done by the system as it goes from the initial to the final state, assuming that the only heat transfer is with the surroundings. It is evident that this concept of reversible work involves both the first and second laws of thermodynamics. It is also evident that the work will be maximum only if the process is entirely reversible.

In developing this concept of reversible work, let us consider the change of state that takes place as a mass δm_i enters the system, a mass δm_e leaves the system, heat δQ is transferred to the system, work δW is done by the system, and the energy of the system changes by the amount $d\left(U + \dfrac{m\overline{V}^2}{2g_c} + m\dfrac{g}{g_c}Z\right)_\sigma$. This is shown in Fig. 10.3.

Let us derive an expression for the reversible work for this change of state. Since the process must be entirely reversible, it is necessary, whenever the temperature of the system is different from the tempera-

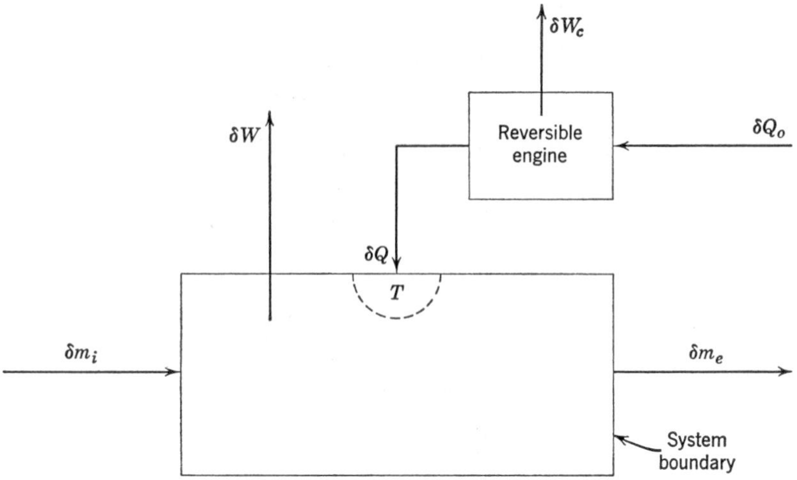

Fig. 10.3 Process in which maximum work is done.

ture of the surroundings, for the heat transfer to take place through a reversible heat engine (i.e., one which operates on the Carnot cycle). Since the temperature of the fluid may be changing as it flows through the system (as in the case of flow through a turbine), the heat transfer is shown for the general case to take place at that point where the temperature of the system is T.

The first law of thermodynamics for this process (Eq. 5.17) is

$$\delta Q + \delta m_i\left(h_i + \frac{\overline{V}_i^2}{2g_c} + \frac{g}{g_c} Z_i\right) = \delta W + \delta m_e\left(h_e + \frac{\overline{V}_e^2}{2g_c} + \frac{g}{g_c} Z_e\right) + d\left(U + m\frac{\overline{V}^2}{2g_c} + m\frac{g}{g_c} Z\right)_\sigma \quad (10.1)$$

Since this process is reversible, the change in entropy of the system (Eq. 7.10) is

$$dS_\sigma = \frac{\delta Q}{T} + \delta m_i s_i - \delta m_e s_e \quad (10.2)$$

The work done by the reversible heat engine, denoted W_c, is given by the expression

$$\delta W_c = \delta Q_o - \delta Q \quad (10.3)$$

Since the heat engine is reversible we can write

$$\frac{\delta Q_o}{T_o} = \frac{\delta Q}{T}$$

and the work of the reversible engine is given by the relation

$$\delta W_c = T_o\left(\frac{\delta Q}{T}\right) - \delta Q \tag{10.4}$$

On substituting Eq. 10.2 into Eq. 10.4, we have

$$\delta W_c = T_o(dS_\sigma - \delta m_i s_i + \delta m_e s_e) - \delta Q \tag{10.5}$$

The reversible work is the sum of the work done by the system, δW, plus the work done by the reversible engine, δW_c. That is,

$$\delta W_{\text{rev}} = \delta W + \delta W_c \tag{10.6}$$

Substituting Eqs. 10.1 and 10.5, we have a relation for the reversible work.

$$\delta W_{\text{rev}} = \delta Q + \delta m_i\left(h_i + \frac{\overline{V}_i^2}{2g_c} + \frac{g}{g_c} Z_i\right) - \delta m_e\left(h_e + \frac{\overline{V}_e^2}{2g_c} + \frac{g}{g_c} Z_e\right)$$

$$- d\left(U + \frac{m\overline{V}^2}{2g_c} + m\frac{g}{g_c} Z\right)_\sigma + T_o(dS_\sigma - \delta m_i s_i + \delta m_e s_e) - \delta Q$$

This simplifies to the following expression:

$$\delta W_{\text{rev}} = \delta m_i\left(h_i - T_o s_i + \frac{\overline{V}_i^2}{2g_c} + \frac{g}{g_c} Z_i\right) - \delta m_e\left(h_e - T_o s_e + \frac{\overline{V}_e^2}{2g_c}\right.$$

$$\left. + \frac{g}{g_c} Z_e\right) - d\left(U - T_o S + \frac{m\overline{V}^2}{2g_c} + \frac{mg}{g_c} Z\right)_\sigma \tag{10.7}$$

This equation can be integrated to give the reversible work for a particular system and a given change of state. As stated previously, this involves heat transfer with the surroundings only. It should be noted that the reversible work is not only a function of the initial and final states of the system, but it also depends on the temperature of the surroundings.

Two cases are of particular interest, namely the steady-flow process and the closed system. The requirements for a steady-flow process were discussed in Chapter 5, and as applied to Eq. 10.7 they mean that

$$d\left(U - T_o S + \frac{m\overline{V}^2}{2g_c} + m\frac{g}{g_c} Z\right)_\sigma = 0$$

and

$$\delta m_i = \delta m_e \equiv \delta m$$

Expressed on the basis of a unit mass of fluid entering and leaving the system the relation for reversible work for a steady-flow process becomes

$$w_{\text{rev}} = (h_i - h_e) - T_o(s_i - s_e) + \frac{\overline{V}_i^2 - \overline{V}_e^2}{2g_c} + (Z_i - Z_e)\frac{g}{g_c} \tag{10.8}$$

This relation can also be expressed on a time basis. Letting $\dot{W}_{\text{rev}}$ designate the power and $\dot{m}$ the mass rate of flow, this equation becomes

$$\dot{W}_{\text{rev}} = \dot{m}\left(h_i - T_o s_i + \frac{\overline{V}_i^2}{2g_c} + \frac{g}{g_c}Z_i\right) - \dot{m}\left(h_e - T_o s_e + \frac{\overline{V}_e^2}{2g_c} + \frac{g}{g_c}Z_e\right) \tag{10.9}$$

Equation 10.7 can be simplified to apply to the closed system by noting that in this case $\delta m_i = 0$ and $\delta m_e = 0$. Therefore, for the closed system

$$\delta W_{\text{rev}} = -d\left(U - T_o S + \frac{m\overline{V}^2}{2g_c} + m\frac{g}{g_c}Z\right)_\sigma$$

This is readily integrated to give

$$_1W_{\text{rev}\,2} = \left(U_1 - T_o S_1 + \frac{m\overline{V}_1^2}{2g_c} + m\frac{g}{g_c}Z_1\right) - \left(U_2 - T_o S_2 + \frac{m\overline{V}_2^2}{2g_c} + m\frac{g}{g_c}Z_2\right) \tag{10.10}$$

or for a unit mass

$$_1w_{\text{rev}\,2} = (u_1 - u_2) - T_o(s_1 - s_2) + \frac{\overline{V}_1^2 - \overline{V}_2^2}{2g_c} + (Z_1 - Z_2)\frac{g}{g_c} \tag{10.11}$$

Example 10.1 Air enters an air turbine at a pressure of 100 lbf/in.^{2}, 150 F, with a velocity of 300 ft/sec, and leaves the turbine at 20 lbf/in.^{2}, 40 F, and 200 ft/sec velocity. There is no heat transfer from the turbine. Assume air to be an ideal gas with constant specific heat.

Calculate the reversible work per pound of air for the change of state that occurs in the turbine, and compare this with the actual work done by the turbine.

Since this is a steady-flow process, Eq. 10.8 can be used to calculate the reversible work. Changes in potential energy are neglected.

$$w_{\text{rev}} = (h_i - h_e) - T_o(s_i - s_e) + \frac{\overline{V}_i^2 - \overline{V}_e^2}{2g_c}$$

$$= C_{po}(T_i - T_e) - T_o\left(C_{po}\ln\frac{T_i}{T_e} - R\ln\frac{P_i}{P_e}\right) + \frac{\overline{V}_i^2 - \overline{V}_e^2}{2g_c}$$

$$= 0.240(610 - 500) - 537\left(0.240\ln\frac{610}{500} - \frac{53.3}{778}\ln\frac{100}{20}\right)$$

$$+ \frac{(300)^2 - (200)^2}{64.34 \times 778}$$

$$= 26.4 + 33.6 + 1.0 = 61 \text{ Btu/lbm}$$

The actual work of the turbine can be found from the steady-flow energy equation,

$$w = (h_i - h_e) + \frac{\overline{V}_i^2 - \overline{V}_e^2}{2g_c}$$

$$= 0.240(610 - 500) + \frac{(300)^2 - (200)^2}{64.34 \times 778}$$

$$= 26.4 + 1.0 = 27.4 \text{ Btu/lbm}$$

Example 10.2 A cylinder fitted with a piston contains 1 lbm of air at 20 lbf/in.^{2}, 120 F. Work is done by the piston on the gas. The final pressure is 80 lbf/in.^{2} and the final temperature is 200 F.

Calculate the reversible work per pound of air for this process.

Since this is a closed system, Eq. 10.11 can be applied.

Neglecting changes in kinetic and potential energy this becomes

$${}_1w_{\text{rev}\,2} = u_1 - u_2 - T_o(s_1 - s_2)$$

$$= C_{v\infty}(T_1 - T_2) - T_o\left(C_{po}\ln\frac{T_1}{T_2} - R\ln\frac{P_1}{P_2}\right)$$

$$= 0.171(580 - 660) - 537\left(0.240\ln\frac{580}{660} - \frac{53.3}{778}\ln\frac{20}{80}\right)$$

$$= -13.7 - 34.2 = -47.9 \text{ Btu/lbm}$$

In this case W_{rev} is negative. This means that the actual work will be less than this (say -60 Btu/lbm). Or, to say it another way, the minimum work of compression is 47.9 Btu/lbm, and the actual work of compression is greater. In general then, when W_{rev} is negative, the numerical value of W_{rev} is the minimum work of compression.

Example 10.3 Steam at a pressure of 200 lbf/in.^{2}, 600 F flows in a pipe. Attached to this pipeline is an evacuated vessel. The valve connecting the vessel to the pipe is opened, and steam flows into the evacuated vessel until the pressure is 200 lbf/in.² The process is adiabatic. Determine the reversible work per pound of steam for this change of state.

It was determined in Example 5.10 that the final temperature of the steam is 885 F.

The reversible work can be determined by use of Eq. 10.7. Since $\delta m_e = 0$ and changes in kinetic and potential energy are not significant, this equation reduces to

$$\delta W_{\text{rev}} = \delta m_i(h_i - T_o s_i) - d(U - T_o S)_\sigma$$

On integrating this becomes

$$W_{\text{rev}} = m_i(h_i - T_o s_i) - m_2(u_2 - T_o s_2)$$

Since (Example 5.10)

$$m_i = m_2 \qquad \text{and} \qquad h_i = u_2$$

$$w_{\text{rev}} = -T_o(s_i - s_2)$$

$$w_{\text{rev}} = -537(1.6767 - 1.7990) = 65.7 \text{ Btu/lbm}$$

Example 10.4 Water at 60 F is available in large quantities. Some of this water is to be made into ice at 0 F. Determine the reversible work per pound of ice for this process, assuming a steady-flow process.

In this case t_o is assumed to be 60 F, the temperature of the water. From Eq. 10.8

$$\begin{aligned} w_{\text{rev}} &= (h_i - h_e) - T_o(s_i - s_e) \\ &= 28.06 - (-158.93) - 520\,[0.0555 - (-0.3241)] \\ &= 186.99 - 197.39 = -10.40 \text{ Btu/lbm} \end{aligned}$$

Since a refrigeration device would be required, the minimum work input to the refrigeration device is 10.40 Btu/lbm of ice produced.

The significance of Eq. 10.8 as used in this problem can be demonstrated by reference to the *T-s* diagram of Fig. 10.4. A reversible refrigerator receives heat from the water that is being frozen and rejects heat to water at 60 F. Such large quantities of cooling water are used that the temperature does not rise significantly above 60 F. Line *i*–1–2–*e* represents the path followed by the water during the process (*i*–1 represents the cooling of the liquid, 1–2 the freezing process, and 2–*e* the cooling of the ice). The area underneath this curve, area *i*–1–2–*e*–4–5–*i*, represents the heat transferred from the ice during this process; i.e., $h_i - h_e$. Area *i*–5–4–3–*i* represents $T_o(s_i - s_e)$, the heat transferred from the working fluid of the reversible refrigerator to the

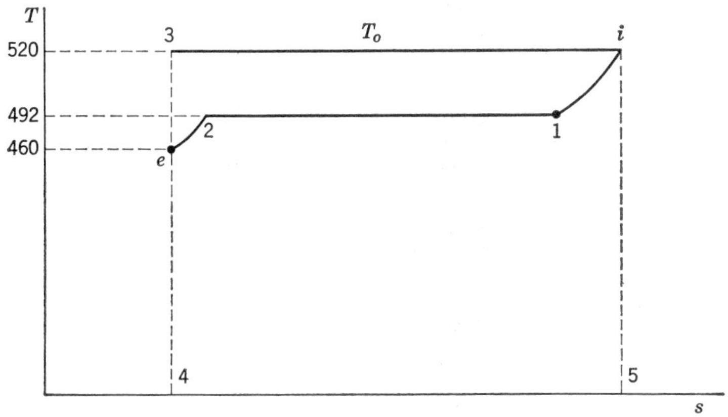

Fig. 10.4 Temperature-entropy diagram for Example 10.4.

water at 60 F. Thus, area i–3–e–2–1–i represents the minimum work necessary to produce 1 lbm of ice, which in this example is 10.40 Btu/lbm.

10.3 Availability

What is the maximum work that can be done by a system in a given state, assuming heat transfer with the surroundings only? We have just developed an expression to calculate the reversible work for a given change of state of a system. But the question that now arises is, for a given system in a given initial state, what final state will give the greatest amount of work, or the maximum reversible work?

The answer to this question is that when a system is in equilibrium with the environment, no spontaneous change of state will occur, and the system will not be capable of doing any work. Therefore, if a system in a given state undergoes a completely reversible process until it reaches a state in which it is in equilibrium with the environment, the maximum possible amount of work will have been done by the system. This amount of work is called the *availability* of the system.

If a system is in equilibrium with the surroundings, it must certainly be in pressure and temperature equilibrium with the surroundings, that is, at pressure P_o and temperature T_o. It must also be in chemical equilibrium with the surroundings, which implies that no further chemical reaction will take place. Equilibrium with the surroundings also requires that the system have zero velocity and minimun potential energy. Similar requirements could be set forth regarding magnetic, electrical, and surface effects if these are relevant to a given problem.

Any of the relations for reversible work that were derived in the previous section can be used to calculate availability provided that the final state is the state in which the system is in equilibrium with the environment. The subscript "o" will designate this state.

The two cases of particular interest are the steady-flow process and the closed system. Consider first the steady-flow process. Let the fluid in a given state enter the open system with velocity $\overline{V}$ and at an elevation Z. If the maximum reversible work is to be done by this fluid, the fluid must leave the system at the pressure and temperature of the surroundings, at essentially zero velocity, and at the minimum elevation. If for this change of state all processes are reversible, then the work done is called the availability.

Let us denote (in the absence of surface, magnetic, and electrical effects) the availability per unit mass flowing in a steady-flow process by the symbol ψ. Let us denote the state in which the fluid enters the system without subscript, and the state of the fluid leaving the system, which is at the pressure and temperature of the surroundings, by the subscript "o." Applying Eq. 10.8 to this process we have

$$\psi = \left(h - T_o s + \frac{\overline{V}^2}{2g_c} + \frac{g}{g_c} Z\right) - \left(h_o - T_o s_o + \frac{g}{g_c} Z_o\right) \quad (10.12)$$

Two important points should be made regarding this equation. The first is that, for a given pressure and temperature of the surroundings, the availability is a property of a substance, because it is determined by the initial state. Thus, given a steady-flow process, the availability of the fluid entering the system can be considered a thermodynamic property, and so can the availability of the fluid leaving.

The second point is that the reversible work for any change of state is equal to the decrease in availability for the two states. This is evident from the fact that:

$$\psi_i - \psi_e = \left[\left(h_i - T_o s_i + \frac{\overline{V}_i^2}{2g_c} + \frac{g}{g_c} Z_i\right) - \left(h_o - T_o s_o + \frac{g}{g_c} Z_o\right)\right]$$

$$- \left[\left(h_e - T_o s_e + \frac{\overline{V}_e^2}{2g_c} + \frac{g}{g_c} Z_e\right) - \left(h_o - T_o s_o + \frac{g}{g_c} Z_o\right)\right]$$

$$= (h_i - h_e) - T_o(s_i - s_e) + \frac{\overline{V}_i^2 - \overline{V}_e^2}{2g_c} + (Z_i - Z_e)\frac{g}{g_c} = w_{\text{rev}}$$

Thus, we can write for a steady-flow process:

$$w_{\text{rev}} = \psi_1 - \psi_2 \quad (10.13)$$

Example 10.5 Calculate the availability of the air entering and leaving the turbine of Example 10.1, and determine the reversible work of the turbine from the decrease in availability.

This is a steady-flow process, and therefore Eq. 10.12 can be used to calculate availability.

$$\psi_i = (h_i - h_o) - T_o(s_i - s_o) + \frac{\overline{V}_i^2}{2g_c}$$

$$= C_{po}(T_i - T_o) - T_o\left(C_{po} \ln \frac{T_i}{T_o} - R \ln \frac{P_i}{P_o}\right) + \frac{\overline{V}_i^2}{2g_c}$$

$$= 0.240(610 - 537) - 537\left(0.240 \ln \frac{610}{537} - \frac{53.34}{778} \ln \frac{100}{14.7}\right) + \frac{(300)^2}{64.34 \times 778}$$

$$= 17.5 + 54.1 + 1.8 = 73.4 \text{ Btu/lbm}$$

The availability of the air leaving the turbine is found in a similar manner:

$$\psi_e = (h_e - h_o) - T_o(s_e - s_o) + \frac{\overline{V}_e^2}{2g_c}$$

$$= C_{po}(T_e - T_o) - T_o\left(C_{po} \ln \frac{T_e}{T_o} - R \ln \frac{P_e}{P_o}\right) + \frac{\overline{V}_e^2}{2g_c}$$

$$= 0.24(500 - 537) - 537\left(0.24 \ln \frac{500}{537} - \frac{53.34}{778} \ln \frac{20}{14.7}\right) + \frac{(200)^2}{64.34 \times 778}$$

$$= -8.9 + 20.5 + 0.8 = 12.4 \text{ Btu/lbm}$$

The reversible work is equal to the decrease in availability.

$$w_{\text{rev}} = \psi_i - \psi_e$$

$$= 73.4 - 12.4 = 61.0 \text{ Btu/lbm}$$

This agrees with the results of Example 10.1.

It is of interest here to note how one might do an amount of work equal to the availability of the air entering the turbine. One way would be to expand the air through a reversible adiabatic turbine until it reaches atmospheric pressure and a very low velocity. This is process 1–b on the T-s diagram of Fig. 10.5. The temperature leaving the turbine is less than T_o and it is therefore possible to do an additional amount of work by transferring heat from the surroundings to a series of Carnot engines which reject heat to the air, causing it to increase in

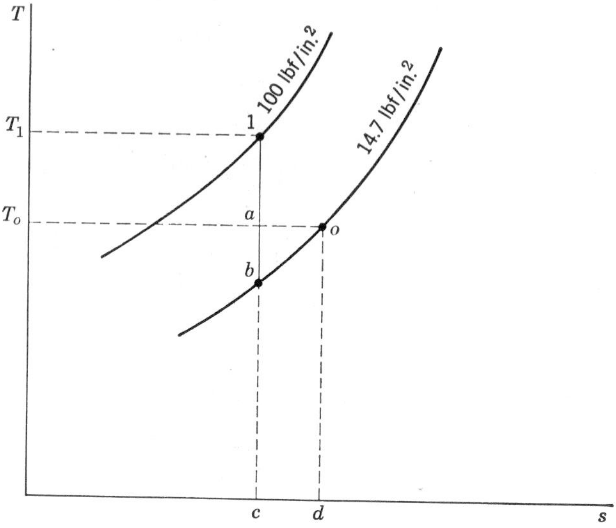

Fig. 10.5 Temperature-entropy diagram for Example 10.5.

temperature from T_b to T_o. (A series of Carnot engines is necessary because the temperature of the air is increasing.) This entire process is shown in the schematic diagram of Fig. 10.6. Area *a–c–d–o–a* on the *T-s* diagram represents the heat transferred to the series of Carnot engines from the atmosphere, and area *b–c–d–o–b* represents the heat

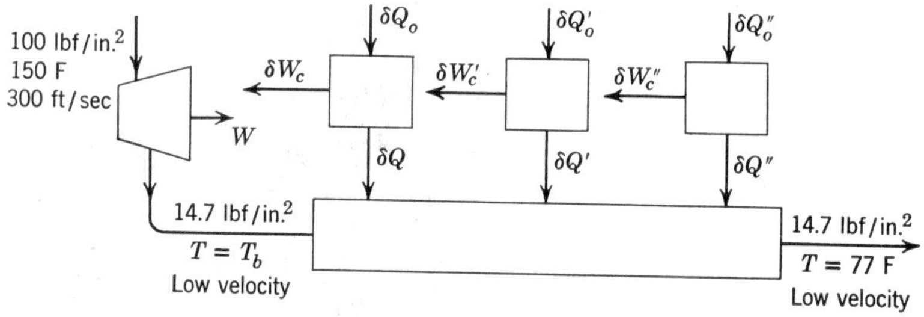

Fig. 10.6 Sketch for Example 10.5.

rejected from the Carnot engines to the air. Therefore, area *a–b–o–a* represents the work done by the series of Carnot engines. It is recommended that the student calculate the availability of the air entering the turbine by calculating the work for the process outlined above.

For a closed system the availability (in the absence of surface, electrical, magnetic effects) per unit mass can be found from Eq. 10.11. Again, the initial state has no subscript and the final state, designated with the subscript o, is the state in which the system is in equilibrium with the surroundings. In this case the availability per unit mass is designated ϕ.

$$\phi = \left(u - T_o s + \frac{\overline{V}^2}{2g_c} + \frac{g}{g_c} Z\right) - \left(u_o - T_o s_o + \frac{g}{g_c} Z_o\right) \tag{10.14}$$

For given conditions of the surroundings the quantity ϕ is a property of a substance.

For a closed system it follows that

$$w_{\text{rev}} = \phi_1 - \phi_2 \tag{10.15}$$

It should be emphasized that the two availability functions, ψ and ϕ, are both properties of a substance (for a given state of the surroundings). However, ψ is significant for steady-flow processes and ϕ is significant for a closed system.

Example 10.6 For the problem given in Example 10.2, calculate the availability in the initial and final state.

This is a closed system, and therefore, we use Eq. 10.14.

$$\begin{aligned}
\phi_1 &= (u_1 - u_o) - T_o(s_1 - s_o)\\
&= C_{v\infty}(T_1 - T_o) - T_o\left(C_{po} \ln \frac{T_1}{T_o} - R \ln \frac{P_1}{P_o}\right)\\
&= 0.171(580 - 537) - 537\left(0.240 \ln \frac{580}{537} - \frac{53.34}{778} \ln \frac{20}{14.7}\right)\\
&= 7.35 + 1.40 = 8.75 \text{ Btu/lbm}\\
\phi_2 &= (u_2 - u_o) - T_o(s_2 - s_o)\\
&= C_{v\infty}(T_2 - T_o) - T_o\left(C_{po} \ln \frac{T_2}{T_o} - R \ln \frac{P_2}{P_o}\right)\\
&= 0.171(660 - 537) - 537\left(0.240 \ln \frac{660}{537} - \frac{53.34}{778} \ln \frac{80}{14.7}\right)\\
&= 21.03 + 35.82 = 56.85 \text{ Btu/lbm}
\end{aligned}$$

The reversible work for this process is equal to the decrease in availability:

$${}_1w_{\text{rev}2} = \phi_1 - \phi_2 = 8.75 - 56.85 = -48.1 \text{ Btu/lbm}$$

This agrees with Example 10.2.

Example 10.7 A steam turbine (see Fig. 10.7) receives 200,000 lbm of steam per hour at 400 lbf/in.^{2}, 600 F. At the point in the turbine where the pressure is 50 lbf/in.^{2}, steam is bled off for use in processing equipment at the rate of 50,000 lbm/hr. The temperature of this bleed steam is 290 F. The balance of the steam leaves the turbine at 2 lbf/in.^{2}, 90 per cent quality. The heat transfer to the surroundings

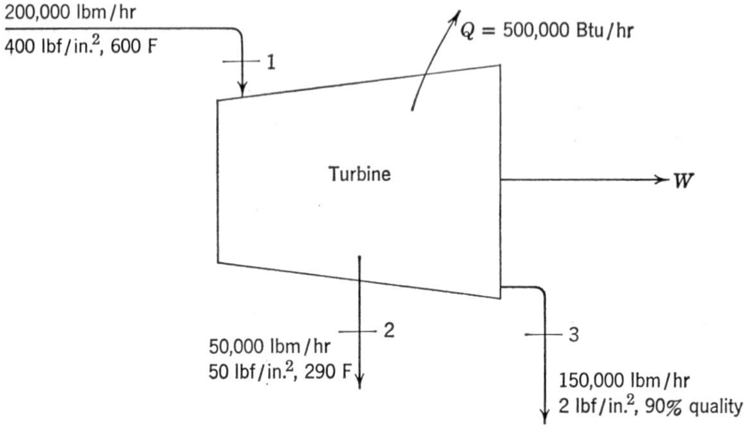

Fig. 10.7 Sketch for Example 10.7.

is 500,000 Btu/hr. Determine the availability per pound of the steam entering and leaving the turbine and the reversible work per pound of steam for the given change of state.

Since the kinetic and potential energies may be assumed to be negligible, the availability of the steam entering and leaving is given by the relation (Eq. 10.12)

$$\psi = (h - h_o) - T_o(s - s_o)$$

At the pressure and temperature of the surroundings, namely, 14.7 lbf/in.^{2}, 77 F, the water is a slightly compressed liquid, and the properties of the water are essentially equal to those for saturated liquid at 77 F.

$$\psi_i = (1306.9 - 45.0) - 537(1.5894 - 0.0876) = 1261.9 - 806.5$$
$$= 455.4 \text{ Btu/lbm}$$

The bleed steam and the exhaust steam must both be considered in determining the availability of the fluid leaving the turbine.

$$\psi_e = 0.25[(h_2 - h_o) - T_o(s_2 - s_o)] + 0.75[(h_3 - h_o) - T_o(s_3 - s_o)]$$

$$= 0.25[(1178.9 - 45.0) - 537(1.6650 - 0.0876)]$$

$$+ 0.75[(1014.0 - 45.0) - 537(1.7455 - 0.0876)]$$

$$= 0.25[286.3] + 0.75[78.7] = 130.6 \text{ Btu/lbm}$$

$$w_{\text{rev}} = \psi_i - \psi_e = 455.4 - 130.6 = 324.8 \text{ Btu/lbm}$$

10.4 Irreversibility

In Section 10.2 the reversible work was considered for a change of state in which the only heat transfer was with the surroundings. This reversible work is the maximum work that can be done during the given change of state (assuming that the only heat transfer is with the surroundings). Since every actual process has irreversibilities associated with it, the actual work W for a given change of state is always less than the corresponding reversible work, W_{rev}.

$$W \leq W_{\text{rev}}$$

This leads to a definition of the *irreversibility* of a given process. The irreversibility I for a given process is defined by the relation

$$\delta I = \delta W_{\text{rev}} - \delta W$$
$$I = W_{\text{rev}} - W \tag{10.16}$$

In words this equation states that the actual work is less than the reversible work by the amount of the irreversibility.

It will be helpful to develop a general expression for the irreversibility of a process. The reversible work is given by Eq. 10.7.

$$\delta W_{\text{rev}} = \delta m_i \left(h_i - T_o s_i + \frac{\overline{V}_i^2}{2g_c} + Z_i \frac{g}{g_c} \right) - \delta m_e \left(h_e - T_o s_e + \frac{\overline{V}_e^2}{2g_c} + Z_e \frac{g}{g_c} \right) - d\left(U - T_o S + \frac{m\overline{V}^2}{2g_c} + mZ \frac{g}{g_c} \right)_\sigma$$

The actual work is given by Eq. 10.1:

$$\delta W = \delta m_i \left(h_i + \frac{\overline{V}_i^2}{2g_c} + Z_i \frac{g}{g_c} \right) - \delta m_e \left(h_e + \frac{\overline{V}_e^2}{2g_c} + Z_e \frac{g}{g_c} \right) - d\left(U + \frac{m\overline{V}^2}{2g_c} + mZ \frac{g}{g_c} \right)_\sigma + \delta Q$$

Therefore,

$$\delta I = \delta W_{\text{rev}} - \delta W$$

$$= \delta m_i(-T_o s_i) - \delta m_e(-T_o s_e) - d(-T_o S)_\sigma - \delta Q \quad (10.17)$$

For a steady-flow process Eq. 10.17 becomes

$$\begin{aligned} \delta I &= T_o(s_e - s_i)\,\delta m - \delta Q \\ i &= T_o(s_e - s_i) - q \end{aligned} \quad (10.18)$$

For a closed system Eq. 10.17 reduces to

$$\begin{aligned} \delta I &= T_o\,(dS)_\sigma - \delta Q \\ {}_1I_2 &= T_o(S_2 - S_1) - {}_1Q_2 \\ {}_1i_2 &= T_o(s_2 - s_1) - {}_1q_2 \end{aligned} \quad (10.19)$$

Example 10.8 Calculate the irreversibility per pound of air that flows through the turbine of Example 10.1.

The reversible work was found to be 61.0 Btu/lbm and the actual work 27.4 Btu/lbm. Therefore,

$$i = w_{\text{rev}} - w$$

$$= 61.0 - 27.4 = 33.6 \text{ Btu/lbm}$$

Example 10.9 Calculate the irreversibility in Btu/hr for Example 10.7.

Let us first do this utilizing Eq. 10.18.

$$\delta I = T_o(\delta m_e s_e - \delta m_i s_i) - \delta Q$$

In this case this integrates to

$$\dot{I} = T_o(\dot{m}_2 s_2 + \dot{m}_3 s_3 - \dot{m}_1 s_1) - \dot{Q}$$

$$= 537[(50{,}000 \times 1.6650) + (150{,}000 \times 1.7455) - (200{,}000 \times 1.5894)] - (-500{,}000)$$

$$= 537 \times 10{,}000(8.3250 + 26.1825 - 31.7880) - (-500{,}000)$$

$$= 15{,}103{,}000 \text{ Btu/hr}$$

This is equivalent to

$$\frac{15{,}103{,}000}{2545} = 5930 \text{ hp}$$

Since W_{rev} was calculated in Example 10.7, we can check this answer by first determining the actual work.

$$w = (h_i - h_e) + q$$

$$= 1306.9 - 0.25(1178.9) - 0.75(1014.0) - 2.5 = 249.2 \text{ Btu/lbm}$$

$$i = w_{rev} - w$$

$$= 324.3 - 249.2 = 75.1 \text{ Btu/lbm}$$

$$\dot{I} = 75.1 \times 200{,}000 = 15{,}020{,}000 \text{ Btu/hr}$$

It is of interest to compare the concept of lost work, as defined by Eq. 7.9 with the concept of irreversibility developed here. Equation 7.9 defined the lost work for a closed system as

$$dS = \frac{\delta Q + \delta LW}{T}$$

$$\delta LW = T\,dS - \delta Q$$

Comparing this equation with Eq. 10.19, we note that the irreversibility of a system is identical to the lost work, except that the T of Eq. 7.9 is replaced by T_o, the temperature of the surroundings, in the expression for availability. An example may help to clarify this matter.

Example 10.10 Air at a temperature of 1000 R and a pressure of 100 lbf/in.^{2} flows across a restriction in a pipe, and the pressure of the air is thus reduced to 20 lbf/in.^{2} The process is adiabatic. Determine the lost work and the irreversibility per unit mass flowing for this process shown schematically in Fig. 10.9*a*.

Since this is a steady-flow process, we use Eq. 7.11 to determine the lost work.

$$dS_\sigma = \frac{\delta Q + \delta LW}{T} + \delta m_i s_i - \delta m_e s_e$$

In this case $dS_\sigma = 0$, $\delta Q = 0$, and $\delta m_i = \delta m_e$. Therefore,

$$\delta LW = T(s_e - s_i)\,\delta m$$

On integrating we have, since T is constant,

$$lw = T(s_e - s_i) = T\left(R \ln \frac{P_i}{P_e}\right)$$

$$= 1000 \times \frac{53.34}{778} \ln \frac{100}{20} = 110.2 \text{ Btu/lbm}$$

The irreversibility can be determined from Eq. 10.18.

$$i = T_o(s_e - s_i) = 537 \times \frac{53.34}{778} \ln \frac{100}{20} = 59.2 \text{ Btu/lbm}$$

The significance of the difference between the lost work and the irreversibility can be best explained by reference to the T-s diagram of Fig. 10.8. Points i and e represent the initial and final states of the air.

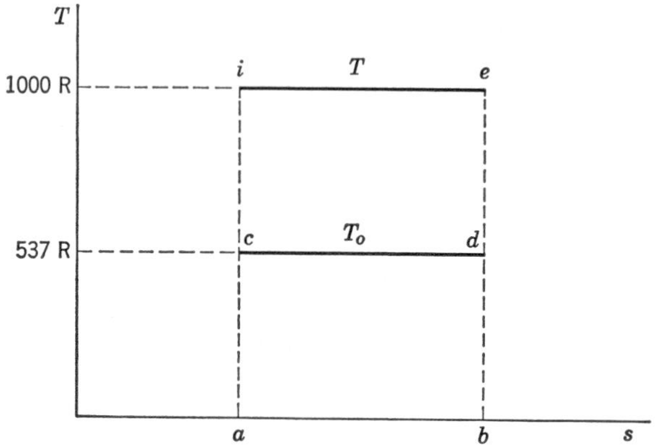

Fig. 10.8 Temperature-entropy diagram for Example 10.10.

If the expansion between these two states had been reversible, the work done would have been 110.2 Btu/lbm. However, since $\Delta h = 0$ for this process, $q = w$ (area a–b–e–i–a) and it would be necessary to transfer 110.2 Btu of heat to the air during this process. Figure 10.9 shows two ways in which the heat transfer could take place. One way would be to have a reservoir at 1000 R; the difference between the work for the ideal case (110.2 Btu/lbm) and the actual case (zero) is equal to the lost work.

The second method involves a heat pump which receives heat from the surroundings and rejects heat at 1000 R. The heat pump requires a work input of 51.0 Btu/lbm (area i–e–c–d–i), and the net work output would be 110.2 − 51.0 = 59.2 Btu/lbm, and the difference between the net work (59.2 Btu/lbm) and the actual work (zero) is equal to the irreversibility.

In summary, the lost-work concept assumes that there is a reservoir available at the temperature T required for the given situation. The concept of irreversibility assumes heat transfer with the surroundings only at temperature T_o.

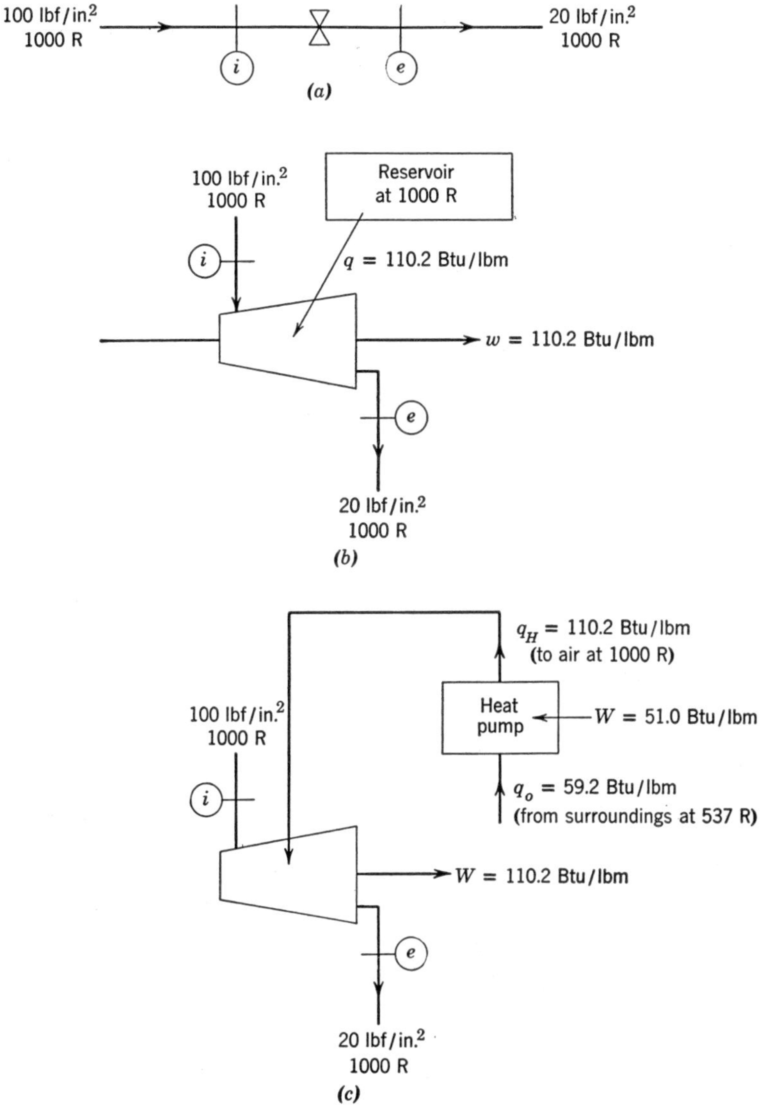

Fig. 10.9 Sketch for Example 10.10.

10.5 Efficiency

In Chapter 6 the concept of thermal efficiency was introduced for an engine that operates in a thermodynamic cycle. In this case the thermal efficiency, η_{th}, was defined by the relation

$$\eta_{th} = \frac{W_{net}}{Q_H} \tag{10.20}$$

where W_{net} is the net work of the cycle and Q_H is the heat transfer from the high-temperature body.

Many times, however, we wish to evaluate the performance of a certain process, rather than a complete cycle. Thus, we might be interested in the efficiency of the turbine in a steam power plant or the efficiency of the compressor in a gas turbine. In this section, therefore, we give some general consideration to efficiency.

In general we can say that efficiency compares the actual performance of a given machine to the performance that would have been achieved in an ideal process. Consider as an example the steam turbine, which is intended to be an adiabatic machine. The only heat transfer is the unavoidable heat transfer to the surroundings. Therefore, the ideal process for the steam turbine is the reversible adiabatic process or the isentropic process. Therefore, if we denote by w_a the actual work done (per unit mass flowing) by a given turbine, and w_s the work that would have been done in a reversible adiabatic process between the same initial state and the same final pressure, the efficiency of the turbine is defined by the relation

$$\eta_{turbine} = \frac{w_a}{w_s} \tag{10.21}$$

The same relation would hold for a gas turbine.

A few other examples may help to clarify this point. In a nozzle the objective is to have the maximum kinetic energy leaving the nozzle for the given inlet conditions. The nozzle is also an adiabatic device and therefore the ideal process is a reversible adiabatic or isentropic process. The efficiency of a nozzle is the ratio of the actual kinetic energy leaving the nozzle, $\overline{V}_a^2/2g_c$, to the kinetic energy for an isentropic process from the same initial conditions, $\overline{V}_s^2/2g_c$.

$$\eta_{nozzle} = \frac{\overline{V}_a^2/2g_c}{\overline{V}_s^2/2g_c} \tag{10.22}$$

In compressors for air or other gases, there are two ideal processes to which the actual performance can be compared. If no effort is made to cool the gas during compression (that is, when the process is adiabatic), the ideal process is a reversible adiabatic or isentropic process. If we denote the work for this isentropic process as w_s, and the actual work w_a (the actual work input will be greater than the work for an isentropic process), the efficiency is defined by the relation

$$\eta_{\text{adiabatic compressor}} = \frac{w_s}{w_a} \tag{10.23}$$

If an attempt is made to cool the air during compression by use of a water jacket or fins, the ideal process is considered a reversible isothermal process. If w_t is the work for the reversible isothermal process and w_a the actual work, the efficiency is defined by the relation

$$\eta_{\text{cooled compressor}} = \frac{w_t}{w_a} \tag{10.24}$$

These examples will suffice to illustrate the way in which the term efficiency is used. Further illustrations will be given when combustion is considered in a later chapter. The point to be made here is that the term efficiency has significance only when its definition for the particular case is clearly understood.

The relationship between the reversible work and efficiency for a given change of state should also be understood. Reversible work always refers to the reversible work between two given states assuming heat transfer with the surroundings only. Efficiency always refers to a comparison of the actual process to some ideal process.

Consider a simple steam turbine as an example. The steam enters the turbine at state 1, and leaves at state 2 (Fig. 10.10). If the steam had undergone a reversible adiabatic process, it would leave the tur-

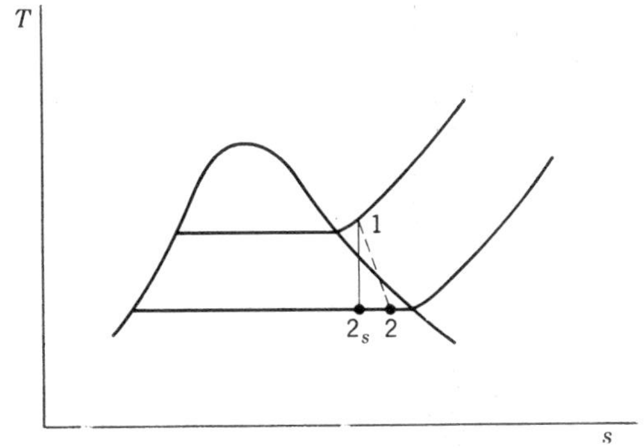

Fig. 10.10 Temperature-entropy diagram showing definition of turbine efficiency.

bine at state 2_s. The efficiency is the ratio of the work for the actual process to the work for the reversible adiabatic process. Note, however, that state 2_s is never reached in the turbine. On the other hand, the reversible work can be calculated for the process involving the actual states achieved, namely states 1 and 2. Note that as the efficiency changes, state 2 changes, and the reversible work also changes. It is advantageous to base efficiency on the reversible adiabatic process, because w_s remains fixed for a given back pressure.

Example 10.11 A steam turbine receives steam at a pressure of 100 lbf/in.^{2}, 500 F and it leaves the turbine at a pressure of 2 lbf/in.^{2} The efficiency of the turbine is 60 per cent, and the process is adiabatic. Determine the actual work of the turbine and the reversible work for the initial and final states of the steam.

$$h_1 = 1279.1 \qquad s_1 = 1.7085$$

$$s_{2s} = s_1 = 1.7085 = 1.9200 - (1 - x)_{2s}1.7451$$

$$(1 - x)_{2s} = \frac{0.2115}{1.7451} = 0.1210$$

$$h_{2s} = 1116.2 - 0.1210(1022.2) = 992.4$$

From the steady-flow energy equation

$$w_s = h_1 - h_{2s} = 1279.1 - 992.4 = 286.7 \text{ Btu/lbm}$$

$$w_a = \eta_{\text{turbine}}(w_s) = 0.6(286.7) = 172.0 \text{ Btu/lbm}$$

The actual enthalpy leaving the turbine, h_2, is found by applying the steady-flow energy equation to the actual turbine.

$$h_1 = h_2 + w_a$$

$$h_2 = 1279.1 - 172.0 = 1107.1$$

The entropy leaving the turbine is found from the steam tables by first finding the final quality.

$$1107.1 = 1116.2 - (1 - x)_2(1022.2)$$

$$(1 - x)_2 = \frac{9.1}{1022.2} = 0.0089$$

$$s_2 = 1.9200 - 0.0089(1.7451) = 1.9045$$

$$\begin{aligned} w_{\text{rev}} &= (h_1 - h_2) - T_o(s_1 - s_2) \\ &= (1279.1 - 1107.1) - 537(1.7085 - 1.9045) \\ &= 172.0 + 105.2 = 277.2 \text{ Btu/lbm} \end{aligned}$$

From this example it is quite evident that it would be impractical to base efficiency on maximum work because the state of the steam leaving the turbine varies as the efficiency varies, whereas for a given inlet condition and given exit pressure the isentropic work remains constant.

10.6 Reheat Factor

The relation between efficiency and availability leads to a consideration of reheat factor, a concept which is important in multistage turbines and turbocompressors.

Consider a turbine having two stages, as represented on the h-s diagram of Fig. 10.11. The fluid enters the turbine at state 1, leaves

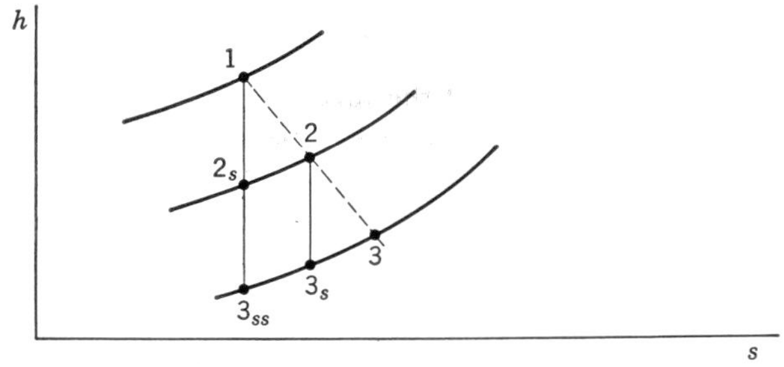

Fig. 10.11 Enthalpy-entropy diagram to show significance of reheat factor.

the first stage and enters the second stage at state 2, and leaves the final stage at state 3. A single isentropic expansion from the initial state to the final pressure is process 1–3_{ss}. However, the isentropic process for the first stage is process 1–2_s and for the second stage is process 2–3_s, and the sum of the isentropic enthalpy drop (i.e., the work) for the two stages is greater than the enthalpy drop of the single isentropic process 1–3_{ss}. The reheat factor $\mathcal{R}$ is defined as the ratio of the sum of the isentropic enthalpy drops of the individual stages to the enthalpy drop in a single isentropic process. In equation form this is written

$$\mathcal{R} = \frac{\Sigma \Delta h_s}{\Delta h_{ss}} \tag{10.25}$$

where Δh_s = isentropic enthalpy drop of the individual stage

Δh_{ss} = isentropic enthalpy drop of a single process from initial state to the given final pressure

The reheat factor has a value greater than unity, usually between 1.0 and 1.1. It is evident from an h-s diagram that the reheat factor is greater than unity, because the constant-pressure lines diverge as the entropy increases. From another point of view, we can say that the reason the reheat factor is greater than unity is that the availability of the state leaving a stage (state 2 for example) is greater than the availability after an isentropic process (state 2_s). The following example illustrates this.

Example 10.12 Steam enters a turbine that has two stages at a pressure of 200 lbf/in.^{2}, 500 F. The efficiency of each stage is 75 per cent (the efficiency of one stage is defined in the same way turbine efficiency is defined, namely, the ratio of actual work to the work of an isentropic process through that stage), and the pressure between stages is 40 lbf/in.^{2} The steam leaves the turbine at a pressure of 2 lbf/in.^{2} The entire process is adiabatic, and kinetic and potential energy changes are not significant. Calculate the availability entering and leaving each stage, the maximum work for each state, the reheat factor, and the over-all efficiency of the turbine.

Table 10.1 is compiled using the designations for the various stages given in Fig. 10.11. In computing the availability one notes that the quantity $(h_o - T_o s_o)$ for 14.7 lbf/in.^{2}, 77 F is given by

$$h_o - T_o s_o = 45.02 - 537(0.0876) = -2.0 \text{ Btu/lbm}$$

The negative sign has no particular significance as it simply depends on the reference state selected for the steam tables. Thus, for any given state

$$\psi = (h - T_o s) - (h_o - T_o s) = h - T_o s + 2.0 \text{ Btu/lbm}$$

TABLE 10.1

State	P, lbf/in.^{2}	h, Btu/lbm	s, Btu/lbm R	$h - T_o s$, Btu/lbm	ψ, Btu/lbm
1	200	1268.9	1.6240	396.8	398.8
2_s	40	1133.2	1.6240	261.1	263.1
2	40	1167.1	1.6727	268.9	270.9
3_{ss}	2	942.9	1.6240	70.8	72.8
3_s	2	971.4	1.6727	73.2	75.2
3	2	1020.3	1.7565	77.1	79.1

$${}_1w_{\max 2} = \psi_1 - \psi_2 = 128.0 \text{ Btu/lbm} \qquad h_1 - h_2 = 101.8 \text{ Btu/lbm}$$

$${}_2w_{\max 3} = \psi_2 - \psi_3 = 191.8 \text{ Btu/lbm} \qquad h_2 - h_3 = 146.8 \text{ Btu/lbm}$$

$$\mathcal{R} = \frac{\Sigma \Delta h_s}{\Delta h_{ss}} = \frac{(1268.9 - 1133.2) + (1167.1 - 971.4)}{1268.9 - 942.9} = 1.018$$

Note in particular that the availability ψ is greater in state 2 than in state 2_s, and also greater in state 3 than in state 3_s.

The turbine efficiency is based on the work of a single isentropic process. That is,

$$\eta_{\text{turb}} = \frac{(h_1 - h_2) + (h_2 - h_3)}{(h_1 - h_{3ss})} = \frac{248.6}{326.0} = 76.3\%$$

Note that the efficiency of the turbine is greater than the efficiency of the individual stages. This is a consequence of the same factor that gives rise to the reheat factor.

10.7 Consideration of Processes That Involve Heat Transfer with a Body Other Than the Atmosphere

Up to this point we have considered processes in which the only heat transfer is with the surroundings. Since many processes do involve heat transfer with bodies at a temperature above or below the temperature of the surroundings, a few remarks should be made regarding them.

The simplest way of handling such processes is to consider two separate systems, one comprising the body from which the heat is transferred, and the other the body to which heat is transferred. Then the reversible work, availability, and irreversibility will be the sum of these quantities for the individual systems. The two examples that follow illustrate this approach. The first involves the irreversible transfer of heat and the second involves reversible heat transfer.

Example 10.13 In a boiler, heat is transferred from the products of combustion to the steam. The temperature of the products of combustion decreases from 2000 F to 1000 F while the pressure remains constant at 1 atm. The average constant-pressure specific heat of the products of combustion is 0.26 Btu/lbm R. The water enters at 100 lbf/in.^{2}, 300 F and leaves at 100 lbf/in.^{2}, 500 F. Determine the reversible work and the irreversibility for this process per pound of water evaporated. In solving this problem we must consider both the products of combustion and the water. Since we are taking 1 lbm of water as our parameter, we must determine the pounds of products per pound of water evaporated.

The heat transferred to the water from the products of combustion is

$$q = h_2 - h_1 = 1279.1 - 269.6 = 1009.5 \text{ Btu/lbm water}$$

The pounds of products required per pound of air is given by

$$q = 1009.5 \text{ Btu/lbm water} = 0.26 \times \frac{\text{lbm products}}{\text{lbm water}}(2000 - 1000)$$

$$\frac{\text{lbm products}}{\text{lbm water}} = \frac{1009.5}{0.26(1000)} = 3.885$$

We first calculate w_{rev} for the change of state of the water.

$$\begin{aligned}(w_{rev})_{water} &= (h_1 - h_2) - T_o(s_1 - s_2)\\ &= (269.6 - 1279.1) - 537(0.4369 - 1.7085)\\ &= -326 \text{ Btu/lbm}\end{aligned}$$

Next we consider the products.

$$(w_{rev})_{products} = \frac{\text{lbm products}}{\text{lbm water}}[(h_3 - h_4) - T_o(s_3 - s_4)]$$

(States 3 and 4 designate the initial and final states of the products.)

$$\begin{aligned}(w_{rev})_{products} &= 3.885[0.26(2000 - 1000) - 537(0.26 \ln \tfrac{2460}{1460})]\\ &= 728 \text{ Btu/lbm water}\end{aligned}$$

$$w_{rev} = 728 - 326 = 402 \text{ Btu/lbm water}$$

$$i = w_{rev} - w = 402 - 0 = 402 \text{ Btu/lbm water}$$

It is also of interest to determine the net change of entropy. The change in the entropy of the water is

$$(s_2 - s_1)_{water} = 1.7085 - 0.4369 = 1.2716 \text{ Btu/lbm water}$$

The change in the entropy of the product is

$$(s_3 - s_4)_{products} = 3.885 \times 0.26 \ln \tfrac{2460}{1460} = 0.524 \text{ Btu/lbm water}$$

Thus, there is a net increase in entropy during the process. The irreversibility could also have been calculated from Eq. 10.18:

$$i = T_o(s_e - s_i) - q$$

Considering a single open system with water and products crossing the boundary, $q = 0$, and therefore,

$$i = 537(1.2716 - 0.524) = 402 \text{ Btu/lbm}$$

These two processes are shown on the T-s diagram of Fig. 10.12. Line 3–4 represents the process for the 3.885 lbm of products. Area

3–4–c–d–3 represents the heat transferred from the 3.885 lbm of products of combustion, and area 3–4–e–f–3 represents the reversible work for the given change in state of these products. Area 1–a–b–2–h–c–1 represents the heat transferred to the water, and this is equal to area 3–4–c–d–3 which represents the heat transferred from the products of combustion. Area 1–a–b–2–g–e–1 represents the reversible work for the given change in state of the water. The difference between area 3–4–e–f–3 (the reversible work for the products) and area 1–a–b–2–g–e–1 (the reversible work for the water) represents the net reversible work. It is readily shown that this net reversible work is equal to area f–g–h–d–f, or $T_o(\Delta s)_{\text{net}}$. Since the actual work is zero, this area also represents the irreversibility, which agrees with our calculation above.

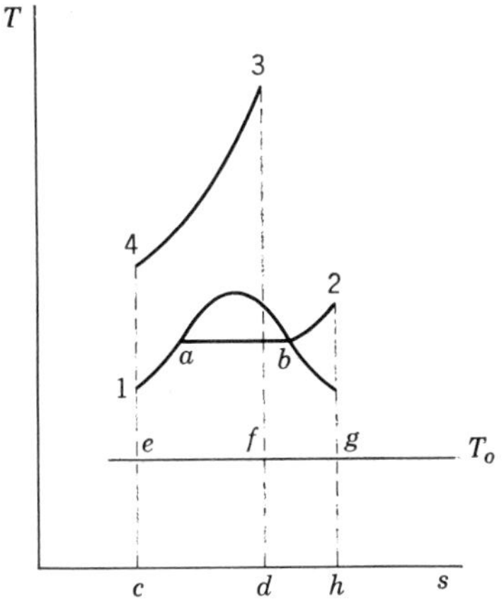

Fig. 10.12 Temperature-entropy diagram for Example 10.13.

It is essential to note that when the change of state involving heat transfer between two systems takes place reversibly, the net change in entropy is zero, and therefore the decrease in the entropy of the body from which heat is transferred must be equal to the increase in entropy of the body to which heat is transferred. This is best demonstrated by considering an example similar to 10.13.

Example 10.14 Repeat Example 10.13, but assume that the heat transfer is entirely reversible, i.e., that the heat transfer takes place through a reversible engine.

Schematically this would involve heat transfer from the products of combustion to reversible engines that reject heat to the water, as shown in Fig. 10.13. Since this is entirely reversible

$$\Delta S_{\text{water}} + \Delta S_{\text{products}} = 0$$

or

$$(s_2 - s_1)_{\text{water}} + \frac{\text{lbm products}}{\text{lbm water}}(s_4 - s_3)_{\text{products}} = 0$$

$$\frac{\text{lbm products}}{\text{lbm water}}(s_3 - s_4)_{\text{products}} = (1.7085 - 0.4369)$$

$$\frac{\text{lbm products}}{\text{lbm water}}\left(0.26 \ln \frac{2440}{1460}\right) = 1.2716$$

$$\frac{\text{lbm products}}{\text{lbm water}} = \frac{1.2716}{0.26 \times 0.521} = 9.39$$

We now calculate the reversible work for the change of state of the water and the products.

$$(w_{\text{rev}})_{\text{water}} = (h_1 - h_2) - T_o(s_1 - s_2)$$

$$(W_{\text{rev}})_{\text{products}} = [(h_3 - h_4) - T_o(s_3 - s_4)] \times \frac{\text{lbm products}}{\text{lbm water}}$$

But, since $(s_2 - s_1)_{\text{water}} = \dfrac{\text{lbm products}}{\text{lbm water}}(s_3 - s_4)_{\text{products}}$

$$(w_{\text{rev}})_{\text{net}} = (h_1 - h_2) + \frac{\text{lbm products}}{\text{lbm water}}(h_3 - h_4)$$

$$= (269.6 - 1279.1) + 9.39 \times 0.26(2000 - 1000)$$

$$= 1431.9 \text{ Btu/lbm water}$$

Since the net change in entropy is zero, the T-s diagram is as shown in Fig. 10.13. Area 3–4–c–d–3 represents the heat transferred from the products to the engines, and area 1–a–b–2–d–c–1 represents the heat received by the water. Area 3–4–1–a–b–2–3 represents the work done by the heat engines, which is equal to the reversible work for this process.

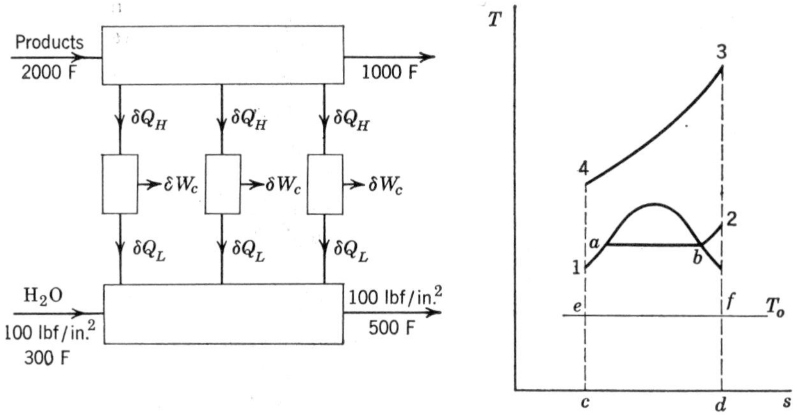

Fig. 10.13 Diagram for Example 10.14.

10.8 Processes Involving Chemical Reactions

Although chemical reactions will not be considered in detail until Chapter 15, some preliminary remarks regarding availability in such processes can be made here. In the first place, we observe that often the reactants are in pressure and temperature equilibrium with the surroundings before the reaction takes place, and the same is true of the products after the reaction. An automobile engine would be an example of such a process if we visualized the products being cooled to atmospheric temperature before being discharged from the engine.

Let us first consider for a closed system the implications of temperature equilibrium with the surroundings during a chemical reaction. The temperature of the system T is equal to T_o, the temperature of the surroundings. Therefore, from Eq. 10.10 we can write (noting that in this case $T_o = T$)

$$\delta W_{\text{rev}} = -d\left(U - TS + \frac{m\overline{V}^2}{2g_c} + m\frac{g}{g_c}Z\right)_\sigma$$

The quantity $(U - TS)$ is a thermodynamic property of a substance and is called the *Helmholtz function*. It is an extensive property and we designate it by the symbol A. Thus,

$$A = U - TS$$
$$a = u - Ts \tag{10.26}$$

Therefore, when a closed system undergoes a change of state while in temperature equilibrium with the surroundings, the reversible work is given by the relation

$$\delta W_{\text{rev}} = -d\left(A + \frac{m\overline{V}^2}{2g_c} + m\frac{g}{g_c}Z\right) \tag{10.27}$$

In most cases the kinetic and potential energy changes are not significant, and in this case

$$\delta W_{\text{rev}} = -dA$$

or

$${}_1W_{\text{rev}\,2} = A_1 - A_2 \tag{10.28}$$

We now consider a closed system that undergoes a chemical reaction while in both pressure and temperature equilibrium with the surroundings. Since the system may undergo a change in volume during the process, work is done by the system on the surroundings if the volume increases during the process, and work is done by the surroundings on the system if the volume decreases. The work done by the system on the surroundings, W_{sur}, is given by the relation (note the pressure of the system P is equal to the pressure of the surroundings P_o)

$$W_{sur} = P_o(V_2 - V_1) = P(V_2 - V_1) = P_2V_2 - P_1V_1 \quad (10.29)$$

It is convenient at this point to define one more term, the maximum useful work, W_{max}. The maximum useful work is the reversible work minus the work done by the system on the surroundings during the process.

$$\begin{aligned} W_{max} &= W_{rev} - W_{sur} \\ &= \left(U_1 - T_1S_1 + \frac{m\overline{V}_1^2}{2g_c} + \frac{mZ_1g}{g_c}\right) \\ &\quad - \left(U_2 - T_2S_2 + \frac{m\overline{V}_2^2}{2g_c} + \frac{mZ_2g}{g_c}\right) - (P_2V_2 - P_1V_1) \\ &= \left(H_1 - T_1S_1 + \frac{m\overline{V}_1^2}{2g_c} + \frac{mZg}{g_c}\right) \\ &\quad - \left(H_2 - T_2S_2 + \frac{m\overline{V}_2^2}{2g_c} + \frac{mZg}{g_c}\right) \end{aligned} \quad (10.30)$$

The quantity $(H - TS)$ is also a thermodynamic property, and is called the *Gibbs function*. It is an extensive property designated by the symbol G. Thus,

$$\begin{aligned} G &= H - TS \\ g &= h - Ts \end{aligned} \quad (10.31)$$

Introducing the Gibbs function Eq. 10.30 can be written

$$\begin{aligned} -W_{max} &= (G_2 - G_1) + \frac{m(\overline{V}_2^2 - \overline{V}_1^2)}{2g_c} + \frac{mg}{g_c}(Z_2 - Z_1) \\ &= \Delta G + \Delta KE + \Delta PE \end{aligned} \quad (10.32)$$

The Gibbs function is also of significance in a steady-flow process that takes place in temperature equilibrium with the surroundings. Since in this case $T = T_o$, Eq. 10.8 reduces to

$$w_{rev} = \left(h_i - T_is_i + \frac{\overline{V}_i^2}{2g_c} + \frac{g}{g_c}Z_i\right) - \left(h_e - T_es_e + \frac{\overline{V}_e^2}{2g_c} + \frac{g}{g_c}Z_e\right)$$

Introducing the Gibbs function we have

$$w_{rev} = \left(g_i + \frac{\overline{V}_i^2}{2g_c} + \frac{g}{g_c}Z_i\right) - \left(g_e + \frac{\overline{V}_e^2}{2g_c} + \frac{g}{g_c}Z_e\right) \quad (10.33)$$

If we define, for the steady-flow process,

$$\Delta G = m(g_e - g_i) \qquad \Delta KE = \frac{m(\overline{V}_e^2 - \overline{V}_i^2)}{2g_c} \qquad \Delta PE = \frac{mg}{g_c}(Z_e - Z_i)$$

Equation 10.33 can be written as follows:

$$-W_{\text{rev}} = \Delta G + \Delta KE + \Delta PE \tag{10.34}$$

Note that this is similar to Eq. 10.32, which applies to a closed system that is in both pressure and temperature equilibrium with the surroundings. For the steady-flow process it was necessary to assume only temperature equilibrium. However, if we wish to make a general statement that applies to both a closed system and a steady-flow process, we can state that for a process that takes place in pressure and temperature equilibrium with the surroundings, and in the absence of surface, magnetic, and electrical effects,

$$-W_{\text{max}} = \Delta G + \Delta KE + \Delta PE \tag{10.35}$$

This equation will be important in Chapter 16 when equilibrium is discussed.

PROBLEMS

Unless otherwise stated assume that the surroundings are at 1 atm pressure, 77 F.

10.1 Consider a steam turbine that has a throttling governor. (That is, the power output of the turbine is controlled by throttling the inlet steam.) The steam in the pipeline flowing to the turbine has a pressure of 400 lbf/in.^{2} and a temperature of 600 F. At a certain load the steam is throttled in an adiabatic process to 300 lbf/in.^{2} Calculate the availability per lbm of steam before and after this process, and the reversible work and irreversibility per lbm of steam for this process. Show the initial and final states of the steam on a T-s diagram.

10.2 Freon-12 enters the expansion valve of a refrigerator at a pressure of 150 lbf/in.^{2} and a temperature of 80 F. It leaves the expansion valve at a temperature of 10 F. Calculate the reversible work and irreversibility for this process. Do you think it would be worthwhile to replace the irreversible throttling process with a reversible process in order to reduce the work required for operating the compressor?

10.3 Air enters the compressor of a gas turbine at 14.0 lbf/in.^{2}, 60 F, with a velocity of 400 ft/sec. The air leaves the compressor at a pressure of 60 lbf/in.^{2}, 400 F, and a velocity of 200 ft/sec. The process is adiabatic. Calculate the reversible work and irreversibility per lbm of air for this process.

10.4 A pressure vessel has a volume of 30 ft^3 and contains air at 200 lbf/in.^{2}, 300 F. The air is cooled to 77 F by heat transfer to the surroundings at 77 F. Calculate the availability in the initial and final states and the irreversibility of this process.

10.5 (*a*) 0.79 mole of nitrogen at 14.7 lbf/in.^{2}, 77 F are separated from 0.21 mole of oxygen at 14.7 lbf/in.^{2}, 77 F by a membrane. The membrane ruptures and the gases mix in an adiabatic process to form a uniform mixture. Determine the reversible work and irreversibility for this process.

(*b*) Determine the minimum work required to separate 1 mole of air (assume composition to be 79% nitrogen and 21% oxygen by volume) at 14.7 lbf/in.^{2}, 77 F into nitrogen and oxygen at 14.7 lbf/in.^{2}, 77 F.

(*c*) How would you evaluate the performance of an air separation plant regarding work input?

10.6 A steam turbine receives 18,000 lbm of steam per hr at 400 lbf/in.^{2}, 600 F and exhausts steam at 2 lbf/in.^{2} The efficiency of the turbine is 78%. Determine the power output of the turbine and the irreversibility per lbm of steam for the actual change of state.

10.7 Air enters the compressor of a gas turbine at 10.0 lbf/in.^{2}, 20 F and leaves the compressor at 35 lbf/in.^{2} The compressor has an efficiency of 82%. Determine the work of compression per lbm of air, the reversible work for the actual change of state, and irreversibility per lbm of air.

10.8 Cooling water is available at 50 F, and ice is to be manufactured from some of this water. The final temperature of the ice is 10 F. The heat rejected from the refrigerator used to manufacture the ice is transferred to a stream of this cooling water, causing the temperature to rise to 80 F. Assuming the entire process to be reversible, determine the work required to produce 1 ton of ice.

10.9 Two identical blocks of metal have a mass of 10 lbm and a specific heat of 0.1 Btu/lbm R. One has an initial temperature of 2000 R and the other an initial temperature of 500 R. The blocks are brought to the same temperature in a reversible process. Determine the final temperature of the blocks and the net work.

10.10 Consider the blocks of metal in Problem 10.9. The blocks are placed in thermal communication and allowed to come to temperature equilibrium. Determine the final temperature of the blocks and the irreversibility for the process. Show the processes of Problems 10.9 and 10.10 on a *T-s* diagram.

10.11 Two blocks of metal having a mass of 10 lbm and a specific heat of 0.1 Btu/lbm F are at a temperature of 100 F. A reversible refrigerator receives heat from one block and rejects heat to the other. Calculate the work required to cause a temperature difference of 200 F between the two blocks.

10.12 A lead storage battery of the type used in an automobile is able to deliver 1440 watt-hours of electrical energy. This energy is available for starting the car.

Suppose we wish to use compressed air for doing an equivilant amount of work in starting the car. The compressed air is to be stored at 1000 lbf/in.^{2}, 77 F. What volume of tank would be required to have the compressed air have an availability of 1440 watt-hours?

10.13 One lbm of water at 14.7 lbf/in.^{2} undergoes a change of state from saturated liquid to saturated vapor in a steady-flow process. Calculate the irreversibility for the following cases.

(*a*) This change of state occurs as a result of heat transfer from a constant-temperature reservoir at 1000 F.

(*b*) This change of state is effected by an electric resistance heater.

10.14 Water is used as the working fluid in a power plant utilizing a nuclear reactor. A schematic diagram for this power plant is shown in Fig. 10.14. The compressed liquid that leaves the reactor enters the

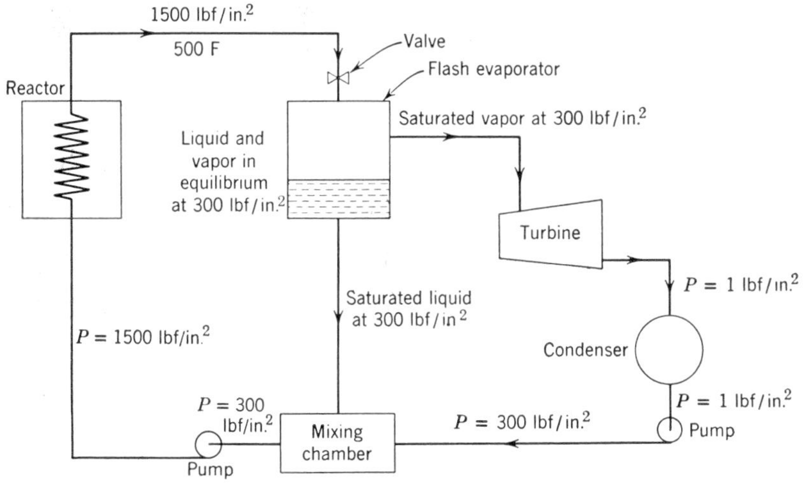

Fig. 10.14 Sketch for Problem 10.14.

flash evaporator, where the pressure is reduced. That fraction of the water which flashes into steam flows to the turbine, while that which remains liquid flows to the mixing chamber. Calculate the following quantities:

(*a*) The work done by the turbine per lbm of water leaving the reactor. Assume a reversible adiabatic turbine.

(*b*) The availability of the water leaving the reactor.

(*c*) The irreversibility in the flash evaporator per lbm of water entering.

(*d*) The irreversibility in the mixing chamber per lbm of water leaving.

10.15 Figure 10.15 shows a closed feedwater heater used in a steam power plant. Determine the irreversibility per lbm of feedwater leaving the heater at 380 lbf/in.^{2}, 200 F.

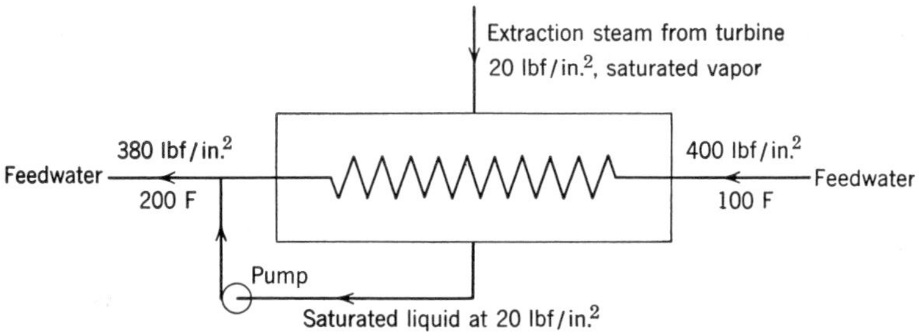

Fig. 10.15 Sketch for Problem 10.15.

10.16 A pressure vessel of 10 ft^3 capacity contains air at 500 lbf/in.^{2}, 77 F. A valve on the vessel is opened and the air escapes as the pressure drops to 200 lbf/in.^{2} Assume no heat transfer during this process and that the gas remaining in the tank has undergone a reversible process. Calculate:

(*a*) The initial availability of the air in the vessel.

(*b*) The availability of the air in the tank immediately after the expansion process.

(*c*) After some time the air that remains in the tank comes to a temperature of 77 F as a result of heat transfer from the surroundings. The volume of the air remains constant during this process. Determine the availability of the air in the tank in this final state.

10.17 An air preheater is used to cool the products of combustion from a furnace while heating the air to be used for combustion. The rate of flow of products is 100,000 lbm/hr, and the products are cooled from 600 F to 400 F, and for the products at this temperature C_p =

0.26 Btu/lbm R. The rate of air flow is 93,000 lbm/hr, the initial air temperature is 100 F, and for the air $C_p = 0.24$ Btu/lbm R.

(*a*) What is the initial and final availability of the products (Btu/hr)?

(*b*) What is the irreversibility for this process?

(*c*) Suppose this heat transfer from the products took place reversibly through heat engines. What would be the final temperature of the air? What power would be developed by the heat engines?

10.18 One mole of carbon dust is burned with 1 mole of oxygen to form 1 mole of carbon dioxide in a steady-flow process. The temperature of the carbon and oxygen before combustion is 77 F and the temperature of the carbon dioxide after combustion is also 77 F. The heat transferred during the process is −169,293 Btu, and the entropy of the carbon dioxide after combustion is 0.713 Btu/lb mole R higher than the entropy of carbon and oxygen before combustion. Calculate the reversible work and the irreversibility for this process.

CHAPTER 11

Reciprocating machines

Many machines that require a thermodynamic analysis are reciprocating machines. Typical ones are reciprocating internal-combustion engines, gas compressors, refrigeration compressors, steam engines and gas-expansion engines. In this chapter we consider a number of terms and definitions that apply to such reciprocating machines. A brief consideration of the compression of gases is included in this chapter.

11.1 Some Definitions

A number of terms are common to all types of reciprocating machines. The cylinder diameter is referred to as the *bore*, and the distance through which the piston moves is the *stroke*. The two extreme positions of the piston are known as *head-end dead center* (also called top dead center) and *crank-end dead center* (also called bottom dead center). The volume occupied by the fluid when the piston is at head-end dead center is known as the *clearance volume*. The volume a piston sweeps through (i.e., the piston area $\times$ the stroke) is called the *piston displacement*. Thus, when the piston is at crank-end dead center the volume of the fluid in the cylinder is equal to the clearance volume plus the piston displacement.

Piston displacement can be expressed either on the basis of piston displacement per stroke or displacement per unit time. The displacement per unit time is the product of the displacement per stroke and the cycles per unit time.

Both the four-stroke cycle and the two-stroke cycle are used in practice in internal-combustion engines. The four-stroke cycle, which is shown schematically in Fig. 11.1, is the one used in the conventional

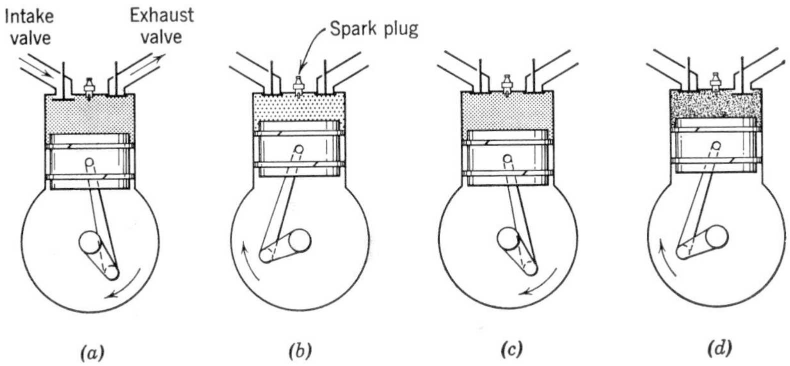

Fig. 11.1 The four-stroke cycle reciprocating internal-combustion engine. (a) Intake. (b) Compression. (c) Power (expansion). (d) Exhaust.

automobile engine, and requires two revolutions of the engine (four strokes) for a complete cycle of events.

In the two-stroke cycle engine a complete cycle occurs with each revolution of the engine, and can be achieved in many ways. One common method, shown schematically in Fig. 11.2, involves crank-

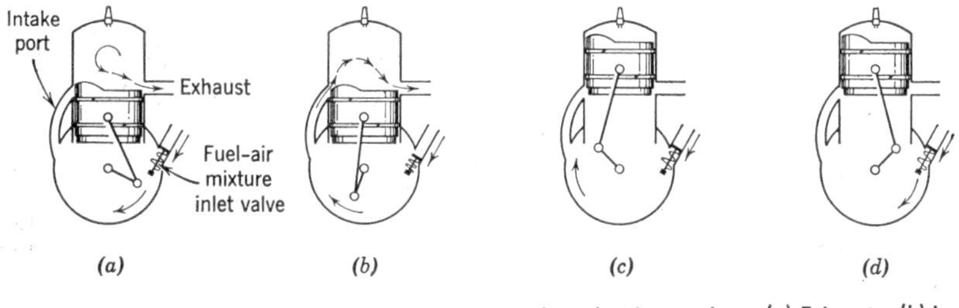

Fig. 11.2 The two-stroke cycle reciprocating internal-combustion engine. (a) Exhaust. (b) Intake. (c) Compression. (d) Expansion.

case compression, in which the piston acts as a valve as it uncovers ports in the cylinder wall.

Compression ratio is an important term in reciprocating internal-combustion engines, and is defined as the ratio of cylinder volume with the piston at crank-end dead center to the cylinder volume with the piston at head-end dead center (i.e., the clearance volume). In reciprocating compressors the clearance volume is usually expressed as percentage of piston displacement. Thus, 4 per cent clearance means that the clearance volume is 4 per cent of the piston displacement.

The difference between a single-acting and a double-acting machine is shown in Fig. 11.3, the difference being whether one or two sides of the piston are utilized. A double-acting machine requires the use of

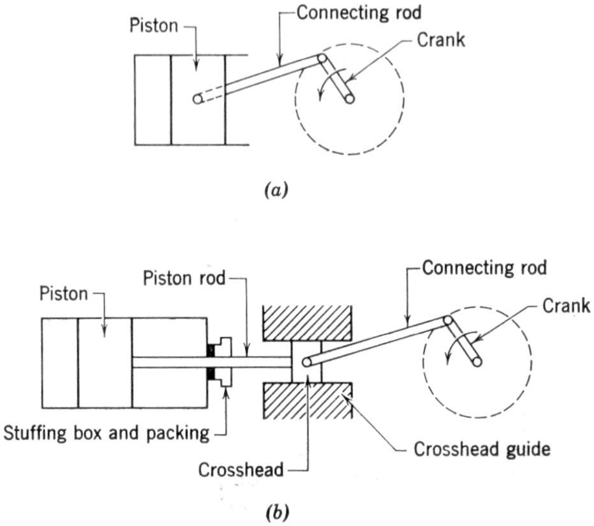

Fig. 11.3 Schematic arrangement of (a) single-acting, and (b) double-acting reciprocating machines.

a stuffing box and crosshead. Some machines, such as expansion engines in air-liquifaction plants, operate only on the rod end. This is to keep the rod in tension so that a smaller diameter rod can be used, thus reducing the heat transfer down the rod.

11.2 The Indicator Diagram and Mechanical Efficiency

A plot of the cylinder pressure vs. cylinder volume of a reciprocating device is a very convenient method of analyzing the performance and determining the actual work done by or on the piston.

Let us consider first the ideal indicator diagram of a reciprocating air compressor. Such a diagram is shown in Fig. 11.4. The piston is at crank-end dead center in position 1. The gas is then compressed in the cylinder until the discharge pressure is reached at state 2. From 2–3 the high-pressure air is forced from the cylinder and through the discharge valve. There is no change in the state of the air during this movement of the piston, but the volume of the air in the cylinder decreases. At point 3 the piston is at head-end dead center and the clearance volume is filled with air at the discharge pressure and temperature. As the piston moves away from head-end dead center the high-pressure air in the clearance volume expands along path 3–4, with a resulting drop in cylinder pressure. At point 4 the pressure in the

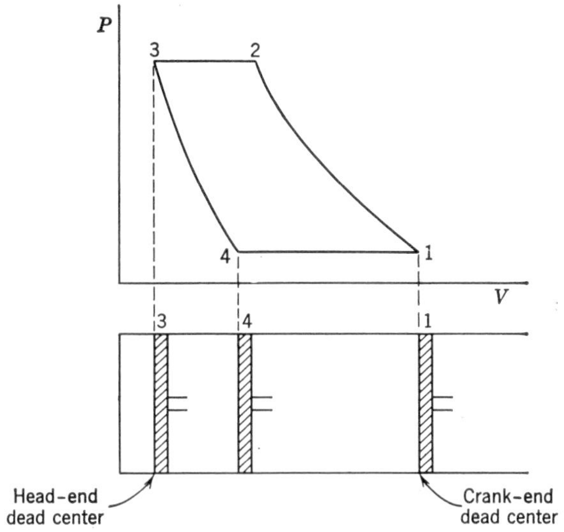

Fig. 11.4 Ideal indicator diagram of a reciprocating air compressor.

cylinder is slightly below the inlet pressure and the inlet valve opens and a new charge of air is drawn into the cylinder. There is no change in the state of the air during the intake stroke, and at point 1 the cylinder is filled with air at the inlet pressure and temperature.

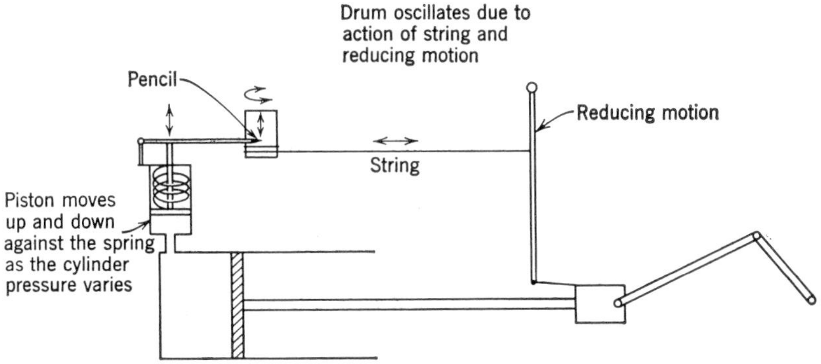

Fig. 11.5 Schematic arrangement of one type of indicator utilized on reciprocating machines.

The actual indicator diagram will differ significantly from this ideal diagram because of such factors as pressure drop through the valves, leakage past the piston and valves, the time required to open the valves, and inertia effects in the inlet and discharge pipes. Therefore, an actual indicator diagram must be made by measuring the actual

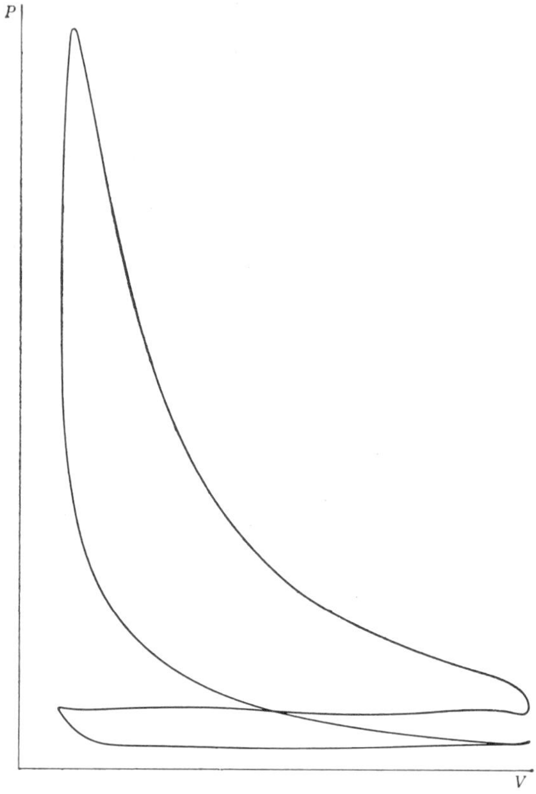

Fig. 11.6 Examples of actual indicator cards of a four-stroke cycle reciprocating internal-combustion engine.

pressure in the cylinder as a function of volume. The instrument used for making these measurements is called an indicator, and is available in many different forms. One of the simplest devices is shown in Fig. 11.5. In this device the cylinder pressure acts on a spring, causing the vertical deflection of the pencil to be proportional to the cylinder pressure. The paper on which the pencil writes is moved in relation to the piston motion, and thus the horizontal position is proportional to cylinder volume. The spring constant of an indicator is defined as the pressure in lbf/in.^{2} required to give a deflection of 1 in. at the pencil.

A typical indicator diagram for a four-stroke cycle internal-combustion engine under road load conditions is shown in Fig. 11.6.

An indicator such as shown in Fig. 11.5 is limited to use in low-speed devices, because at high speeds the inertia of the moving parts of the indicator prevents accurate measurements. Other devices are available which overcome these difficulties.

The work done during one complete cycle (two revolutions of a four-stroke cycle engine) is equal to the net area of the indicator diagram. This is usually found in the following manner. Let the area of the diagram be measured, either by counting squares or by use of a planimeter. Let this area be divided by the length of the indicator diagram, to give a mean ordinate, which when multiplied by the spring constant gives an average pressure in the cylinder. This pressure is called the *mean effective pressure* (abbreviated mep), which is defined as the pressure which, if it acted on the piston during the entire power stroke, would do an amount of work equal to that actually done on the piston. The work for one cycle is found by multiplying this mean effective pressure by the area of the piston (minus the area of the rod on the crank end of a double-acting engine) and by the stroke. This is summarized in Fig. 11.7.

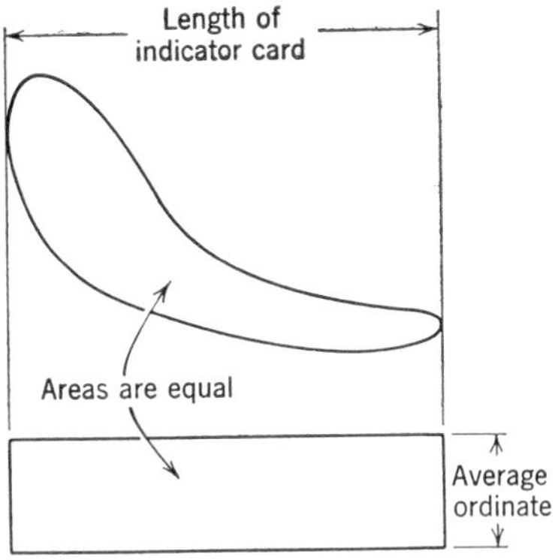

Fig. 11.7 Diagram to show procedure for determining mean effective pressure from an indicator card.

Example 11.1 A single-cylinder air compressor has a 3-in. bore and a 4-in. stroke. An indicator card from this compressor has an area of 0.620 in.^{2} and a length of 2.45 in. The spring constant is 100 lbf/in.^{2} per in. Determine the work per cycle done by the piston on the air.

$$\text{average ordinate} = \frac{\text{area}}{\text{length}} = \frac{0.620}{2.45} = 0.253 \text{ in.}$$

$$\text{mep} = \text{average ordinate} \times \text{spring constant}$$

$$= 0.253 \times 100 = 25.3 \text{ lbf/in.}^2$$

$$\frac{\text{work}}{\text{cycle}} = \text{mep} \times \text{piston area} \times \text{stroke}$$

$$= 25.3 \text{ lbf/in.}^2 \times \frac{(3)^2\pi \text{ in.}^2}{4} \times \frac{4}{12} \text{ ft} = 59.6 \text{ ft-lbf/cycle}$$

Since this is an air compressor the indicator diagram would represent work done by the piston on the air.

Two things should be pointed out regarding the use of indicator cards. First of all, when work is found from the area on the indicator diagram, the assumption involved is that the pressure throughout the cylinder is uniform; that is, that we have a quasistatic process. Most processes in reciprocating machinery occur at such a rate that the assumption of a quasistatic process is justified.

The second point is that the work found from an indicator diagram is the work done on the face of the piston. This is frequently referred to as the indicated work. Due to such losses as friction between the piston and cylinder, and friction in the bearings, the work put into the driving shaft, which is called the brake work, will be greater than indicated work in a compressor; in an engine, the indicated work is greater than the brake work.

This leads to a definition of mechanical efficiency of a reciprocating machine.

For a compressor or pump:

$$\eta_{\text{mech}} = \frac{\text{indicated work}}{\text{brake work}} \tag{11.1}$$

For an engine:

$$\eta_{\text{mech}} = \frac{\text{brake work}}{\text{indicated work}} \tag{11.2}$$

Thus, the mechanical efficiency gives an indication of the losses between the piston and the driving shaft.

A final point should be made regarding mean effective pressure. Mean effective pressure can be based on either the brake work or indicated work, and the terms brake mean effective pressure (bmep) and indicated mean effective pressure (imep) are used. The product of the brake mean effective pressure, the piston area, and the stroke gives the brake work, and similarly the product of the indicated mean effective pressure, the piston area, and the stroke gives the indicated work.

11.3 Volumetric Efficiency

The *volumetric efficiency* of a reciprocating machine involving a compression process is a measure of the effectiveness of the machine regarding its gas handling capacity, and is defined by the relation

$$\eta_{\text{vol}} = \frac{\begin{array}{c}\text{volume of the gas actually compressed and delivered}\\ \text{as measured at the inlet pressure and temperature}\end{array}}{\text{piston displacement}} \tag{11.3}$$

An alternate definition can be given in terms of the mass of gas delivered.

$$\eta_{\text{vol}} = \frac{\text{mass of gas actually compressed and delivered}}{\text{mass of gas occupying the piston displacement at inlet pressure and temperature}} \tag{11.4}$$

The significance of volumetric efficiency may be better understood by considering some reasons why it is usually less than 100 per cent. Consider first the re-expansion of the clearance gas. This is process 4–1 on the ideal indicator diagram of Fig. 11.8. Note that as the piston

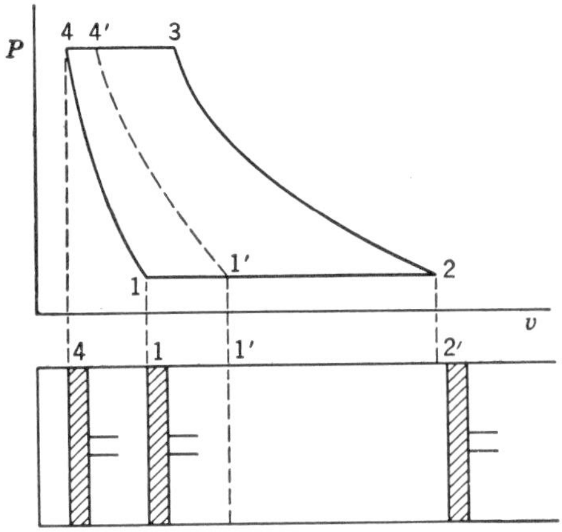

Fig. 11.8 Diagram showing the influence of clearance on volumetric efficiency.

moves from 4 to 1 on the suction stroke, the pressure is dropping from the discharge pressure at 4 to the inlet pressure at 1. Thus, gas is drawn into the cylinder only between points 1 and 2, and the volume of the gas drawn in is less than the volume displaced by the piston. Note also that increasing the clearance volume from 4 to 4′ causes the volumetric efficiency to decrease. In air compressors therefore the clearance volume is kept small.

Another factor influencing volumetric efficiency is the pressure drop in the inlet passages and across the inlet valve. This pressure drop means that the gas in the cylinder has a higher specific volume, and therefore the mass of gas in the cylinder when the piston is at crank-end dead center is decreased as a result of the pressure drop, thus causing a decrease in volumetric efficiency.

Similarly, heat transfer to the incoming gas influences volumetric efficiency. If the gas is heated up in the inlet passage and in the cylinder during the suction stroke, its specific volume is increased and the volumetric efficiency decreases. Conversely, the volumetric efficiency would be increased by cooling during the suction stroke, though only rarely is this possible or feasible.

Finally, any leakage through the valves or past the piston causes a decrease in volumetric efficiency, since this decreases the mass of gas delivered.

In compressors a lower volumetric efficiency means that a larger machine is required for a given gas handling capacity. In a reciprocating internal-combustion engine a lower volumetric efficiency means that less air is present in the cylinder and therefore less fuel can be burned, and a lower output results.

Example 11.2 The volumetric efficiency of a reciprocating refrigeration compressor is to be determined experimentally. The refrigerant is Freon-12, and the inlet conditions are 20 lbf/in.^{2}, 80 F. The discharge pressure is 175 lbf/in.^{2} The compressor has a single cylinder; the bore is 1.25 in.; the stroke is 1.50 in.; the rpm is 1740. The mass rate of flow of the refrigerant is experimentally determined to be 35.2 lbm/hr. What is the volumetric efficiency of the compressor?

The piston displacement is

$$\text{P.D.} = \frac{\pi(1.25)^2}{4} \times 1.5 \times \frac{1740}{1728} = 1.854 \text{ ft}^3/\text{min}$$

At 20 lbf/in.^{2}, 80 F,

$$v = 2.329 \text{ ft}^3/\text{lbm}$$

Therefore, if the piston displacement had been filled with Freon-12 at the inlet conditions, the mass of refrigerant pumped per hour would have been

$$\text{Mass of refrigerant/hr} = \frac{1.854}{2.329} \times 60 = 47.7 \text{ lbm/hr}$$

Since the actual mass of refrigerant pumped is 35.2 lbm/hr, the volumetric efficiency is

$$\eta_{\text{vol}} = \frac{35.2}{47.7} = 73.7\%$$

11.4 Work of Compression for Reciprocating Compressors

This section and the following one deal specifically with compressors, and involve a consideration of the work of compression.

The first point is that for the ideal reciprocating compressor (reversible process, no leakage, etc.) the work of compression is not influenced by clearance volume. The reason is that the work done by the clearance gas on the piston during the re-expansion is exactly equal to the work done in compressing the clearance gas originally. Thus, in the ideal case the only influence of volumetric efficiency is on the piston displacement required for a given amount of compressed gas. In the actual compressor however, a larger piston displacement means more frictional losses, and thus, if the clearance volume is increased, more work is required to compress a given amount of gas. If the volumetric efficiency of an actual compressor decreases due to leakage, the amount of work increases.

The second consideration involves the fact that by transferring heat from the gas during compression, the amount of work can be reduced. This is shown qualitatively in Fig. 11.9, which shows a P-v diagram of

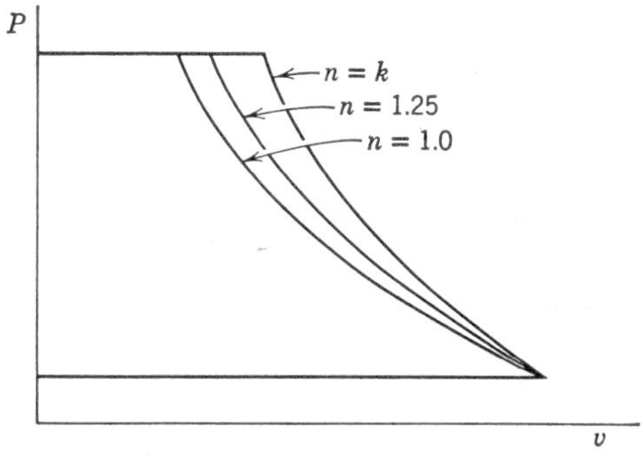

Fig. 11.9 A P-v diagram showing the influence of heat transfer on the work required for steady-flow compression.

an ideal air compressor with zero clearance and with negligible changes in kinetic and potential energy. Three compression paths are shown, namely, $n = k$ (reversible adiabatic), $n = 1.25$ (which involves some heat transfer during compression), and $n = 1$ (the isothermal case which involves considerable heat transfer during compression). Since the work of compression under these conditions is equal to the net area $(-\int v\, dP)$, it is evident that the work of compression can be decreased by transferring heat from the gas during the compression process. This applies to a rotary type of compressor as well as to a reciprocating compressor. However, it is very difficult to transfer heat from the gas during compression in a rotary compressor. In a reciprocating compressor the heat transfer is accomplished by use of a water jacket or by providing fins on the cylinder and using air cooling. From actual indicator cards on compressors it is found that the polytropic exponent during the actual compression stroke may be as low as $n = 1.20$ on a well-cooled, slow-speed compressor.

The final point to be made involves the effect of heat transfer to the gas during the intake stroke. This also may be demonstrated qualitatively by considering the P-v diagram, Fig. 11.10 for a compressor with zero clearance. Assume that the compressor takes in 1 lbm of air per

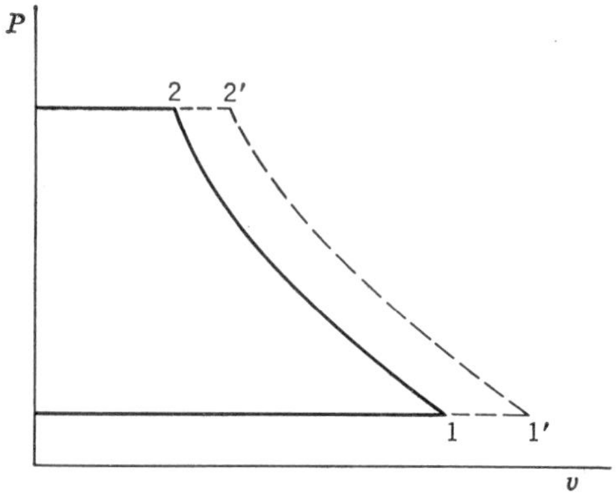

Fig. 11.10 A P-v diagram showing the influence of heat transfer during the inlet stroke on work required for steady-flow compression.

stroke. If the air is heated during the intake stroke the specific volume changes from 1 to 1′, and a larger piston displacement is necessary. Note that if this takes place the compression work per pound of air is greater. This follows from the expression for the net work of compression for an ideal gas in a steady-flow reversible polytropic process with negligible changes in kinetic and potential energy.

$$-w = \int_1^2 v\,dP = \frac{n}{n-1}(P_2v_2 - P_1v_1) = \frac{nP_1v_1}{n-1}\left[\left(\frac{P_2}{P_1}\right)^{(n-1)/n} - 1\right]$$

Note that the work is proportional to v_1, the initial specific volume. Thus, we conclude that it is desirable to avoid heat transfer to the gas during the inlet stroke. It also follows that it is desirable to have the inlet of a compressor so located that cool gas is drawn into the compressor. This of course applies to both rotary and reciprocating compressors.

11.5 Multistage Compression

When gases are compressed to reasonably high pressures, it is advantageous to use multistage compression with intercooling between the stages. This applies to both reciprocating and rotary compressors, although there are certain aspects of this matter that apply particularly to reciprocating compressors. A schematic arrangement for a two-stage reciprocating compressor with a water-cooled intercooler is shown in Fig. 11.11.

There are at least three gains to be achieved by multistage compression in a reciprocating compressor. First of all, if the gas is compressed

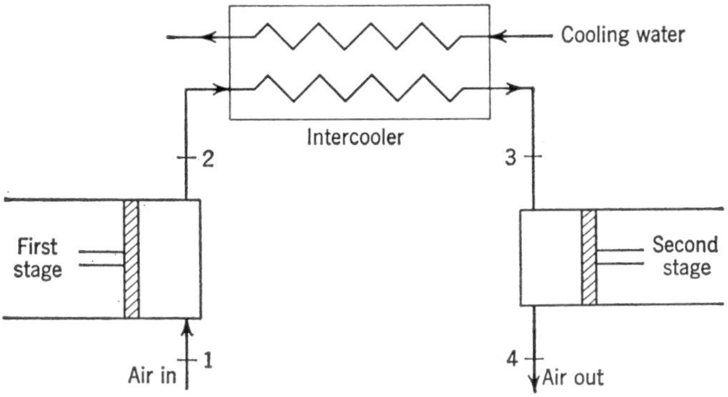

Fig. 11.11 Schematic arrangement for multistage compression with intercooling.

to a very high pressure in a single stage, the gas in the clearance volume expands to a large volume during the inlet stroke, and a very low volumetric efficiency results, necessitating a large piston displacement. By using several stages and cooling the gas between stages a high volumetric efficiency can be achieved in each stage.

Second, if a gas is compressed to a high pressure in a single stage, the final temperature of the gas is high. This leads to lubrication problems, and also means a high average cylinder wall temperature. Under these conditions a significant heat transfer to the incoming gas would take place during the suction stroke, with resulting increase in work and decrease in volumetric efficiency.

The third point is that a saving in work can be achieved by using multistage compression with intercooling to achieve a given discharge pressure. This applies to both reciprocating and rotating compressors, and is shown schematically for the ideal case on the *P-v* and *h-s* diagrams of Fig. 11.12.

The ideal case assumes reversible compression, no clearance, no pressure drop in the intercooler, and that the temperature leaving the intercooler is equal to the initial temperature. If the compression had taken place in a single stage, the compression line would be 1–x, whereas by intercooling the path followed is 1–2–3–4. Therefore, it is evident that area 2–3–4–x–2 represents the work saved by using two stages with intercooling. This saving is evident on the *h-s* diagram because the distance 2–x (which represents the work of compression between the interstage pressure and final pressure without intercooling) is greater than the distance 3–4 (which represents the second-stage work with intercooling), due to the curvature of the constant-pressure lines. In effect, intercooling is a means of approaching an isothermal process more closely than is possible in a single-stage compressor. A significant reduction in the work can be effected by multistage compression with intercooling in actual compressors, even though they deviate significantly from the ideal case of Fig. 11.12.

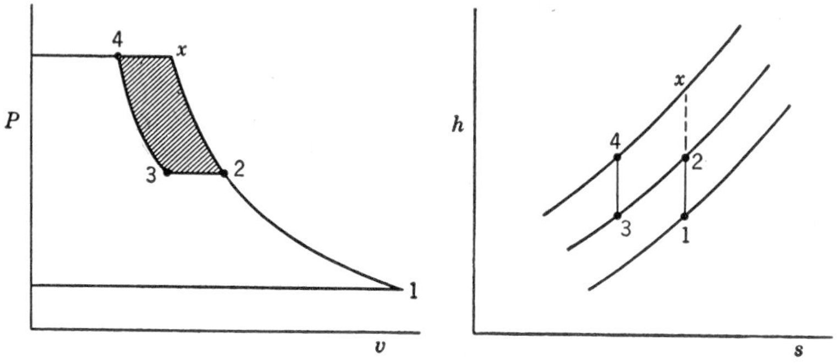

Fig. 11.12 Demonstration of the work saved by use of intercooling.

What interstage pressure will give the minimum total work of compression? This optimum interstage pressure may have to be determined by trial and error for substances that deviate significantly from an ideal gas, but for an ideal gas and an ideal two-stage compressor as defined in the last paragraph, a simple relationship can be found. Using the symbols of Fig. 11.12, the work per pound of gas for the first stage is

$$-w_{\text{1st stage}} = \int_{P_1}^{P_2} v\,dP = \frac{n}{n-1}(P_2v_2 - P_1v_1) = \frac{nRT_1}{n-1}\left(\frac{T_2}{T_1} - 1\right)$$

$$= \frac{nRT_1}{n-1}\left[\left(\frac{P_2}{P_1}\right)^{(n-1)/n} - 1\right]$$

Similarly, for the second stage

$$-w_{\text{2nd stage}} = \frac{nRT_1}{n-1}\left[\left(\frac{P_4}{P_2}\right)^{(n-1)/n} - 1\right]$$

$$-w_{\text{total}} = \frac{nRT_1}{n-1}\left[\left(\frac{P_2}{P_1}\right)^{(n-1)/n} + \left(\frac{P_2}{P_4}\right)^{(1-n)/n} - 2\right]$$

If we differentiate the work with respect to P_2, and set the derivative equal to zero, we can determine the interstage pressure for minimum work.

$$\frac{\delta w}{dP_2} = \frac{nRT_1}{n-1}\left[\left(\frac{n-1}{n}\right) \times \left(\frac{1}{P_1}\right)^{(n-1)/n} \times P_2^{-1/n} + \left(\frac{1-n}{n}\right) \times \left(\frac{1}{P_4}\right)^{(1-n)/n} \times P_2^{(1-2n)/n}\right] = 0$$

Therefore

$$\frac{P_2}{P_1} = \frac{P_4}{P_2} \qquad \text{and} \qquad P_2 = \sqrt{P_1 P_4} \tag{11.5}$$

That is, the work will be minimum when the pressure ratio across each stage is equal. This also applies with good accuracy to actual compressors and it also applies to more than two stages. Thus, in a compressor of three stages the first interstage pressure would be $P_{\text{initial}}\,(P_{\text{final}}/P_{\text{initial}})^{1/3}$ and the second interstage pressure would be $P_{\text{initial}}\,(P_{\text{final}}/P_{\text{initial}})^{2/3}$.

From a practical point of view the pressure ratio across each stage of a reciprocating compressor is usually maintained between 3 and 7, because the saving in work by having more stages is offset by the additional cost of the equipment.

Example 11.3 Consider the compression of air from 14.7 lbf/in.^{2}, 80 F to 500 lbf/in.^{2} in an ideal two-stage compressor with intercooling. Assume that the temperature of the air leaving the intercooler is 80 F, and that the optimum interstage pressure is used. The compressor is water jacketed and the polytropic exponent n is 1.30 for both stages. Determine the work of compression per pound of air, and compare this with the work of compression for a single-stage compressor with $n = 1.30$.

Using the notation of Fig. 11.12, the two-stage compressor will be considered first.

The interstage pressure for minimum work is

$$P_2 = \sqrt{P_1 P_4} = \sqrt{14.7 \times 500} = 85.5 \text{ lbf/in.}^2$$

The temperature leaving the first stage is

$$\frac{T_2}{T_1} = \left(\frac{P_2}{P_1}\right)^{(n-1)/n} = \left(\frac{85.5}{14.7}\right)^{(1.3-1)/1.3} = 1.501$$

$$T_2 = 540(1.501) = 810 \text{ R}$$

$$-w_{\text{1st stage}} = \frac{nR}{n-1}(T_2 - T_1)$$

$$= \frac{1.3 \times 53.3}{0.3 \times 778}(810 - 540) = 80.1 \text{ Btu/lbm}$$

It is evident from Eq. 11.5 that the work of the second stage is equal to the work of the first stage. Thus, the total work of compression is

$$-w_{\text{total}} = 160.2 \text{ Btu/lbm}$$

If this had been done in a single stage the final temperature, T_f, would have been

$$\frac{T_f}{T_i} = \left(\frac{P_f}{P_i}\right)^{(n-1)/n} = \left(\frac{500}{14.7}\right)^{(1.3-1)/1.3} = 2.26$$

$$T_f = 540(2.26) = 1220 \text{ R}$$

$$-w = \frac{nR}{n-1}(T_f - T_i) = \frac{1.3 \times 53.3}{0.3 \times 778}(1220 - 540)$$

$$= 202 \text{ Btu/lbm}$$

PROBLEMS

11.1 A four-stroke cycle V-8 internal combustion engine has a bore of 4.125 in. and a stroke of 3.40 in. The engine delivers a maximum power of 250 hp at 4400 rpm. The compression ratio is 9.5 to 1.

(*a*) What is the cylinder volume of one cylinder when the piston is at crank-end and at head-end dead center?

(*b*) What is the bmep of this engine?

11.2 An indicator diagram of a reciprocating air-expansion engine used in an air-liquifaction plant has an area of 2.73 in.^{2} and a length of 2.40 in. The spring scale of the indicator is 100 (lbf/in.^{2})/in. The expansion engine has a bore of 3.75 in. and a stroke of 4.25 in. and operates at 320 rpm. Determine the indicated power output of the expansion engine.

11.3 A reciprocating engine has a 3-in. bore and a 4-in. stroke and operates at 800 rpm. The engine is double acting, and the diameter of the rod is 0.75 in. The area of the head-end indicator card is 1.45 in.^{2} and the area of the crank-end card is 1.32 in.^{2} The length of each card is 1.80 in., and the spring scale is 120 (lbf/in.^{2})/in. Determine the power output of the engine. Assume a two-stroke cycle.

11.4 A refrigeration compressor must be designed to handle 125 lbm of Freon-12 per hr. The inlet conditions to the compressor are 25 lbf/in.^{2}, 60 F. The discharge pressure is 80 lbf/in.^{2} From previous experience it is estimated that a volumetric efficiency of 75% can be achieved. It is planned to make the stroke of the compressor equal to the bore. A four-pole induction motor will be used to drive the compressor. What bore should the compressor have?

11.5 If the engine of Problem 11.1 has a volumetric efficiency of 70% at maximum power, what is the air handling capacity of the engine in lbm/min? Atmospheric conditions are 60 F, 14.7 lbf/in.^{2}

11.6 Consider an air compressor that receives air at 14.7 lbf/in.^{2}, 60 F and delivers it at 100 lbf/in.^{2} The air compressor delivers 100 lbm/min of air. The polytropic exponent for both compression and expansion processes is $n = 1.32$.

(*a*) Determine the hp and piston displacement required, assuming an ideal compressor with no clearance.

(*b*) Repeat (*a*) for a clearance of 8%.

(*c*) Repeat (*a*) assuming that the pressure drops to 14.0 lbf/in.^{2} through the inlet valve, and that the pressure drop across the discharge valve is 2 lbf/in.^{2}

11.7 Consider an air compressor that receives air at 14.7 lbf/in.^{2}, 60 F and delivers air at 100 lbf/in.^{2} The compressor handles 600 ft^3/min of air (at inlet conditions). Determine the hp required for the following conditions, assuming reversible processes:

(*a*) Isothermal compression.

(*b*) $n = 1.30$

(*c*) Adiabatic compression.

11.8 A compressor is available with a 3-in. bore and a 4-in. stroke. The compressor operates at 800 rpm. Normally this compressor handles air, and it receives the air at 14.7 lbf/in.^{2}, 80 F, and delivers it at 60 lbf/in.^{2} Assume the compression process to be reversible and adiabatic, and that the volumetric efficiency is 80%.

(*a*) What hp motor would be required to drive the compressor under the conditions outlined above?

(*b*) Suppose this compressor were used to compress helium at the same conditions. What hp motor should be used?

11.9 Determine the work of compression for the compressor of Example 11.3, assuming an ideal three-stage compressor with intercooling.

How many stages would you recommend for this compressor?

11.10 A two-stage compressor is to be designed which receives air at 14.2 lbf/in.^{2}, 80 F and delivers air at 150 lbf/in.^{2} From previous experience it is believed that a volumetric efficiency of 80% can be achieved in each stage. Water jacketing will be used and it is estimated that the temperature leaving the intercooler will be 100 F. The capacity of the compressor is to be 200 ft^3/min of free air ("free air" is air measured at the ambient conditions).

(*a*) What will be the piston displacement of each stage?

(*b*) What hp motor would you recommend to drive this compressor?

CHAPTER 12

Power and Refrigeration Cycles

Some power plants, such as the simple steam power plant which we have considered several times, operate in a cycle. That is, the working fluid undergoes a series of processes and finally returns to the initial state. In other power plants, such as the internal-combustion engine and gas turbine, the working fluid does not go through a cycle, even though the engine itself may operate in a mechanical cycle. In this case the working fluid has a different composition or is in a different state at the conclusion of the process than at the beginning. Such equipment is sometimes said to operate on the open cycle (the word cycle is really a misnomer), whereas the steam power plant operates on a closed cycle. The same distinction between open and closed cycles can be made regarding refrigeration devices. For both the open- and closed-cycle type of apparatus, however, it is advantageous to analyze the performance of an idealized closed cycle similar to the actual cycle. Such a procedure is particularly advantageous in determining the influence of various variables on performance. For example, the spark-ignition internal-combustion engine is usually approximated by the Otto cycle. From an analysis of the Otto cycle one concludes that increasing the compression ratio increases the efficiency. This is also true for the actual engine, even though the Otto-cycle efficiencies may deviate significantly from the actual efficiencies.

This chapter is concerned with these idealized cycles, both for power and refrigeration apparatus. The working fluids considered are both vapors and ideal gases. An attempt will be made to point out how the processes in actual apparatus deviate from the ideal. The order in which the cycles will be considered is:

Vapor power cycles.
Vapor refrigeration cycles.
Air-standard power cycles.
Air-standard refrigeration cycles.

VAPOR POWER CYCLES

12.1 The Rankine Cycle

The ideal cycle for a simple steam power plant is the Rankine cycle, shown in Fig. 12.1.

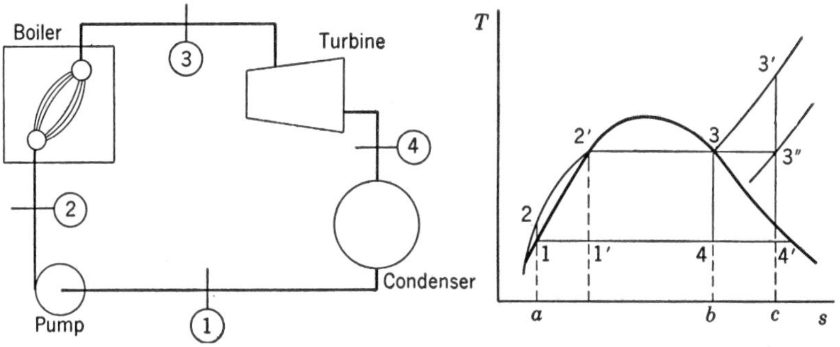

Fig. 12.1 Simple steam power plant which operates on the Rankine cycle.

The processes that comprise the cycle are:

1–2: Reversible adiabatic pumping process in the pump.
2–3: Constant-pressure transfer of heat in the boiler.
3–4: Reversible adiabatic expansion in the turbine (or other prime mover such as a steam engine).
4–1: Constant-pressure transfer of heat in the condenser.

The Rankine cycle also includes the possibility of superheating the vapor, as cycle 1–2–3′–4′–1.

If changes of kinetic and potential energy are neglected, heat transfer and work may be represented by various areas on the T-s diagram. The heat transferred to the working fluid is represented by area a–2–2′–3–b–a, and the heat transferred from the working fluid by area

a–1–4–b–a. From the first law we conclude that the area representing the work is the difference between these two areas, namely, area 1–2–2′–3–4–1. The thermal efficiency is defined by the relation

$$\eta_{th} = \frac{w_{net}}{q_H} = \frac{\text{area } 1\text{–}2\text{–}2'\text{–}3\text{–}4\text{–}1}{\text{area } a\text{–}2\text{–}2'\text{–}3\text{–}b\text{–}a} \tag{12.1}$$

In analyzing the Rankine cycle it is helpful to think of efficiency as depending on the average temperature at which heat is supplied and the average temperature at which heat is rejected. Any changes that increase the average temperature at which heat is supplied or decrease the average temperature at which heat is rejected will increase the Rankine cycle efficiency.

It should be stated that in analyzing the ideal cycles in this chapter the changes in kinetic and potential energies from one point in the cycle to another are neglected. In general this is a reasonable assumption for the actual cycles.

It is readily evident that the Rankine cycle has a lower efficiency than a Carnot cycle with the same maximum and minimum temperatures as a Rankine cycle, because the average temperature between 2 and 2′ is less than the temperature during evaporation. The question might well be asked, why choose the Rankine cycle as the ideal cycle? Why not rather select the Carnot cycle 1′–2′–3–4–1′? At least two reasons can be given. The first involves the pumping process. State 1′ is a mixture of liquid and vapor, and great difficulties are encountered in building a pump that will handle the mixture of liquid and vapor at 1′ and deliver saturated liquid at 2′. It is much easier to completely condense the vapor and handle only liquid in the pump, and the Rankine cycle is based on this fact. The second reason involves superheating the vapor. In the Rankine cycle the vapor is superheated at constant pressure, process 3–3′. In the Carnot cycle all the heat transfer is at constant temperature, and therefore the vapor is superheated in process 3–3″. Note, however, that during this process the pressure is dropping, which means that the heat must be transferred to the vapor as it undergoes an expansion process in which work is done. This also is very difficult to achieve in practice. Thus, the Rankine cycle is the ideal cycle that can be approximated in practice. In the sections that follow we will consider some variations on the Rankine cycle that enable one to more closely approach the Carnot-cycle efficiency.

Before discussing the influence of various variables on the performance of the Rankine cycle, an example is given.

Example 12.1 Determine the efficiency of a Rankine cycle utilizing steam as the working fluid in which the condenser pressure is 1 lbf/in.[2] The boiler pressure is 300 lbf/in.[2] The steam leaves the boiler as saturated vapor.

In solving Rankine-cycle problems we will let w_p denote the work into the pump per pound of fluid flowing and q_L the heat rejected from the working fluid per pound of fluid flowing.

Pump work (see Section 7.9):

$$w_p = h_2 - h_1 = \int v\, dP$$

Assuming the fluid to be incompressible

$$w_p = v(P_2 - P_1) = 0.01614(300 - 1)\tfrac{144}{778} = 0.893 \text{ Btu/lbm}$$

One could also use Figure 3 of Keenan and Keyes' steam tables to determine pump work.

$$h_2 = h_1 + w_p = 69.7 + 0.9 = 70.6 \text{ Btu/lbm}$$

Heat transfer in boiler:

$$q_H = h_3 - h_2 = 1202.8 - 70.6 = 1132.2 \text{ Btu/lbm}$$

Turbine work:

$$w_T = h_3 - h_4$$

$$s_3 = s_4 = 1.5104 = [s_g - (1 - x)s_{fg}]_4$$

$$= 1.9782 - (1 - x)_4\, 1.8456$$

$$(1 - x)_4 = \frac{0.4678}{1.8456} = 0.2535$$

$$h_4 = [h_g - (1 - x)h_{fg}]_4$$

$$= 1106.0 - 0.2535(1036.3)$$

$$= 843.3 \text{ Btu/lbm}$$

$$w_T = 1202.8 - 843.3 = 359.5 \text{ Btu/lbm}$$

Heat transfer from steam in condenser:

$$q_L = h_4 - h_1$$

$$= 843.3 - 69.7 = 773.6 \text{ Btu/lbm}$$

$$\eta_{th} = \frac{w_{net}}{q_H} = \frac{q_H - q_L}{q_H} = \frac{w_T - w_p}{q_H}$$

$$\eta_{th} = \frac{(h_3 - h_2) - (h_4 - h_1)}{h_3 - h_2} = \frac{(h_3 - h_4) - (h_2 - h_1)}{h_3 - h_2}$$

$$= \frac{1132.2 - 773.6}{1132.2} = \frac{359.5 - 0.9}{1132.2} = 31.7\%$$

12.2 Effect of Pressure and Temperature on Rankine Cycle

Let us first consider the effect of exhaust pressure and temperature on the Rankine cycle. This effect is shown on the T-s diagram of Fig. 12.2.

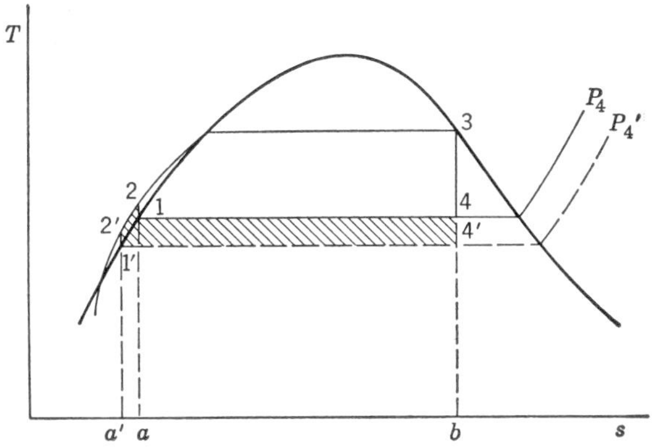

Fig. 12.2 Effect of exhaust pressure on Rankine-cycle efficiency.

Let the exhaust pressure drop from P_4 to P_4', with the corresponding decrease in temperature at which heat is rejected. The net work is increased by area 1–4–4′–1′–2′–2–1 (shown by the cross hatching). The heat transferred to the steam is increased by area a'–2′–2–a–a'. Since these two areas are approximately equal, the net result is an increase in cycle efficiency. This is also evident from the fact that the average temperature at which heat is rejected is decreased. Note however, that lowering the back pressure causes an increase in the moisture content in the steam leaving the turbine. This is a significant factor because if the moisture in the low-pressure stages of the turbine exceeds about 10 per cent, not only is there a decrease in turbine efficiency, but also the erosion of the turbine blades may be a very serious problem.

Next consider the effect of superheating the steam in the boiler, as shown in Fig. 12.3. It is readily evident that the work is increased by area 3–3′–4′–4–3, and the heat transferred in the boiler is increased by area 3–3′–b'–b–3. Since this ratio of these two areas is greater than the ratio of net work to heat supplied for the rest of the cycle, it is evident that for given pressures, superheating the steam increases the Rankine-cycle efficiency. This would also follow from the fact that the

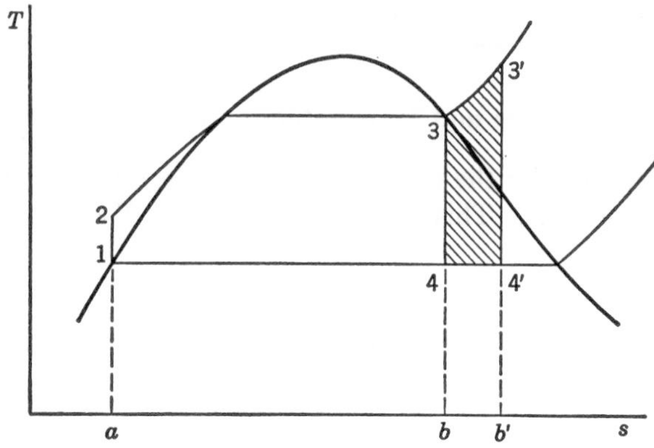

Fig. 12.3 Effect of superheating on Rankine-cycle efficiency.

average temperature at which heat is transferred to the steam is increased. Note also that when the steam is superheated the quality of the steam leaving the turbine increases.

Finally, the influence of the maximum pressure of the steam must be considered, and this is shown in Fig. 12.4. In this analysis the maximum temperature of the steam, as well as the exhaust pressure, is held constant. The heat rejected decreases by area b'–$4'$–4–b–b'. The

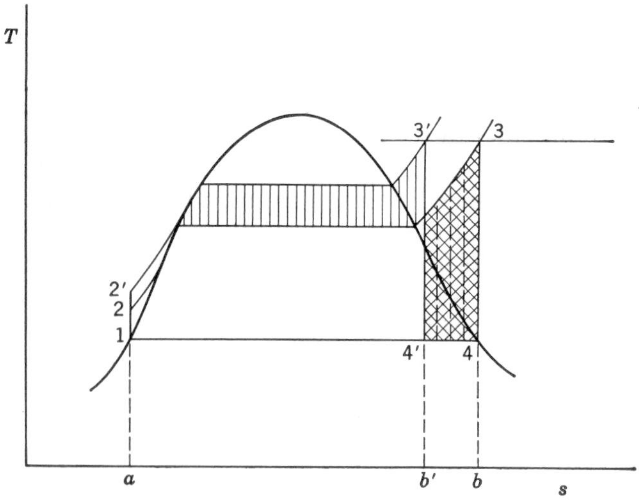

Fig. 12.4 Effect of boiler pressure on Rankine-cycle efficiency.

net work increases by the amount of the single crosshatching and decreases by the amount of the double crosshatching. Therefore the net work tends to remain the same, but the heat rejected decreases, and therefore the Rankine-cycle efficiency increases with an increase in maximum pressure. Note that in this case also the average temperature at which heat is supplied increases with an increase in pressure. The quality of the steam leaving the turbine decreases as the maximum pressure increases.

To summarize this section we can say that the Rankine-cycle efficiency can be increased by lowering the exhaust pressure, increasing the pressure during heat addition, and by superheating the steam. The quality of the steam leaving the turbine is increased by superheating the steam, and decreased by lowering the exhaust pressure and by increasing the pressure during heat addition.

Example 12.2 In a Rankine cycle steam leaves the boiler and enters the turbine at 600 lbf/in.^{2}, 800 F. The condenser pressure is 1 lbf/in.^{2} Determine the cycle efficiency.

$$h_3 = 1407.7 \qquad s_3 = 1.6343$$

$$s_3 = s_4 = 1.6343 = 1.9783 - (1 - x)_4\, 1.8456$$

$$(1 - x)_4 = 0.1865$$

$$h_4 = 1106.0 - 0.1865(1036.3) = 913.3$$

$$h_1 = 69.70$$

$$w_p = v(P_2 - P_1) = 0.01614(600 - 1) \times \tfrac{144}{778} = 1.8 \text{ Btu/lbm}$$

$$h_2 = 69.7 + 1.8 = 71.5$$

$$q_H = h_3 - h_2 = 1407.7 - 71.5 = 1336.2 \text{ Btu/lbm}$$

$$w_t = h_3 - h_4 = 1407.7 - 913.3 = 494.4 \text{ Btu/lbm}$$

$$w_n = w_t - w_p = 494.4 - 1.8 = 492.6 \text{ Btu/lbm}$$

$$\eta_{\text{th}} = \frac{w_n}{q_H} = \frac{492.6}{1336.2} = 36.9\%$$

The net work could also be determined by calculating the heat rejected in the condenser, q_L, and noting, from the first law, that the net work for the cycle is equal to the net heat transfer.

$$q_L = h_4 - h_1 = 913.3 - 69.7 = 843.6 \text{ Btu/lbm}$$

$$w_n = q_H - q_L = 1336.2 - 843.6 = 492.6 \text{ Btu/lbm}$$

12.3 The Reheat Cycle

In the last paragraph we noted the efficiency of the Rankine cycle could be increased by increasing the pressure during the addition of heat. However, this also increases the moisture content of the steam in the low-pressure end of the turbine. The reheat cycle has been developed to take advantage of the increased efficiency with higher pressures, and yet avoid excessive moisture in the low-pressure stages of the turbine. This cycle is shown schematically and on a T-s diagram in Fig. 12.5. The unique feature of this cycle is that the steam is expanded to some intermediate pressure in the turbine, and is then reheated in the boiler, after which it expands in the turbine to the exhaust pressure. It is evident from the T-s diagram that there is very little gain in efficiency from reheating the steam, because the average temperature at which heat is supplied is not greatly changed. The chief advantage is in decreasing the moisture content in the low-pressure stages of the turbine to a safe value. Note also, that if metals could be found that would enable one to superheat the steam to 3′, the simple Rankine cycle would be more efficient than the reheat cycle, and there would be no need for the reheat cycle.

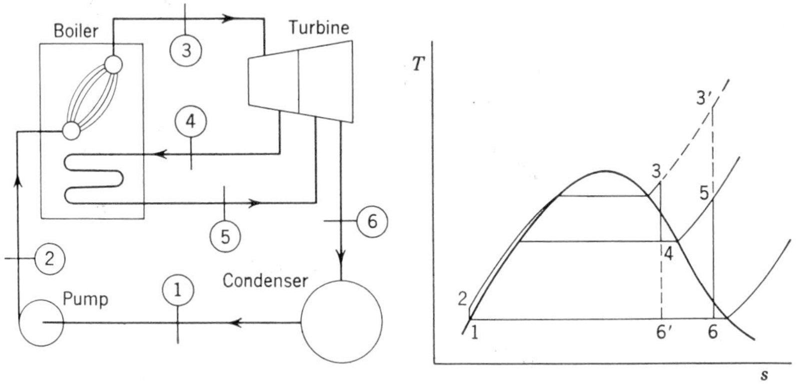

Fig. 12.5 The ideal reheat cycle.

Example 12.3 Consider a reheat cycle utilizing steam. Steam leaves the boiler and enters the turbine at 600 lbf/in.^{2}, 800 F. After expansion in the turbine to 60 lbf/in.^{2}, the steam is reheated to 800 F and then expanded in the low-pressure turbine to 1 lbf/in.^{2} Determine the cycle efficiency.

Designating the states in accordance with Fig. 12.5 we proceed as follows:

$$h_3 = 1407.7 \qquad s_3 = 1.6343$$

$$s_4 = s_3 = 1.6343 = 1.6438 - (1-x)_4\, 1.2168$$

$$(1-x)_4 = 0.0078$$

$$h_4 = 1177.6 - 0.0078(915.5) = 1170.5$$

$$h_5 = 1430.5 \qquad s_5 = 1.9015$$

$$s_5 = s_6 = 1.9015 = 1.9782 - (1-x)_6\, 1.8456$$

$$(1-x)_6 = 0.0416$$

$$h_6 = 1106.0 - 0.0416(1036.8) = 1062.9$$

$$w_p = v(P_2 - P_1) = 0.01614(600 - 1)\tfrac{144}{778} = 1.8 \text{ Btu/lbm}$$

$$h_2 = 69.7 + 1.8 = 71.5$$

$$q_H = (h_3 - h_2) + (h_5 - h_4)$$

$$q_H = (1407.7 - 71.5) + (1430.5 - 1170.5) = 1596.2 \text{ Btu/lbm}$$

$$w_T = (h_3 - h_4) + (h_5 - h_6)$$

$$w_T = (1407.7 - 1170.5) + (1430.5 - 1062.9) = 604.8 \text{ Btu/lbm}$$

$$w_n = w_T - w_p = 604.8 - 1.8 = 603.0 \text{ Btu/lbm}$$

$$\eta_{\text{th}} = \frac{w_n}{q_H} = \frac{603.0}{1596.2} = 37.8\%$$

Note by comparison with Example 12.2 that the gain in efficiency from reheating is relatively small, but that the moisture content leaving the turbine is decreased as a result of reheating from 18.6 per cent to 4.2 per cent.

12.4 The Regenerative Cycle

Another important variation from the Rankine cycle is the regenerative cycle, which involves the use of feedwater heaters. The basic concepts of this cycle can be demonstrated by considering the Rankine cycle without superheat as shown in Fig. 12.6. During the process between states 2 and 2′ the working fluid is heated while in the liquid phase, and the average temperature of the working fluid is much lower during this process than during the vaporization process 2′–3. This causes the average temperature at which heat is supplied in the Rankine cycle to be lower than in the Carnot cycle 1′–2′–3–4–1′, and consequently the efficiency of the Rankine cycle is less than that of the cor-

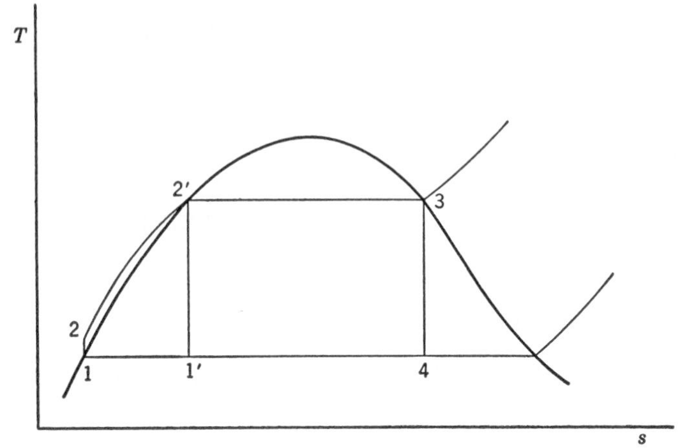

Fig. 12.6 Temperature-entropy diagram showing the relationship between Carnot-cycle efficiency and Rankine-cycle efficiency.

responding Carnot cycle. In the regenerative cycle the working fluid enters the boiler at some state between 2 and 2′, and consequently the average temperature at which heat is supplied is increased.

Consider first an idealized regenerative cycle, shown in Fig. 12.7. The unique feature of this cycle compared to the Rankine cycle is that

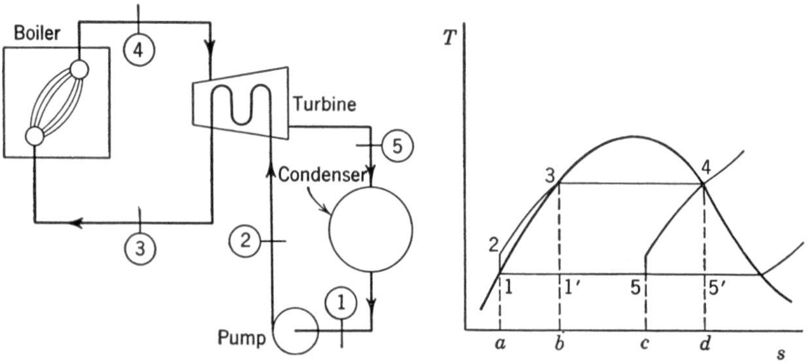

Fig. 12.7 The ideal regenerative cycle.

after leaving the pump, the liquid circulates around the turbine casing, counterflow to the direction of vapor flow in the turbine. Thus, it is possible to transfer heat from the vapor as it flows through the turbine to the liquid flowing around the turbine. Let us assume for the moment that this is a reversible heat transfer; that is, at each point the temperature of the vapor is only infinitesimally higher than the temperature of the liquid. In this case line 4–5 on the *T-s* diagram of Fig. 12.7, which represents the states of the vapor flowing through the turbine, is exactly parallel to line 1–2–3, which represents the pumping process (1–2) and the states of the liquid flowing around the turbine. Consequently areas 2–3–*b*–*a*–2 and 5–4–*d*–*c*–5 are not only equal but congruous, and these areas respectively represent the heat transferred to the liquid and from the vapor. Note also that heat is transferred to the working fluid at constant temperature in process 3–4, and area 3–4–*d*–*b*–3 represents this heat transfer. Heat is transferred from the working fluid in process 5–1, and area 1–5–*c*–*a*–1 represents this heat transfer. Note that this area is exactly equal to area 1′–5′–*d*–*b*–1′, which is the heat rejected in the related Carnot cycle 1′–3–4–5′–1′. Thus, this idealized regenerative cycle has an efficiency exactly equal to the efficiency of the Carnot cycle with the same heat-supply and heat-rejection temperatures.

Quite obviously this idealized regenerative cycle is not practical. First of all, it would not be possible to effect the necessary heat transfer from the vapor in the turbine to the liquid feedwater. Furthermore, the moisture content of the vapor leaving the turbine is considerably increased as a result of the heat transfer, and the disadvantage of this has been noted previously. The practical regenerative cycle involves

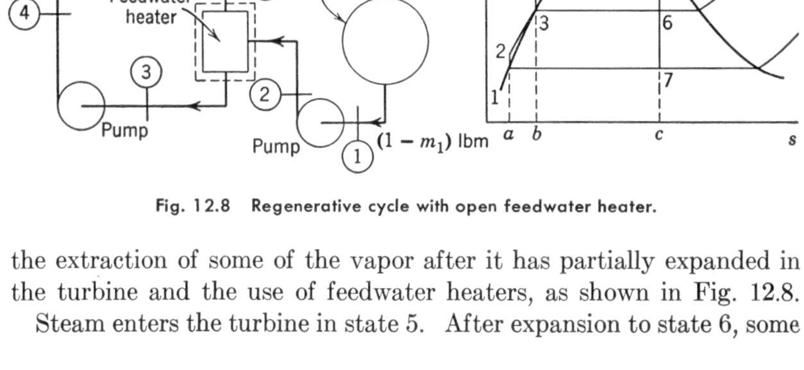

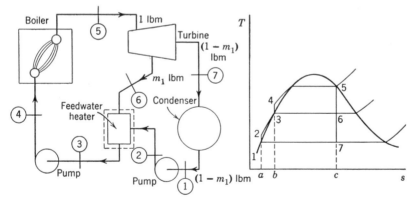

Fig. 12.8 Regenerative cycle with open feedwater heater.

the extraction of some of the vapor after it has partially expanded in the turbine and the use of feedwater heaters, as shown in Fig. 12.8.

Steam enters the turbine in state 5. After expansion to state 6, some of the steam is extracted and enters the feedwater heater. The steam that is not extracted is expanded in the turbine to state 7 and is then condensed in the condenser. This condensate is pumped into the feedwater heater where it mixes with the steam extracted from the turbine. The proportion of steam extracted is just sufficient to cause the liquid leaving the feedwater heater to be saturated at state 3. Note that the liquid has not been pumped to the boiler pressure, but only to the intermediate pressure corresponding to state 6. Another pump is required to pump the liquid leaving the feedwater heater to boiler pressure. The significant point is that the average temperature at which heat is supplied has been increased.

This cycle is somewhat difficult to show on a T-s diagram because the mass of steam flowing through the various components is not the same. The T-s diagram of Fig. 12.7 simply shows the state of the fluid at the various points.

Area 4–5–c–b–4 in Fig. 12.8 represents the heat transferred per pound mass of working fluid. Process 7–1 is the heat-rejection process, but since not all the steam passes through the condenser, area 1–7–c–a–1 represents the heat transfer per pound mass flowing through the condenser, which does not represent the heat transfer per pound mass of working fluid entering the turbine. Note also that between states 6 and 7 only part of the steam is flowing through the turbine. The example that follows illustrates the calculations involved in the regenerative cycle.

Example 12.4 Consider a regenerative cycle utilizing steam as the working fluid. Steam leaves the boiler and enters the turbine at 600 lbf/in.^{2}, 800 F. After expansion to 60 lbf/in.^{2} some of the steam is extracted from the turbine for the purpose of heating the feedwater in an open feedwater heater. The pressure in the feedwater heater is 60 lbf/in.^{2} and the water leaving it is saturated liquid at 60 lbf/in.^{2} The steam not extracted expands to 1 lbf/in.^{2} Determine the cycle efficiency.

The line diagram and T-s diagram for this cycle are shown in Fig. 12.8.

From Examples 12.2 and 12.3 we have the following properties:

$$h_5 = 1407.7 \qquad h_6 = 1170.5 \qquad h_7 = 913.3 \qquad h_1 = 69.70$$

Between states 1 and 2 the pump work is

$$w_p = v(P_2 - P_1) = 0.01614(60 - 1)\,\tfrac{144}{778} = 0.2 \text{ Btu/lbm}$$

$$h_2 = h_1 + w_p = 69.7 + 0.2 = 69.9$$

$$h_3 = 262.1$$

Between states 3 and 4 the pump work is

$$w_p = v(P_4 - P_3) = 0.01738(600 - 60)\,\tfrac{144}{778} = 1.7 \text{ Btu/lbm}$$

$$h_4 = h_3 + w_p = 262.1 + 1.7 = 263.8$$

The amount of steam extracted per pound of steam entering the turbine can be found by considering the feedwater as the system and applying the first law. Denoting by m_1 the fraction of steam extracted we have

$$m_1(h_6) + (1 - m_1)h_2 = h_3$$

$$m_1(1170.5) + (1 - m_1)70.1 = 262.1$$

$$m_1 = \frac{262.1 - 69.9}{1170.5 - 69.9} = 0.1747$$

$$q_H = h_5 - h_4 = 1407.7 - 263.8 = 1143.9$$

$$w_T = (h_5 - h_6) + (1 - m_1)(h_6 - h_7)$$

$$= (1407.7 - 1170.5) + (1 - 0.1747)(1170.5 - 913.3) = 449.4$$

$$w_n = w_T - w_{p1} - w_{p2} = 449.4 - 0.2(0.825) - 1.7 = 447.5$$

$$\eta_{\text{th}} = \frac{w_n}{q_H} = \frac{447.5}{1143.9} = 39.1\%$$

Note the increase in efficiency over the Rankine cycle of Example 12.2.

Up to this point the discussion and example have tacitly assumed that the extraction steam and feedwater are mixed in the feedwater heater. Another much-used type of feedwater heater, known as a closed heater, is one in which the steam and feedwater do not mix, but rather heat is transferred from the extracted steam as it condenses on the outside of tubes as the feedwater flows through the tubes. In a closed heater, a schematic sketch of which is shown in Fig. 12.9, the steam and feedwater may be at considerably different pressures. The condensate may be pumped into the feedwater line, or it may be removed through a trap (a device that permits liquid but no vapor to flow to a region of lower pressure) to a lower pressure heater or to the main condenser.

Open feedwater heaters have the advantage of being less expensive and having better heat-transfer characteristics compared to closed feedwater heaters. They have the disadvantage of requiring a pump to handle the feedwater between each heater.

In many power plants a number of stages of extraction are used, though only rarely more than five. The number is, of course, determined by economic considerations. It is evident that by using a very large number of extraction stages and feedwater heaters, the cycle efficiency would approach that of the idealized regenerative cycle of Fig. 12.7, where the feedwater enters the boiler as saturated liquid at the maximum pressure. However, in practice this could not be eco-

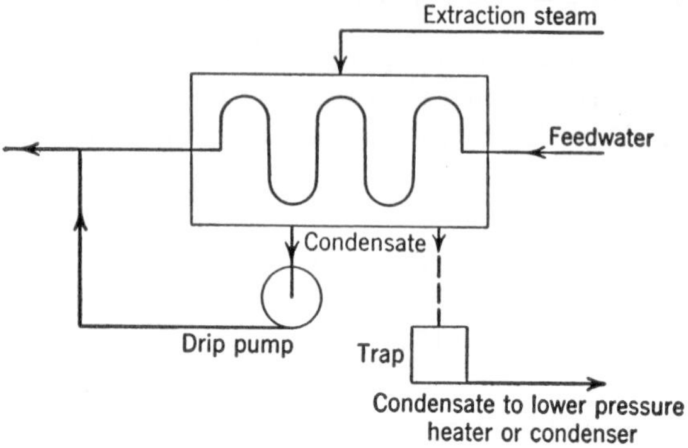

Fig. 12.9 **Schematic arrangement for a closed feedwater heater.**

nomically justified because the savings effected by the increase in efficiency would be more than offset by the cost of additional equipment (feedwater heaters, piping, etc.).

A typical arrangement of the main components in an actual power plant is shown in Fig. 12.10. Note that one open feedwater heater is

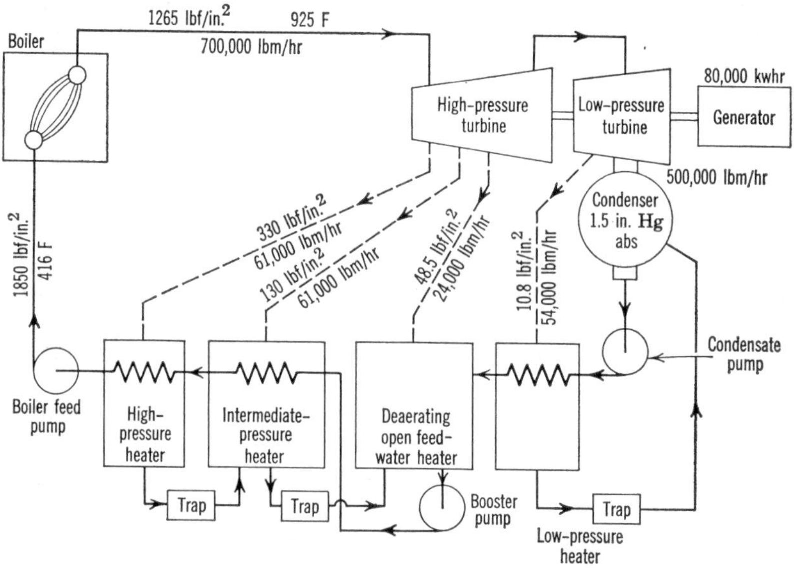

Fig. 12.10 **Arrangement of heaters in an actual power plant utilizing regenerative feedwater heaters.**

a deaerating feedwater heater, and this has the dual purpose of heating and removing the air from the feedwater. Unless the air is removed excessive corrosion occurs in the boiler. Note also that the condensate from the high-pressure heater drains (through a trap) to the intermediate heater, and the intermediate heater drains to the deaerating feedwater heater. The low-pressure heater drains to the condenser.

In many cases an actual power plant combines one reheat stage with a number of extraction stages. The principles already considered are readily applied to such a cycle.

12.5 The Binary Vapor Cycle

The purpose and concept of the binary cycle can be best demonstrated by evaluating water as the working fluid in a Rankine cycle. We have noted that the cycle efficiency can be increased by increasing the average temperature at which heat is supplied and by decreasing the average temperature at which heat is rejected. Let us consider the characteristics of water at these upper and lower temperatures.

Consider the upper temperature first. The maximum temperature is determined by metallurgical considerations; that is, by the availability of metals that can contain the steam in the boiler and turbine where the highest pressures and temperatures prevail. The higher the pressure the lower the maximum temperature a given metal is suited for. Thus, line ab in Fig. 12.11 might represent the states of

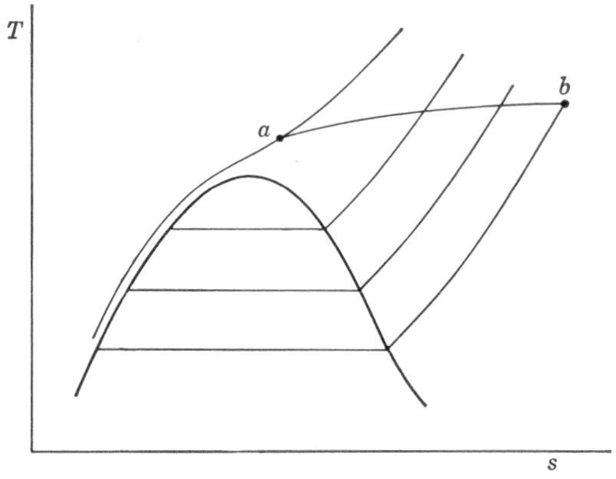

Fig. 12.11 Temperature-entropy diagram showing pressure limitations for the Rankine cycle.

steam associated with this metallurgical limit. Note that it is well above the critical temperature for steam (705 F); at present this limit would be about 1200 F.

It is evident from Fig. 12.11 that because of the constant temperature during vaporization, a higher average temperature during the heat-supply process is attained by utilizing the high pressures. However, three factors must be considered. First of all, the enthalpy of evaporation decreases as the pressure increases, and above the critical pressure there is no vaporization process. Therefore it is impossible to have an isothermal vaporization process near the metallurgical limit. Second, using higher pressures means a lower maximum temperature. Third, the moisture problem in the turbine becomes more severe at these higher pressures.

At the lower temperatures at which heat is rejected, steam has one main disadvantage, namely, that at the temperatures at which heat can be rejected (generally between 40 F and 120 F) the saturation pressure of steam is much less than 1 atm, and therefore the condenser must operate at a vacuum. This necessitates rather elaborate steps to minimize leakage of air into the condenser (for example, leakage along the turbine shaft) and to remove the air that does get into the condenser. (The air is usually removed by a steam-jet ejector. If the air is not removed, the pressure in the condenser will build up, thus increasing the temperature at which the steam condenses.) At these low pressures the steam has a relatively large specific volume, which necessitates large flow areas in the low-pressure stages of the turbine, and large condensers.

This leads one to suggest the characteristics of an ideal working fluid for the Rankine cycle. The *T*-*s* diagram for such a fluid is shown in Fig. 12.12. It would have a relatively low saturation pressure at the

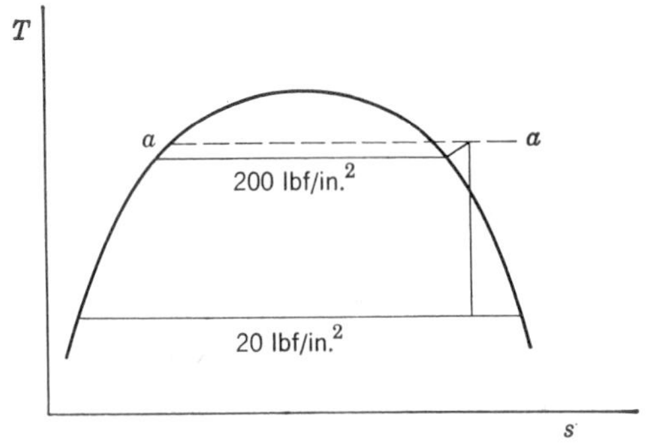

Fig. 12.12 Temperature-entropy diagram showing the characteristics of an ideal fluid for the Rankine cycle.

metallurgical limit so that most of the heat transfer would be at this upper limit. The saturation pressure would be above atmospheric pressure at the heat-rejection temperatures. The specific heat of the liquid would be relatively small, and the saturated vapor curve of the T-s diagram would be steep, thus minimizing the moisture problem. Added to these would be the requirement that the working fluid be nontoxic, noncorrosive, not excessively viscous, and low in cost.

As might be expected, no single fluid meets all these requirements. The binary cycle utilizes two different fluids in such a way that more nearly optimum conditions prevail than could be attained by using a single fluid. The mercury-steam binary cycle has been proposed for power plants, and a few plants utilizing these two fluids have been built. The ideal binary vapor cycle utilizing mercury and steam is shown in Fig. 12.13 on a line diagram and a corresponding T-s diagram.

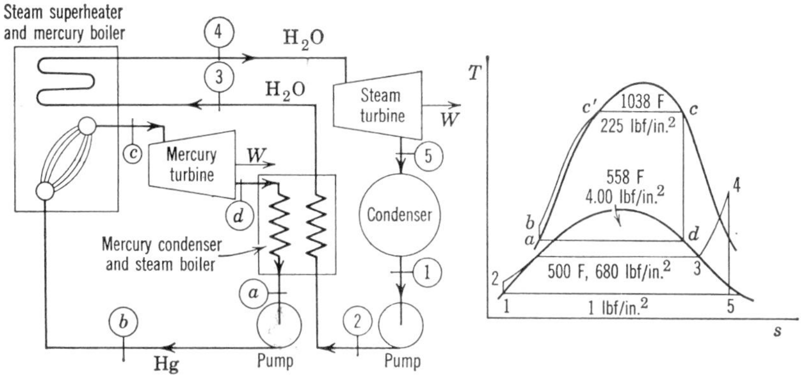

Fig. 12.13 The binary cycle.

Heat is transferred from an external source to the mercury in process $bc'c$, during which the liquid mercury is first heated to saturation temperature and then vaporized. After expanding isentropically in the turbine in process cd, the mercury vapor is condensed in process da, as heat is transferred from the mercury to the steam. The steam leaves the mercury condenser–steam boiler as saturated vapor, and it is then superheated by heat transfer from the external source. In essence then, the mercury and steam each operate on the Rankine cycle. The important thing to note is that the average temperature at which heat is supplied is much higher than the Rankine cycle utilizing steam only, and the maximum pressure is relatively low. The main disadvantage of the mercury-steam cycle is that mercury is very expensive and highly toxic. Therefore much more effort has been expended in developing steam cycles utilizing higher pressures and temperatures. Perhaps the most important point to note here is the concept of utilizing more than one fluid to accomplish a given objective more efficiently, for this principle has wide application.

Example 12.5 Determine the thermal efficiency of the binary cycle shown in Fig. 12.13. Assume saturated vapor at points c and 3, and saturated liquid at points a and 1, and $t_4 = 900$ F.

Consider the mercury cycle first. From the tables of thermodynamic properties of mercury (Table A.4 in the Appendix) we have

$$h_c = 156.319 \qquad s_c = 0.11852$$

$$s_c = s_d = 0.11852 = 0.14787 - (1 - x)_d\, 0.12434$$

$$(1 - x)_d = \frac{0.02353}{0.12434} = 0.189$$

$$h_d = 143.436 - 0.189(126.275) = 119.51$$

$$h_a = 17.16$$

Because of the high density of mercury the pump work will be very low, so let us assume that $h_b = h_a$. Then the heat supplied to the mercury,

$$q_H = h_c - h_a = 156.32 - 17.16 = 139.16 \text{ Btu/lbm Hg}$$

$$q_L = h_d - h_a = 119.51 - 17.16 = 102.35 \text{ Btu/lbm Hg}$$

Now consider the steam cycle.

$$h_3 = 1201.7$$

$$h_4 = 1459.7 \qquad s_4 = 1.6609$$

$$s_5 = s_4 = 1.6609 = 1.9782 - (1 - x)_5\, 1.8456$$

$$(1 - x)_5 = \frac{0.3173}{1.8456} = 0.172$$

$$h_5 = 1106.0 - 0.172(1036.3) = 929 \text{ Btu/lbm steam}$$

$$h_1 = 69.70$$

$$w_p = 0.01614(680 - 1)\, \tfrac{144}{778} = 2.0 \text{ Btu/lbm steam}$$

$$h_2 = 69.70 + 2.0 = 71.7 \text{ Btu/lbm steam}$$

$$q_H \text{ (from Hg)} = h_3 - h_2 = 1201.7 - 71.7$$

$$= 1130.0 \text{ Btu/lbm steam}$$

$$q_H \text{ (from external source)} = h_4 - h_3 = 1459.7 - 1201.7$$

$$= 258.0 \text{ Btu/lbm steam}$$

By applying the first law to the mercury condenser–steam boiler the pounds of mercury circulated per pound of steam can be determined.

$$\frac{\text{lbm Hg}}{\text{lbm steam}}(h_d - h_a) = h_3 - h_2$$

$$\frac{\text{lbm Hg}}{\text{lbm steam}} = \frac{1130.0}{102.35} = 11.02$$

The total heat supply from the external source is the sum of the heat transferred to the mercury and to the steam. Using 1 lbm of steam as our parameter we have

$$q_H\,(\text{total}) = \frac{\text{lbm Hg}}{\text{lbm steam}}(h_c - h_a) + (h_4 - h_3)$$

$$= 11.02(139.16) + 258.0 = 1795 \text{ Btu/lbm steam}$$

The only heat rejected is in the steam condenser.

$$q_L = h_5 - h_1 = 929 - 69.7 = 859.3 \text{ Btu/lbm steam}$$

From the first law we conclude that

$$w_n = q_H - q_L = 1795 - 859 = 936 \text{ Btu/lbm steam}$$

$$\eta = \frac{w_n}{q_H} = \frac{936}{1795} = 52.1\%$$

12.6 Deviation of Actual Cycles from Ideal Cycles

Before leaving the matter of vapor power cycles a few comments regarding the ways in which an actual cycle deviates from an ideal cycle are in order. (The losses associated with the combustion process are considered in a later chapter.) The most important of these are as follows:

Piping losses Pressure drop due to frictional effects and heat transfer to the surroundings are the most important piping losses. Consider for example the pipe connecting the turbine to the boiler. If only frictional effects occurred the states a and b in Fig. 12.14 would represent the states of the steam leaving the boiler and entering the turbine respectively. Note that this causes an increase in entropy. Heat transferred to the surroundings at constant pressure can be represented by process bc. This effect causes a decrease in entropy. Both the pressure drop and heat transfer cause a decrease in the availability of the steam entering the turbine, and the irreversibility of this process can be calculated by the methods outlined in Chapter 10.

A similar loss is the pressure drop in the boiler. Because of this pressure drop the water entering the boiler must be pumped to a

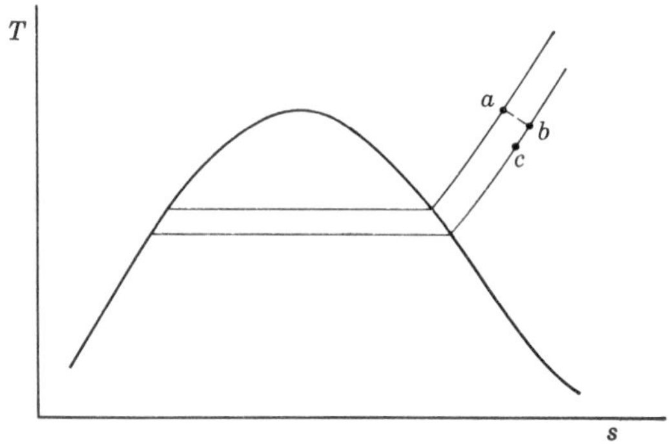

Fig. 12.14 Temperature-entropy diagram showing effect of losses between boiler and turbine.

much higher pressure than the desired steam pressure leaving the boiler, and this requires additional pump work.

Turbine losses The losses in the turbine are primarily those associated with the flow of the working fluid through the turbine. Heat transfer to the surroundings also represents a loss, but this is usually of secondary importance. The effects of these two losses are the same as these outlined for piping losses, and the process might be as represented in Fig. 12.15, where 4_s represents the state after an isentropic

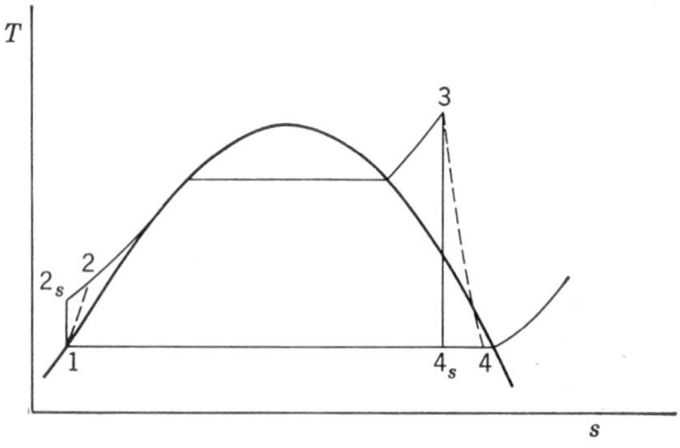

Fig. 12.15 Temperature-entropy diagram showing effect of turbine and pump inefficiencies on cycle performance.

expansion and state 4 represents the actual state leaving the turbine. The governing procedures may also cause a loss in the turbine, particularly if a throttling process is used to govern the turbine.

The efficiency of the turbine has been defined (in Chapter 10) as

$$\eta_T = \frac{w_T}{h_3 - h_{4s}}$$

where the states are as designated in Fig. 12.15.

Pump losses The losses in the pump are similar to those of the turbine, and are primarily due to the irreversibilities associated with the fluid flow. Heat transfer is usually a minor loss.

The pump efficiency is defined as

$$\eta_p = \frac{h_{2s} - h_1}{w_p}$$

where the states are as shown in Fig. 12.15, and w_p is the actual work input per pound of fluid.

Condenser losses The losses in the condenser are relatively small. One of these minor losses is the cooling below the saturation temperature of the liquid leaving the condenser. This represents a loss because additional heat transfer is necessary to bring the water to saturation temperature.

The influence of these losses on the cycle is illustrated in the following example, which should be compared to Example 12.2.

Example 12.6 A steam power plant operates on a cycle with pressures and temperatures as designated in Fig. 12.16. The efficiency of

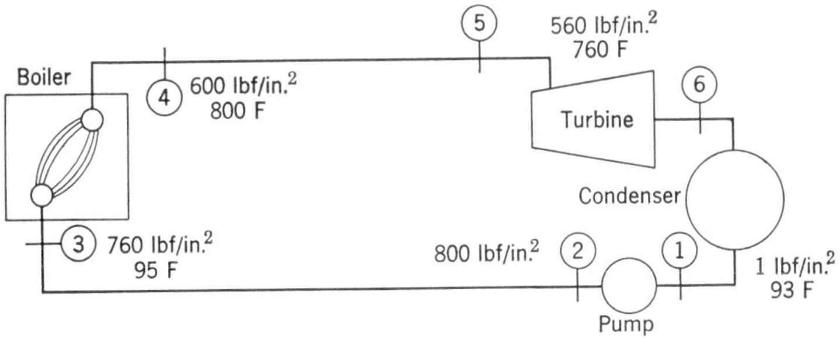

Fig. 12.16 Schematic diagram for Example 12.6.

the turbine is 86 per cent and the efficiency of the pump is 80 per cent. Determine the thermal efficiency of this cycle.

From the steam tables

$$h_1 = 61.0$$

$$h_3 = 63.0 + 2.1 = 65.1$$

$$h_4 = 1407.7 \qquad s_4 = 1.6343$$

$$h_5 = 1387.3 \qquad s_5 = 1.6251$$

This cycle is shown on the T-s diagram of Fig. 12.17.

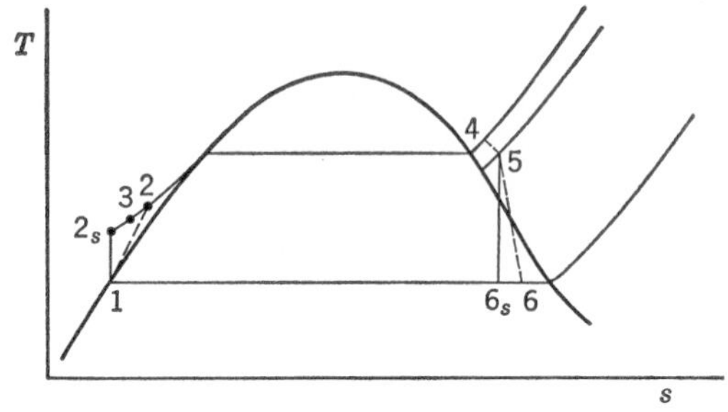

Fig. 12.17 Temperature-entropy diagram for Example 12.6.

$$s_{6s} = s_5 = 1.6251 = 1.9782 - (1-x)_{6s}\, 1.8456$$

$$(1-x)_{6s} = \frac{0.3531}{1.8456} = 0.1912$$

$$h_{6s} = 1106.0 - 0.1912(1036.3) = 907.8$$

$$w_T = \eta_T(h_5 - h_{6s}) = 0.86(1387.3 - 907.8)$$

$$= 0.86(479.5) = 412.3 \text{ Btu/lbm}$$

$$w_p = \frac{h_{2s} - h_1}{\eta_p} = \frac{v(P_2 - P_1)}{\eta_p} = \frac{0.01615(800 - 1)144}{0.8 \times 778}$$

$$= \frac{2.4}{0.8} = 3.0 \text{ Btu/lbm}$$

$$w_n = w_T - w_p = 412.3 - 3.0 = 409.3 \text{ Btu/lbm}$$

$$q_H = h_4 - h_3 = 1407.7 - 65.1 = 1342.6 \text{ Btu/lbm}$$

$$\eta_{\text{th}} = \frac{409.3}{1342.6} = 30.4\%$$

This compares to an efficiency of 36.9 per cent for the Rankine efficiency of the similar cycle of Example 12.2.

VAPOR REFRIGERATION CYCLES

12.7 Vapor-Compression Refrigeration Cycles

The ideal cycle for vapor-compression refrigeration is shown in Fig. 12.18 as cycle 1–2–3–4–1. Saturated vapor at low pressure enters the compressor and undergoes a reversible adiabatic compression, 1–2.

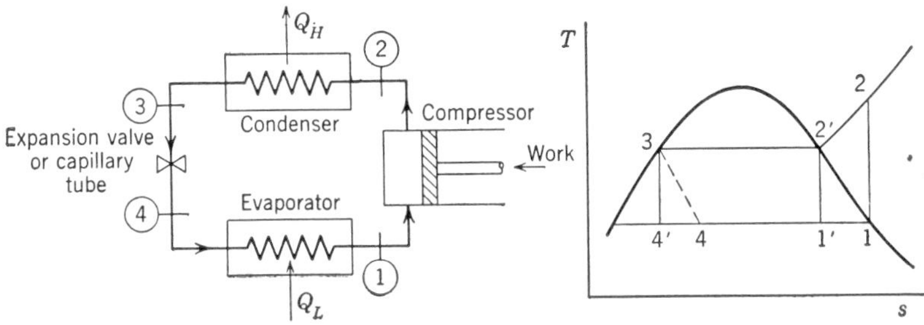

Fig. 12.18 The ideal vapor-compression refrigeration cycle.

Heat is then rejected at constant pressure in process 2–3, and the working fluid leaves the condenser as saturated liquid. An adiabatic throttling process follows, process 3–4, and the working fluid is then evaporated at constant pressure, process 4–1, to complete the cycle.

The similarity between this cycle and the Rankine cycle is evident, for it is essentially the same cycle in reverse, except that an expansion valve replaces the pump. This throttling process is irreversible, whereas the pumping process of the Rankine cycle is reversible. The deviation of this ideal cycle from the Carnot cycle 1′–2′–3–4′–1′ is evident from the T-s diagram. The reason for the deviation is that it is much more expedient to have a compressor handle only vapor than a mixture of liquid and vapor as would be required in process 1′–2′ of the Carnot cycle. It is virtually impossible to compress (at a reasonable rate) a mixture such as that represented by state 1′ and maintain equilibrium between the liquid and vapor, because there must be a heat and mass transfer across the phase boundary. It is also much simpler to have the expansion process take place irreversibly through an expansion valve than to have an expansion device that receives saturated liquid and discharges a mixture of liquid and vapor, as would be required in process 3–4′. For these reasons the ideal cycle for vapor-compression refrigeration is as shown in Fig. 12.18 by cycle 1–2–3–4–1.

As was pointed out in Chapter 6 the performance of a refrigeration cycle is given in terms of the coefficient of performance, β, which is defined for a refrigeration cycle as

$$\beta = \frac{q_L}{w} \tag{12.2}$$

The capacity of a refrigeration plant is usually given in terms of tons of refrigeration. This term had its origin in the ice-making industry, where cooling capacity was given in terms of tons of ice melting per day. The unit is now defined as

$$\begin{aligned} 1 \text{ ton refrigeration} &= 288{,}000 \text{ Btu of refrigeration/day} \\ &= 12{,}000 \text{ Btu of refrigeration/hr} \end{aligned}$$

Example 12.7 Consider an ideal refrigeration cycle that utilizes Freon-12 as the working fluid. The temperature of the refrigerant in the evaporator is 0 F and in the condenser it is 100 F. The refrigerant is circulated at the rate of 200 lbm/hr. Determine the coefficient of performance and the capacity of the plant in tons of refrigeration.

From the thermodynamic tables for Freon-12 we find the following properties for the states as designated in Fig. 12.18.

$$h_1 = 77.271 \qquad s_1 = 0.16888$$

$$P_2 = 131.86 \qquad s_2 = s_1 = 0.16888 \qquad h_2 = 90.306$$

$$h_3 = 31.100 = h_4$$

$$q_L = h_1 - h_4 = 77.271 - 31.100 = 46.171$$

$$-w = h_2 - h_1 = 90.306 - 77.271 = 13.035$$

$$\beta = \frac{q_L}{w} = \frac{46.171}{13.035} = 3.55$$

$$\text{capacity} = \frac{46.171 \times 200}{12{,}000} = 0.768 \text{ tons}$$

12.8 Working Fluids for Vapor-Compression Refrigeration Systems

A much larger number of different working fluids (refrigerants) are utilized in vapor-compression refrigeration systems than in vapor power cycles. Ammonia and sulfur dioxide were important in the early days of vapor-compression refrigeration. Today, however, the main refrigerants are the halogenated hydrocarbons, which are marketed under the trade names of Freon and Genatron. For example, dichlorodifluoromethane (CCl_2F_2) is known as Freon-12 and Genatron-12. Two important considerations in selecting a refrigerant are the temperature at which refrigeration is desired and the type of equipment to be used.

Since the refrigerant undergoes a change of phase during the heat-transfer process, the pressure of the refrigerant will be the saturation pressure during the heat-supply and heat-rejection processes. Low pressures mean large specific volumes and correspondingly large equipment. High pressures mean smaller equipment but it must be designed to withstand higher pressure. In particular, the pressures should be well below the critical pressure. For extremely low temperature applications a binary fluid system may be used in a manner analogous to the mercury-steam cycle described previously.

The type of compressor used has a particular bearing on the refrigerant. Reciprocating compressors are best adapted to low specific volumes, which means higher pressures, whereas centrifugal compressors are most suitable for low pressures and high specific volumes.

It is also important that the refrigerants used in domestic appliances be nontoxic. Other important characteristics are tendency to cause corrosion, miscibility with compressor oil, dielectric strength, stability, and cost. It should also be noted that for given temperatures during evaporation and condensation, not all refrigerants have the same coefficient of performance for the ideal cycle. It is, of course, desirable to utilize the refrigerant with the highest coefficient of performance, other factors permitting.

12.9 Deviation of the Actual Vapor-Compression Refrigeration Cycle from the Ideal Cycle

The actual refrigeration cycle deviates from the ideal cycle primarily because of pressure drops associated with fluid flow and heat transfer to or from the surroundings. The actual cycle might approach the one shown in Fig. 12.19.

The vapor entering the compressor will probably be superheated. During the compression process there are irreversibilities and heat transfer either to or from the surroundings, depending on the temperature of the refrigerant and the surroundings. Therefore, the entropy might increase or decrease during this process, for the irreversibility and heat transfer to the refrigerant cause an increase in entropy, and heat transfer from the refrigerant causes a decrease in entropy. These possibilities are represented by the two dotted lines 1–2 and 1–2′.

The pressure of the liquid leaving the condenser will be less than that of the vapor entering, and the temperature of the refrigerant in the condenser will be somewhat above that of the surroundings to which heat is being transferred. Usually the temperature of the liquid leaving the condenser is lower than the saturation temperature, and might drop somewhat more in the piping between the condenser and expansion valve. This represents a gain, however, because as a result of this

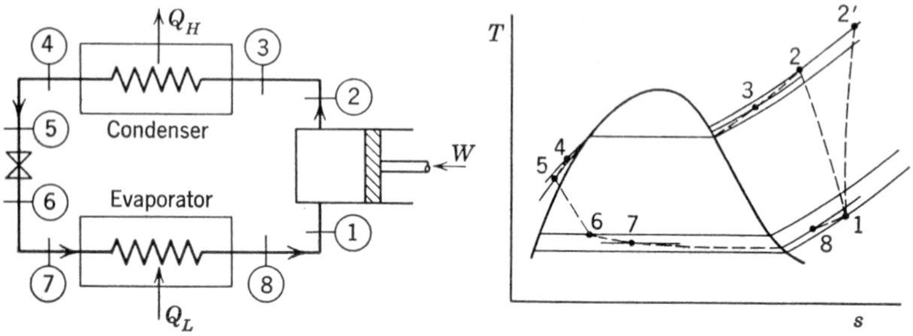

Fig. 12.19 The actual vapor-compression refrigeration cycle.

heat transfer the refrigerant enters the evaporator with a lower enthalpy, thus permitting more heat transfer to the refrigerant in the evaporator.

There is some drop in pressure as the refrigerant flows through the evaporator. It may be slightly superheated as it leaves the evaporator, and due to heat transfer from the surroundings the temperature will increase in the piping between the evaporator and compressor. This heat transfer represents a loss, because it increases the work of the compressor as a result of the increased specific volume of the fluid entering it.

Example 12.8 A refrigeration cycle utilizes Freon-12 as the working fluid. Following are the properties of the various points of the cycle designated in Fig. 12.19.

$$P_1 = 18 \text{ lbf/in.}^2 \qquad t_1 = 20 \text{ F}$$
$$P_2 = 180 \text{ lbf/in.}^2 \qquad t_2 = 220 \text{ F}$$
$$P_3 = 175 \text{ lbf/in.}^2 \qquad t_3 = 180 \text{ F}$$
$$P_4 = 170 \text{ lbf/in.}^2 \qquad t_4 = 110 \text{ F}$$
$$P_5 = 168 \text{ lbf/in.}^2 \qquad t_5 = 104 \text{ F}$$
$$P_6 = p_7 = 20 \text{ lbf/in.}^2 \qquad x_6 = x_7$$
$$P_8 = 19 \text{ lbf/in.}^2 \qquad t_8 = 0 \text{ F}$$

The heat transfer from the Freon-12 during the compression process is 2.0 Btu/lbm. Determine the coefficient of performance of this cycle.

From the Freon-12 tables the following properties are found.

$$h_1 = 80.527 \qquad h_2 = 106.896$$

$$h_5 = h_6 = h_7 = 32.067 \qquad h_8 = 77.621$$

The work of compression is found by applying the steady-flow energy equation to the compressor.

$$q + h_1 = h_2 + w$$

$$-w = h_2 - h_1 - q$$

$$= 106.896 - 80.527 - (-2) = 26.4 + 2.0 = 28.4 \text{ Btu/lbm}$$

The refrigeration effect q_L is found in a similar manner.

$$q_L = h_8 - h_7 = 77.621 - 32.067 = 45.6$$

$$\beta = \frac{q_L}{w} = \frac{45.6}{28.4} = 1.64$$

12.10 The Ammonia-Absorption Refrigeration Cycle

The ammonia-absorption refrigeration cycle differs from the vapor-compression cycle in the manner in which compression is achieved. In the absorption cycle the low-pressure ammonia vapor is absorbed in water and the liquid solution is pumped to a high pressure by a liquid pump. Figure 12.20 shows a schematic arrangement of the essential elements of such a system.

The low-pressure ammonia vapor leaving the evaporator enters the absorber where it is absorbed in the weak ammonia solution. This process takes place at a temperature slightly above that of the surroundings and heat must be transferred to the surroundings during this process. The strong ammonia solution is then pumped through a heat exchanger to the generator where a higher pressure and temperature are maintained. Under these conditions ammonia vapor is driven from the solution as a result of heat transfer from a high-temperature source. The ammonia vapor goes to the condenser where it is condensed, as in a vapor-compression system, and thence to the expansion valve and evaporator. The weak ammonia solution is returned to the absorber through the heat exchanger.

The distinctive feature of the absorption system is that very little work input is required because the pumping process involves a liquid. This follows from the fact that for a reversible steady-flow process with negligible changes in kinetic and potential energy, the work is equal to $-\int v\,dP$, and the specific volume of the liquid is much less than the specific volume of the vapor. On the other hand, a relatively high-temperature source of heat must be available (200 F to 400 F). The

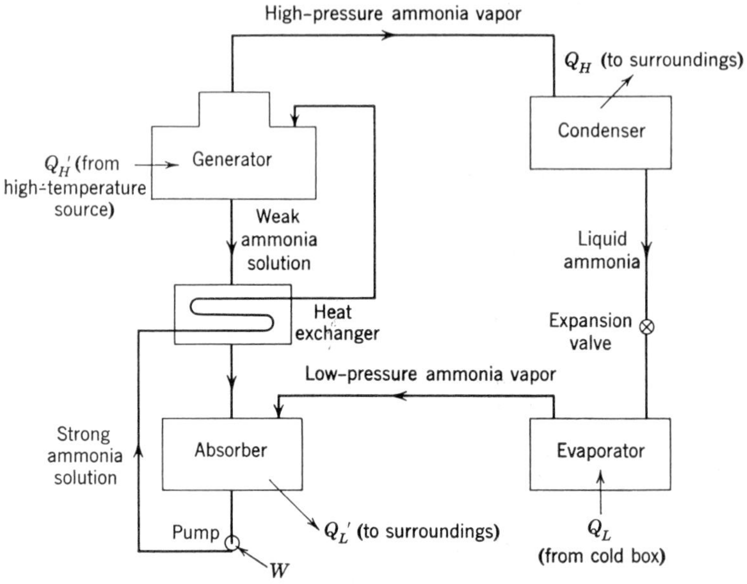

Fig. 12.20 The ammonia-absorption refrigeration cycle.

equipment involved in an absorption system is somewhat greater than in a vapor-compression system and it can usually be economically justified only in those cases where a suitable source of heat is available that would otherwise be wasted.

This cycle brings out the important principle that since the work in a reversible steady-flow process with negligible changes in kinetic and potential energy is $-\int v\,dP$, a compression process should take place with the smallest possible specific volume.

AIR-STANDARD POWER CYCLES

12.11 Air-Standard Cycles

Many work-producing devices (engines) utilize a working fluid that is always a gas. The spark-ignition automotive engine is a familiar example, and the same is true of the Diesel engine and conventional gas turbine. In all of these engines there is a change in the composition of the working fluid, because during combustion it changes from air and fuel to combustion products. For this reason these engines are called internal-combustion engines. In contrast to this the steam power plant may be called an external-combustion engine, because heat is transferred from the products of combustion to the working fluid. External-combustion engines using a gaseous working fluid (usually air) have been built. These are known as hot-air engines, but they have had very limited application.

Because the working fluid does not go through a complete cycle in the engine (even though the engine operates in a mechanical cycle) the internal-combustion engine operates on the so-called open cycle. However, in order to analyze internal-combustion engines it is advantageous to devise closed cycles that closely approximate the open cycles. One such approach is the air-standard cycle, which is based on the following assumptions:

(*a*) A fixed mass of air is the working fluid throughout the entire cycle, and the air is always an ideal gas. Thus there is no inlet process or exhaust process.

(*b*) The combustion process is replaced by a heat-transfer process from an external source.

(*c*) The cycle is completed by heat transfer to the surroundings (in contrast to the exhaust and intake process of an actual engine).

(*d*) All processes are internally reversible.

(*e*) The additional assumption is usually made that air has a constant specific heat.

The main value of the air-standard cycle is to enable one to examine qualitatively the influence of a number of variables on performance. The results obtained from the air-standard cycle, such as efficiency and mean effective pressure, will differ a great deal from those of the actual engine. The emphasis, therefore, in our consideration of the air-standard cycle will be primarily on the qualitative aspects.

12.12 The Air-Standard Carnot Cycle

The air-standard Carnot cycle is shown on the P-v and T-s diagrams of Fig. 12.21. Such a cycle could be achieved in either a reciprocating or steady-flow device, as shown in the same figure.

In Chapter 7 it was noted that the efficiency of a Carnot cycle depends only on the temperatures at which heat is supplied and rejected, and is given by the relation

$$\eta_{\text{th}} = 1 - \frac{T_L}{T_H} = 1 - \frac{T_4}{T_1} = 1 - \frac{T_3}{T_2}$$

where the subscripts refer to Fig. 12.21. The efficiency may also be

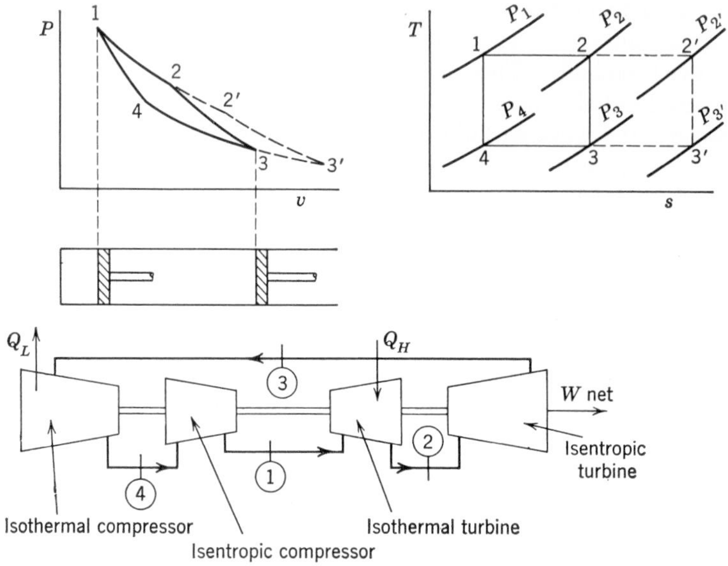

Fig. 12.21 The air-standard Carnot cycle.

expressed by the pressure ratio or compression ratio during the isentropic processes. This follows from the fact that

$$\text{Isentropic pressure ratio} = r_{ps} = \frac{P_1}{P_4} = \frac{P_2}{P_3} = \left(\frac{T_3}{T_2}\right)^{k/(1-k)}$$

$$\text{Isentropic compression ratio} = r_{vs} = \frac{V_4}{V_1} = \frac{V_3}{V_2} = \left(\frac{T_3}{T_2}\right)^{1/(1-k)}$$

Therefore

$$\eta_{\text{th}} = 1 - r_{ps}^{(1-k)/k} = 1 - r_{vs}^{1-k} \tag{12.3}$$

One other important variable in the air-standard Carnot cycle is the amount of heat transferred to the working fluid per cycle. It is evident from Fig. 12.21 that increasing the heat transfer per cycle at a given T_H causes a larger change in volume per cycle, and this causes a lower mean effective pressure in a reciprocating engine. In fact, for air-standard Carnot cycles having a minimum pressure between 1 and 10 atm and a reasonable heat transfer per cycle, the mean effective pressure is so low that it will scarcely overcome friction forces.

Another practical difficulty of the Carnot cycle, which applies to both the reciprocating and steady-flow types of cycle, is the difficulty of transferring heat during the isothermal expansion and compression processes. It is virtually impossible to even approach this in an actual machine operating at a reasonable rate of speed. Thus, the air-standard Carnot cycle is not practical. None the less it is of value as a standard for comparison with other cycles, and in a later paragraph we will consider the ideal-gas turbine cycle with intercooling and reheating and note how its efficiency approaches that of the corresponding Carnot cycle.

Example 12.9 In an air-standard Carnot cycle heat is transferred to the working fluid at 2000 R, and heat is rejected at 500 R. The heat transfer to the working fluid at 2000 R is 50 Btu/lbm. The minimum pressure in the cycle is 1 atm. Assuming constant specific heat of air, determine the cycle efficiency and the mean effective pressure.

Designating the states as in Fig. 12.21 we have

$$P_3 = 14.7 \text{ lbf/in.}^2 \qquad T_3 = T_4 = 500 \text{ R}$$

$$T_1 = T_2 = 2000 \text{ R}$$

$$\frac{T_2}{T_3} = \left(\frac{P_2}{P_3}\right)^{(k-1)/k} = 4 \qquad \therefore \quad \frac{P_2}{P_3} = 128$$

$$P_2 = 14.7(128) = 1882 \text{ lbf/in.}^2$$

$$_1q_2 = RT \ln \frac{V_2}{V_1} = RT \ln \frac{P_1}{P_2} = \frac{53.34 \times 2000}{778} \ln \frac{P_1}{P_2} = 50 \text{ Btu/lbm}$$

$$\frac{P_1}{P_2} = 1.44 \qquad P_1 = 1882(1.44) = 2710 \text{ lbf/in.}^2$$

$$\frac{T_1}{T_4} = \left(\frac{P_1}{P_4}\right)^{(k-1)/k} = 4 \qquad \therefore \quad \frac{P_1}{P_4} = 128$$

$$P_4 = \frac{2710}{128} = 21.2 \text{ lbf/in.}^2$$

$$\eta_{\text{th}} = 1 - \frac{T_L}{T_H} = 1 - \frac{500}{2000} = 0.75$$

The product of mean effective pressure and the piston displacement is equal to the net work.

$$\text{mep}\,(v_3 - v_1) = w_{\text{net}}$$

$$v_3 = \frac{RT_3}{P_3} = \frac{53.34 \times 500}{14.7 \times 144} = 12.6 \text{ ft}^3/\text{lbm}$$

$$v_1 = \frac{RT_1}{P_1} = \frac{53.34 \times 2000}{2710 \times 144} = 0.273 \text{ ft}^3/\text{lbm}$$

$$\text{mep} = \frac{37.5 \times 778}{(12.6 - 0.273)} = 2360 \text{ lbf/ft}^2 = 16.4 \text{ lbf/in.}^2$$

This is a much lower mean effective pressure than can be effectively utilized in a reciprocating engine.

12.13 The Air-Standard Otto Cycle

The air-standard Otto cycle is an ideal cycle that approximates a spark-ignition internal-combustion engine. This cycle is shown on the P-v and T-s diagrams of Fig. 12.22. Process 1–2 is an isentropic com-

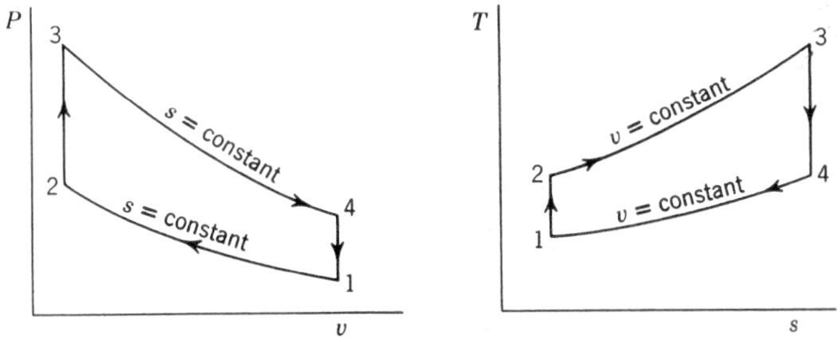

Fig. 12.22 The air-standard Otto cycle.

pression of the air as the piston moves from crank-end dead center to head-end dead center. Heat is then added at constant volume while the piston is momentarily at rest at head-end dead center. (This process corresponds to the ignition of the fuel-air mixture by the spark and the subsequent burning in the actual engine.) Process 3–4 is an isentropic expansion, and process 4–1 is the rejection of heat from the air while the piston is at crank-end dead center.

The thermal efficiency of this cycle is found as follows, assuming constant specific heat of air.

$$\eta_{th} = \frac{Q_H - Q_L}{Q_H} = 1 - \frac{Q_L}{Q_H} = 1 - \frac{mC_v(T_4 - T_1)}{mC_v(T_3 - T_2)}$$

$$= 1 - \frac{T_1}{T_2}\frac{(T_4/T_1 - 1)}{(T_3/T_2 - 1)}$$

We note further then

$$\frac{T_2}{T_1} = \left(\frac{V_1}{V_2}\right)^{k-1} = \left(\frac{V_4}{V_3}\right)^{k-1} = \frac{T_3}{T_4}$$

Therefore,

$$\frac{T_3}{T_2} = \frac{T_4}{T_1}$$

and

$$\eta_{th} = 1 - \frac{T_1}{T_2} = 1 - (r_v)^{1-k} = 1 - \frac{1}{r_v^{\,k-1}} \tag{12.4}$$

where r_v = compression ratio $= \dfrac{V_1}{V_2} = \dfrac{V_4}{V_3}$

The important thing to note is that the efficiency of the air-standard Otto cycle is a function only of the compression ratio, and that the efficiency is increased by increasing the compression ratio. Figure 12.23

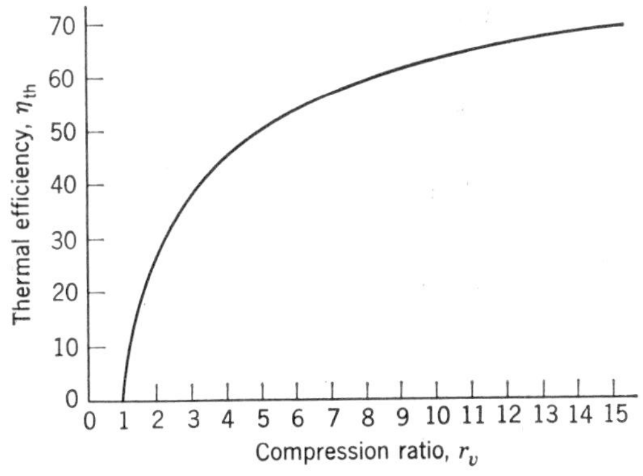

Fig. 12.23 Thermal efficiency of the Otto cycle as a function of compression ratio.

is a plot of the air-standard cycle thermal efficiency vs. compression ratio. It is also true of an actual spark-ignition engine that the efficiency can be increased by increasing the compression ratio. The trend toward higher compression ratios is prompted by the effort to obtain higher thermal efficiency. In the actual engine there is an increased tendency towards detonation of the fuel as compression ratio is increased. Detonation is characterized by an extremely rapid burning of the fuel and strong pressure waves present in the engine cylinder which give rise to the so-called spark knock. Therefore, the maximum compression ratio that can be used is fixed by the fact that detonation must be avoided. The advance in compression ratios over the years in the actual engine has been made possible by developing fuels with better antiknock characteristics.

Some of the most important ways in which the actual open-cycle spark-ignition engine deviates from the air-standard cycle are as follows:

(*a*) The specific heats of the actual gases increase with an increase in temperature.

(*b*) The combustion process replaces the heat-transfer process at high temperature, and combustion may be incomplete.

(*c*) Each mechanical cycle of the engine involves an inlet and an exhaust process, and due to the pressure drop through the valves a certain amount of work is required to charge the cylinder with air and exhaust the products of combustion.

(*d*) There will be considerable heat transfer between the gases in the cylinder and the cylinder walls.

(*e*) There will be irreversibilities associated with pressure and temperature gradients.

Example 12.10 The compression ratio in an air-standard Otto cycle is 8. At the beginning of the compression stroke the pressure is 14.7 lbf/in.² and the temperature is 60 F. The heat transfer to the air per cycle is 800 Btu/lbm air. Determine:

(*a*) The pressure and temperature at the end of each process of the cycle.

(*b*) The thermal efficiency.

(*c*) The mean effective pressure.

Designating the states as in Fig. 12.22 we have

$$P_1 = 14.7 \text{ lbf/in.}^2 \qquad T_1 = 520 \text{ R}$$

$$v_1 = \frac{53.34 \times 520}{14.7 \times 144} = 13.08 \text{ ft}^3\text{/lbm}$$

$$\frac{T_2}{T_1} = \left(\frac{V_1}{V_2}\right)^{k-1} = 8^{0.4} = 2.3 \qquad T_2 = 2.3(520) = 1197 \text{ R}$$

$$\frac{P_2}{P_1} = \left(\frac{V_1}{V_2}\right)^{k} = 8^{1.4} = 18.4 \qquad P_2 = 18.4(14.7) = 270.3 \text{ lbf/in.}^2$$

$$v_2 = \frac{13.08}{8} = 1.637 \text{ ft}^3/\text{lbm}$$

$$_2q_3 = C_V(T_3 - T_2) = 800 \text{ Btu/lbm}$$

$$T_3 - T_2 = \frac{800}{0.171} = 4690 \text{ R} \qquad T_3 = 1197 + 4690 = 5887 \text{ R}$$

$$\frac{T_3}{T_2} = \frac{P_3}{P_2} = \frac{5887}{1197} = 4.92 \qquad P_3 = 4.92(270.3) = 1331 \text{ lbf/in.}^2$$

$$\frac{T_3}{T_4} = \left(\frac{V_4}{V_3}\right)^{k-1} = 8^{0.4} = 2.3 \qquad T_4 = \frac{5887}{2.3} = 2558 \text{ R}$$

$$\frac{P_3}{P_4} = \left(\frac{V_4}{V_3}\right)^{k} = 8^{1.4} = 18.4 \qquad P_4 = \frac{1331}{18.4} = 73.0 \text{ lbf/in.}^2$$

$$\eta_{\text{th}} = 1 - \frac{1}{r_v^{\,k-1}} = 1 - \frac{1}{8^{0.4}} = 1 - \frac{1}{2.3} = 1 - 0.435 = 0.565$$

This can be checked by finding heat rejected

$$_4q_1 = C_V(T_1 - T_4) = 0.171(520 - 2558) = -348 \text{ Btu/lbm}$$

$$\therefore \eta_{\text{th}} = 1 - \frac{348}{800} = 1 - 0.435 = 0.565$$

$$w_{\text{net}} = 800 - 348 = 452 \text{ Btu/lbm} = (v_1 - v_2)\text{mep}$$

$$\text{mep} = \frac{452 \times 778}{(13.08 - 1.637)144} = 213.5 \text{ lbf/in.}^2$$

Note how much higher this mean effective pressure is than for the Carnot cycle of Example 12.9. A low mean effective pressure means a large piston displacement for a given power output, and a large piston displacement means high frictional losses in an actual engine. In fact, the mean effective pressure of the Carnot cycle of Example 12.9 would scarcely overcome the friction in an actual engine.

12.14 The Air-Standard Diesel Cycle

The air-standard Diesel cycle is shown in Fig. 12.24. This is the ideal cycle for the Diesel engine, which is also called the compression-ignition engine.

In this cycle the heat is transferred to the working fluid at constant pressure. This process corresponds to the injection and burning of the fuel in the actual engine. Since the gas is expanding during the heat addition in the air-standard cycle, the heat transfer must be just sufficient to maintain constant pressure. When state 3 is reached the heat addition ceases and the gas undergoes an isentropic expansion, process 3–4, until the piston reaches crank-end dead center. As in the

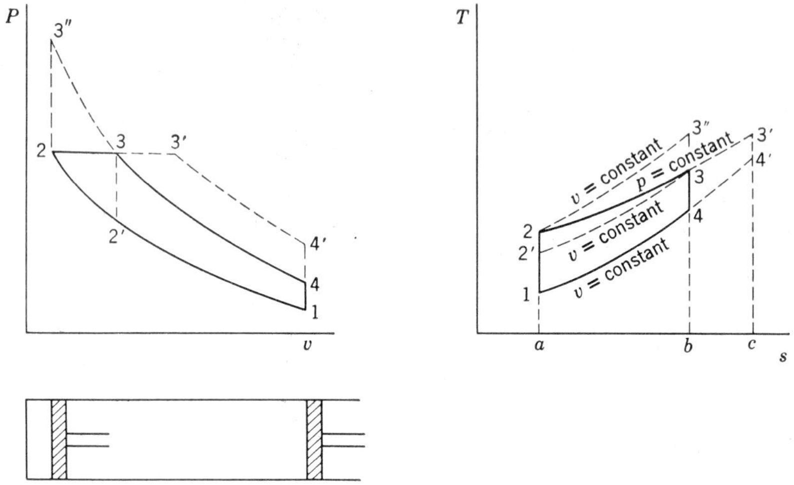

Fig. 12.24 The air-standard Diesel cycle.

air-standard Otto cycle, a constant-volume rejection of heat at crank-end dead center replaces the exhaust and intake processes of the actual engine.

The efficiency of the Diesel cycle is given by the relation

$$\eta_{\text{th}} = 1 - \frac{Q_L}{Q_H} = 1 - \frac{C_v(T_4 - T_1)}{C_p(T_3 - T_2)} = 1 - \frac{T_1(T_4/T_1 - 1)}{kT_2(T_3/T_2 - 1)} \qquad (12.5)$$

It is important to note that the isentropic compression ratio is greater than the isentropic expansion ratio in the Diesel cycle. Also, for a given state before compression and a given compression ratio (i.e., given states 1 and 2) the cycle efficiency decreases as the maximum temperature increases. This is evident from the T-s diagram, because the constant-pressure and constant-volume lines converge, and increasing the temperature from 3 to 3′ requires a large addition of heat (area 3–3′–c–b–3) and results in a relatively small increase in work (area 3–3′–4′–4–3).

There are a number of comparisons between the Otto cycle and the Diesel cycle, but here we will note only two. Consider Otto cycle

1–2–3″–4–1 and Diesel cycle 1–2–3–4–1, which have the same state at the beginning of the compression stroke and the same piston displacement and compression ratio. It is evident from the T-s diagram that the Otto cycle has the higher efficiency. In practice, however, the Diesel engine can operate on a higher compression ratio than the spark-ignition engine. The reason is that in the spark-ignition engine an air-fuel mixture is compressed, and detonation (spark knock) becomes a serious problem if too high a compression ratio is used. This problem does not exist as such in the Diesel engine because only air is compressed during the compression stroke. The development of higher octane fuels has permitted the use of higher compression ratios in spark-ignition engines.

Therefore, we might compare an Otto cycle with a Diesel cycle and in each case select a compression ratio that might be achieved in practice. Such a comparison can be made by considering Otto cycle 1–2′–3–4–1 and Diesel cycle 1–2–3–4–1. The maximum pressure and temperature is the same for both cycles, which means that the Otto cycle has a lower compression ratio than the Diesel cycle. It is evident from the T-s diagram that in this case the Diesel cycle has the higher efficiency. Thus, the conclusions drawn from a comparison of these two cycles must always be related to the basis on which the comparison has been made.

The actual compression-ignition open cycle differs from the air-standard Diesel cycle in much the same way that the spark-ignition open cycle differs from the air-standard Otto cycle.

Example 12.11 An air-standard Diesel cycle has a compression ratio of 15, and the heat transferred to the working fluid per cycle is 800 Btu/lbm. At the beginning of the compression process the pressure is 14.7 lbf/in.^{2} and the temperature is 60 F. Determine:

(*a*) The pressure and temperature at each point in the cycle.

(*b*) The thermal efficiency.

(*c*) The mean effective pressure.

Designating the cycle as in Fig. 12.24 we have

$$P_1 = 14.7 \text{ lbf/in.}^2 \qquad T_1 = 520 \text{ R}$$

$$v_1 = \frac{53.34 \times 520}{14.7 \times 144} = 13.08 \text{ ft}^3/\text{lbm}$$

$$v_2 = \frac{v_1}{15} = \frac{13.08}{15} = 0.872 \text{ ft}^3/\text{lbm}$$

$$\frac{T_2}{T_1} = \left(\frac{V_1}{V_2}\right)^{k-1} = 15^{0.4} = 2.955$$

$$T_2 = 2.955(520) = 1535 \text{ R}$$

$$\frac{P_2}{P_1} = \left(\frac{V_1}{V_2}\right)^{k} = 15^{1.4} = 44.2$$

$$P_2 = 44.2(14.7) = 650 \text{ lbf/in.}^2$$

$$q_H = {}_2q_3 = C_p(T_3 - T_2) = 800 \text{ Btu/lbm}$$

$$T_3 - T_2 = \frac{800}{0.24} = 3333 \text{ R}$$

$$T_3 = 3333 + 1535 = 4868 \text{ R}$$

$$\frac{V_3}{V_2} = \frac{T_3}{T_2} = \frac{4868}{1535} = 3.17$$

$$v_3 = 3.17(0.872) = 2.77 \text{ ft}^3\text{/lbm}$$

$$\frac{T_3}{T_4} = \left(\frac{V_4}{V_3}\right)^{k-1} = \left(\frac{13.08}{2.83}\right)^{0.4} = 1.860 \qquad T_4 = \frac{4868}{1.860} = 2620 \text{ R}$$

$$q_L = {}_4q_1 = C_v(T_1 - T_4) = 0.171(520 - 2620) = -359 \text{ Btu/lbm}$$

$$w_n = 800 - 359 = 441 \text{ Btu/lbm}$$

$$\eta_{\text{th}} = \frac{w_n}{q_H} = \frac{441}{800} = 0.551$$

$$\text{mep} = \frac{w_{\text{net}}}{v_1 - v_2} = \frac{441 \times 778}{(13.08 - 0.87)144} = 195 \text{ lbf/in.}^2$$

12.15 Some Other Cycles

Many other cycles have been proposed that use a gaseous working fluid, and engines have been built that approximate these cycles. Most of these cycles can be analyzed qualitatively by means of an air-standard cycle. A number are presented in this section. The analysis of these cycles is similar to the analysis already made of the air-standard Carnot, Otto, and Diesel cycles, and therefore only a P-v and T-s diagram and a few comments are given for each.

The limited-pressure cycle The combustion process in a spark-ignition engine does not occur exactly at constant volume, nor does the combustion process in an actual compression-ignition engine occur exactly at constant pressure. In order to more closely approximate these actual cycles, the limited-pressure air-standard cycle (also called the dual cycle) has been developed. This cycle is shown in Fig. 12.25.

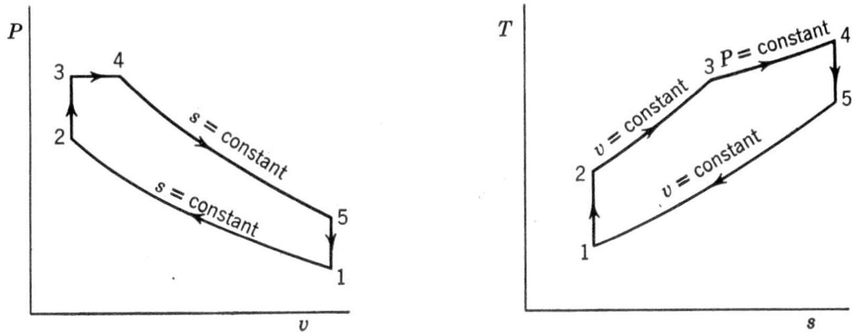

Fig. 12.25 The air-standard limited-pressure cycle.

Some heat is first transferred to the working fluid at constant volume and the remainder is transferred at constant pressure.

Stirling cycle The Stirling cycle is shown on the P-v and T-s diagrams of Fig. 12.26. Heat is transferred to the working fluid during the

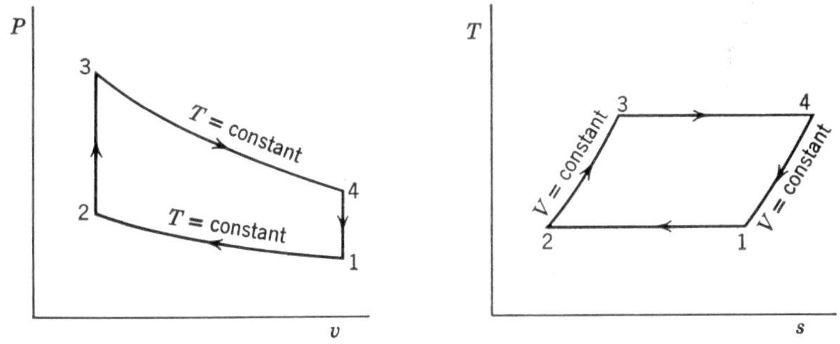

Fig. 12.26 The air-standard Stirling cycle.

constant-volume process 2–3 and during the isothermal expansion process 3–4. Heat is rejected during the constant-volume process 4–1 and during the isothermal compression process 1–2. The significance of this cycle in conjunction with a regenerator is discussed in the next paragraph.

Ericsson cycle The Ericsson cycle is shown on the P-v and T-s diagrams of Fig. 12.27. This cycle differs from the Stirling cycle in that the constant-volume processes of the Stirling cycle are replaced by constant-pressure processes. In both cycles there is an isothermal compression and expansion.

The importance of both cycles is the possibility of including a regenerator; by so doing the air-standard Stirling and Ericsson cycles may

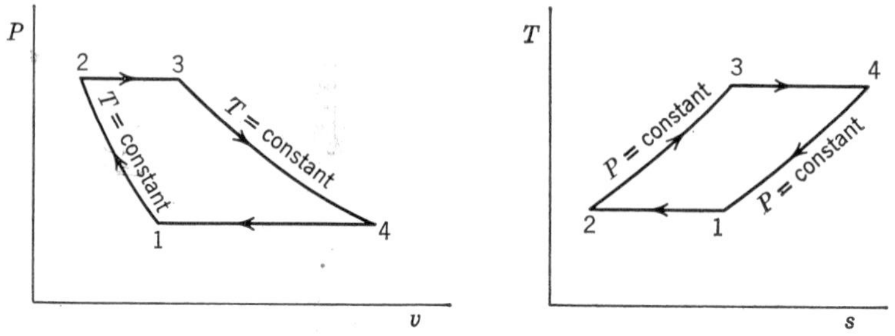

Fig. 12.27 The air-standard Ericsson cycle.

have an efficiency equal to that of a Carnot cycle operating between the same temperatures. This may be demonstrated by considering Fig. 12.28, in which the Ericsson cycle is accomplished in a device that is essentially a gas turbine. If we assume an ideal heat-transfer process in the regenerator, i.e., no pressure drop and an infinitesimal temperature difference between the two streams, and reversible compression and expansion processes, then this device operates on the Ericsson cycle.

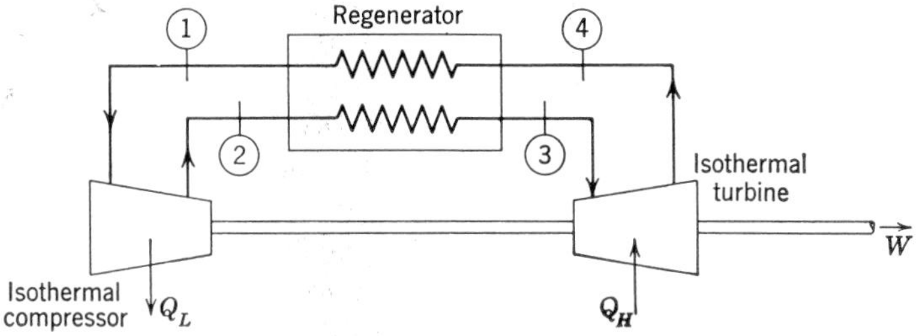

Fig. 12.28 Schematic arrangement of an engine operating on the Ericsson cycle and utilizing a regenerator.

Since all the heat is supplied and rejected isothermally, the efficiency of this cycle will equal the efficiency of a Carnot cycle operating between the same temperatures. A similar cycle could be developed which would approximate the Stirling cycle.

The difficulties in achieving such a cycle are primarily those associated with heat transfer. It is difficult to achieve an isothermal compression or expansion in a machine operating at a reasonable speed, and there will be pressure drops in the regenerator and a temperature difference between the two streams flowing through the regenerator. However, the gas turbine with intercooling and regenerators, which is described in Section 12.17, is a practical attempt to approach the Ericsson cycle. There have also been attempts to approach the Stirling cycle with the use of regenerators.

Atkinson cycle The Atkinson cycle is shown on the P-v and T-s diagrams of Fig. 12.29. It is the ideal cycle for an Otto cycle exhaust-

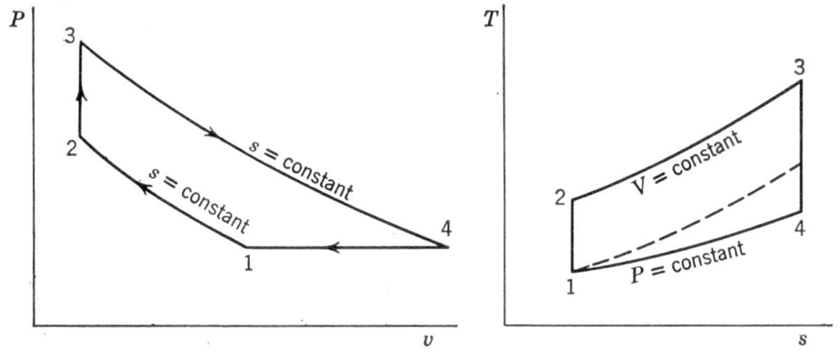

Fig. 12.29 The air-standard Atkinson cycle.

ing to a gas turbine. Compound engines that approach this cycle have been built and successfully operated.

Lenoir cycle This cycle, shown in Fig. 12.30, was developed about 1860. It is unique in that the engine operates without compressing the working fluid. Rather, the heat transfer (combustion in the actual engine) takes place at constant volume, thus causing an increase in

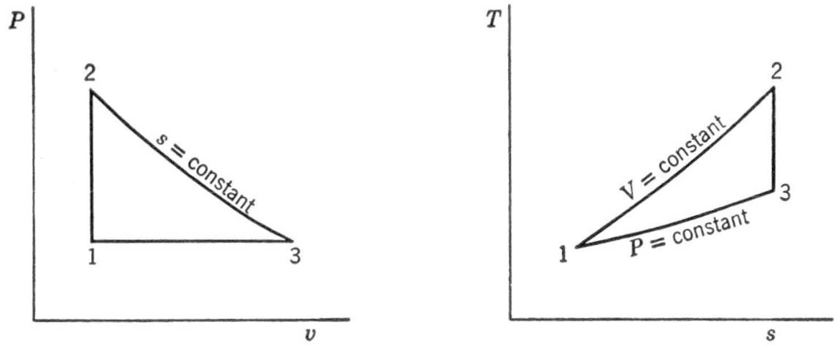

Fig. 12.30 The air-standard Lenoir cycle.

pressure. Work is then done by expanding the gas, and the heat rejection takes place at constant pressure. This cycle is obsolete as far as a piston machine is concerned, but it is approximated by the pulse jet which was developed in World War II.

12.16 The Brayton Cycle

The air-standard Brayton cycle is the ideal cycle for the simple gas turbine. The simple open-cycle gas turbine utilizing an internal-combustion process and the simple closed-cycle gas turbine, which

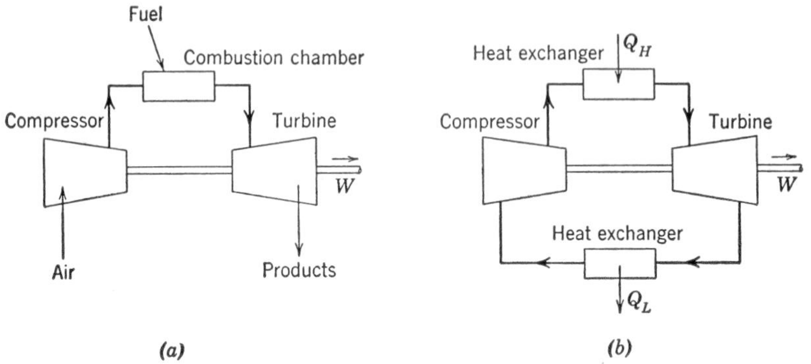

Fig. 12.31 A gas turbine operating on the Brayton cycle. (a) Open cycle. (b) Closed cycle.

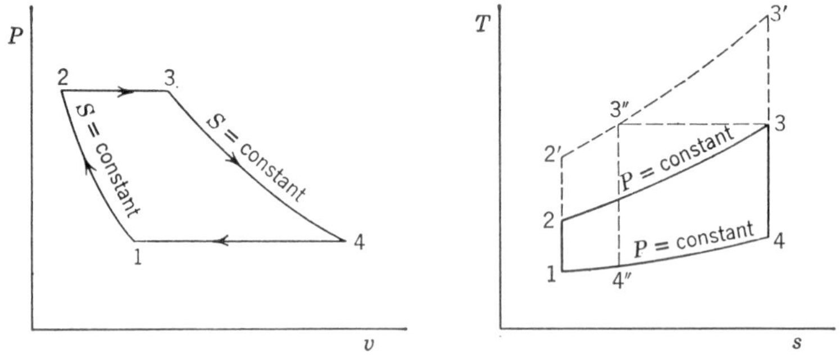

Fig. 12.32 The air-standard Brayton cycle.

utilizes heat-transfer processes, are both shown schematically in Fig. 12.31. The air-standard Brayton cycle is shown on the P-v and T-s diagrams of Fig. 12.32.

The efficiency of the air-standard Brayton cycle is found as follows:

$$\eta_{th} = 1 - \frac{Q_L}{Q_H} = 1 - \frac{C_p(T_4 - T_1)}{C_p(T_3 - T_2)} = 1 - \frac{T_1(T_4/T_1 - 1)}{T_2(T_3/T_2 - 1)}$$

We note, however, that

$$\frac{P_3}{P_4} = \frac{P_2}{P_1} \qquad \therefore \frac{P_3}{P_2} = \frac{P_4}{P_1}$$

$$\frac{P_2}{P_1} = \left(\frac{T_2}{T_1}\right)^{k/(k-1)} = \frac{P_3}{P_4} = \left(\frac{T_3}{T_4}\right)^{k/(k-1)}$$

$$\frac{T_3}{T_4} = \frac{T_2}{T_1} \qquad \therefore \frac{T_3}{T_2} = \frac{T_4}{T_1} \quad \text{and} \quad \frac{T_3}{T_2} - 1 = \frac{T_4}{T_1} - 1$$

$$\eta_{\text{th}} = 1 - \frac{T_1}{T_2} = 1 - \frac{1}{(P_2/P_1)^{(k-1)/k}} \tag{12.6}$$

The efficiency of the air-standard Brayton cycle is therefore a function of isentropic pressure ratio; Fig. 12.33 shows a plot of efficiency vs.

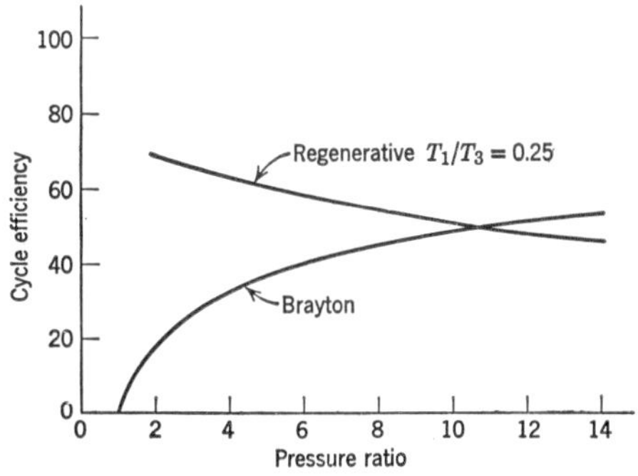

Fig. 12.33 Cycle efficiency as a function of pressure ratio for the Brayton and regenerative cycles.

pressure ratio. The fact that efficiency increases with pressure ratio is evident from the T-s diagram of Fig. 12.32, because increasing the pressure ratio will change the cycle from 1–2–3–4–1 to 1–2′–3′–4–1. The latter cycle has a greater heat supply and the same heat rejected as the original cycle, and therefore it has a greater efficiency. Note further that the latter cycle has a higher maximum temperature (T_3') than the original cycle (T_3). In an actual gas turbine the maximum temperature of the gas entering the turbine is fixed by metallurgical considerations. Therefore, if we fix the temperature T_3 and increase the pressure ratio, the resulting cycle is 1–2′–3″–4″–1. This cycle would have a higher efficiency than the original cycle, but the work per pound of working fluid is thereby changed.

With the advent of nuclear reactors the closed-cycle gas turbine has become more important. Heat is transferred, either directly or via a second fluid, from the fuel in the nuclear reactor to the working fluid in the gas turbine. Heat is rejected from the working fluid to the surroundings.

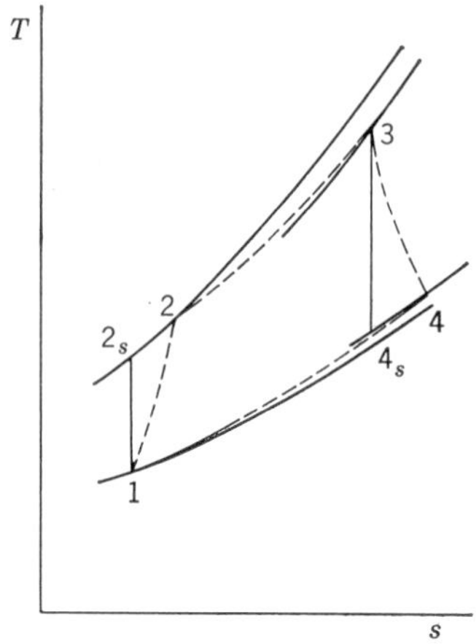

Fig. 12.34 Effect of inefficiencies on the gas-turbine cycle.

The actual gas-turbine engine differs from the ideal cycle primarily because of irreversibilities in the compressor and turbine, and because of pressure drop in the flow passages and combustion chamber (or in the heat exchanger of a closed-cycle turbine). Thus, the state points in a simple open-cycle gas turbine might be as shown in Fig. 12.34.

The efficiencies of the compressor and turbine are defined in relation to isentropic processes. Designating the states as in Fig. 12.34, the definitions of compressor and turbine efficiencies are as follows:

$$\eta_{\text{comp}} = \frac{h_{2s} - h_1}{h_2 - h_1} \tag{12.7}$$

$$\eta_{\text{turb}} = \frac{h_3 - h_4}{h_3 - h_{4s}} \tag{12.8}$$

One other important feature to note about the Brayton cycle is the large amount of compressor work (also called back work) as compared to the turbine work. Thus, the compressor might require from 40 per cent to 80 per cent of the output of the turbine. This is particularly important when the actual cycle is considered, because the effect of the losses is to require a larger amount of compression work from a smaller amount of turbine work, and thus the over-all efficiency drops very rapidly with a decrease in the efficiencies of the compressor and turbine. In fact, if these efficiencies drop much below 60 per cent, all the work of the turbine will be required to drive the compressor, and the over-all efficiency will be zero. This is in sharp contrast to the Rankine cycle, where only 1 or 2 per cent of the turbine work is required to drive the pump. The reason for this is that for a reversible steady-flow process with negligible change in kinetic and potential energy, the work is equal to $-\int v\,dP$. Because we are pumping a liquid in the Rankine cycle, the specific volume is very low compared to the specific volume of the gas in a gas turbine. This matter is illustrated in the following examples:

Example 12.12 In an air-standard Brayton cycle the air enters the compressor at 14.7 lbf/in.^{2}, 60 F. The pressure leaving the compressor is 70 lbf/in.^{2} and the maximum temperature in the cycle is 1600 F. Determine:

(*a*) The pressure and temperature at each point in the cycle.

(*b*) The compressor work, turbine work, and cycle efficiency.

Designating the points as in Fig. 12.32 we have

$$P_1 = P_4 = 14.7 \text{ lbf/in.}^2 \qquad T_1 = 520 \text{ R}$$

$$P_2 = P_3 = 70 \text{ lbf/in.}^2$$

$$\left(\frac{P_2}{P_1}\right)^{(k-1)/k} = \left(\frac{70}{14.7}\right)^{0.286} = \frac{T_2}{T_1} = 1.563 \qquad T_2 = 1.563(520) = 814 \text{ R}$$

$$q_H = C_p(T_3 - T_2) = 0.24(2060 - 814) = 299.0 \text{ Btu/lbm}$$

$$\left(\frac{P_3}{P_4}\right)^{(k-1)/k} = \left(\frac{70}{14.7}\right)^{0.286} = \frac{T_3}{T_4} = 1.563 \qquad T_4 = \frac{2060}{1.563} = 1318 \text{ R}$$

$$q_L = C_p(T_4 - T_1) = 0.24(1318 - 520) = 191.4 \text{ Btu/lbm}$$

$$w_c = h_2 - h_1 = C_p(T_2 - T_1) = 0.24(814 - 520)$$

$$= 70.6 \text{ Btu/lbm}$$

Note that the compressor work w_c is here defined as work input to the compressor.

$$w_T = h_3 - h_4 = C_p(T_3 - T_4) = 0.24(2060 - 1318) = 177.8 \text{ Btu/lbm}$$

$$w_n = w_T - w_c = 177.8 - 70.6 = 107.2 \text{ Btu/lbm}$$

We can also write

$$w_n = q_H - q_L = 299.0 - 191.4 = 107.6 \text{ Btu/lbm}$$

$$\eta_{\text{th}} = \frac{w_n}{q_H} = \frac{107.6}{299.0} = 35.95\%$$

This may be checked by using Eq. 12.6

$$\eta = 1 - \frac{1}{(P_2/P_1)^{(k-1)/k}} = 1 - \frac{1}{(70/14.7)^{0.286}} = 1 - \frac{1}{1.563}$$

$$= 1 - 0.64 = 36\%$$

Example 12.13 Consider a gas turbine with air entering the compressor under the same conditions as in Example 12.12, and leaving at a pressure of 70 lbf/in.^{2} The maximum temperature is 1600 F. Assume a compressor efficiency of 80 per cent, a turbine efficiency of 85 per cent, and a pressure drop between the compressor and turbine of 2 lbf/in.^{2} Determine the compressor work, turbine work, and cycle efficiency.

Designating the states as in Fig. 12.34 we have

$$\left(\frac{P_2}{P_1}\right)^{(k-1)/k} = \frac{T_{2s}}{T_1} = \left(\frac{70}{14.7}\right)^{0.286} = 1.563 \qquad T_{2s} = 1.563(520) = 814\ \text{R}$$

$$\eta_c = \frac{h_{2s} - h_1}{h_2 - h_1} = \frac{T_{2s} - T_1}{T_2 - T_1} = \frac{814 - 520}{T_2 - T_1} = 0.80$$

$$T_2 - T_1 = \frac{294}{0.80} = 367\ \text{R} \qquad T_2 = 520 + 367 = 887\ \text{R}$$

$$w_c = h_2 - h_1 = C_p(T_2 - T_1) = 0.24(887 - 520)$$

$$= 87.9\ \text{Btu/lbm}$$

$$q_H = C_p(T_3 - T_2) = 0.24(2060 - 887) = 281.6\ \text{Btu/lbm}$$

$$P_3 = P_2 - \text{pressure drop} = 70 - 2 = 68\ \text{lbf/in.}^2$$

$$\left(\frac{P_3}{P_4}\right)^{(k-1)/k} = \frac{T_3}{T_{4s}} = \left(\frac{68}{14.7}\right)^{0.286} = 1.550 \qquad T_{4s} = \frac{2060}{1.550}$$

$$\eta_T = \frac{h_3 - h_4}{h_3 - h_{4s}} = \frac{T_3 - T_4}{T_3 - T_{4s}} = 0.85$$

$$T_3 - T_4 = 0.85(2060 - 1330) = 620\ \text{R}$$

$$T_4 = 2060 - 620 = 1440\ \text{R}$$

$$w_T = h_3 - h_4 = C_p(T_3 - T_4) = 0.24(2060 - 1440) = 148.8$$

$$w_n = w_T - w_c = 148.8 - 87.9 = 60.9\ \text{Btu/lbm}$$

$$\eta_{\text{th}} = \frac{w_n}{q_H} = \frac{60.9}{281.6} = 21.6\%$$

The following comparisons can be made between Examples 12.12 and 12.13.

	w_c	w_T	w_n	q_H	η_{th}
Example 12.12 (Ideal)	70.6	177.8	107.6	299	36.0
Example 12.13 (Actual)	87.9	148.8	60.9	281.6	21.6

As stated previously the result of the irreversibilities is to decrease the turbine work and increase the compressor work. Since the net work is the difference between these two it decreases very rapidly as compressor and turbine efficiencies decrease. The development of compressors and turbines of high efficiency is therefore an important aspect of the development of gas turbines.

Note also that in the ideal cycle (Example 12.12) about 40 per cent of the turbine work is required to drive the compressor and 60 per cent is delivered as net work. In the actual turbine (Example 12.13) 59 per cent of the turbine work is required to drive the compressor and 41 per cent is delivered as net work. Thus, if the net power of this unit is to be 10,000 hp, a 25,000-hp turbine and a 15,000-hp compressor are required. This demonstrates the statement that a gas turbine has a high back-work ratio.

12.17 The Simple Gas-Turbine Cycle with Regenerator

The efficiency of the gas-turbine cycle may be improved by introducing a regenerator. The simple open-cycle gas-turbine cycle with regenerator is shown in Fig. 12.35, and the corresponding ideal air-

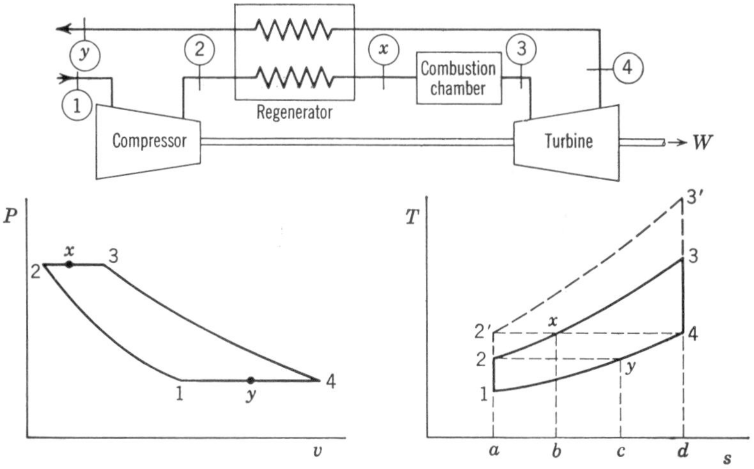

Fig. 12.35 The ideal regenerative cycle.

standard cycle with regenerator is shown on the P-v and T-s diagrams. Note that in cycle 1–2–x–3–4–y–1, the temperature of the exhaust gas leaving the turbine in state 4 is higher than the temperature of the gas leaving the compressor. Therefore, heat can be transferred from the exhaust gases to the high-pressure gases leaving the compressor. If this is done in a counterflow heat exchanger, which is known as a regenerator, the temperature of the high-pressure gas leaving the regenerator, T_x, may, in the ideal case, have a temperature equal to T_4, the temperature of the gas leaving the turbine. In this case heat transfer from the external source is necessary only to increase the temperature from T_x to T_3, and this heat transfer is represented by area x–3–d–b–x. Area y–1–a–c–y represents the heat rejected.

The influence of pressure ratio on the simple gas-turbine cycle with regenerator is shown by considering cycle 1–2′–3′–4–1. In this cycle the temperature of the exhaust gas leaving the turbine is just equal to the temperature of the gas leaving the compressor; therefore there is no possibility of utilizing a regenerator. This may be shown more exactly by determining the efficiency of the ideal gas-turbine cycle with regenerator.

The efficiency of this cycle with regeneration is found as follows, where the states are as given in Fig. 12.35.

$$\eta_{\text{th}} = \frac{w_n}{q_H} = \frac{w_T - w_c}{q_H}$$

$$q_H = C_p(T_3 - T_x)$$

$$w_T = C_p(T_3 - T_4)$$

But for the ideal regenerator $T_4 = T_x$, and therefore $q_H = w_T$. Therefore,

$$\eta_{\text{th}} = 1 - \frac{w_c}{w_T} = 1 - \frac{C_p(T_2 - T_1)}{C_p(T_3 - T_4)}$$

$$= 1 - \frac{T_1(T_2/T_1 - 1)}{T_3(1 - T_4/T_3)} = 1 - \frac{T_1\,[(P_2/P_1)^{(k-1)/k} - 1]}{T_3\,[1 - (P_1/P_2)^{(k-1)/k}]}$$

$$\eta_{\text{th}} = 1 - \frac{T_1}{T_3}\left(\frac{P_2}{P_1}\right)^{(k-1)/k}$$

Thus, we see that for the ideal cycle with regeneration the thermal efficiency depends not only upon the pressure ratio, but also upon the ratio of the minimum to maximum temperature. We also note, in contrast to the Brayton cycle, that the efficiency decreases with an increase in pressure ratio. The thermal efficiency vs. pressure ratio for this cycle is plotted in Fig. 12.33 for a value of

$$\frac{T_1}{T_3} = 0.25$$

The effectiveness or efficiency of a regenerator is given by the term regenerator efficiency. This may best be defined by reference to Fig. 12.36. State x represents the high-pressure gas leaving the regenerator.

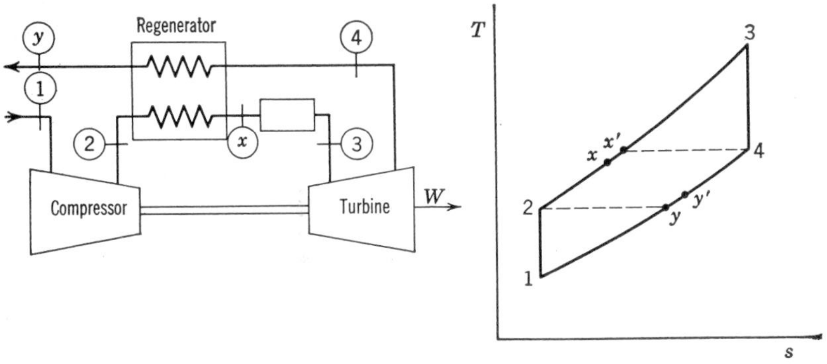

Fig. 12.36 Temperature-entropy diagram to illustrate the definition of regenerator efficiency.

In the ideal regenerator there would be only an infinitesimal temperature difference between the two streams, and the high-pressure air would leave the regenerator at temperature T_x', and $T_x' = T_4$. In an actual regenerator, which must operate with a finite temperature difference, T_x, the actual temperature leaving the regenerator is therefore less than T_x'. The regenerator efficiency is defined by

$$\eta_{\text{reg}} = \frac{h_x - h_2}{h_x' - h_2}$$

If the specific heat is assumed to be constant, the regenerator efficiency is also given by the relation

$$\eta_{\text{reg}} = \frac{T_x - T_2}{T_x' - T_2}$$

It should also be pointed out that a higher efficiency can be achieved by using a regenerator with greater heat-transfer area. However, this also increases the pressure drop, which represents a loss, and both the pressure drop and the regenerator efficiency must be considered in determining which regenerator gives maximum thermal efficiency for the cycle. From an economic point of view, the cost of the regenerator must be weighed against the saving that can be effected by its use.

Example 12.14 If an ideal regenerator is incorporated into the cycle of Example 12.12, determine the thermal efficiency of the cycle.

Designating the states as in Fig. 12.35,

$$T_x = T_4 = 1318 \text{ R}$$

$$q_H = h_3 - h_x = C_p(T_3 - T_x) = 0.24(2060 - 1318) = 178 \text{ Btu/lbm}$$

$$w_n = 107.6 \text{ Btu/lbm} \qquad \text{(from Example 12.12)}$$

$$\eta_{\text{th}} = \frac{107.6}{178} = 60.4\%$$

12.18 The Ideal Gas-Turbine Cycle Using Multistage Compression with Intercooling, Multistage Expansion with Reheating, and Regenerator

In Section 12.15 it was pointed out that when an ideal regenerator was incorporated into the Ericsson cycle, an efficiency equal to the efficiency of the corresponding Carnot cycle could be attained. It was also pointed out that it is impossible to even closely approach in practice the reversible isothermal compression and expansion required in the Ericsson cycle.

The practical approach to this cycle is to use multistage compression with intercooling between stages, multistage expansion with reheat between stages, and a regenerator. Figure 12.37 shows a cycle with two stages of compression and two stages of expansion. The air-standard cycle is shown on the corresponding T-s diagram. It may be shown that for this cycle the maximum efficiency is obtained if equal pressure ratios are maintained across the two compressors and the two turbines. In this ideal cycle it is assumed that the temperature of the air leaving the intercooler, T_3, is equal to the temperature of the air entering the first stage of compression, T_1, and that the temperature after reheating, T_8, is equal to the temperature entering the first turbine, T_6. Further, in the ideal cycle it is assumed that the temperature of the high-pressure air leaving the regenerator, T_5, is equal to the temperature of the low-pressure air leaving the turbine, T_9.

If a large number of stages of compression and expansion are used, it is evident that the Ericsson cycle is approached. This is shown in Fig. 12.38. In practice the economical limit to the number of stages is usually two or three. The turbine and compressor losses and pressure drops which have already been discussed would be involved in any actual unit employing this cycle.

There are a variety of ways in which the turbines and compressors using this cycle can be utilized. Two possible arrangements for closed

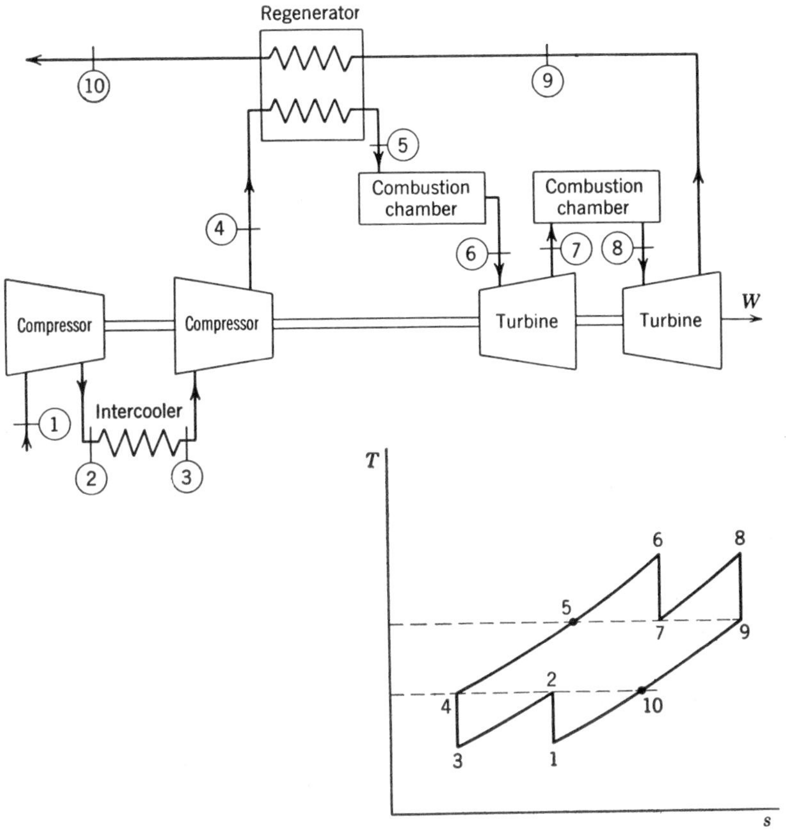

Fig. 12.37 The ideal gas-turbine cycle utilizing intercooling, reheat, and a regenerator.

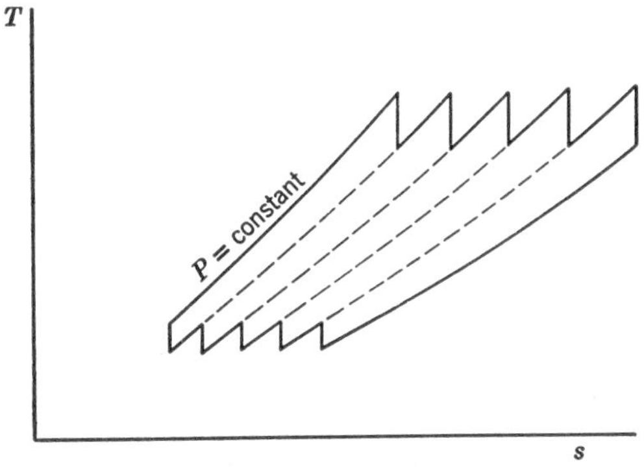

Fig. 12.38 Temperature-entropy diagram which shows how the gas-turbine cycle with many stages approaches the Ericsson cycle.

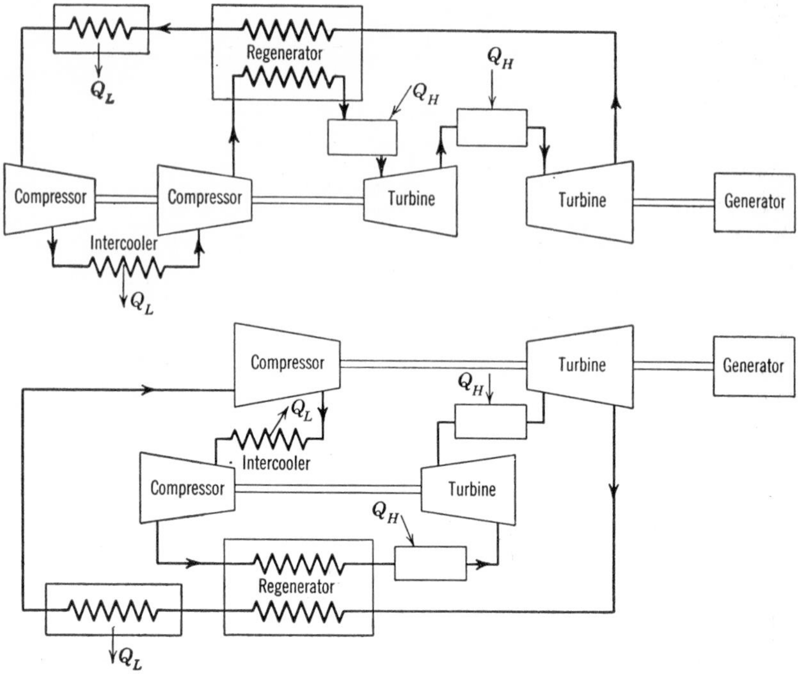

Fig. 12.39 Some arrangements of components that may be utilized in stationary gas-turbine power plants.

cycles are shown in Fig. 12.39. One advantage frequently sought in a given arrangement is ease of control of the unit under various loads. Detailed discussion of this point, however, is beyond the scope of this text.

12.19 The Air-Standard Cycle for Jet Propulsion

The last air-standard power cycle we will consider is utilized in jet propulsion. In this cycle the work done by the turbine is just sufficient to drive the compressor—the gases are expanded in the turbine to a pressure such that the turbine work is just equal to the compressor work. The exhaust pressure of the turbine will then be above that of the surroundings, and the gas can be expanded in a nozzle to the pressure of the surroundings. Since the gases leave at a high velocity, the change in momentum the gases undergo results in a thrust upon the aircraft in which the engine is installed. The air-standard cycle for this is shown in Fig. 12.40.

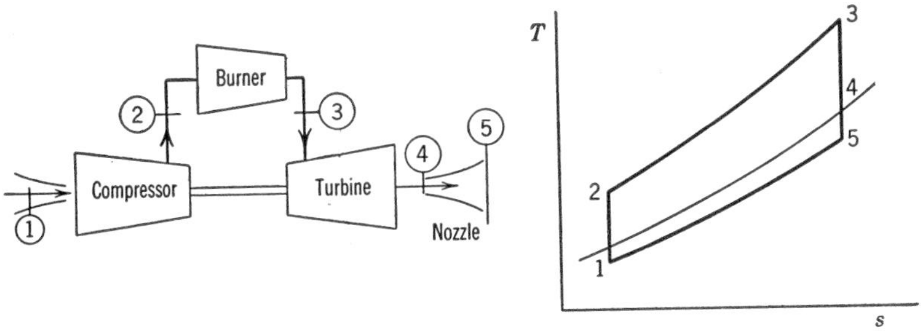

Fig. 12.40 The ideal gas-turbine cycle for a jet engine.

Since all the principles involved in this cycle have already been covered, the example that follows will conclude our consideration of air-standard power cycles.

Example 12.15 Consider an ideal cycle in which air enters the compressor at 14.7 lbf/in.^{2}, 60 F. The pressure leaving the compressor is 70 lbf/in.^{2} and the maximum temperature is 1600 F. The air expands in the turbine to such a pressure that the turbine work is just equal to the compressor work. On leaving the turbine the air expands in a reversible adiabatic process in a nozzle to 14.7 lbf/in.^{2} Determine the velocity of the air leaving the nozzle.

From Example 12.12 we have the following, where the states are as designated in Fig. 12.40:

$$P_1 = 14.7 \text{ lbf/in.}^2 \qquad T_1 = 520 \text{ R}$$

$$P_2 = 70 \text{ lbf/in.}^2 \qquad T_2 = 814 \text{ R}$$

$$w_c = 70.6 \text{ Btu/lbm}$$

$$P_3 = 70 \text{ lbf/in.}^2 \qquad T_3 = 2060 \text{ R}$$

$$w_c = w_T = C_p(T_3 - T_4)$$

$$= 70.6 \text{ Btu/lbm}$$

$$T_3 - T_4 = \frac{70.6}{0.24} = 294 \text{ R}$$

$$T_4 = 2060 - 294 = 1766 \text{ R}$$

$$\frac{T_3}{T_4} = \left(\frac{P_3}{P_4}\right)^{(k-1)/k} = \frac{2060}{1766} = 1.167$$

$$\frac{P_3}{P_4} = 1.715 \qquad P_4 = \frac{70}{1.715} = 40.8 \text{ lbf/in.}^2$$

$$T_5 = 1318 \text{ R} \qquad \text{(from Example 12.12)}$$

If we assume that the velocity entering the nozzle is low, we can write the first law for the nozzle.

$$h_4 = h_5 + \frac{\overline{V}_5^{\,2}}{2g_c} \qquad \overline{V}_5^{\,2} = 2g_cC_p(T_4 - T_5)$$

$$\overline{V}_5^{\,2} = 2 \times 32.17 \times 778 \times 0.24(1766 - 1318)$$

$$\overline{V}_5 = 2318 \text{ ft/sec}$$

AIR-STANDARD REFRIGERATION CYCLE

12.20 The Air-Standard Refrigeration Cycle

The final section of this chapter concerns the air-standard refrigeration cycle. Its main use in practice is in the liquifaction of air and other gases and in certain special situations that require refrigeration.

The simplest form of the air-standard refrigeration cycle, which is essentially the reverse of the Brayton cycle, is shown in Fig. 12.41.

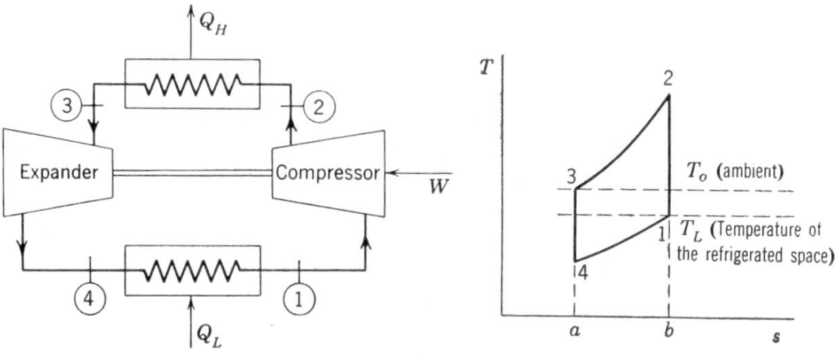

Fig. 12.41 The air-standard refrigeration cycle.

The compressor and expander might be either reciprocating or rotary. After compression from 1 to 2, the air is cooled as a result of heat transfer to the surroundings at temperature T_0. The air is then expanded in process 3–4 to the pressure entering the compressor, and the temperature drops to T_4 in the expander. Heat may then be transferred to the air until temperature T_L is reached. The work for this cycle is represented by area 1–2–3–4–1, and the refrigeration effect is represented by area 4–1–b–a–4. The coefficient of performance is the ratio of these two areas.

In practice this cycle has been utilized for the cooling of aircraft in an open cycle, a simplified form of which is shown in Fig. 12.42. Upon

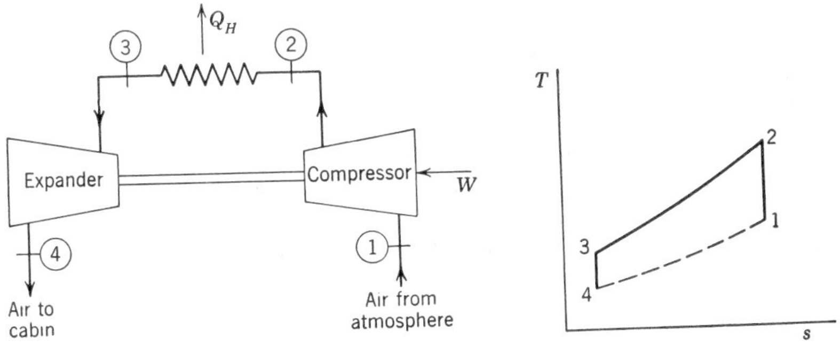

Fig. 12.42 An air refrigeration cycle that might be utilized for aircraft cooling.

leaving the expander the cool air is blown directly into the cabin thus providing the cooling effect where needed.

When counterflow heat exchangers are incorporated, very low temperatures can be obtained. This is essentially the cycle used in low-pressure air liquifaction plants and in other liquifaction devices such as the Collins helium liquifier. The ideal cycle in this case is as shown in Fig. 12.43. It is evident that the expander operates at very low

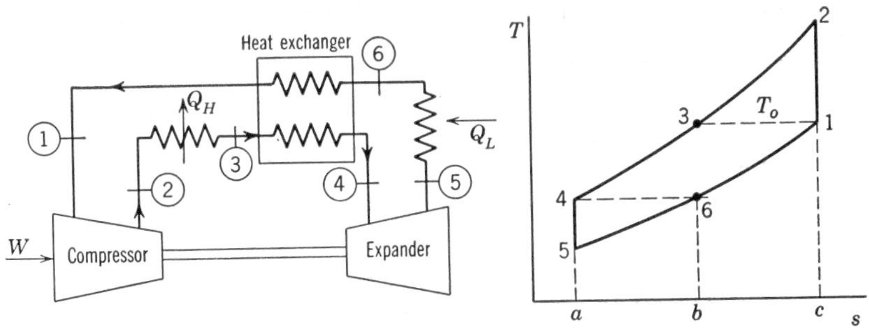

Fig. 12.43 The air-refrigeration cycle utilizing a heat exchanger.

temperature, which presents unique problems to the designer in regard to lubrication and materials.

Example 12.16 Consider the simple air-standard refrigeration cycle of Fig. 12.41. Air enters the compressor at 14.7 lbf/in.^{2}, 0 F, and leaves at 80 lbf/in.^{2} Air enters the expander at 60 F. Determine:

(a) The coefficient of performance for this cycle.

(*b*) The rate at which air must enter the compressor in order to provide one ton of refrigeration.

$$T_1 = 460 \text{ R} \qquad P_1 = 14.7 \text{ lbf/in.}^2$$

$$\frac{T_2}{T_1} = \left(\frac{P_2}{P_1}\right)^{(k-1)/k} = \left(\frac{80}{14.7}\right)^{0.286} = 1.624$$

$$T_2 = 460(1.624) = 747 \text{ R}$$

$$T_3 = 520 \text{ R}$$

$$\frac{T_3}{T_4} = \left(\frac{P_3}{P_4}\right)^{(k-1)/k} = \left(\frac{80}{14.7}\right)^{0.286} = 1.624$$

$$T_4 = \frac{520}{1.624} = 320 \text{ R}$$

$$w_c = h_2 - h_1 = C_p(T_2 - T_1) = 0.24(747 - 460) = 68.9 \text{ Btu/lbm}$$

(w_c designates work into the compressor)

$$q_H = h_2 - h_3 = C_p(T_2 - T_3) = 0.24(747 - 520) = 54.5 \text{ Btu/lbm}$$

$$w_T = h_3 - h_4 = 0.24(520 - 320) = 48.0 \text{ Btu/lbm}$$

(w_T designates work done by expander)

$$q_L = h_1 - h_4 = C_p(T_1 - T_4) = 0.24(460 - 320) = 33.6 \text{ Btu/lbm}$$

$$w_n = w_c - w_T = 68.9 - 48.0 = 20.9 \text{ Btu/lbm}$$

$$\beta = \frac{q_L}{w_n} = \frac{33.6}{20.9} = 1.61$$

1 ton of refrigeration = 12,000 Btu/hr = 200 Btu/min

$$\frac{\text{lbm air/min}}{\text{ton refrigeration}} = \frac{200}{33.6} = 5.95$$

PROBLEMS

12.1 In a Rankine cycle utilizing steam the steam enters the turbine at 400 lbf/in.², 600 F. Determine the thermal efficiency of the cycle and the moisture content leaving the turbine for exhaust pressures of 14.7 lbf/in.², 5 lbf/in.², 2 lbf/in.², 1 lbf/in.² Plot thermal efficiency vs. exhaust pressure for the given inlet conditions to the turbine.

12.2 A Rankine cycle has an exhaust pressure of 2 lbf/in.^{2} Determine the thermal efficiency and moisture content leaving the turbine for an inlet temperature of 600 F and the following inlet pressures: 100 lbf/in.^{2}, 200 lbf/in.^{2}, 400 lbf/in.^{2}, saturated vapor at 600 F. Plot thermal efficiency vs. turbine inlet pressure for the given turbine inlet temperature and exhaust pressure.

12.3 A Rankine cycle has an exhaust pressure of 2 lbf/in.^{2} and a pressure entering the turbine of 400 lbf/in.^{2} Determine the thermal efficiency and moisture content leaving the turbine for the following temperatures entering the turbine: saturated vapor at 400 lbf/in.^{2}, 600 F, 1000 F, 1400 F. Plot thermal efficiency vs. turbine inlet temperature for the given turbine inlet and exhaust pressures.

12.4 Consider the reheat cycle. Steam enters the turbine at 400 lbf/in.^{2}, 600 F and expands to 90 lbf/in.^{2} It is then reheated to 600 F, after which it expands to 2 lbf/in.^{2} Determine the thermal efficiency and the moisture content leaving the low-pressure turbine.

12.5 Consider the ideal regenerative cycle. Steam enters the turbine at 400 lbf/in.^{2}, 600 F. Condenser pressure is 2 lbf/in.^{2} Steam is extracted at 90 lbf/in.^{2} and 16 lbf/in.^{2} for purposes of heating the feedwater. The feedwater heaters are open heaters, and the feedwater leaves at the temperature of the condensing steam. The appropriate pumps are used for the water leaving the condenser and the two feedwater heaters. Determine the thermal efficiency and the net work per lbm of steam.

12.6 Repeat Problem 12.5 assuming closed feedwater heaters. A single pump is used which pumps the water leaving the condenser to 400 lbf/in.^{2} The condensate from the high-pressure heater is drained through a trap to the low-pressure heater, and the low-pressure heater is drained through a trap to the main condenser.

12.7 An ideal cycle combines the reheat and regenerative cycles. Steam enters the turbine at 400 lbf/in.^{2}, 600 F. Steam is extracted at 90 lbf/in.^{2} for purposes of feedwater heating. The steam not extracted is reheated to 600 F. As this reheated steam expands through the turbine, steam is extracted at 16 lbf/in.^{2} for feedwater heating. The condenser pressure is 2 lbf/in.^{2} Both the feedwater heaters are open heaters. Determine the thermal efficiency and the net work per pound.

12.8 Steam leaves the boiler of a steam power plant at 400 lbf/in.^{2}, 600 F and when it enters the turbine it is at 370 lbf/in.^{2}, 560 F. The turbine has an efficiency of 80% and the condenser pressure is 2 lbf/in.^{2} The condensate leaves the condenser at 2 lbf/in.^{2}, 100 F. The pump has an efficiency of 70% and the pressure of the water leaving the pump is 500 lbf/in.^{2} Water enters the boiler at 460 lbf/in.^{2}, 96 F. Determine:

(*a*) The irreversibility for the process between the boiler and turbine. (Assume ambient temperature = 77 F.)

(*b*) The thermal efficiency of the cycle.

12.9 Consider an ideal binary cycle similar to that of Fig. 12.13. One of the variables is the temperature difference between the mercury and steam in the mercury condenser–steam boiler. To show the influence of this temperature difference on the cycle efficiency, consider a mercury cycle with a maximum pressure of 250 lbf/in.^{2} and a minimum pressure of 4 lbf/in.^{2}, a maximum steam temperature of 1000 F, and a steam-condenser pressure of 1 lbf/in.^{2} Plot a curve of cycle efficiency vs. temperature difference in the mercury condenser–steam boiler from 0 F to 200 F.

12.10 Consider an ideal steam cycle that combines the reheat cycle and the regenerative cycle. The net power output of the turbine is 100,000 kw. Steam enters the high-pressure turbine at 1200 lbf/in.^{2}, 1000 F. After expansion to 90 lbf/in.^{2}, some of the steam goes to an open feedwater heater and the balance is reheated to 700 F, after which it expands to 1 lbf/in.^{2}

(*a*) Draw a line diagram of the unit and show the state points on a *T-s* diagram.

(*b*) What is the steam flow rate to the high-pressure turbine?

(*c*) What hp motor is required to drive each of the pumps?

(*d*) If there is a 20 F rise in the temperature of the cooling water, what is the rate of flow of cooling water through the condenser?

(*e*) The velocity of the steam flowing from the turbine to the condenser is limited to a maximum of 400 ft/sec. What is the diameter of this connecting pipe?

12.11 In an ideal refrigeration cycle the temperature of the condensing vapor is 100 F and the temperature during evaporation is 0 F. Determine the coefficient of performance of this cycle for the working fluids Freon-12 and ammonia.

12.12 In an actual refrigeration cycle using Freon-12 as a working fluid the rate of flow of refrigerant is 300 lbm/hr. The refrigerant enters the compressor at 25 lbf/in.^{2}, 20 F and leaves at 175 lbf/in.^{2}, 170 F. The power input to the compressor is 2.5 hp. The refrigerant enters the expansion valve at 165 lbf/in.^{2}, 100 F and leaves the evaporator at 27 lbf/in.^{2}, 10 F. Determine:

(*a*) The irreversibility during the compression process.

(*b*) The refrigeration capacity in tons.

(*c*) The coefficient of performance for this cycle.

12.13 A study is to be made of the influence on power requirements of the difference between the temperature of the refrigerant in the condenser and the temperature of the surroundings. For this purpose consider the ideal cycle of Fig. 12.18 with Freon-12 as the refrigerant. The temperature of the surroundings is 100 F and the temperature of the refrigerant in the evaporator is 10 F. Plot a curve of hp per ton of refrigeration for temperature differences of 0 to 100 F between the refrigerant in the condenser and the surroundings.

12.14 A study is to be made of the influence on power requirements of the unit of the temperature difference between the cold box and the refrigerant in the evaporator. For this purpose consider the ideal cycle of Fig. 12.18 with Freon-12 as the refrigerant. The temperature of the cold box is 10 F and the temperature of the refrigerant in the condenser is 160 F. Plot a curve of hp per ton of refrigeration for temperature differences of 0 to 60 F between the cold box and the refrigerant in the evaporator.

12.15 An ammonia-absorption system has an evaporator temperature of 10 F and a condenser temperature of 120 F. The generator temperature is 302 F. In this cycle, 0.42 Btu is transferred to the ammonia in the evaporator for each Btu transferred to the ammonia solution in the generator from the high-temperature source.

We wish to compare the performance of this cycle with the performance of a similar vapor-compression cycle. To do this, assume that a reservoir is available at 302 F, and that heat is transferred from this reservoir to a reversible engine which rejects heat to the surroundings at 77 F. This work is then used to drive an ideal vapor-compression system with ammonia as the refrigerant. Compare the amount of refrigeration that can be achieved per Btu from the high-temperature source in this case with the 0.42 Btu that can be achieved in the absorption system.

12.16 A stoichiometric mixture of fuel and air has an enthalpy of combustion of approximately -1200 Btu/lbm of mixture. In order to approximate an actual spark-ignition engine using such a mixture, consider an air-standard Otto cycle that has a heat addition of 1200 Btu per lbm of air, a compression ratio of 8, and a pressure and temperature at the beginning of the compression process of 14.7 lbf/in.^{2}, 60 F. Determine:

(*a*) The maximum pressure and temperature for this cycle.
(*b*) The thermal efficiency.
(*c*) The mean effective pressure.

12.17 Repeat Problem 12.16, but take into account the variation of the specific heat of air with temperature. The *Gas Tables* are recommended for this calculation.

12.18 An air-standard Diesel cycle has a compression ratio of 14. The pressure at the beginning of the compression stroke is 14.7 lbf/in.^{2}, and the temperature is 60 F. The maximum temperature is 4500 R. Determine the thermal efficiency and the mean effective pressure for this cycle.

12.19 Consider an air-standard Ericsson cycle with an ideal regenerator incorporated. The temperature at which heat is supplied is 2000 R. Heat is rejected at 520 R. The pressure at the beginning of the isothermal compression process is 14.7 lbf/in.^{2} Determine the compressor and turbine work per lbm of air and the thermal efficiency of the cycle. Q_H = 200 Btu/lbm.

12.20 The pressure ratio across the compressor of an air-standard Brayton cycle is 4 to 1. The pressure of the air entering the compressor is 14.7 lbf/in.^{2}, and the temperature is 60 F. The maximum temperature in the cycle is 1500 F. The rate of air flow is 20 lbm/sec. Determine:

(*a*) The compressor work, turbine work, and thermal efficiency of the cycle.

(*b*) If this cycle was utilized for a reciprocating machine, what would be the mean effective pressure? Would you recommend this cycle for a reciprocating machine?

12.21 Repeat Problem 12.20 using the *Gas Tables,* which take into account the variation of specific heat with temperature.

12.22 An ideal regenerator is incorporated into the air-standard Brayton cycle of Problem 12.20. Determine the thermal efficiency of the cycle with this modification.

12.23 Consider a gas-turbine cycle with two stages of compression and two stages of expansion. The pressure ratio across each turbine and each compressor is 2. The temperature entering each compressor is 60 F and the temperature entering each turbine is 1500 F. An ideal regenerator is incorporated into the cycle. Determine the compressor work, turbine work, and thermal efficiency. P_1 = 14.7 lbf/in.^{2}

12.24 Repeat Problem 12.23 assuming that each compressor has an efficiency of 80%, each turbine has an efficiency of 85%, and the regenerator has an efficiency of 60%.

12.25 Consider the air-standard cycle for a gas turbine–jet propulsion unit. The pressure and temperature entering the compressor are

14.7 lbf/in.^{2}, 60 F respectively. After expansion in the turbine, the air expands to 14.7 lbf/in.^{2} Determine the velocity of the air leaving the nozzle. The temperature at the turbine inlet is 1500 F. $P_2/P_1 = 4$.

12.26 A stationary gas-turbine power plant operates on the Brayton cycle and delivers 20,000 hp to an electric generator. The maximum temperature is 1540 F and the minimum temperature is 60 F. The minimum pressure is 14.0 lbf/in.^{2} and the maximum pressure is 60 lbf/in.^{2}

(*a*) What is the power output of the turbine?

(*b*) What fraction of the output of the turbine is used to drive the compressor?

(*c*) What is the mass rate of air flow to the compressor per min? What is the volume rate of air flow to the compressor per min?

12.27 Repeat Problem 12.26 assuming that a regenerator of 75% efficiency is added to the cycle, and the efficiency of the compressor is 80% and the efficiency of the turbine is 85%.

12.28 (*a*) A Brayton cycle has a minimum pressure of 15 lbf/in.^{2} and a minimum temperature of 40 F. The maximum pressure is 60 lbf/in.^{2} Plot a curve of thermal efficiency as a function of maximum cycle temperature.

(*b*) Repeat (*a*) assuming a compressor efficiency of 80%, a turbine efficiency of 85%, and a 2 lbf/in.^{2} pressure drop between the compressor and turbine.

12.29 A jet aircraft is flying at an altitude of 16,000 ft, where the ambient pressure is approximately 8 lbf/in.^{2} and the ambient temperature is 0 F. The velocity of the aircraft is 500 ft/sec, and the pressure ratio across the compressor is 4.

Devise an air-standard cycle that approximates this cycle, and determine the velocity (relative to the aircraft) of the air leaving the engine, assuming that it has been expanded to the ambient pressure. Assume a maximum gas temperature of 1600 F.

12.30 A Brayton cycle and Rankine cycle are to be combined in a power plant in such a way that heat transferred from the gas leaving the turbine is used to evaporate the water. The Brayton cycle, which utilizes air, has a minimum pressure of 15 lbf/in.^{2} and a maximum pressure of 70 lbf/in.^{2}, and a minimum temperature of 60 F and a maximum temperature of 1540 F. The Rankine cycle, which utilizes steam, has a maximum pressure of 400 lbf/in.^{2}, a maximum temperature of 600 F, and a minimum pressure of 1 lbf/in.^{2} All the heat transferred to the steam comes from the air leaving the turbine of the Brayton

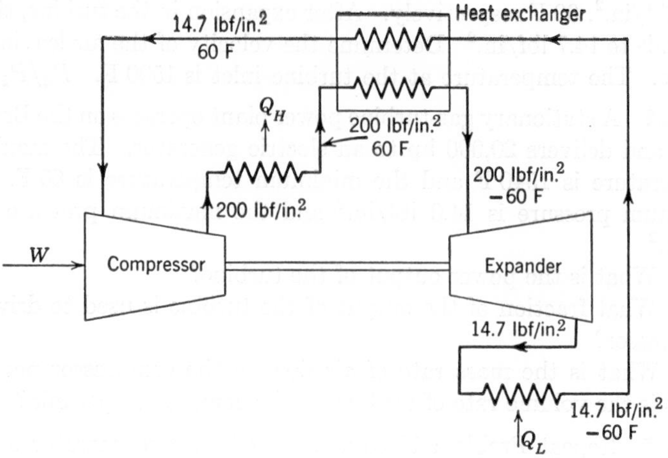

Fig. 12.44 **Schematic arrangement of Problem 12.31.**

cycle, and after this heat-transfer process the air temperature is 600 F.

(*a*) Draw a line diagram of this power plant.

(*b*) Determine the thermal efficiency of this cycle.

12.31 A heat exchanger is incorporated into an air-standard refrigeration cycle as shown in Fig. 12.44. Assume that both the compression and expansion are reversible adiabatic processes. Determine the coefficient of performance for this cycle.

12.32 Repeat Problem 12.31 assuming an isentropic efficiency for the compressor and expander of 75%.

CHAPTER 13

FLOW THROUGH NOZZLES AND BLADE PASSAGES

This chapter involves the flow of fluids through nozzles and blade passages. A complete analysis of these problems involves not only thermodynamics, but also other topics, such as boundary-layer theory, turbulence, Reynolds number, and similarity laws, which are usually covered in courses in fluid flow. Our purpose in this chapter is to present a thermodynamic analysis of one-dimensional fluid-flow problems, which the student is encouraged to integrate with his fluid-flow courses, though there will, no doubt, be some overlap.

13.1 Stagnation Properties

In dealing with problems involving flow, many discussions and equations can be simplified by introducing the concept of the isentropic stagnation state and the properties associated with it. The *isentropic stagnation state* is the state a flowing fluid would attain if it underwent a reversible adiabatic deceleration to zero velocity. This state is designated in this chapter with the subscript 0. From the steady-flow energy equation we conclude that

$$h + \frac{\overline{V}^2}{2g_c} = h_0 \tag{13.1}$$

The actual and the isentropic stagnation states for a typical gas or vapor are shown on the h-s diagram of Fig. 13.1. Sometimes it is advantageous to make a distinction between the actual and the isentropic stagnation states. The actual stagnation state is the state achieved after an actual deceleration to zero velocity (as at the nose of a body placed in a fluid stream), and there may be irreversibilities associated with the deceleration process. Therefore, the term stagnation property is sometimes reserved for the properties associated with the actual state, and the term total property is used for the isentropic stagnation state.

It is evident from Fig. 13.1 that the enthalpy is the same for both the actual and isentropic stagnation states (assuming that the actual

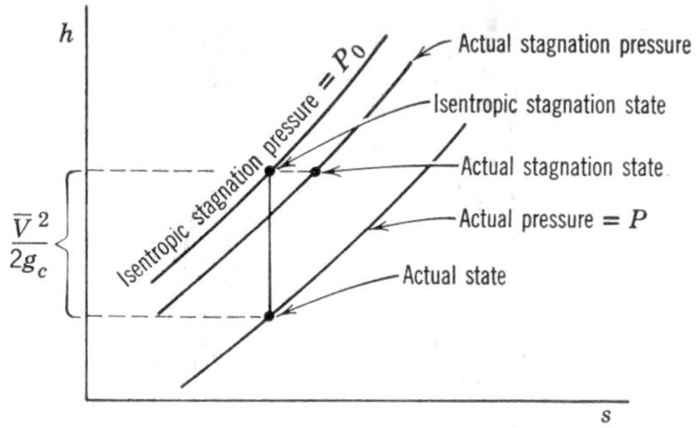

Fig. 13.1 Enthalpy-entropy diagram illustrating the definition of stagnation state.

process is adiabatic). Therefore, in an ideal gas, the actual stagnation temperature is the same as the isentropic stagnation temperature. However, the actual stagnation pressure may be less than the isentropic stagnation pressure and for this reason the term total pressure (meaning isentropic stagnation pressure) has particular meaning compared to the actual stagnation pressure.

Example 13.1 Air flows in a duct at a pressure of 20 lbf/in.^{2} with a velocity of 600 ft/sec. The temperature of the air is 80 F. Determine the isentropic stagnation pressure and temperature.

If we assume that the air is a perfect gas with constant specific heat as given in Table A.6, the calculation is as follows. From Eq. 13.1

$$\frac{\overline{V}^2}{2g_c} = h_0 - h = C_{p0}(T_0 - T)$$

$$\frac{(600)^2}{64.34 \times 778} = 0.240(T_0 - T)$$

$$T_0 = 570 \text{ R} = 110 \text{ F}$$

The stagnation pressure can be found from the relation

$$\frac{T_0}{T} = \left(\frac{P_0}{P}\right)^{(k-1)/k}$$

$$\frac{570}{540} = \left(\frac{P_0}{P}\right)^{0.286}$$

$$\frac{P_0}{P} = 1.210$$

$$P_0 = 20(1.210) = 24.2 \text{ lbf/in.}^2$$

Keenan and Kaye's *Gas Tables* could also have been used, and then the variation of specific heat with temperature would have been taken into account. Since the actual and stagnation states have the same entropy, we proceed as follows: Using Table 1 of the *Gas Tables* (Appendix Table A.8),

$$T = 540 \text{ R} \qquad h = 129.06 \qquad P_r = 1.3860$$

$$h_0 = h + \frac{\overline{V}^2}{2g_c} = 129.06 + \frac{(600)^2}{64.34 \times 778} = 136.26$$

$$T_0 = 570 \text{ R} \qquad P_{r0} = 1.6748$$

$$P_0 = P \times \frac{P_{r0}}{P_r} = 20.0 \times \frac{1.6748}{1.3860} = 24.2 \text{ lbf/in.}^2$$

13.2 The Momentum Equation for the Open System

Before proceeding further it will be of value to develop the momentum equation for the open system. Newton's second law states that the force on a body of mass m is proportional to the rate of change of momentum.

$$F \propto \frac{d\,(m\overline{V})}{d\tau}$$

For the system of units used in this text it is convenient to introduce the constant g_c.

$$F = \frac{1}{g_c}\frac{d\,(m\overline{V})}{d\tau} \qquad (13.2)$$

It should be pointed out that both force and momentum are vector quantities. Therefore it would be more precise to say that the force in a given direction is proportional to the rate of change of momentum in that direction.

Equation 13.2 has been written for a body of a given mass, which in our thermodynamic parlance is a closed system. In using the momentum equation for an open system it is convenient to use the concept of momentum flux. Consider the system shown in Fig. 13.2. The flow is

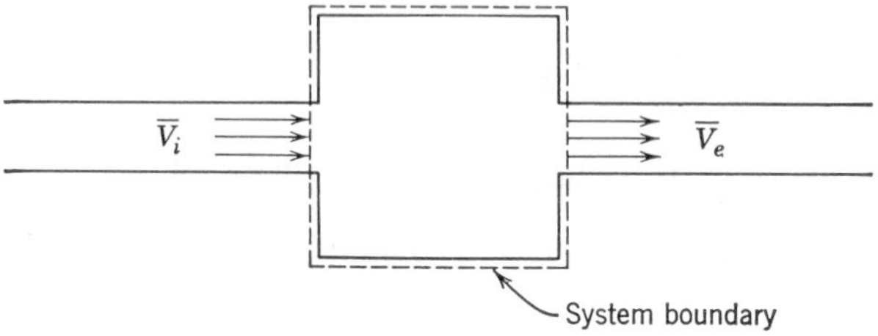

Fig. 13.2 Diagram showing the concept of momentum flux.

assumed to be uniform, one-dimensional flow. The momentum of a mass of fluid, δm_i, entering the system, is $\delta(m_i\overline{V}_i)$. Let the time required for δm_i to cross the boundary of the system be $d\tau$. Then the rate at which momentum crosses the boundary of the system is

$$\frac{\delta(m_i\overline{V}_i)}{g_c\,d\tau}$$

If we consider the velocity $\overline{V}_i$ constant, and denote the mass rate of flow as $\dot{m}$ (where $\dot{m} = \delta m/d\tau$), the rate at which momentum crosses the boundary (momentum flux) is

$$(\text{Momentum flux})_{\text{in}} = \frac{\dot{m}_i\overline{V}_i}{g_c} \tag{13.3}$$

Similarly, for constant exit velocity, the momentum flux at the exit is

$$(\text{Momentum flux})_{\text{exit}} = \frac{\dot{m}_e\overline{V}_e}{g_c} \tag{13.4}$$

We can now write the momentum equation for an open system with constant inlet and exit fluid velocity by adding to Eq. 13.2, which applies to a closed system, the momentum flux across the boundary of the system. The momentum equation for the open system then becomes

$$F = \frac{d\,(m\overline{V})_\sigma}{g_c\,d\tau} + \frac{\dot{m}_e\overline{V}_e}{g_c} - \frac{\dot{m}_i\overline{V}_i}{g_c}$$

The first term on the right represents the rate of change of momentum within the boundary of the system, the second term the momentum flux leaving the system, and the third term the momentum flux entering the system. F is the net force on the system. The addition of the quantities on the right side must be done by the laws of vectors, since momentum is a vector quantity. We can also write the momentum equations for the x, y, and z directions as follows:

$$
\begin{aligned}
F_x &= \frac{d\,(m\overline{V}_x)_\sigma}{g_c\,d\tau} + \frac{(\dot{m}_e\overline{V}_e)_x}{g_c} - \frac{(\dot{m}_i\overline{V}_i)_x}{g_c} \\
F_y &= \frac{d\,(m\overline{V}_y)_\sigma}{g_c\,d\tau} + \frac{(\dot{m}_e\overline{V}_e)_y}{g_c} - \frac{(\dot{m}_i\overline{V}_i)_y}{g_c} \\
F_z &= \frac{d\,(m\overline{V}_z)_\sigma}{g_c\,d\tau} + \frac{(\dot{m}_e\overline{V}_e)_z}{g_c} - \frac{(\dot{m}_i\overline{V}_i)_z}{g_c}
\end{aligned}
\qquad (13.5)
$$

These equations assume that $\overline{V}_e$ and $\overline{V}_i$ are constant.

For a steady-flow process there is no change of momentum within the boundary of the system, and the mass rate of flow at the inlet is equal to the mass rate of flow at the exit. Therefore, for the steady-flow process, the momentum equation reduces to

$$F = \frac{\dot{m}}{g_c}(\overline{V}_e - \overline{V}_i) \qquad (13.6)$$

The remarks made above regarding the fact that momentum is a vector quantity apply to this equation also. Therefore, for the directions x, y, and z we write

$$
\begin{aligned}
F_x &= \frac{\dot{m}}{g_c}(\overline{V}_e - \overline{V}_i)_x \\
F_y &= \frac{\dot{m}}{g_c}(\overline{V}_e - \overline{V}_i)_y \\
F_z &= \frac{\dot{m}}{g_c}(\overline{V}_e - \overline{V}_i)_z
\end{aligned}
\qquad (13.7)
$$

Example 13.2 A man is pushing a wheel barrow (Fig. 13.3) on a level floor into which sand is falling at the rate of 1 lbm/sec. The man is walking at the rate of 5 ft/sec and the sand has a velocity of 40 ft/sec as it falls into the wheel barrow. Determine the force the man must exert on the wheel barrow and the force the floor exerts on the wheel barrow due to the falling sand.

Let us consider the wheel barrow as our system. Consider first the horizontal (x) direction. The momentum equation, Eq. 13.5, can be applied in this case because we are considering constant velocities.

$$F_x = \frac{d\,(m\overline{V}_x)_\sigma}{g_c\,d\tau} + \frac{(\dot{m}_e\overline{V}_e)_x}{g_c} - \frac{(\dot{m}_i\overline{V}_i)_x}{g_c}$$

The second term drops out because there is no mass leaving the system; i.e., $\dot{m}_e = 0$. The third term drops out because the sand entering

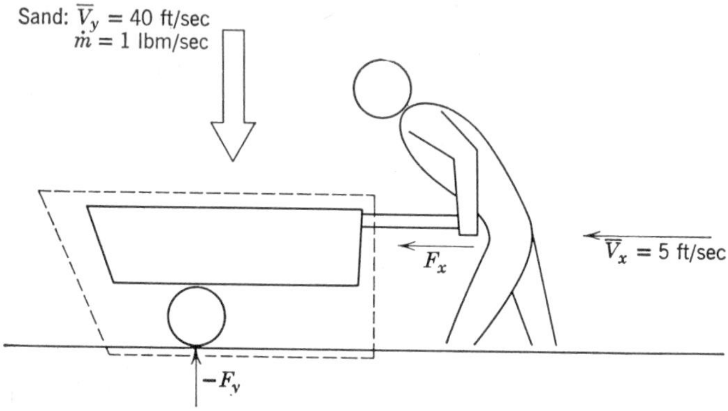

Fig. 13.3 Sketch for Example 13.2.

has no horizontal component of velocity; i.e., $\overline{V}_{ix} = 0$. Therefore,

$$F_x = \frac{d\,(m\overline{V}_x)_\sigma}{g_c\,d\tau} = \frac{1\text{ lbm/sec} \times 5\text{ ft/sec}}{32.17\text{ lbm-ft/lbf-sec}^2} = 0.155\text{ lbf}$$

Next consider the vertical (y) direction:

$$F_y = \frac{d\,(m\overline{V}_y)_\sigma}{g_c\,d\tau} + \frac{(\dot{m}_e\overline{V}_e)_y}{g_c} - \frac{(\dot{m}_i\overline{V}_i)_y}{g_c}$$

The second term again drops out because $\dot{m}_e = 0$. The first term drops out because the system has no velocity in the vertical direction; i.e., $(\overline{V}_y)_\sigma = 0$. Therefore, the momentum equation reduces to

$$F_y = -\frac{(\dot{m}_i\overline{V}_i)_y}{g_c} = -\frac{1\text{ lbm/sec} \times 40\text{ ft/sec}}{32.17\text{ lbm-ft/lbf-sec}^2} = -1.24\text{ lbf}$$

The minus sign indicates that the force is in the opposite direction of $\overline{V}_y$. The sign of F_x is positive, indicating that it is in the same direction as the velocity V_x.

13.3 Forces Acting on an Open System

In the last section we considered the momentum equation for the open system. We now wish to evaluate the net force on a system which causes this change in momentum. Let us do this by considering the system shown in Fig. 13.4, which involves a pipe bend. The boundaries of the system are designated by the dotted lines, and are so chosen that at the point where the fluid crosses the system boundary the flow is perpendicular to the boundary of the system. The shear forces at the section where the fluid crosses the boundary of the system are negligible. Part (*a*) of Fig. 13.4 shows the velocities and part (*b*) shows the forces involved. The force R is the result of all external forces on the system, except for the pressure of the surroundings. The pressure of the surroundings, P_0, acts on the entire boundary except at A_i and A_e, where the fluid crosses the boundary. P_i and P_e represent the absolute pressures at these points.

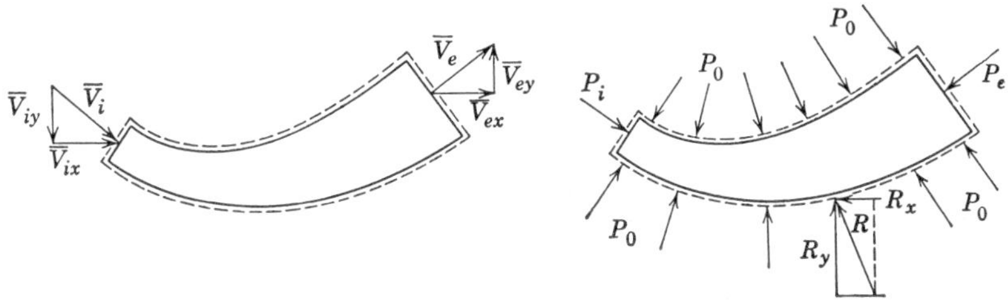

Fig. 13.4 Forces acting on an open system.

The net forces acting on the system in the x and y directions, F_x and F_y, are the sum of the pressure forces and the external force R in their respective directions. The influence of the pressure of the surroundings, P_0, is most easily taken into account by noting that it acts over the entire system boundary except at A_i and A_e. Therefore, we can write

$$\begin{aligned} F_x &= (P_iA_i)_x - (P_0A_i)_x + (P_eA_e)_x - (P_0A_e)_x + R_x \\ F_y &= (P_iA_i)_y - (P_0A_i)_y + (P_eA_i)_y - (P_0A_e)_y + R_y \end{aligned} \tag{13.8}$$

Equation 13.8 may be simplified by combining the pressure terms.

$$\begin{aligned} F_x &= [(P_i - P_0)A_i]_x + [(P_e - P_0)A_e]_x + R_x \\ F_y &= [(P_i - P_0)A_i]_y + [(P_e - P_0)A_e]_y + R_y \end{aligned} \tag{13.9}$$

The proper sign for each pressure and force must of course be used in all calculations.

Equations 13.7 and 13.9 may now be combined as follows:

$$F_x = \frac{\dot{m}}{g_c}(\overline{V}_e - \overline{V}_i)_x = [(P_i - P_0)A_i]_x + [(P_e - P_0)A_e]_x + R_x$$

$$F_y = \frac{\dot{m}}{g_c}(\overline{V}_e - \overline{V}_i)_y = [(P_i - P_0)A_i]_y + [(P_e - P_0)A_e]_y + R_y \tag{13.10}$$

These equations are very useful in analyzing the forces involved in open systems.

Example 13.3 A jet engine is being tested on a test stand (Fig. 13.5). The inlet area to the compressor is 1.8 ft^2, and air enters the compressor

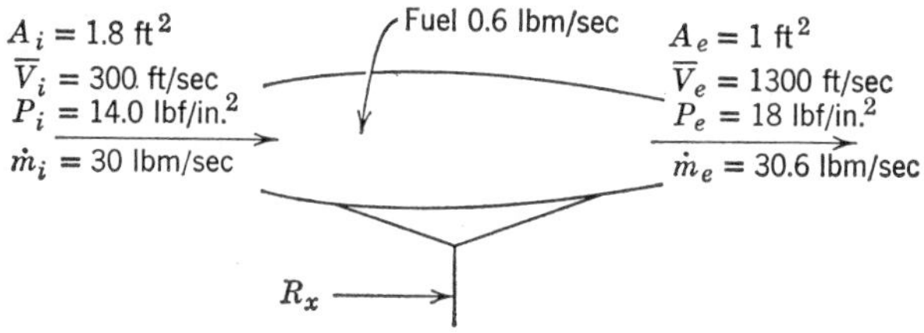

Fig. 13.5 Sketch for Example 13.3.

at 14.0 lbf/in.^{2}, 300 ft/sec. The pressure of the atmosphere is 14.7 lbf/in.^{2} The exit area of the engine is 1 ft^2, and the products of combustion leave the exit plane at a pressure of 18 lbf/in.^{2} and a velocity of 1300 ft/sec. The air-fuel ratio is 50 lbm air/lbm fuel, and the fuel enters with a low velocity. The rate of air flow entering the engine is 30 lbm/sec. Determine the thrust on the engine.

In the solution that follows it is assumed that forces and velocities to the right are positive.

Using Eq. 13.10

$$R_x + [(P_i - P_0)A_i]_x + [(P_e - P_0)A_e]_x = \frac{\dot{m}}{g_c}(\overline{V}_e - \overline{V}_i)_x$$

$$R_x + [(14.0 - 14.7)144 \text{ lbf/ft}^2 \times 1.8 \text{ ft}^2] - [(18.0 - 14.7)144 \text{ lbf/ft}^2 \times 1.0 \text{ ft}^2] = \frac{(30.6 \times 1300 - 30 \times 300) \text{ lbm ft/sec}^2}{32.17 \text{ lbm-ft/lbf-sec}^2}$$

$$R_x = 957 + 180 + 475 = 1612 \text{ lbf}$$

(Note that the momentum of the fuel entering has been neglected.)

13.4 Reversible, Adiabatic, One-Dimensional Steady Flow of an Incompressible Fluid Through a Nozzle

A nozzle is a device whose function is to increase the kinetic energy of a fluid in an adiabatic process. This involves a decrease in pressure and is accomplished by the proper change in flow area. A diffuser is a device that has the opposite function, namely to increase the pressure by decelerating the fluid. This paragraph discusses both nozzles and diffusers, but to minimize the use of words only the term nozzle is used.

Since there is no work involved in flow through nozzles, the steady-flow energy equation for the adiabatic flow through nozzles in differential form is

$$dh + \frac{d\overline{V}^2}{2g_c} + \frac{g}{g_c}\,dZ = 0 \tag{13.11}$$

In integrated form this equation is

$$(h_2 - h_1) + \frac{\overline{V}_2^{\,2} - \overline{V}_1^{\,2}}{2g_c} + (Z_2 - Z_1)\frac{g}{g_c} = 0 \tag{13.12}$$

For a reversible adiabatic process $ds = 0$, and therefore,

$$dh = v\,dp$$

It follows that Eq. 13.11 can be written for a reversible process in the form

$$v\,dp + \frac{d\overline{V}^2}{2g_c} + \frac{g}{g_c}\,dZ = 0 \tag{13.13}$$

This is readily integrated for an incompressible fluid to give Bernoulli's equation.

$$v(P_2 - P_1) + \frac{\overline{V}_2^{\,2} - \overline{V}_1^{\,2}}{2g_c} + (Z_2 - Z_1)\frac{g}{g_c} = 0 \tag{13.14}$$

Equation 5.26, which relates specific volume, velocity, mass rate of flow, and cross-sectional area is also an important relation when considering flow through nozzles. This equation is

$$\dot{m}v = A\overline{V}$$

Equations 13.12, 13.14 (which applies only to reversible processes) and 5.26 are the important relations governing the flow of an incompressible fluid through a nozzle or diffuser.

Example 13.4 Water enters the diffuser in a pump casing with a velocity of 100 ft/sec, a pressure of 50 lbf/in.^{2}, and a temperature of 80 F. It leaves the diffuser with a velocity of 20 ft/sec and a pressure of 90 lbf/in.^{2} Determine the exit pressure for a reversible diffuser with these inlet conditions and exit velocity. Determine the increase in enthalpy, internal energy, and entropy for the actual diffuser.

Considering a reversible diffuser with no change in elevation, Bernoulli's equation, Eq. 13.14, reduces to

$$v(P_2 - P_1) + \frac{\overline{V}_2^2 - \overline{V}_1^2}{2g_c} = 0$$

From the Keenan and Keyes' steam tables, $v = 0.01608$ ft^3/lbm.

$$P_2 - P_1 = \frac{[(100)^2 - (20)^2]\text{ ft}^2/\text{sec}^2}{0.01608 \text{ ft}^3/\text{lbm} \times 64.34 \text{ lbm-ft/lbf-sec}^2 \times 144 \text{ in.}^2/\text{ft}^2}$$

$$= 64.5 \text{ lbf/in.}^2$$

Therefore, for the reversible diffuser, the final pressure P_{2s} is 50 + 64.5 = 114.5 lbf/in.^{2}

The change in enthalpy can be found from Eq. 13.12.

$$h_2 - h_1 = \frac{\overline{V}_1^2 - \overline{V}_2^2}{2g_c}$$

$$= \frac{[(100)^2 - (20)^2]\text{ ft}^2/\text{sec}^2}{64.34 \text{ lbm-ft/lbf-sec}^2 \times 778 \text{ ft-lbf/Btu}} = 0.192 \text{ Btu/lbm}$$

The change in internal energy can be found from the definition of enthalpy, $h_2 - h_1 = (u_2 - u_1) + (P_2v_2 - P_1v_1)$.

Thus, for an incompressible fluid

$$u_2 - u_1 = h_2 - h_1 - v(P_2 - P_1)$$

$$= 0.192 \text{ Btu/lbm} - 0.01608 \text{ ft}^3/\text{lbm} \times \frac{(90 - 50)144 \text{ lbf/ft}^2}{778 \text{ ft-lbf/Btu}}$$

$$= 0.192 - 0.119 = 0.073 \text{ Btu/lbm}$$

The change of entropy can be approximated from the familiar relation

$$T\,ds = du + P\,dv$$

by assuming that the temperature is constant (which is approximately true in this case) and noting that for an incompressible fluid $dv = 0$. With these assumptions

$$s_2 - s_1 = \frac{u_2 - u_1}{T} = \frac{0.073}{540} = 0.000135 \text{ Btu/lbm R}$$

Since this is an irreversible adiabatic process, the entropy will increase, as the above calculation indicates.

13.5 Velocity of Sound in an Ideal Gas

When a pressure disturbance occurs in a compressible fluid, the disturbance travels with a velocity that depends on the state of the fluid. A sound wave is a very small pressure disturbance; the velocity of sound, also called the *sonic velocity,* is an important parameter in compressible-fluid flow. We proceed now to determine an expression for the sonic velocity of an ideal gas in terms of the properties of the gas.

Let a disturbance be set up by the movement of the piston at the end of the tube, Fig. 13.6*a*. A wave travels down the tube with a

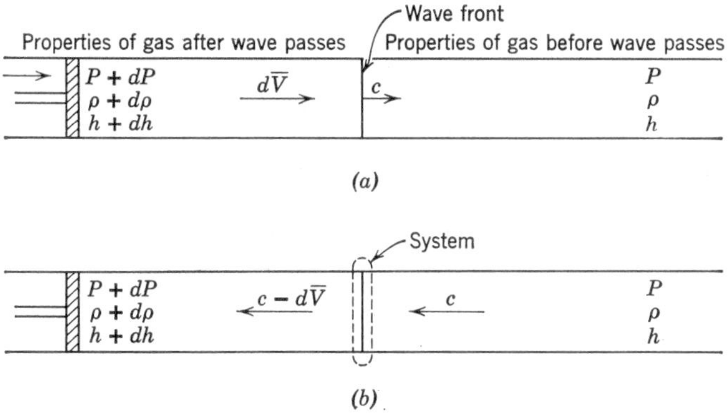

Fig. 13.6 Diagram illustrating sonic velocity. (*a*) Stationary observer. (*b*) Observer travelling with wave front.

velocity c, which is the sonic velocity. Assume that after the wave has passed the properties of the gas have changed an infinitesimal amount and that the gas is moving with the velocity $d\bar{V}$ toward the wave front.

In Fig. 13.6*b* this process is shown from the point of view of an observer who travels with the wave front. Consider the open system shown in Fig. 13.6*b*. From the energy equation we can write

$$h + \frac{c^2}{2g_c} = (h + dh) + \frac{(c - d\overline{V})^2}{2g_c}$$

$$dh - \frac{c\,d\overline{V}}{g_c} = 0 \tag{13.15}$$

From the continuity equation we can write

$$\rho A c = (\rho + d\rho) A (c - d\overline{V})$$

$$c\,d\rho - \rho\,d\overline{V} = 0 \tag{13.16}$$

Consider also the relation between properties

$$T\,ds = dh - \frac{dP}{\rho}$$

If the process is isentropic, $ds = 0$, and this equation can be combined with Eq. 13.15 to give the relation

$$\frac{dP}{\rho} - \frac{c\,d\overline{V}}{g_c} = 0 \tag{13.17}$$

This can be combined with Eq. 13.16 to give the relation

$$\frac{dP}{d\rho} = \frac{c^2}{g_c}$$

Since we have assumed the process to be isentropic this is better written as a partial derivative.

$$\left(\frac{\partial P}{\partial \rho}\right)_s = \frac{c^2}{g_c} \tag{13.18}$$

An alternate derivation is to introduce the momentum equation. For the open system of Fig. 13.6*b* the momentum equation is

$$PA - (P + dP)A = \frac{\dot{m}}{g_c}(c - d\overline{V} - c) = \frac{\rho A c}{g_c}(c - d\overline{V} - c)$$

$$d\overline{V} = \frac{g_c}{\rho c}\,dP \tag{13.19}$$

On combining this with Eq. 13.16 we obtain Eq. 13.18

$$\left(\frac{\partial P}{\partial \rho}\right)_s = \frac{c^2}{g_c}$$

It will be of particular advantage to solve Eq. 13.18 for the velocity of sound in an ideal gas.

When an ideal gas undergoes an isentropic change of state, we found in Chapter 8 that, for this process, assuming constant specific heat,

$$\frac{dP}{P} - k\frac{d\rho}{\rho} = 0$$

or

$$\left(\frac{\partial P}{\partial \rho}\right)_s = \frac{kP}{\rho}$$

Substituting this equation in Eq. 13.18 we have an equation for the velocity of sound in an ideal gas

$$c^2 = \frac{kPg_c}{\rho} \qquad (13.20)$$

Since for an ideal gas

$$\frac{P}{\rho} = RT$$

this equation may also be written

$$c^2 = kg_cRT \qquad (13.21)$$

Example 13.5 Determine the velocity of sound in air at 80 F and at 1000 F.

Using Eq. 13.21

$$\begin{aligned} c &= \sqrt{kg_cRT} \\ &= \sqrt{1.4 \times 32.17 \text{ lbm-ft/lbf-sec}^2 \times 53.34 \text{ ft-lbf/lbm-R} \times 540 \text{ R}} \\ &= 1138 \text{ ft/sec} \end{aligned}$$

Similarly, at 1000 F

$$\begin{aligned} c &= \sqrt{1.4 \times 32.17 \times 53.34 \times 1460} \\ &= 1862 \text{ ft/sec} \end{aligned}$$

Note the significant increase in sonic velocity as the temperature increases.

The *Mach number*, $\boldsymbol{M}$, is defined as the ratio of the actual velocity $\overline{V}$ to the sonic velocity c.

$$\boldsymbol{M} = \frac{\overline{V}}{c} \qquad (13.22)$$

When $\boldsymbol{M} > 1$ the flow is supersonic; when $\boldsymbol{M} < 1$ the flow is subsonic;

and when $M = 1$ the flow is sonic. The importance of the Mach number as a parameter in fluid-flow problems will be evident in the paragraphs which follow.

13.6 Reversible, Adiabatic, One-Dimensional, Steady Flow of an Ideal Gas Through a Nozzle

A nozzle or diffuser with both a converging and diverging section is shown in Fig. 13.7. The minimum cross-sectional area is called the throat.

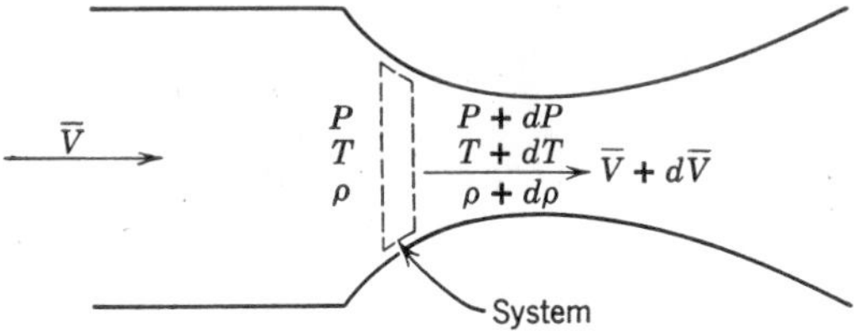

Fig. 13.7 One-dimensional reversible adiabatic steady flow through a nozzle.

Our first consideration concerns the conditions that determine whether a nozzle or diffuser should be converging or diverging, and the conditions that prevail at the throat. For the system shown the following relations can be written.

First law:

$$dh + \frac{\overline{V}\,d\overline{V}}{g_c} = 0 \tag{13.23}$$

Property relation:

$$T\,ds = dh - \frac{dP}{\rho} = 0 \tag{13.24}$$

Continuity equation:

$$\rho A \overline{V} = \dot{m} = \text{constant}$$

$$\frac{d\rho}{\rho} + \frac{dA}{A} + \frac{d\overline{V}}{\overline{V}} = 0 \tag{13.25}$$

Combining Eqs. 13.23 and 13.24 we have

$$dh = \frac{dP}{\rho} = \frac{-\overline{V}\,d\overline{V}}{g_c}$$

$$d\overline{V} = -\frac{g_c}{\rho\overline{V}}\,dP$$

Substituting this in Eq. 13.25

$$\frac{dA}{A} = \left(-\frac{d\rho}{\rho} - \frac{d\overline{V}}{\overline{V}}\right) = -\frac{d\rho}{\rho}\left(\frac{dP}{dP}\right) + \frac{g_c}{\rho \overline{V}^2}\,dP$$

$$= \frac{-dP}{\rho}\left(\frac{d\rho}{dP} - \frac{g_c}{\overline{V}^2}\right) = \frac{dP}{\rho}\left[-\frac{1}{(dP/d\rho)} + \frac{g_c}{\overline{V}^2}\right]$$

Since the flow is isentropic

$$\frac{dP}{d\rho} = \frac{c^2}{g_c} = \frac{\overline{V}^2}{\boldsymbol{M}^2 g_c}$$

and therefore,

$$\frac{dA}{A} = \frac{dP}{\rho \overline{V}^2/g_c}(1 - \boldsymbol{M}^2) \tag{13.26}$$

This is a very significant equation, for from it we can draw the following conclusions regarding the proper shape for nozzles and diffusers:

For a nozzle, $dP < 0$. Therefore,
for a subsonic nozzle, $\boldsymbol{M} < 1$, $dA < 0$, and the nozzle is converging.
for a supersonic nozzle, $\boldsymbol{M} > 1$, $dA > 0$, and the nozzle is diverging.
For a diffuser, $dP > 0$. Therefore,
for a subsonic diffuser, $\boldsymbol{M} < 1$, $dA > 0$, and the diffuser is diverging.
for a supersonic diffuser, $\boldsymbol{M} > 1$, $dA < 0$, and the diffuser is converging.

When $\boldsymbol{M} = 1$, $dA = 0$, which means that sonic velocity can be achieved only at the throat of a nozzle or diffuser. These conclusions are summarized in Fig. 13.8.

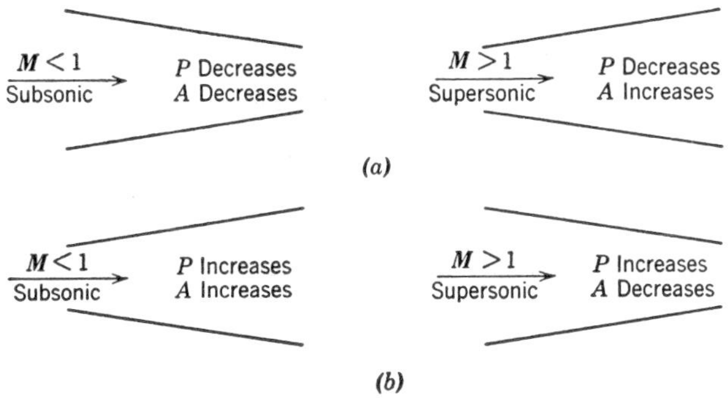

Fig. 13.8 Required area changes for (a) nozzles, and (b) diffusers.

We will now develop a number of relations between the actual properties, stagnation properties, and Mach number. These relations are very useful in dealing with isentropic flow of an ideal gas in a nozzle, and a number of these ratios have been tabulated in the *Gas Tables*, a summary of which is given in the Appendix.

Equation 13.1 gives the relation between enthalpy, stagnation enthalpy, and kinetic energy.

$$h + \frac{\overline{V}^2}{2g_c} = h_0$$

For an ideal gas with constant specific heat Eq. 13.1 can be written

$$\overline{V}^2 = 2g_c C_{P0}(T_0 - T) = 2g_c \frac{kRT}{k-1}\left(\frac{T_0}{T} - 1\right)$$

Since

$$c^2 = kg_c RT$$

$$\overline{V}^2 = \frac{2c^2}{k-1}\left(\frac{T_0}{T} - 1\right)$$

$$\frac{\overline{V}^2}{c^2} = \boldsymbol{M}^2 = \frac{2}{k-1}\left(\frac{T_0}{T} - 1\right)$$

$$\frac{T_0}{T} = 1 + \frac{(k-1)}{2}\boldsymbol{M}^2 \tag{13.27}$$

For an isentropic process,

$$\left(\frac{T_0}{T}\right)^{k/(k-1)} = \frac{P_0}{P} \qquad \left(\frac{T_0}{T}\right)^{1/(k-1)} = \frac{\rho_0}{\rho}$$

Therefore,

$$\frac{P_0}{P} = \left[1 + \frac{(k-1)}{2}\boldsymbol{M}^2\right]^{k/(k-1)} \tag{13.28}$$

$$\frac{\rho_0}{\rho} = \left[1 + \frac{(k-1)}{2}\boldsymbol{M}^2\right]^{1/(k-1)} \tag{13.29}$$

Values of P/P_0, ρ/ρ_0, and T/T_0 are given as a function of $\boldsymbol{M}$ in Tables 30 to 35 of the *Gas Tables*, each table being for a given value of k. Table A.10 of the Appendix has been abstracted from Table 30 of the *Gas Tables*, and applies to an ideal gas with $k = 1.4$.

The conditions at the throat of the nozzle can be found by noting that $\boldsymbol{M} = 1$ at the throat. The properties at the throat are denoted by an asterisk *. Therefore,

$$\frac{T^*}{T_0} = \frac{2}{k+1} \tag{13.30}$$

$$\frac{P^*}{P_0} = \left(\frac{2}{k+1}\right)^{k/(k-1)} \tag{13.31}$$

$$\frac{\rho^*}{\rho_0} = \left(\frac{2}{k+1}\right)^{1/(k-1)} \tag{13.32}$$

These properties at the throat of a nozzle when $\boldsymbol{M} = 1$ are frequently referred to as critical pressure, critical temperature, and critical density, and the ratios given by Eqs. 13.30, 13.31, and 13.32 are referred to as the critical-temperature ratio, critical-pressure ratio, and critical-density ratio. Table 13.1 gives these ratios for various values of k.

TABLE 13.1

Critical Pressure, Density, and Temperature Ratios for Isentropic Flow of an Ideal Gas

	$k = 1.1$	$k = 1.2$	$k = 1.3$	$k = 1.4$	$k = 1.67$
P^*/P_0	0.5847	0.5644	0.5457	0.5283	0.4867
ρ^*/ρ_0	0.6139	0.6209	0.6276	0.6340	0.6497
T^*/T_0	0.9524	0.9091	0.8696	0.8333	0.7491

13.7 Mass Rate of Flow of an Ideal Gas Through an Isentropic Nozzle

We now turn our attention to a consideration of the mass rate of flow per unit area, $\dot{m}/A$, in a nozzle. From the continuity equation we proceed as follows:

$$\frac{\dot{m}}{A} = \rho\overline{V} = \frac{P\overline{V}}{RT}\sqrt{\frac{kg_cT_0}{kg_cT_0}}$$

$$= \frac{P\overline{V}}{\sqrt{kg_cRT}}\sqrt{\frac{g_ck}{R}}\sqrt{\frac{T_0}{T}}\sqrt{\frac{1}{T_0}}$$

$$= \frac{PM}{\sqrt{T_0}}\sqrt{\frac{kg_c}{R}}\sqrt{1+\frac{k-1}{2}M^2} \tag{13.33}$$

By substituting Eq. 13.28 into Eq. 13.32 the flow per unit area can be expressed in terms of stagnation pressure, stagnation temperature, Mach number, and gas properties.

$$\frac{\dot{m}}{A} = \frac{P_0}{\sqrt{T_0}} \sqrt{\frac{kg_c}{R}} \times \frac{\boldsymbol{M}}{\left(1 + \frac{k-1}{2}\boldsymbol{M}^2\right)^{(k+1)/2(k-1)}} \tag{13.34}$$

At the throat, $\boldsymbol{M} = 1$, and therefore the flow per unit area at the throat, $\dot{m}/A^*$, can be found by setting $\boldsymbol{M} = 1$ in Eq. 13.34.

$$\frac{\dot{m}}{A^*} = \frac{P_0}{\sqrt{T_0}} \sqrt{\frac{kg_c}{R}} \times \frac{1}{\left(\frac{k+1}{2}\right)^{(k+1)/2(k-1)}} \tag{13.35}$$

The area ratio A/A^* can be obtained by dividing Eq. 13.35 by Eq. 13.34.

$$\frac{A}{A^*} = \frac{1}{\boldsymbol{M}}\left[\left(\frac{2}{k+1}\right)\left(1 + \frac{k-1}{2}\boldsymbol{M}^2\right)\right]^{(k+1)/2(k-1)} \tag{13.36}$$

The area ratio A/A^* is the ratio of the area at the point where the Mach number is $\boldsymbol{M}$ to the throat area, and values of A/A^* as a function of Mach number are given in Tables 30 through 35 of the *Gas Tables* and in Table A.10 in the Appendix. Figure 13.9 shows a plot

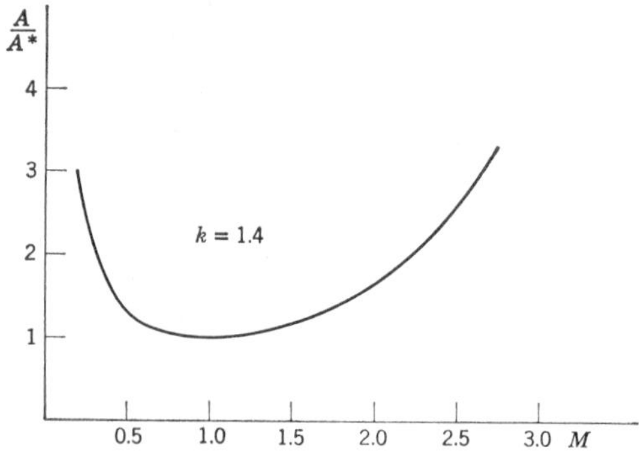

Fig. 13.9 Area ratio as a function of Mach number for a reversible adiabatic nozzle.

of A/A^* vs. $\boldsymbol{M}$, which is in accordance with our previous conclusion that a subsonic nozzle is converging and a supersonic nozzle is diverging.

The final point to be made regarding the isentropic flow of an ideal gas through a nozzle involves the effect of varying the back pressure (the pressure outside the nozzle exit) on the mass rate of flow.

Consider first a convergent nozzle as shown in Fig. 13.10, which also shows the pressure ratio P/P_0 along the length of the nozzle. The

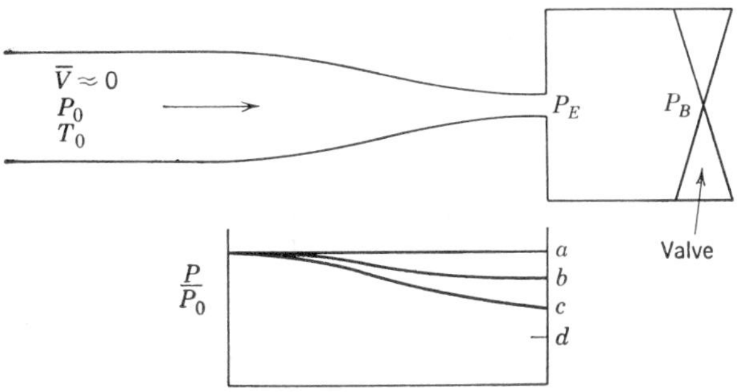

Fig. 13.10 **Pressure ratio as a function of back pressure for a convergent nozzle.**

conditions upstream are the stagnation conditions, which are assumed to be constant. The pressure at the exit plane of the nozzle is designated P_E, and the back pressure P_B. Let us consider how the mass rate of flow $\dot{m}$ and the exit plane pressure P_E/P_0, vary as the back pressure P_B is decreased. These quantities are plotted in Fig. 13.11.

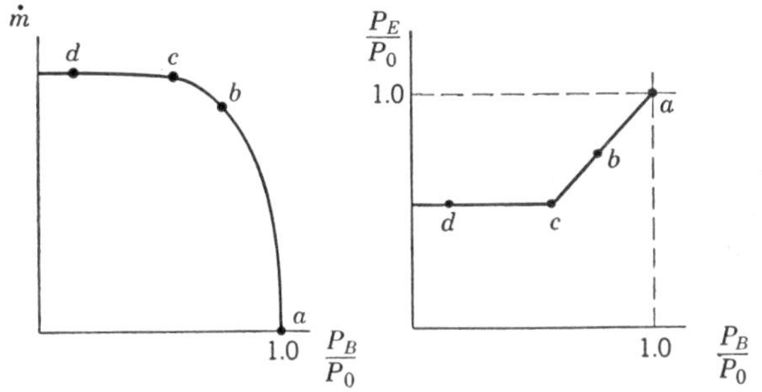

Fig. 13.11 **Mass rate of flow and exit pressure as a function of back pressure for a convergent nozzle.**

When $P_B/P_0 = 1$ there is of course no flow, and $P_E/P_0 = 1$ as designated by point a. Next let the back pressure P_B be lowered to that designated by point b, so that P_B/P_0 is greater than the critical-pressure ratio. The mass rate of flow has a certain value and $P_E = P_B$. The exit Mach number is less than 1. Next let the back pressure be lowered to the critical pressure, designated by point c. The Mach number at the exit is now unity, and P_E is equal to P_B. When P_B is decreased below the critical pressure, designated by point d, there is no further increase in the mass rate of flow, and P_E remains constant at a value equal to the critical pressure, and the exit Mach number is unity. The drop in pressure from P_E to P_B takes place outside the nozzle exit. Under these conditions the nozzle is said to be choked, which means that for given stagnation conditions the nozzle is passing the maximum possible mass flow.

Consider next a convergent-divergent nozzle in a similar arrangement, Fig. 13.12. Point a designates the condition when $P_B = P_0$ and

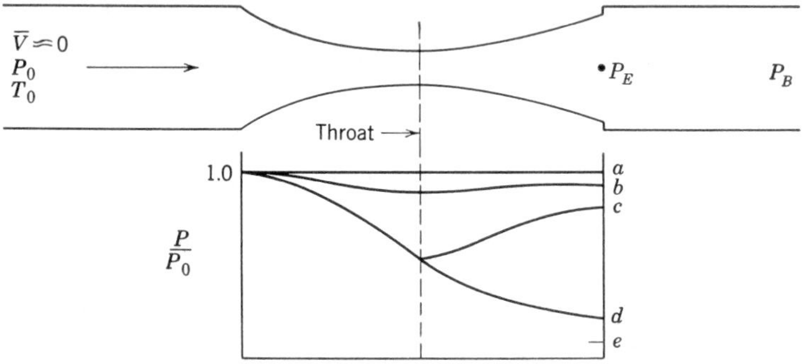

Fig. 13.12 Nozzle pressure ratio as a function of back pressure for a reversible convergent-divergent nozzle.

there is no flow. When P_B is decreased to the pressure indicated by point b, so that P_B/P_0 is less than 1 but considerably greater than the critical-pressure ratio, the velocity increases in the convergent section, but $\boldsymbol{M} < 1$ at the throat. Therefore, the diverging section acts as a subsonic diffuser in which the pressure increases and velocity decreases. Point c designates the back pressure at which $\boldsymbol{M} = 1$ at the throat, but the diverging section acts as a subsonic diffuser (with $\boldsymbol{M} = 1$ at the inlet) in which the pressure increases and velocity decreases. Point d designates one other back pressure that permits isentropic flow, and in this case the diverging section acts as a supersonic nozzle, with a decrease in pressure and an increase in velocity. Between the back pressures designated by points c and d, an isentropic solution is not possible, and shock waves will be present. This matter is discussed in the section that follows. When the back pressure is decreased below that designated by point d, the exit-plane pressure P_E remains constant, and the drop in pressure from P_E to P_B takes place outside the nozzle. This is designated by point e.

Example 13.6 A convergent nozzle has an exit area of 1 in.^{2} Air enters the nozzle with a stagnation pressure of 100 lbf/in.^{2} and a stagnation temperature of 200 F. Determine the mass rate of flow for back pressures of 80 lbf/in.^{2}, 52.8 lbf/in.^{2}, and 30 lbf/in.^{2}, assuming isentropic flow.

For air $k = 1.4$ and Table 30 of the *Gas Tables* (or Table A.10 in the Appendix) may be used. The critical-pressure ratio, P^*/P_0, is 0.528. Therefore, for a back pressure of 52.8 lbf/in.^{2} $\mathbf{M} = 1$ at the nozzle exit and the nozzle is choked. Decreasing the back pressure below 52.8 lbf/in.^{2} will not increase the flow.

For a back pressure of 52.8 lbf/in.^{2}

$$\frac{T^*}{T_0} = 0.833 \qquad T^* = 0.833(660) = 550 \text{ R}$$

At the exit

$$\bar{V} = c = \sqrt{kg_cRT}$$

$$= \sqrt{1.4 \times 32.17 \times 53.34 \times 550} = 1152 \text{ ft/sec}$$

$$\rho^* = \frac{P^*}{RT^*} = \frac{52.8 \times 144}{53.34 \times 550} = 0.259 \text{ lbm/ft}^3$$

$$\dot{m} = \rho A \bar{V}$$

Applying this relation to the throat section

$$\dot{m} = 0.259 \text{ lbm/ft}^3 \times \tfrac{1}{144} \text{ ft}^2 \times 1152 \text{ ft/sec} = 2.07 \text{ lbm/sec}$$

For a back pressure of 80 lbf/in.^{2}, $P_E/P_0 = 0.8$ (subscript E designates the properties in the exit plane). From Table 30 of the *Gas Tables*,

$$\mathbf{M}_E = 0.573 \qquad T_E/T_0 = 0.938$$

$$T_E = 0.938(660) = 619 \text{ R}$$

$$c_E = \sqrt{kg_cRT_E} = \sqrt{1.4 \times 32.17 \times 53.34 \times 619} = 1222 \text{ ft/sec}$$

$$\bar{V}_E = \mathbf{M}_E c_E = 0.573(1222) = 700 \text{ fps}$$

$$\rho_E = \frac{P_E}{RT_E} = \frac{80 \times 144}{53.34 \times 619} = 0.35 \text{ lbm/ft}^3$$

$$\dot{m} = \rho A \bar{V}$$

Applying this relation to the exit section

$$\dot{m} = 0.35 \text{ lbm/ft}^3 \times \tfrac{1}{144} \text{ ft}^2 \times 700 \text{ fps} = 1.69 \text{ lbm/sec}$$

In solving this problem we could also have used Table 1 of the *Gas Tables*, noting that this is an isentropic process. However, for a back pressure below the critical pressure we must realize that this convergent nozzle is choked and will pass only the mass flow that corresponds to a back pressure equal to the critical pressure.

Example 13.7 A converging-diverging nozzle has an exit area to throat area ratio of 2. Air enters this nozzle with a stagnation pressure of 100 lbf/in.^{2} and a stagnation temperature of 200 F. The throat area is 1 in.^{2} Determine the mass rate of flow, exit pressure, exit temperature, exit Mach number, and exit velocity for the following conditions:

(*a*) Sonic velocity at the throat, diverging section acting as a nozzle. (Corresponds to point *d* in Fig. 13.12.)

(*b*) Sonic velocity at the throat, diverging section acting as a diffuser. (Corresponds to point *c* in Fig. 13.12.)

(*a*) In Table 30 of the *Gas Tables* we find that there are two Mach numbers listed for $A/A^* = 2$. One of these is greater than unity and one is less than unity. When the diverging section acts as a supersonic nozzle we use the value for $\boldsymbol{M} > 1$. The following are from Table 30:

$$\frac{A_E}{A^*} = 2.0 \qquad \boldsymbol{M}_E = 2.197 \qquad \frac{P_E}{P_0} = 0.0939 \qquad \frac{T_E}{T_0} = 0.5089$$

Therefore,

$$P_E = 0.0939(100) = 9.39 \text{ lbf/in.}^2$$

$$T_E = 0.5089(660) = 336 \text{ R}$$

$$c_E = \sqrt{kg_cRT_E} = \sqrt{1.4 \times 32.17 \times 53.34 \times 336} = 900 \text{ ft/sec}$$

$$V_E = \boldsymbol{M}_E c_E = 2.197(900) = 1977 \text{ ft/sec}$$

The mass rate of flow can be determined by considering either the throat section or the exit section. However, in general it is preferable to determine the mass rate of flow from conditions at the throat. Since in this case $\boldsymbol{M} = 1$ at the throat, the calculation is identical to the calculation for the flow in the convergent nozzle of Example 13.6 when it is choked.

(*b*) The following are from Table 30 of the *Gas Tables:*

$$\frac{A_E}{A^*} = 2.0 \qquad \boldsymbol{M}_E = 0.306 \qquad \frac{P_E}{P_0} = 0.9371 \qquad \frac{T_E}{T_0} = 0.9816$$

$$P_E = 0.9371(100) = 93.7 \text{ lbf/in.}^2$$

$$T_E = 0.9816(660) = 649 \text{ R}$$

$$c_E = \sqrt{kg_cRT} = \sqrt{1.4 \times 32.17 \times 53.34 \times 649} = 1250 \text{ ft/sec}$$

$$\overline{V}_E = M_E c_E = 0.306(1250) = 383 \text{ ft/sec}$$

Since $M = 1$ at the throat, the mass rate of flow is the same as in (*a*), which is also equal to the flow in the convergent nozzle of Example 13.6 when it is choked.

In the example above a solution assuming isentropic flow is not possible if the back pressure is between 93.7 lbf/in.^{2} and 9.34 lbf/in.^{2} If the back pressure is in this range there will either be a normal shock in the nozzle or oblique shock waves outside the nozzle. The matter of normal shock waves is considered in the following section.

13.8 Normal Shock in an Ideal Gas Flowing Through a Nozzle

A shock wave involves an extremely rapid and abrupt change of state. In a *normal shock* this change of state takes place across a plane normal to the direction of the flow. Figure 13.13 shows an open system

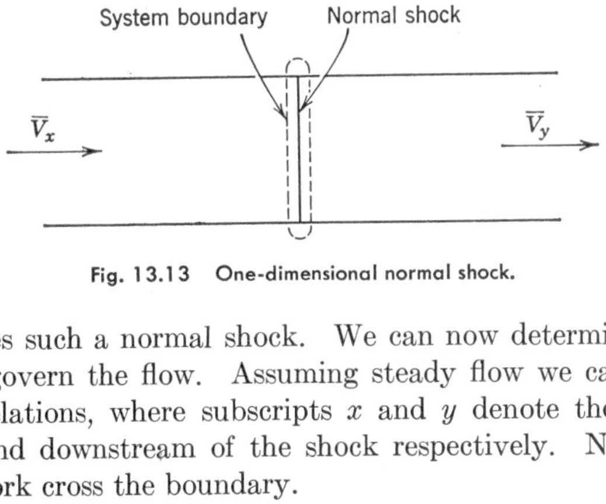

Fig. 13.13 One-dimensional normal shock.

that includes such a normal shock. We can now determine the relations that govern the flow. Assuming steady flow we can write the following relations, where subscripts x and y denote the conditions upstream and downstream of the shock respectively. Note that no heat and work cross the boundary.

First law:

$$h_x + \frac{\overline{V}_x^2}{2g_c} = h_y + \frac{\overline{V}_y^2}{2g_c} = h_{0x} = h_{0y} \tag{13.37}$$

Continuity equation:

$$\frac{\dot{m}}{A} = \rho_x \overline{V}_x = \rho_y \overline{V}_y \tag{13.38}$$

Momentum equation:

$$A(P_x - P_y) = \frac{\dot{m}}{g_c}(\overline{V}_y - \overline{V}_x) \tag{13.39}$$

Second law: Since the process is adiabatic

$$s_y - s_x \geq 0 \tag{13.40}$$

The energy and continuity equations can be combined to give an equation which when plotted on the *h-s* diagram is called the *Fanno line*. Similarly, the momentum and continuity equations can be combined to give an equation, the plot of which on the *h-s* diagram is known as the *Rayleigh line*. Both of these lines are shown on the *h-s* diagram of Fig. 13.14. It can be shown that the point of maximum

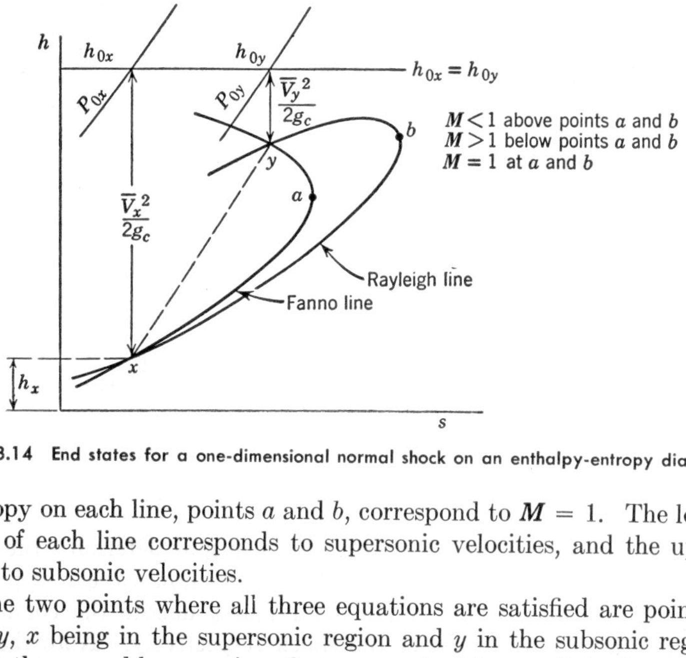

Fig. 13.14 End states for a one-dimensional normal shock on an enthalpy-entropy diagram.

entropy on each line, points a and b, correspond to $M = 1$. The lower part of each line corresponds to supersonic velocities, and the upper part to subsonic velocities.

The two points where all three equations are satisfied are points x and y, x being in the supersonic region and y in the subsonic region. Since the second law requires that $s_y - s_x \geq 0$ in an adiabatic process, we conclude that the normal shock can proceed only from x to y. This means that the velocity changes from supersonic ($M > 1$) before the shock to subsonic ($M < 1$) after the shock.

The equations governing normal shock waves will now be developed. If we assume constant specific heats we conclude from Eq. 13.37, the energy equation, that

$$T_{0x} = T_{0y} \tag{13.41}$$

That is, there is no change in stagnation temperature across a normal shock. Introducing Eq. 13.27

$$\frac{T_{0x}}{T_x} = 1 + \frac{k-1}{2}\boldsymbol{M}_x^2 \qquad \frac{T_{0y}}{T_y} = 1 + \frac{k-1}{2}\boldsymbol{M}_y^2$$

and substituting into Eq. 13.41 we have

$$\frac{T_y}{T_x} = \frac{1 + \dfrac{k-1}{2}\boldsymbol{M}_x^2}{1 + \dfrac{k-1}{2}\boldsymbol{M}_y^2} \tag{13.42}$$

The equation of state, the definition of Mach number, and the relation $c = \sqrt{kg_cRT}$ can be introduced into the continuity equation as follows:

$$\rho_x \overline{V}_x = \rho_y \overline{V}_y$$

But

$$\rho_x = \frac{P_x}{RT_x} \qquad \rho_y = \frac{P_y}{RT_y}$$

$$\frac{T_y}{T_x} = \frac{P_y\overline{V}_y}{P_x\overline{V}_x} = \frac{P_y\boldsymbol{M}_y c_y}{P_x\boldsymbol{M}_x c_x} = \frac{P_y\boldsymbol{M}_y\sqrt{T_y}}{P_x\boldsymbol{M}_x\sqrt{T_x}}$$

$$= \left(\frac{P_y}{P_x}\right)^2\left(\frac{\boldsymbol{M}_y}{\boldsymbol{M}_x}\right)^2 \tag{13.43}$$

Combining Eqs. 13.42 and 13.43, which involves combining the energy equation and the continuity equation, gives the equation of the Fanno line.

$$\frac{P_y}{P_x} = \frac{\boldsymbol{M}_x\sqrt{1 + \dfrac{k-1}{2}\boldsymbol{M}_x^2}}{\boldsymbol{M}_y\sqrt{1 + \dfrac{k-1}{2}\boldsymbol{M}_y^2}} \tag{13.44}$$

The momentum and continuity equations can be combined as follows to give the equation of the Rayleigh line.

$$P_x - P_y = \frac{\dot{m}}{Ag_c}(\overline{V}_y - \overline{V}_x) = \frac{\rho_y \overline{V}_y^2 - \rho_x \overline{V}_x^2}{g_c}$$

$$P_x g_c + \rho_x \overline{V}_x^2 = P_y g_c + \rho_y \overline{V}_y^2$$

$$P_x g_c + \rho_x \boldsymbol{M}_x^2 c_x^2 = P_y g_c + \rho_y \boldsymbol{M}_y^2 c_y^2$$

$$P_x g_c + \frac{P_x \boldsymbol{M}_x^2}{RT_x}(kg_cRT_x) = P_y g_c + \frac{P_y \boldsymbol{M}_y^2}{RT_y}(kg_cRT_y)$$

$$P_x(1 + k\boldsymbol{M}_x^2) = P_y(1 + k\boldsymbol{M}_y^2)$$

$$\frac{P_y}{P_x} = \frac{1 + k\boldsymbol{M}_x^2}{1 + k\boldsymbol{M}_y^2} \tag{13.45}$$

Equations 13.44 and 13.45 can be combined to give the following equation relating $\boldsymbol{M}_x$ and $\boldsymbol{M}_y$.

$$\boldsymbol{M}_y^2 = \frac{\boldsymbol{M}_x^2 + \dfrac{2}{k-1}}{\dfrac{2k}{k-1}\boldsymbol{M}_x^2 - 1} \tag{13.46}$$

Tables 48 through 53 of the *Gas Tables* give the normal shock functions, which include $\boldsymbol{M}_y$ as a function of $\boldsymbol{M}_x$. Table A.11 of the Appendix has been abstracted from Table 48 of the *Gas Tables*, and applies to an ideal gas with $k = 1.4$. Note that $\boldsymbol{M}_x$ is always supersonic and $\boldsymbol{M}_y$ is always subsonic, which agrees with the previous statement that in a normal shock the velocity changes from supersonic to subsonic. These tables also give the pressure, density, temperature, and stagnation pressure ratios across a normal shock as a function of $\boldsymbol{M}_x$. These are found from Eqs. 13.42, 13.43, and the equation of state. Note that there is always a drop in stagnation pressure across a normal shock and an increase in the static pressure.

Example 13.8 Consider the convergent-divergent nozzle of Example 13.7 in which the diverging section acts as a supersonic nozzle (Fig. 13.15). Assume that a normal shock stands in the exit plane of the nozzle. Determine the static pressure and temperature and the stagnation pressure just downstream of the normal shock.

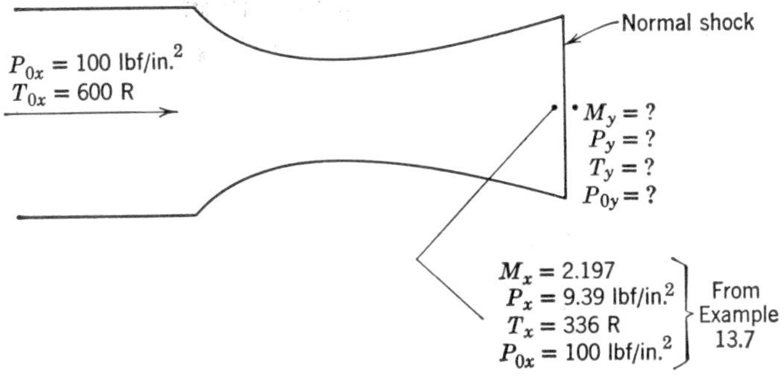

Fig. 13.15 Sketch for Example 13.8.

From Table 48 of the *Gas Tables* (Table A.11 of the Appendix)

$$M_x = 2.197 \qquad M_y = 0.548 \qquad \frac{P_y}{P_x} = 5.46 \qquad \frac{T_y}{T_x} = 1.854 \qquad \frac{P_{0y}}{P_{0x}} = 0.631$$

$$P_y = 5.46 \times P_x = 5.46 \times 9.39 = 51.2 \text{ lbf/in.}^2$$

$$T_y = 1.854 \times T_x = 1.854 \times 336 = 622 \text{ R}$$

$$P_{0y} = 0.631 \times P_{0x} = 0.631 \times 100 = 63.1 \text{ lbf/in.}^2$$

In the light of this example we can conclude the discussion concerning the flow through a convergent-divergent nozzle. Figure 13.12 is repeated here as Fig. 13.16 for convenience, except that points *f*, *g*,

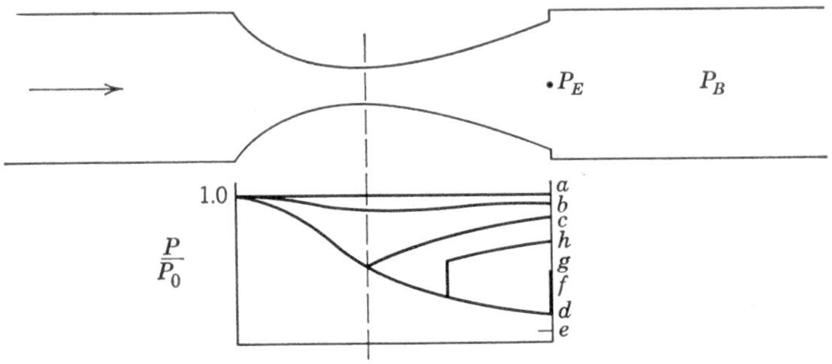

Fig. 13.16 Nozzle-pressure ratio as a function of back pressure for a convergent-divergent nozzle.

and *h* have been added. Consider point *d*. We have already noted that with this back pressure the exit plane pressure P_E is just equal to the back pressure P_B, and isentropic flow is maintained in the nozzle. Let the back pressure be raised to that designated by point *f*. The exit-plane pressure P_E is not influenced by this increase in back pressure, and the increase in pressure from P_E to P_B takes place outside the nozzle. Let the back pressure be raised to that designated by point g, which is just sufficient to cause a normal shock to stand in the exit plane of the nozzle. The exit-plane pressure P_E (downstream of the shock) is equal to the back pressure P_B, and $\boldsymbol{M} < 1$ leaving the nozzle. This is the case in Example 13.8. Now let the back pressure be raised to that corresponding to point h. As the back pressure is raised from g to h the normal shock moves into the nozzle as indicated. Since $\boldsymbol{M} < 1$ downstream of the normal shock, the diverging part of the nozzle which is downstream of the shock acts as a subsonic diffuser. As the back pressure is increased from h to c the shock moves further upstream and disappears at the nozzle throat where the back pressure corresponds to c. This is reasonable since there are no supersonic velocities involved when the back pressure corresponds to c, and hence no shock waves are possible.

Example 13.9 Consider the convergent-divergent nozzle of Examples 13.7 and 13.8. Assume that there is a normal shock wave standing at the point where $\boldsymbol{M} = 1.5$. Determine the exit-plane pressure, temperature, and Mach number. Assume isentropic flow except for the normal shock (Fig. 13.17).

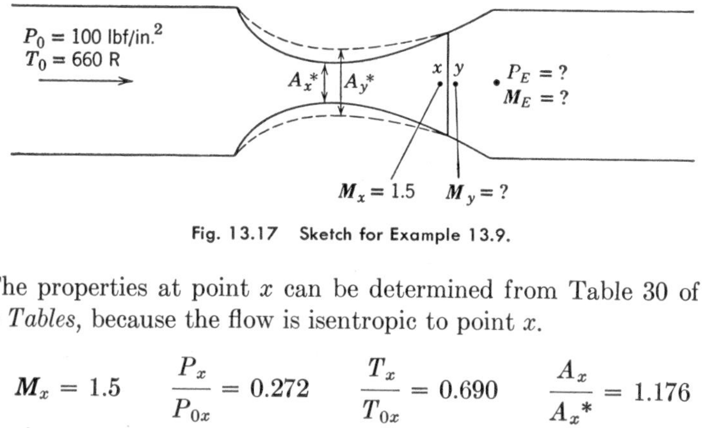

Fig. 13.17 Sketch for Example 13.9.

The properties at point x can be determined from Table 30 of the *Gas Tables*, because the flow is isentropic to point x.

$$\boldsymbol{M}_x = 1.5 \qquad \frac{P_x}{P_{0x}} = 0.272 \qquad \frac{T_x}{T_{0x}} = 0.690 \qquad \frac{A_x}{A_x^*} = 1.176$$

Therefore,

$$P_x = 0.272(100) = 27.2 \text{ lbf/in.}^2$$

$$T_x = 0.690(660) = 455 \text{ R}$$

The properties at point y can be determined from the normal shock functions, Table 48 (Appendix Table A.11).

$$\boldsymbol{M}_y = 0.701 \qquad \frac{P_y}{P_x} = 2.458 \qquad \frac{T_y}{T_x} = 1.320 \qquad \frac{P_{0y}}{P_{0x}} = 0.9298$$

$$P_y = 2.458P_x = 2.458(27.2) = 66.9 \text{ lbf/in.}^2$$

$$T_y = 1.320T_x = 1.320(455) = 601 \text{ R}$$

$$P_{0y} = 0.9298P_{0x} = 0.9298(100) = 93.0 \text{ lbf/in.}^2$$

Since there is no change in stagnation temperature across a normal shock,

$$T_{0x} = T_{0y} = 660 \text{ R}$$

From y to E the diverging section acts as a subsonic diffuser. In solving this problem it is convenient to think of the flow at y as having come from an isentropic nozzle having a throat area A_y^*. Such a hypothetical nozzle is shown by the dotted line. From the table of isentropic flow functions, Table 30 in the *Gas Tables*, we find the following for $\boldsymbol{M}_y = 0.701$:

$$\boldsymbol{M}_y = 0.701 \qquad \frac{A_y}{A_y^*} = 1.094 \qquad \frac{P_y}{P_{0y}} = 0.721 \qquad \frac{T_y}{T_{0y}} = 0.911$$

From the statement of the problem

$$\frac{A_E}{A_x^*} = 2.0$$

Also, since the flow from y to E is isentropic,

$$\frac{A_E}{A_E^*} = \frac{A_E}{A_y^*} = \frac{A_E}{A_x^*} \times \frac{A_x^*}{A_x} \times \frac{A_x}{A_y} \times \frac{A_y}{A_y^*}$$

$$= \frac{A_E}{A_y^*} = 2.0 \times \frac{1}{1.176} \times 1 \times 1.094 = 1.897$$

From the table of isentropic flow functions for $A/A^* = 1.897$ and $\boldsymbol{M} < 1$

$$\boldsymbol{M}_E = 0.343 \qquad \frac{P_E}{P_{0E}} = 0.922 \qquad \frac{T_E}{T_{0E}} = 0.977$$

$$\frac{P_E}{P_{0E}} = \frac{P_E}{P_{0y}} = 0.922$$

$$P_E = 0.922(P_{0y}) = 0.922(93.0) = 85.7 \text{ lbf/in.}^2$$

$$T_E = 0.977(T_{0E}) = 0.977(660) = 645 \text{ R}$$

In conclusion it should be pointed out that in considering the normal shock we have ignored the effect of viscosity and thermal conductivity, which are certain to be present. The actual shock wave will occur over some finite thickness. However, the development as given here gives a very good qualitative picture of normal shocks, and also provides a basis for fairly accurate quantitative results.

13.9 Flow of a Vapor Through a Nozzle

In this section we consider a vapor to be a substance that is in the gaseous phase but with limited superheat. Therefore, the vapor will probably deviate significantly from the ideal-gas relations, and the possibility of condensation must be considered. The most familiar example is the flow of steam through the nozzle of a steam turbine.

The principles that have been developed for the isentropic flow of an ideal gas apply also to the isentropic flow of a vapor. However, because the vapor deviates from the ideal-gas relationships, the appropriate tables of thermodynamic properties must be used. Further, the possibility of condensation must be borne in mind, as indicated in Fig. 13.18. If steam expands isentropically from state 1 to state 2, and if equilibrium is maintained throughout the nozzle, condensation would begin at point a, and at pressures below this a mixture of liquid droplets and vapor would be present. In an actual nozzle the formation of the liquid tends to be delayed due to an effect known as supersaturation. This will be considered in a later paragraph.

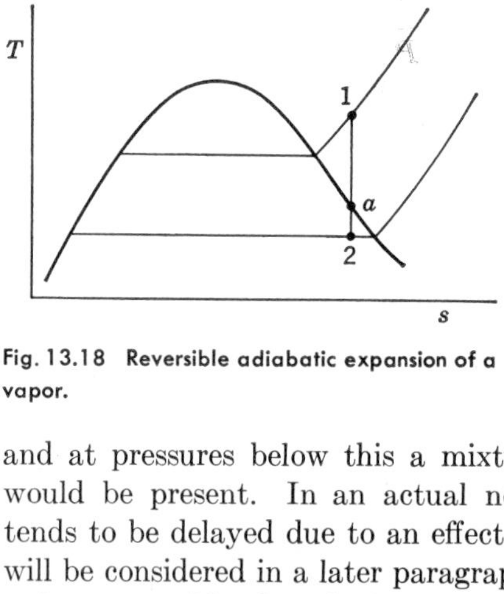

Fig. 13.18 Reversible adiabatic expansion of a vapor.

Let us consider first the isentropic flow of steam in a nozzle without condensation. The value of the specific-heat ratio k for steam varies, but $k = 1.3$ is a good approximation over a considerable range. Therefore, the critical pressure ratio, P^*/P_0, can be found by the relation

$$\frac{P^*}{P_0} = \left(\frac{2}{k+1}\right)^{k/(k-1)} = 0.545 \qquad (13.47)$$

This is also the value given in Table 13.1. Knowing, therefore, the critical-pressure ratio, the throat area for a given flow can be calculated. The exit area of the nozzle can be calculated in a similar manner. The example below illustrates this procedure.

Example 13.10 Steam at a stagnation pressure of 100 lbf/in.^{2} and a stagnation temperature of 600 F expands in a nozzle to 20 lbf/in.^{2} Determine the throat area and exit area required for a flow of 20,000 lbm/hr, assuming reversible adiabatic flow.

First consider the throat.

$$\frac{P^*}{P_0} = 0.545 \qquad P^* = 54.5 \text{ lbf/in.}^2$$

$$s^* = s_0 = 1.7581 \text{ Btu/lbm R} \qquad h_0 = 1329.1 \text{ Btu/lbm}$$

$$t^* = 460 \text{ F} \qquad h^* = 1264.0 \text{ Btu/lbm}$$

$$h_0 = h^* + \frac{\overline{V}^{*2}}{2g_c} \qquad \overline{V}^* = \sqrt{2g_c}\sqrt{h_0 - h^*}$$

$$\overline{V}^* = 223.7\sqrt{1329.1 - 1264.0} = 223.7\sqrt{65.1} = 1800 \text{ ft/sec}$$

$$v^* = 9.961 \text{ (from the steam tables)}$$

$$\dot{m}v^* = A^*\overline{V}^*$$

$$A^* = \frac{20{,}000 \times 9.961}{3600 \times 1800} = 0.0307 \text{ ft}^2 = 4.42 \text{ in.}^2$$

At the nozzle exit

$$P_E = 20 \text{ lbf/in.}^2 \qquad s_E = s_0 = 1.7581 \text{ Btu/lbm R}$$

$$h_E = 1174.9 \text{ Btu/lbm} \qquad v_E = 21.28 \text{ lbf/in.}^2$$

$$V_E = \sqrt{2g_c}\sqrt{h_0 - h_E} = 223.7\sqrt{1329.1 - 1174.4} = 223.7\sqrt{154.2}$$

$$= 2780 \text{ ft/sec}$$

$$\dot{m}v_E = A_E\overline{V}_E$$

$$A_E = \frac{20{,}000 \times 21.28}{3600 \times 2780} = 0.0426 \text{ ft}^2 = 6.13 \text{ in.}^2$$

For steam initially saturated the critical-pressure ratio is usually taken as

$$\frac{P^*}{P_0} = 0.577 \tag{13.48}$$

If the flow with initially saturated steam is assumed to be isentropic there will be a certain amount of liquid entrained with the vapor. Calculation of throat and exit areas under most conditions is similar to the example above.

When the steam flowing through a nozzle is initially superheated, there will be a certain point in the nozzle where the steam becomes saturated. If this point occurs in the diverging section of the nozzle a phenomenon known as supersaturation occurs. The formation of

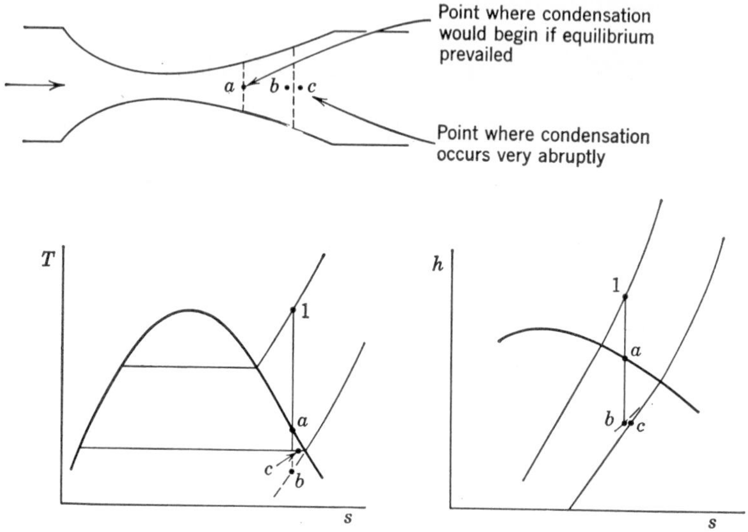

Fig. 13.19 Illustration of the phenomenon of supersaturation in a nozzle.

the droplets of liquid is delayed, and the vapor temperature is less than saturation temperature for the given pressure, as indicated in Fig. 13.19.

In the nozzle the formation of liquid droplets would begin at point a if equilibrium conditions prevailed. However, no liquid is formed until b is reached, and then it occurs rather abruptly in what is sometimes known as a condensation shock. Between points a and b the steam is in a supersaturated, or metastable, state. The metastable state b can be approximated by extending the constant-pressure line from the superheat region into the saturation region, as shown on the T-s and h-s diagrams. This phenomenon of supersaturation is observed only in the diverging portions of the nozzle. In a nozzle in which supersaturation occurs the flow might be slightly greater than that obtained in an isentropic flow without supersaturation.

Supersaturation may also be present in a supersonic wind tunnel. An abrupt condensation of the moisture in the air would disturb the flow pattern, and for this reason most of the moisture is removed from the air to be used in the tunnel.

A similar phenomenon may occur in the flow of a saturated liquid through a nozzle or valve. When a saturated liquid flows through a nozzle or valve some of it will vaporize as the pressure is reduced. In practice the formation of the vapor is delayed, because a metastable state exists.

A detailed consideration of metastable states is beyond the scope of this text. It might be pointed out, however, that it is associated with the effects of surface tension when very small droplets of liquid and very small bubbles of vapor are formed.

13.10 Nozzle and Diffuser Coefficients

Up to this point we have considered only isentropic flow and normal shocks. As was pointed out in Chapter 10, isentropic flow through a nozzle provides a standard to which the performance of an actual nozzle can be compared. For nozzles, the three important parameters by which actual flow can be compared to the ideal flow are nozzle efficiency, velocity coefficient, and discharge coefficient. These are defined as follows:

The *nozzle efficiency* η_N is defined as

$$\eta_N = \frac{\text{Actual kinetic energy at nozzle exit}}{\text{Kinetic energy at nozzle exit with isentropic flow to same exit pressure}} \tag{13.49}$$

The efficiency can be defined in terms of properties. On the h-s diagram of Fig. 13.20 state $0i$ represents the stagnation state of the

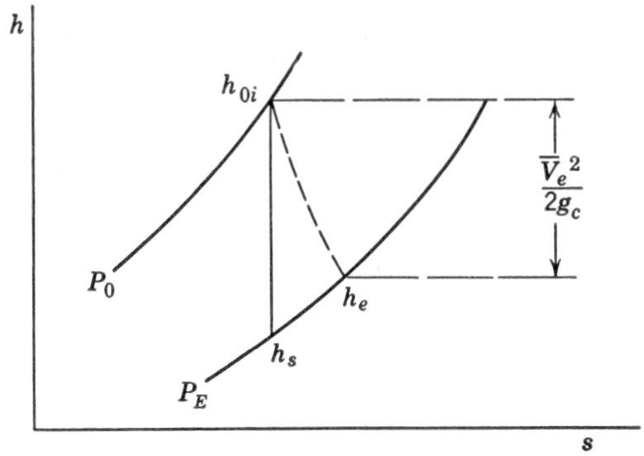

Fig. 13.20 Enthalpy-entropy diagram showing the effects of irreversibility in a nozzle.

fluid entering the nozzle; state e represents the actual state at the nozzle exit; and state s represents the state that would have been achieved at the nozzle exit if the flow had been reversible and adiabatic to the same exit pressure. Therefore, in terms of these states the nozzle efficiency is

$$\eta_N = \frac{h_{0i} - h_e}{h_{0i} - h_s}$$

Nozzle efficiencies vary in general from 90 to 99 per cent. Large nozzles usually have higher efficiencies than small nozzles, and nozzles with straight axes have higher efficiencies than nozzles with curved axes. The irreversibilities, which cause the departure from isentropic flow, are primarily due to frictional effects, and are confined largely to the boundary layer. The rate of change of cross-sectional area along the nozzle axis (i.e., the nozzle contour) is an important parameter in the design of an efficient nozzle, particularly in the divergent section. Detailed consideration of this matter is beyond the scope of this text, and the reader is referred to standard references on the subject.

The *velocity coefficient* $C_{\bar{V}}$ is defined as

$$C_{\bar{V}} = \frac{\text{Actual velocity at nozzle exit}}{\text{Velocity at nozzle exit with isentropic flow and same exit pressure}} \tag{13.50}$$

It follows that the velocity coefficient is equal to the square root of the nozzle efficiency

$$C_{\bar{V}} = \sqrt{\eta_N} \tag{13.51}$$

The *coefficient of discharge* C_D is defined by the relation

$$C_D = \frac{\text{Actual mass rate of flow}}{\text{Mass rate of flow with isentropic flow}}$$

In determining the mass rate of flow with isentropic conditions, the actual back pressure is used if the nozzle is not choked. If the nozzle is choked, the isentropic mass rate of flow is based on isentropic flow and sonic velocity at the minimum section (i.e., sonic velocity at the exit of a convergent nozzle and at the throat of a convergent-divergent nozzle).

The performance of a diffuser is usually given in terms of diffuser efficiency, which is best defined with the aid of an h-s diagram. On the

h-s diagram of Fig. 13.21 states 1 and 01 are the actual and stagnation states of the fluid entering the diffuser. States 2 and 02 are the actual and stagnation states of the fluid leaving the diffuser. State 3 is not attained in the diffuser, but it is the state that has the same entropy as

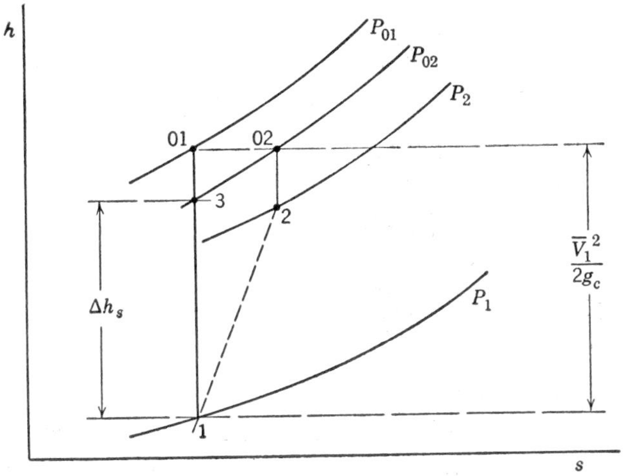

Fig. 13.21 Enthalpy-entropy diagram showing the definition of diffuser efficiency.

the initial state and the pressure of the isentropic stagnation state leaving the diffuser. The efficiency of the diffuser η_D is defined as

$$\eta_D = \frac{\Delta h_s}{\overline{V}_1^2/2g_c} = \frac{h_3 - h_1}{h_{01} - h_1} = \frac{h_3 - h_1}{h_{02} - h_1} \tag{13.52}$$

If we assume an ideal gas with constant specific heat this reduces to

$$\eta_D = \frac{T_3 - T_1}{T_{02} - T_1} = \frac{\dfrac{(T_3 - T_1)}{T_1} T_1}{\dfrac{\overline{V}_1^2}{2g_c C_{p0}}}$$

$$C_{p0} = \frac{kR}{k-1} \qquad T_1 = \frac{c_1^2}{kg_cR} \qquad \overline{V}_1^2 = \boldsymbol{M}_1^2 c_1^2 \qquad \frac{T_3}{T_1} = \left(\frac{P_{02}}{P_1}\right)^{(k-1)/k}$$

Therefore,

$$\eta_D = \frac{\left(\dfrac{P_{02}}{P_1}\right)^{(k-1)/k} - 1}{\dfrac{k-1}{2}\boldsymbol{M}_1^2}$$

$$\left(\frac{P_{02}}{P_1}\right)^{(k-1)/k} = \left(\frac{P_{01}}{P_1}\right)^{(k-1)/k} \times \left(\frac{P_{02}}{P_{01}}\right)^{(k-1)/k}$$

$$= \left(1 + \frac{k-1}{2}\boldsymbol{M}_1^2\right)\left(\frac{P_{02}}{P_{01}}\right)^{(k-1)/k}$$

$$\eta_D = \frac{\left(1 + \dfrac{k-1}{2}\boldsymbol{M}_1^2\right)\left(\dfrac{P_{02}}{P_{01}}\right)^{(k-1)/k} - 1}{\dfrac{k-1}{2}\boldsymbol{M}_1^2} \tag{13.53}$$

13.11 Nozzles and Orifices as Flow-Measuring Devices

The mass rate of flow of a fluid flowing in a pipe is frequently determined by measuring the pressure drop across a nozzle or orifice in the line, as shown in Fig. 13.22. The ideal process for such a nozzle or orifice is assumed to be isentropic flow through a nozzle which has the measured pressure drop from inlet to exit and a minimum cross-sectional area equal to the minimum area of the nozzle or orifice. The actual flow is related to the ideal flow by the coefficient of discharge which is defined by Eq. 13.52.

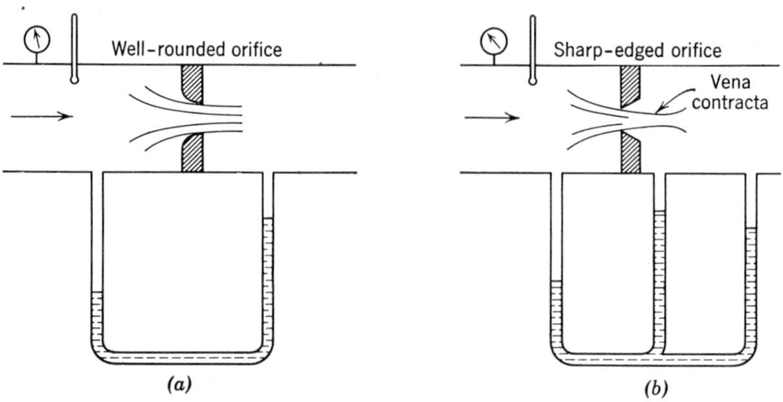

Fig. 13.22 Nozzles and orifices as flow-measuring devices.

The pressure difference measured across an orifice depends upon the location of the pressure taps as indicated in Fig. 13.22. Since the ideal flow is based on the measured pressure difference, it follows that the coefficient of discharge depends on the locations of the pressure taps. Also, the coefficient of discharge for a sharp-edged orifice is considerably less than that for a well-rounded nozzle, primarily due to a contraction of the stream, known as the vena contracta, as it flows through a sharp-edged orifice.

There are two approaches to determining the discharge coefficient of a nozzle or orifice. One is to follow a standard design procedure, such as the ones established by the American Society of Mechanical Engineers,* and use the coefficient of discharge given for a particular design. A more accurate method is to calibrate a given nozzle or orifice, and determine the discharge coefficient for a given installation by accurately measuring the actual mass rate of flow. The procedure to be followed will depend on the accuracy desired and other factors involved (such as time, expense, availability of calibration facilities) in a given situation.

For incompressible fluids flowing through an orifice the ideal flow for a given pressure drop can be found by the procedure outlined in Section 13.4. Actually, it is advantageous to combine Eqs. 13.14 and 5.26 to give the following relation, which is valid for reversible flow.

$$v(P_2 - P_1) + \frac{\overline{V}_2^2 - \overline{V}_1^2}{2g_c} = v(P_2 - P_1) + \frac{\overline{V}_2^2 - (A_2/A_1)^2\overline{V}_2^2}{2g_c} = 0 \tag{13.54}$$

or

$$v(P_2 - P_1) + \frac{\overline{V}_2^2}{2g_c}\left[1 - \left(\frac{A_2}{A_1}\right)^2\right] = 0$$

$$\overline{V}_2 = \sqrt{\frac{2g_c v(P_1 - P_2)}{[1 - (A_2/A_1)^2]}} \tag{13.55}$$

For an ideal gas it is frequently advantageous to use the following simplified procedure when the pressure drop across an orifice or nozzle is small. Consider the nozzle shown in Fig. 13.23. From the first law we conclude that

$$h_i + \frac{\overline{V}_i^2}{2g_c} = h_e + \frac{\overline{V}_e^2}{2g_c}$$

* *Fluid Meters, Their Theory and Application,* ASME, 1937. *Fluid Meters, Their Selection and Installation,* ASME, 1933.

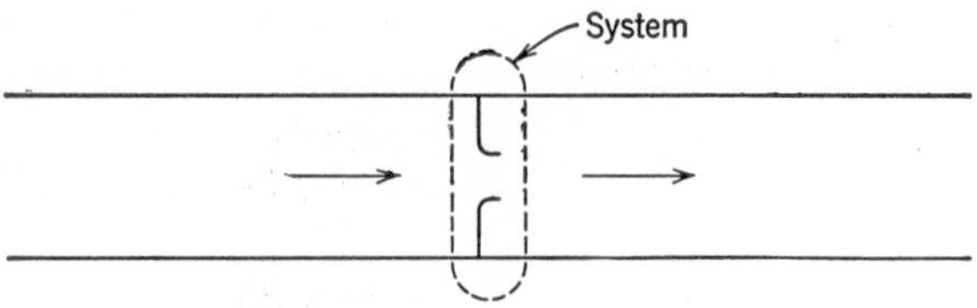

Fig. 13.23 Analysis of a nozzle as a flow-measuring device.

Assuming constant specific heat, this reduces to

$$\frac{\overline{V}_e^2 - \overline{V}_i^2}{2g_c} = h_i - h_e = C_{p0}(T_i - T_e)$$

Let ΔP and ΔT be the decrease in pressure and temperature across the nozzle. Since we are considering reversible adiabatic flow we note that

$$\frac{T_e}{T_i} = \left(\frac{P_e}{P_i}\right)^{(k-1)/k}$$

or

$$\frac{T_i - \Delta T}{T_i} = \left(\frac{P_i - \Delta P}{P_i}\right)^{(k-1)/k}$$

$$1 - \frac{\Delta T}{T_i} = \left(1 - \frac{\Delta P}{P_i}\right)^{(k-1)/k}$$

Using the binomial expansion on the right side of the equation we have

$$1 - \frac{\Delta T}{T_i} = 1 - \frac{k-1}{k}\frac{\Delta P}{P_i} - \frac{k-1}{2k^2}\frac{\Delta P^2}{P_i^2} \cdots$$

If $\Delta P/P_i$ is small this reduces to

$$\frac{\Delta T}{T_i} = \frac{k-1}{k}\frac{\Delta P}{P_i}$$

Substituting this into the first-law equation we have

$$\frac{\overline{V}_e^2 - \overline{V}_i^2}{2g_c} = C_{p0}\frac{k-1}{k}\Delta P\frac{T_i}{P_i}$$

But for an ideal gas

$$C_{p0} = \frac{kR}{k-1} \qquad \text{and} \qquad v_i = R\frac{T_i}{P_i}$$

Therefore,

$$\frac{\overline{V}_e^2 - \overline{V}_i^2}{2g_c} = v_i\,\Delta P$$

which is the same as Eq. 13.54, which was developed for incompressible flow. Therefore, when the pressure drop across a nozzle or orifice is small, the flow can be calculated with high accuracy by assuming incompressible flow.

The Pitot tube, Fig. 13.24, is an important instrument for measuring the velocity of a fluid. In calculating the flow with a Pitot tube it is as-

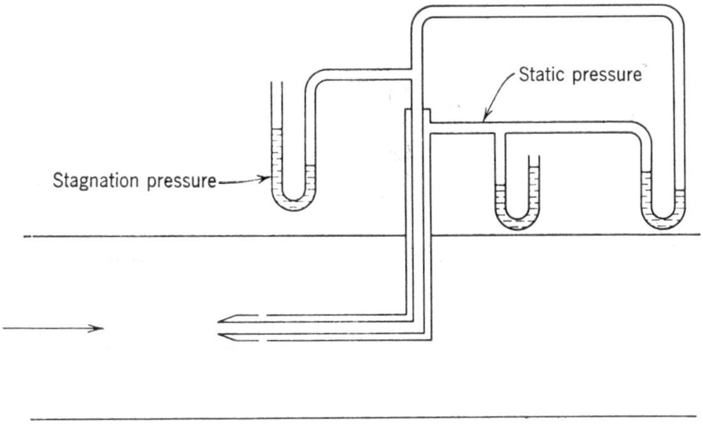

Fig. 13.24 Schematic arrangement of a Pitot tube.

sumed that the fluid is decelerated isentropically in front of the Pitot tube, and therefore the stagnation pressure of the free stream can be measured.

Applying the first law to this process we have

$$h + \frac{\bar{V}^2}{2g_c} = h_0$$

If we assume incompressible flow for this isentropic process, the first law reduces to (because $T\,ds = dh - v\,dp$)

$$\frac{\bar{V}^2}{2g_c} = h_0 - h = v(P_0 - P)$$

or

$$\bar{V} = \sqrt{2g_c v(P_0 - P)} \tag{13.56}$$

If we consider the compressible flow of an ideal gas with constant specific heat, the velocity can be found from the relations

$$\frac{\bar{V}^2}{2g_c} = h_0 - h = C_{p0}(T_0 - T) = C_{p0}T\left(\frac{T_0}{T} - 1\right)$$

$$= C_{p0}T\left[\left(\frac{P_0}{P}\right)^{(k-1)/k} - 1\right] \tag{13.57}$$

It is of interest to know the error introduced by assuming incompressible flow when using the Pitot tube to measure the velocity of an ideal gas. In order to do so we introduce Eq. 13.28 and rearrange it as follows:

$$\frac{P_0}{P} = \left(1 + \frac{k-1}{2}M^2\right)^{k/(k-1)} = \left[1 + \left(\frac{k-1}{2}\right)\left(\frac{\overline{V}^2}{c^2}\right)\right]^{k/(k-1)} \tag{13.58}$$

But

$$\frac{\overline{V}^2}{2g_c} + C_{p0}T = C_{p0}T_0$$

$$\frac{\overline{V}^2}{2g_c} + \frac{kRc^2}{(k-1)kg_cR} = \frac{kR{c_0}^2}{(k-1)kg_cR}$$

$$1 + \frac{2c^2}{(k-1)\overline{V}^2} = \frac{2{c_0}^2}{(k-1)\overline{V}^2}, \quad \text{where } c_0 = \sqrt{kg_cRT_0}$$

$$\frac{c^2}{\overline{V}^2} = \frac{k-1}{2}\left[\left(\frac{2}{k-1}\right)\left(\frac{{c_0}^2}{\overline{V}^2}\right) - 1\right] = \frac{{c_0}^2}{\overline{V}^2} - \frac{k-1}{2}$$

or

$$\frac{c^2}{\overline{V}^2} = \frac{{c_0}^2}{\overline{V}^2} - \frac{k-1}{2} \tag{13.59}$$

Substituting this into Eq. 13.58 and rearranging

$$\frac{P}{P_0} = \left[1 - \frac{k-1}{2}\left(\frac{\overline{V}}{c_0}\right)^2\right]^{k/(k-1)} \tag{13.60}$$

Expanding this equation by the binomial theorem, and including terms through $(\overline{V}/c_0)^4$ we have

$$\frac{P}{P_0} = 1 - \frac{k}{2}\left(\frac{\overline{V}}{c_0}\right)^2 + \frac{k}{8}\left(\frac{\overline{V}}{c_0}\right)^4$$

On rearranging this we have

$$\frac{P_0 - P}{\rho_0\overline{V}^2/2g_c} = 1 - \frac{1}{4}\left(\frac{\overline{V}}{c_0}\right)^2 \tag{13.61}$$

For incompressible flow the corresponding equation is

$$\frac{P_0 - P}{\rho_0\overline{V}^2/2g_c} = 1$$

Therefore, the second term on the right side of Eq. 13.61 represents the error involved if incompressible flow is assumed. The error in pressure for a given velocity, and the error in velocity for a given pressure which would result from assuming incompressible flow are given in Table 13.2.

TABLE 13.2

$\overline{V}/c_0$	Approximate Room-Temperature Velocity, ft/sec	Error in Pressure for a Given Velocity, %	Error in Velocity for a Given Pressure, %
0.0	0	0	0
0.1	110	0.25	−0.13
0.2	220	1.0	−0.50
0.3	330	2.5	−1.1
0.4	440	4.0	−2.4
0.5	550	6.25	−3.3

13.12 Flow Through Blade Passages

We will now apply the principles we have been considering in this chapter to the flow of fluids through blade passages. We will limit our consideration to uniform, parallel, one-dimensional flow.

There are a number of different devices that involve flow through blade passages. Some of the familiar ones are steam and gas turbines, axial-flow compressors, and certain types of blowers, pumps, and torque converters. Some of these devices, such as the steam turbine, are made up of a number of stages. One stage consists of a nozzle (or blade passages which act as a nozzle) and the moving row of blades that follows. Sometimes the blades are also called buckets, but in this text the term blade will be used.

In the analysis of such problems it is necessary to discuss both absolute and relative velocities. By absolute velocity we mean the velocity a stationary observer would detect. The relative velocity is the velocity an observer traveling with the blade would detect, which is designated with a subscript R. The velocity of the blade is designated $\overline{V}_B$.

It is advantageous to show these velocities on a vector diagram. Figure 13.25 shows typical vector diagrams for both a turbine and a compressor. In each case the first sketch shows the velocity vectors in relation to the blades, and the second sketch shows how these vector diagrams are usually drawn. In these diagrams $\overline{V}_1$ represents the velocity of the fluid entering the blade passage, and α designates the angle at which it enters. $\overline{V}_{1R}$ represents the relative velocity of the fluid entering the blade passage and β the angle at which it enters. Similarly, $\overline{V}_2$ and $\overline{V}_{2R}$ represent the velocity and relative velocity of the fluid leaving at the angles are δ and γ, respectively.

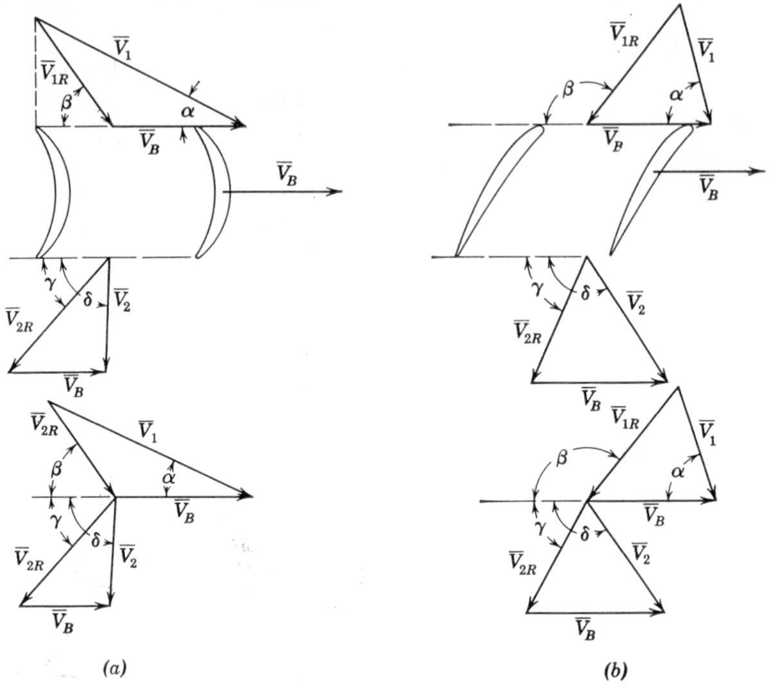

Fig. 13.25 Velocity vector diagrams for (a) a turbine, and (b) a compressor.

In analyzing the flow through a blade passage it is convenient to consider the system shown in Fig. 13.26, which is shown both from a stationary and a moving observer's point of view.

Let us begin the analysis by writing the first law for both cases, assuming steady, adiabatic flow through the blade passage. The stationary observer is aware that work is being done by the system on the blade, and the steady-flow energy equation is

$$h_1 + \frac{\overline{V}_1^2}{2g_c} = h_2 + \frac{\overline{V}_2^2}{2g_c} + w \tag{13.62}$$

The moving observer is not aware of any work being done on the blade, and the steady-flow energy equation is

$$h_1 + \frac{\overline{V}_{1R}^2}{2g_c} = h_2 + \frac{\overline{V}_{2R}^2}{2g_c} \tag{13.63}$$

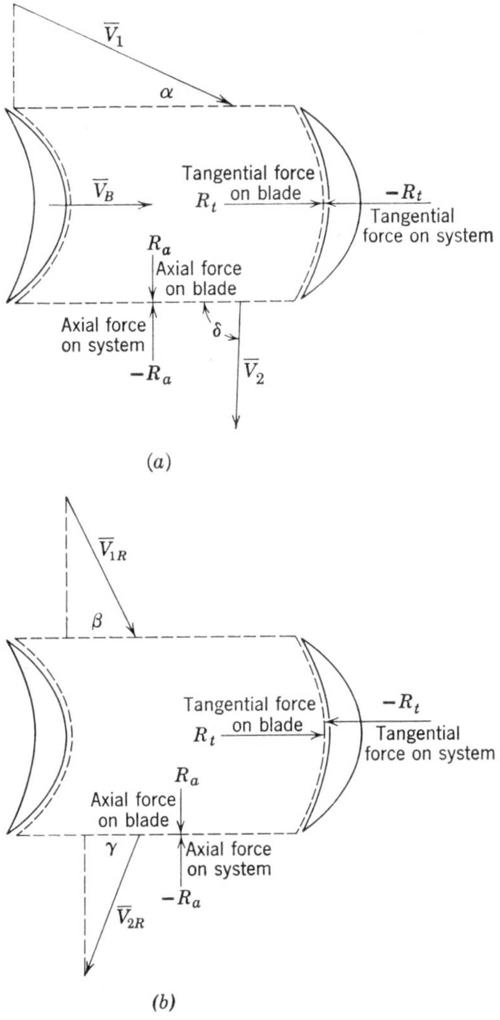

Fig. 13.26 Analysis of forces on a turbine blade. (*a*) Stationary observer. (*b*) Observer moving with the blade.

By combining Eqs. 13.59 and 13.60, we can derive the relation

$$w = \frac{(\overline{V}_1^2 - \overline{V}_{1R}^2) - (\overline{V}_2^2 - \overline{V}_{2R}^2)}{2g_c} \tag{13.64}$$

Applying the second law of thermodynamics to this process we conclude, since the process is adiabatic, that

$$s_2 \geq s_1$$

We will next consider the momentum equation, and evaluate both the tangential force R_t and the axial force R_a on the blade. For our sign convention we assume that tangential velocities and forces in the direction of the movement of the blade are positive and axial velocities and forces in the direction of the axial component of the velocity are positive.

Let us first apply the momentum equation in the tangential direction to the system as viewed by the stationary observer. In doing so we make the reasonable assumption that the pressure forces in the tangential direction balance. Then, using Eq. 13.10

$$-R_t = \frac{\dot{m}}{g_c}(-\overline{V}_2 \cos \delta - \overline{V}_1 \cos \alpha)$$

or

$$R_t = \frac{\dot{m}}{g_c}(\overline{V}_1 \cos \alpha + \overline{V}_2 \cos \delta) \tag{13.65}$$

Similarly, for the system as viewed by the moving observer we have

$$-R_t = \frac{\dot{m}}{g_c}(-\overline{V}_{2R} \cos \gamma - \overline{V}_{1R} \cos \beta)$$

or

$$R_t = \frac{\dot{m}}{g_c}(\overline{V}_{1R} \cos \beta + \overline{V}_{2R} \cos \gamma) \tag{13.66}$$

Equation 13.66 could have been developed from Eq. 13.65 and from the geometry of the vector diagram by noting that

$$\overline{V}_1 \cos \alpha - \overline{V}_B = \overline{V}_{1R} \cos \beta$$

and

$$\overline{V}_2 \cos \delta + \overline{V}_B = \overline{V}_{2R} \cos \gamma$$

Therefore,

$$\overline{V}_1 \cos \alpha + \overline{V}_2 \cos \delta = \overline{V}_{1R} \cos \beta + \overline{V}_{2R} \cos \gamma \tag{13.67}$$

The work done by the system on the blade in time $d\tau$ as the blade moves a distance dx is

$$\delta W = R_t \, dx$$

The rate at which work is done, or the power, is

$$\frac{\delta W}{d\tau} = R_t \frac{dx}{d\tau} = R_t \overline{V}_B$$

Substituting for R_t from Eqs. 13.65 and 13.66

$$\frac{\delta W}{d\tau} = \dot{W} = \frac{\dot{m}}{g_c} \overline{V}_B(\overline{V}_1 \cos \alpha + \overline{V}_2 \cos \delta)$$

$$\frac{\delta W}{d\tau} = \dot{W} = \frac{\dot{m}}{g_c} \overline{V}_B(\overline{V}_{1R} \cos \beta + \overline{V}_{2R} \cos \gamma) \qquad (13.68)$$

The work w per pound of fluid flowing is therefore

$$w = \frac{\overline{V}_B}{g_c} (\overline{V}_1 \cos \alpha + \overline{V}_2 \cos \delta) \qquad (13.69)$$

or

$$w = \frac{\overline{V}_B}{g_c} (\overline{V}_{1R} \cos \beta + \overline{V}_{2R} \cos \gamma) \qquad (13.70)$$

The axial thrust on the blade can be found by applying the momentum equation in the axial direction. We must recognize that the pressure at the inlet section may be different from the pressure at the exit section. Consider first the system as viewed by the stationary observer. Again we use Eq. 13.10

$$R_a = \frac{\dot{m}}{g_c} (\overline{V}_2 \sin \delta - \overline{V}_1 \sin \alpha) - (P_1A_1 - P_2A_2) \qquad (13.71)$$

As viewed by the moving observer we have

$$R_a = \frac{\dot{m}}{g_c} (\overline{V}_{2R} \sin \gamma - \overline{V}_{1R} \sin \beta) - (P_1A_1 - P_2A_2) \qquad (13.72)$$

Since there is no motion in the axial direction there is no work involved. However, adequate thrust bearings must be provided to handle the axial force. It is important to note in Eqs. 13.69 and 13.70 that if the pressure changes across the moving blade, the axial forces will be significantly influenced. This will be discussed more fully in the next section when the difference between an impulse and a reaction stage is considered.

Example 13.11 Steam enters the blade passage of one stage of a steam turbine with a velocity of 1800 ft/sec and at an angle (α) of 20°. The steam leaves the blade (as seen by the moving observer) at an angle (γ) of 50°. There is no pressure change across the blade and there are no irreversibilities (which means that there is no change in the magnitude of the relative velocity during the flow through the blade passage). Determine the work per pound of steam and the axial thrust.

We first draw the vector diagram as shown in Fig. 13.27.

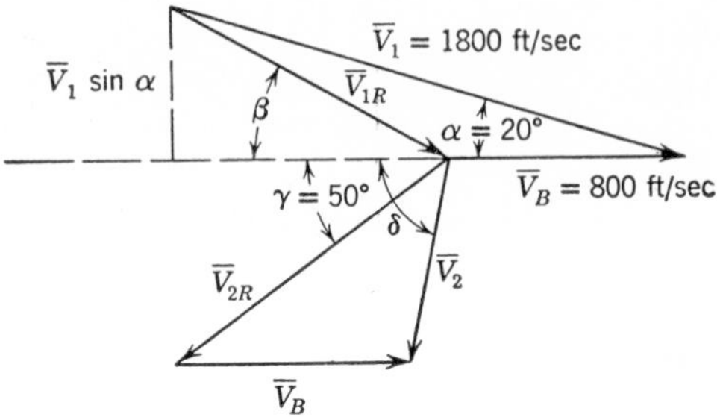

Fig. 13.27 Sketch for Example 13.11.

$$\overline{V}_{1R}\sin\beta = \overline{V}_1\sin\alpha = 1800\sin 20° = 1800(0.3240) = 616\ \text{ft/sec}$$

$$\overline{V}_{1R}\cos\beta + \overline{V}_B = \overline{V}_1\cos\alpha = 1800(0.9397) = 1691\ \text{ft/sec}$$

$$\overline{V}_{1R}\cos\beta = 1691 - 800 = 891\ \text{ft/sec}$$

$$\tan\beta = \frac{\overline{V}_{1R}\sin\beta}{\overline{V}_{1R}\cos\beta} = \frac{616}{891} = 0.692$$

$$\beta = 34°\,41'$$

$$\overline{V}_{1R}\sin\beta = 616$$

$$\overline{V}_{1R} = \frac{616}{\sin 34°\,41'} = \frac{616}{0.569} = 1084\ \text{ft/sec}$$

$$\overline{V}_{2R} = \overline{V}_{1R} = 1084\ \text{ft/sec}$$

$$\overline{V}_2\sin\delta = \overline{V}_{2R}\sin\gamma = 1084\sin 50° = 1084(0.766)$$

$$= 830\ \text{ft/sec}$$

$$\overline{V}_2\cos\delta + \overline{V}_B = \overline{V}_{2R}\cos\gamma = 1084\cos 50° = 1084(0.6428)$$

$$= 696\ \text{ft/sec}$$

$$\overline{V}_2\cos\delta = 696 - 800 = -104\ \text{ft/sec}$$

This implies that we drew our vector diagram slightly wrong and it should have been as shown in Fig. 13.28.

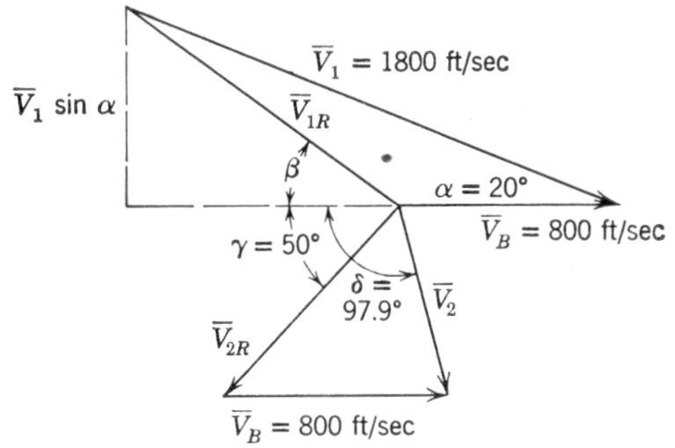

Fig. 13.28 Sketch for Example 13.11.

$$\cot \delta = \frac{\overline{V}_2 \cos \delta}{\overline{V}_2 \sin \delta} = \frac{-104}{830} = -0.1253$$

$$\delta = 97^\circ\, 9'$$

$$\overline{V}_2 = \sqrt{(\overline{V}_2 \sin \delta)^2 + (\overline{V}_2 \cos \delta)^2} = \sqrt{(830)^2 + (104)^2}$$

$$= 842 \text{ ft/sec}$$

13.13 Impulse and Reaction Stages for Turbines

Two terms used in connection with turbines are impulse stage and reaction stage. A turbine stage is defined as the fixed nozzle or blade and the moving blades that follow it. Figure 13.29*a* shows an impulse stage. In an impulse stage the entire pressure drop takes place in a stationary nozzle, and the pressure remains constant as the fluid flows through the blade passage. There is a decrease in the kinetic energy of the fluid as it flows through the blade passage, and the enthalpy will increase due to irreversibilities associated with the fluid flow.

In the pure reaction stage the entire pressure drop occurs as the fluid flows through the moving blades. Thus, the moving blade acts as a nozzle, and the blade passage must have the proper contour for a nozzle (converging if the exit pressure is greater than the critical pressure and converging-diverging if the exit pressure is less than the critical pressure). In the pure reaction stage the only purpose of the stationary blade is to direct the fluid into the moving blade at the proper angle and velocity.

The pure reaction stage is used relatively infrequently. Rather, most turbines that are classified as reaction turbines have a pressure and enthalpy drop in both the fixed and moving blades. The degree of reaction is defined as the fraction of the enthalpy drop that occurs in the moving blades. Thus, in the 50 per cent reaction stage, which is very commonly used, half of the enthalpy drop across the stage occurs in the fixed blade and the other half in the moving blade.

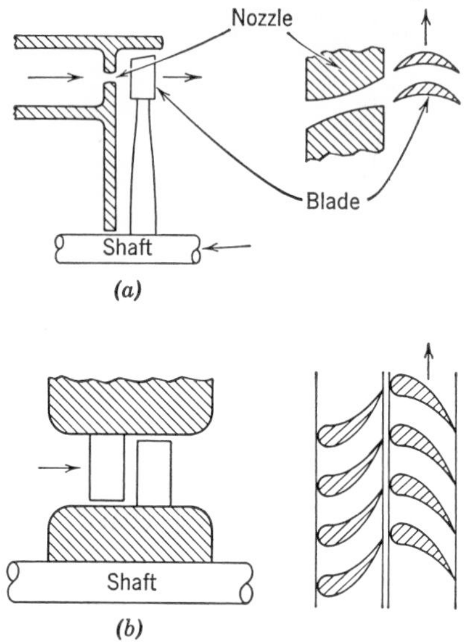

Fig. 13.29 Schematic arrangement for (a) an impulse stage, and (b) a reaction stage.

A few comparisons between the impulse and reaction blade may be noted here. Since there is a pressure drop across both the fixed and moving blades in the reaction stage, there is a tendency for leakage to occur across the tip of the blade. Therefore, very close clearances must be maintained at the blade tips. Further, the pressure drop across the moving blade of the reaction stage gives rise to axial forces that must be balanced in order to prevent axial motion in reaction turbines. In an impulse stage it is possible to utilize only part of the periphery for admission of steam. This is usually termed partial admission. In fact, in turbines that utilize an impulse stage for the first stage the power output can be controlled by opening and closing nozzles. The main advantage of the reaction stage is that lower fluid velocities are utilized, and higher efficiencies can be attained at lower velocities. This will be discussed more fully in the following sections.

13.14 Some Further Considerations of Impulse Stages

Figure 13.30 shows a typical velocity diagram for an impulse stage. The blade velocity coefficient k_B is defined as the ratio of the relative velocity leaving the blade to the relative velocity entering the blade. That is,

$$k_B = \frac{\overline{V}_{2R}}{\overline{V}_{1R}} \tag{13.73}$$

In the ideal impulse stage $\overline{V}_{2R} = \overline{V}_{1R}$ and $k_B = 1$, but in the actual impulse stage $\overline{V}_{2R}$ will be less than $\overline{V}_{1R}$ due to irreversibilities in the fluid as it flows through the blade passage.

The blade efficiency η_B is defined as the ratio of the actual work per pound of fluid flowing to the kinetic energy of the fluid entering the blade passage.

$$\eta_B = \frac{w}{\overline{V}_1^2/2g_c} \tag{13.74}$$

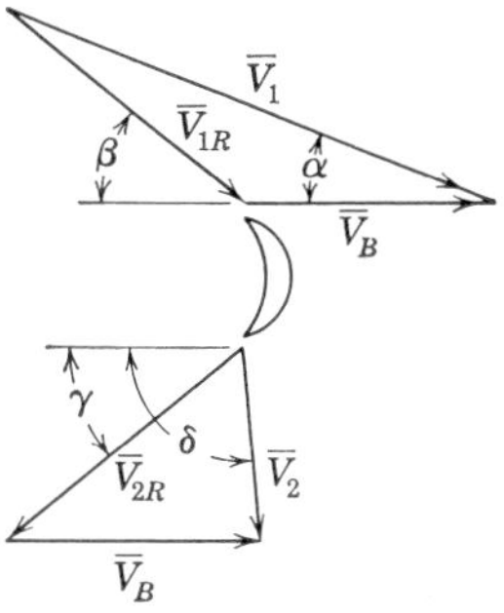

Fig. 13.30 A velocity vector diagram for an impulse blade.

Thus, a blade efficiency of 100 per cent means that the work is exactly equal to the kinetic energy of the fluid entering the blade, and the kinetic energy of the fluid leaving the blade is zero.

Quite obviously, the fluid must have some axial velocity if it is to flow out the blade passage. However, it is of interest to consider a turbine having a zero axial velocity and to determine the blade speed ratio that will give an efficiency of 100 per cent. The blade-speed ratio is the ratio of the blade speed $\overline{V}_B$ to the velocity of the fluid leaving the nozzle $\overline{V}_1$ and the ratio is designated $\overline{V}_B/\overline{V}_1$.

Figure 13.31*a* shows a reversible impulse stage that has a very small entrance and blade exit angle, and a blade-speed ratio of 0.5. It is evident that as the angles α and γ (the inlet and blade exit angles) approach zero that the exit velocity $\overline{V}_2$ will also approach zero. Such a stage would have an efficiency of 100 per cent. Figures 13.31*b* and *c* show reversible impulse stages with essentially zero angles, but with blade-speed ratios considerably less and considerably greater than 0.5.

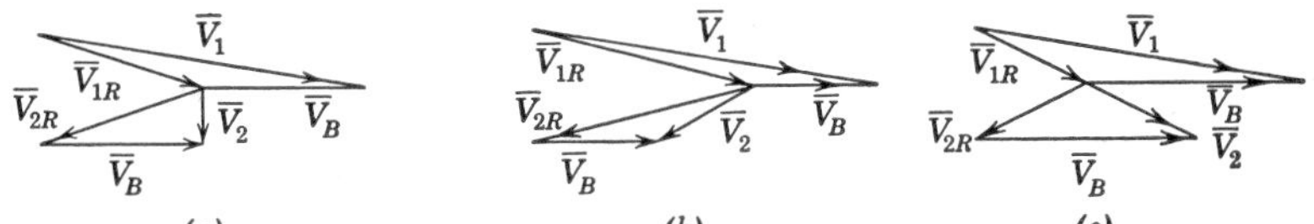

Fig. 13.31 A velocity vector diagram illustrating optimum blade-speed ratio.

In both of these the exit velocity is large, and therefore the blade efficiency is considerably less than 100 per cent. We conclude, therefore, that for a reversible, zero-angle impulse turbine the optimum blade-speed ratio is 0.5. The curve marked impulse stage on Fig. 13.34 shows the variation of blade efficiency with blade-speed ratio for a reversible stage.

An expression for the blade efficiency as a function of the angles α and γ (the inlet angle and blade exit angles), the blade-speed ratio, and the blade velocity coefficient can be derived.

From Eq. 13.69

$$w = \frac{\bar{V}_B}{g_c}(\bar{V}_1 \cos \alpha + \bar{V}_2 \cos \delta)$$

$$\bar{V}_{2R} = k_B \bar{V}_{1R}$$

From the vector diagram we note that

$$\bar{V}_2 \cos \delta = \bar{V}_{2R} \cos \gamma - \bar{V}_B = k_B \bar{V}_{1R} \cos \gamma - \bar{V}_B$$

$$\bar{V}_{1R} = \frac{\bar{V}_1 \cos \alpha - \bar{V}_B}{\cos \beta}$$

Therefore,

$$w = \frac{\bar{V}_B}{g_c}\left(\bar{V}_1 \cos \alpha + k_B \cos \gamma \frac{\bar{V}_1 \cos \alpha - \bar{V}_B}{\cos \beta} - \bar{V}_B\right)$$

$$= \frac{\bar{V}_B}{g_c}\left[(\bar{V}_1 \cos \alpha - \bar{V}_B)\left(1 + \frac{k_B \cos \gamma}{\cos \beta}\right)\right]$$

$$\eta_B = \frac{w}{\bar{V}_1^2/2g_c} = \frac{2\bar{V}_B}{\bar{V}_1^2}(\bar{V}_1 \cos \alpha - \bar{V}_B)\left(1 + k_B \frac{\cos \gamma}{\cos \beta}\right)$$

$$= \frac{2\bar{V}_B^2}{\bar{V}_1^2}\left(\frac{\bar{V}_1}{\bar{V}_B} \cos \alpha - 1\right)\left(1 + k_B \frac{\cos \gamma}{\cos \beta}\right)$$

But

$$\cos \beta = \frac{\bar{V}_1 \cos \alpha - \bar{V}_B}{\sqrt{(\bar{V}_1 \cos \alpha - \bar{V}_B)^2 + (\bar{V}_1 \sin \alpha)^2}}$$

$$= \frac{\cos \alpha - \bar{V}_B/\bar{V}_1}{\sqrt{(\cos \alpha - \bar{V}_B/\bar{V}_1)^2 + \sin^2 \alpha}}$$

$$\eta_B = 2\frac{\bar{V}_B}{\bar{V}_1}\left(\cos \alpha - \frac{\bar{V}_B}{\bar{V}_1}\right)\left[1 + \frac{k_B \cos \gamma}{\dfrac{\cos \alpha - \bar{V}_B/\bar{V}_1}{\sqrt{(\cos \alpha - \bar{V}_B/\bar{V}_1)^2 + \sin^2 \alpha}}}\right] \tag{13.75}$$

This relation gives the blade efficiency as a function of the blade speed-ratio $\bar{V}_B/\bar{V}_1$, the angles α and γ, and the blade velocity coefficient k_B. From this relation it can be shown that for the usual values of angles and blade velocity coefficient, the optimum efficiency is obtained with blade velocity coefficient of approximately 0.5, which was the value obtained for the reversible zero-angle impulse stage.

This leads to consideration of the velocity-compounded turbine. Suppose one was attempting to build a turbine having one impulse stage, and that it was to receive steam at 100 lbf/in.^{2}, 500 F and exhaust at a pressure of 2 lbf/in.^{2} The velocity of the steam leaving the nozzle would be about 3500 ft/sec. In order to have maximum efficiency a blade-speed ratio of 0.5 must be used, and this would require a blade speed of 1750 ft/sec. Blade speeds of this magnitude result in very high stresses due to centrifugal force, and further, the irreversibilities associated with fluid flow increase as the fluid velocity increases.

For these reasons velocity compounding is used, and a schematic diagram and a vector diagram for a velocity-compounded stage are shown in Fig. 13.32. The stationary blade serves to change the direc-

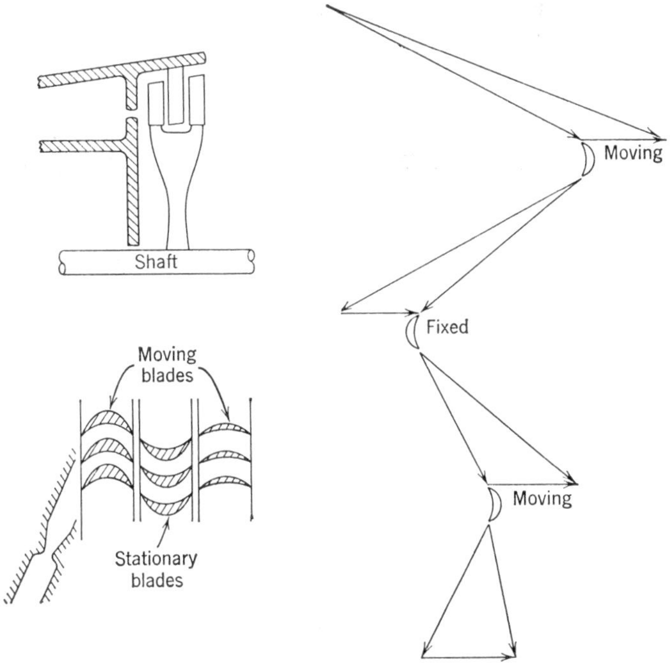

Fig. 13.32 A velocity-compounded impulse stage.

tion of the fluid so that it will enter the second row of moving blades properly. It can be shown that for a reversible zero-angle turbine having two moving rows the blade-speed ratio for optimum efficiency is 1/4. Sometimes velocity-compounded turbines are built with as many as three moving rows.

Velocity-compounded turbines have a lower efficiency than comparable turbines that have a number of impulse or reaction stages. However, they are relatively inexpensive and are therefore frequently used in small sizes where initial cost is more important than efficiency. A velocity-compounded stage is also frequently used as the first stage in large steam turbines in order to have a large pressure and temperature drop before the steam enters the moving blades.

13.15 Some Further Considerations of Reaction Stages

First a remark should be made regarding the blade efficiency of a reaction stage. Both the fixed blades and moving blades act as a nozzle. Therefore, the efficiency of the moving blades can be defined in the same way as the efficiency of a nozzle if the relative velocities are used.

The second matter to consider is the blade-speed ratio for maximum efficiency of a reaction stage. Let us consider a reversible reaction stage in which the isentropic enthalpy drops across the fixed and moving blades are equal (i.e., a 50 per cent reaction stage). For a zero-angle turbine it is evident from Fig. 13.33 that for zero kinetic energy leaving the blade, a blade-speed ratio of 1.0 is required. Thus, maximum efficiency in a reaction stage is obtained for a blade-speed ratio of 1.0.

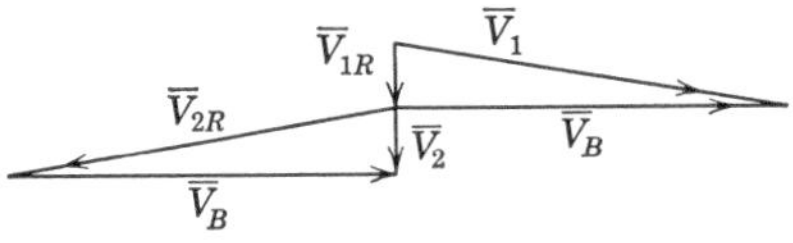

Fig. 13.33 A velocity vector diagram for a reaction stage.

This leads us to a comparison of the impulse stage with the reaction stage regarding the blade-speed ratio for maximum efficiency.

If we compare reversible impulse and reaction stages for a given velocity from the stationary nozzle or blade, it is evident that for maximum efficiency the blade velocity for the reaction stage would be twice the blade velocity of the impulse stage. However, this is not a good comparison because the reaction stage would have a greater enthalpy drop (due to the enthalpy drop in the moving row) than the impulse stage.

A better comparison can be made by considering equal enthalpy drops across the stage. Let this enthalpy drop be denoted Δh_s, and let the velocity $\overline{V}_0$ be defined by the relation

$$\overline{V}_0 = \sqrt{2g_c \, \Delta h_s}$$

That is, $\overline{V}_0$ would be the velocity leaving the nozzle of the reversible impulse stage that has this enthalpy drop. That is, for the impulse stage

$$\overline{V}_1 = \overline{V}_0$$

For the reversible 50 per cent reaction stage the velocity leaving the fixed blades would be

$$\overline{V}_1 = \sqrt{\tfrac{1}{2} \times 2g_c\, \Delta h_s} = \frac{\overline{V}_0}{\sqrt{2}}$$

For maximum efficiency in the impulse stage, $\overline{V}_B/\overline{V}_1 = 0.5$. Therefore, for the impulse stage

$$\frac{\overline{V}_B}{\overline{V}_1} = \frac{\overline{V}_B}{\overline{V}_0} = 0.5$$

For maximum efficiency in the reaction stage, $\overline{V}_B/\overline{V}_1 = 1.0$. Therefore, for the reaction stage

$$\frac{\overline{V}_B}{\overline{V}_1} = \frac{\overline{V}_B\sqrt{2}}{\overline{V}_0} \qquad \frac{\overline{V}_B}{\overline{V}_0} = \frac{1}{\sqrt{2}}$$

Thus, for a given enthalpy drop per stage, the reaction stage requires a higher blade speed for maximum efficiency than the impulse stage. Conversely, for a given blade speed the reaction turbine requires more stages than an impulse stage, and in this case the fluid velocity is lower in the reaction stage than in the impulse stage.

Figure 13.34 shows the blade efficiency as a function of $\overline{V}_B/\overline{V}_0$ for an impulse stage and reaction stage.

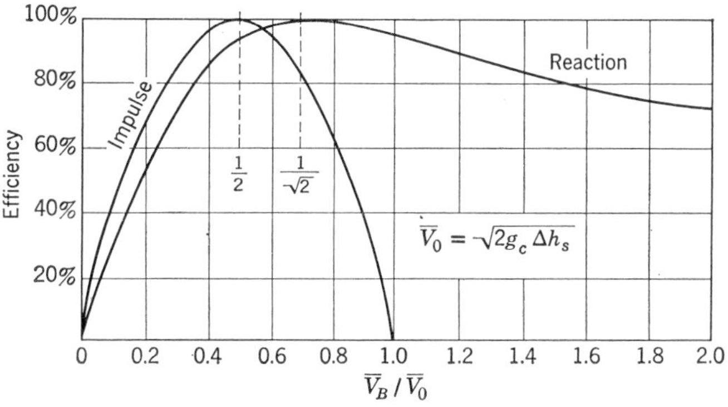

Fig. 13.34 Efficiency of an ideal impulse and reaction stage as a function of $\overline{V}_B/\overline{V}_0$.

PROBLEMS

13.1 Air leaves the compressor of a jet engine at a temperature of 300 F, a pressure of 20 lbf/in.^{2}, and a velocity of 400 ft/sec. Determine the isentropic stagnation temperature and pressure.

13.2 The products of combustion of a jet engine leave the engine with a velocity relative to the plane of 1200 ft/sec, a temperature of 900 F, and a pressure of 12 lbf/in.^{2} Assuming that 400% theoretical air was used for combustion ($k = 1.34$, $C_p = 0.271$ Btu/lbm for the products), determine the stagnation pressure and temperature of the products (relative to the airplane).

13.3 Steam leaves a nozzle with a velocity of 800 ft/sec. The stagnation pressure is 100 lbf/in.^{2}, and the stagnation temperature is 500 F. What is the static pressure and temperature?

13.4 Water is pumped from a lake and discharged through a nozzle as shown in Fig. 13.35. At the pump discharge the pressure is 100

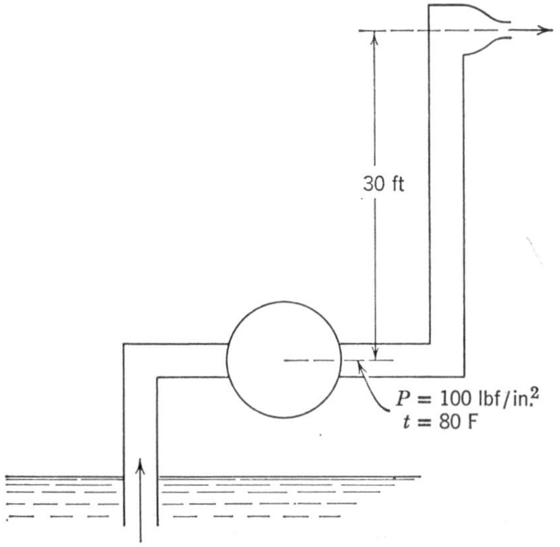

Fig. 13.35 Sketch for Problem 13.4.

lbf/in.^{2}, and the temperature is 80 F. The nozzle is located 30 ft above the pump. Assume standard atmospheric pressure and reversible flow throughout. Determine the velocity of the water leaving the nozzle.

13.5 In a water turbine that utilizes nozzles the water enters the nozzle at 50 lbf/in.^{2}, 80 F and leaves at standard atmospheric pressure.

If the flow through the nozzle is reversible and adiabatic, determine the velocity and kinetic energy per lbm of water leaving the nozzle.

13.6 Air is expanded in a nozzle from 200 lbf/in.^{2}, 400 F, to 20 lbf/in.^{2} The mass rate of flow through the nozzle is 600 lbm/min. Assume the flow to be reversible and adiabatic.

(*a*) Determine specific volume, velocity, Mach number, and cross-sectional area for each 20 lbf/in.^{2} decrease of pressure, and plot these as a function of pressure.

(*b*) Determine the throat and exit areas for the nozzle.

13.7 Consider the nozzle of Problem 13.6 and determine what back pressure will cause a normal shock to stand in the exit plane of the nozzle. What is the mass rate of flow under these conditions?

13.8 At what Mach number will the normal shock occur in the nozzle of Problem 13.6 if the back pressure is 140 lbf/in.^{2}?

13.9 Consider the nozzle of Problem 13.6. What back pressure will be required to cause subsonic flow throughout the entire nozzle with $M = 1$ at the throat?

13.10 Determine the mass rate of flow through the nozzle of Problem 13.6 for a back pressure of 190 lbf/in.^{2}

13.11 A jet plane travels through the air with a speed of 600 mi/hr at an altitude of 20,000 ft, where the pressure is 6.75 lbf/in.^{2} and the temperature is -12 C. Consider the diffuser of the engine. The air leaves the diffuser with a velocity of 300 ft/sec. Determine the pressure and temperature leaving the diffuser, and the ratio of inlet to exit area of the diffuser, assuming the flow to be reversible and adiabatic.

13.12 The products of combustion enter the nozzle of a jet engine at a total pressure of 18 lbf/in.^{2}, and a total temperature of 1200 F. The atmospheric pressure is 6.75 lbf/in.^{2} The nozzle is convergent, and the rate of flow is 50 lbm/sec. Assume the flow to be reversible and adiabatic. Determine the exit area of the nozzle.

13.13 Air is expanded in a nozzle from 200 lbf/in.^{2}, 400 F, to 20 lbf/in.^{2} in a nozzle having an efficiency of 90%. The mass rate of flow is 600 lbm/min. Determine the exit area of the nozzle, the exit velocity, and the increase of entropy per lbm of fluid. Compare these results with those of the reversible adiabatic nozzle of Problem 13.6.

13.14 Repeat Problem 13.11 assuming a diffuser efficiency of 80%.

13.15 Consider the diffuser of a supersonic aircraft flying at $M = 1.4$ at such an altitude that the temperature is -10 F, and the atmosphere pressure is 7 lbf/in.^{2} Consider two possible ways in which this diffuser might operate, and for each case calculate the throat area required for a flow of 100 lbm/sec.

(*a*) The diffuser operates as a reversible adiabatic diffuser with subsonic exit velocity.

(*b*) A normal shock stands at the entrance to the diffuser. Except for the normal shock the flow is reversible and adiabatic, and the exit velocity is subsonic. This is shown in Fig. 13.36. Assume a convergent-divergent diffuser with $M = 1$ at the throat.

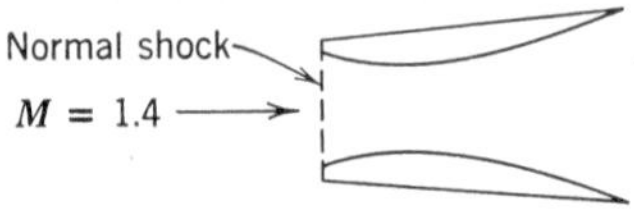

Fig. 13.36 Sketch for Problem 13.15.

13.16 Air enters a diffuser with a velocity of 600 ft/sec, a static pressure of 10 lbf/in.^{2}, and a temperature of 20 F. The velocity leaving the diffuser is 200 ft/sec and the static pressure at the diffuser exit is 11.7 lbf/in.^{2} Determine the static temperature at the diffuser exit and the diffuser efficiency. Compare the stagnation pressures at the diffuser inlet and exit.

13.17 Steam at a pressure of 100 lbf/in.^{2} and a temperature of 600 F expands to a pressure of 20 lbf/in.^{2} in a nozzle having an efficiency of 90%. The mass rate of flow is 20 lbm/sec. Determine the nozzle exit area.

13.18 Steam at an initial pressure of 100 lbf/in.^{2} and a temperature of 600 F flows through a convergent-divergent nozzle having a throat area of 0.5 in.^{2} The pressure at the exit plane is 20 lbf/in.^{2} and the exit velocity is 2600 ft/sec. The flow from the nozzle entrance to the throat is reversible and adiabatic. Determine the exit area of the nozzle, the over-all nozzle efficiency, and the increase in entropy per lbm of fluid.

13.19 A sharp-edged orifice is used to measure the flow of air in a pipe. The pipe diameter is 4 in. and the diameter of the orifice is 1 in. Upstream of the orifice the absolute pressure is 20 lbf/in.^{2} and the temperature is 100 F. The pressure drop across the orifice is 4 in. of mercury, and the coefficient of discharge is 0.62. Determine the mass rate of flow in the pipeline.

13.20 A steam turbine utilizes convergent nozzles. An estimate of the rate of steam flow is to be made from the pressure drop across the nozzles of one stage. The inlet conditions to these nozzles are 80 lbf/in.^{2}, 500 F. The exit pressure is 50 lbf/in.^{2} The coefficient of discharge is estimated to be 0.94. The total exit area of the nozzles in this stage is 5 in.^{2} Determine the rate of flow under these conditions.

13.21 The coefficient of discharge of a sharp-edged orifice is determined at one set of conditions by use of an accurately calibrated gasometer. The orifice has a diameter of 0.75 in. and the pipe diameter is 2.0 in. The absolute upstream pressure is 30 lbf/in.^{2} and the pressure drop across the orifice is 3.2 in. of mercury. The temperature of the air entering the orifice is 80 F. The mass rate of flow as measured by the gasometer is 5.30 lbm/min. What is the coefficient of discharge of the orifice under these conditions?

13.22 Consider the flow of air through an impulse blade passage. The air enters the blade passage at an angle of 18° with a velocity of 1500 ft/sec, a pressure of 16 lbf/in.^{2}, and a temperature of 200 F. The blade velocity is 800 ft/sec and the air leaves the blade passage at an angle of 45° relative to the blade. The mass rate of flow is 20 lbm/sec, and it is assumed that the flow is reversible and adiabatic.

(*a*) Draw the velocity diagram to scale.

(*b*) What is the hp output of the turbine?

(*c*) If the blade wheel to which the blades are attached has a diameter of 1 ft, what is the length of each blade?

13.23 Steam enters an impulse turbine in which all processes are assumed to be reversible and adiabatic. The inlet pressure is 100 lbf/in.^{2} and the inlet temperature is 800 F. The exhaust pressure is 16 lbf/in.^{2} The steam leaves the nozzle and enters the turbine at an angle of 20°. The blade-speed ratio is 0.5 and the blade exit angle is 50°. Determine the blade efficiency of this turbine.

13.24 Consider an impulse turbine having the same inlet conditions as Problem 13.23. The nozzle efficiency is 92% and the blade velocity coefficient is 0.96. The blade-speed ratio is 0.5. Determine the work per lbm of steam and draw the velocity diagram to scale.

13.25 Consider a two-stage, reversible, velocity-compounded impulse turbine that has the same inlet conditions as Problem 13.23. Assume a blade-speed ratio of 0.25, and an inlet angle of 20°. Both the moving and fixed blades have equal inlet and exit angles (i.e., the angle at which the steam enters relative to the blade is equal to the relative angle at which it leaves). Determine the blade efficiency of the turbine and draw the velocity diagram to scale.

13.26 Consider a single-stage reaction turbine having equal enthalpy drops across the fixed and moving blades. Consider the same inlet conditions and exit pressure as Problem 13.23. All processes are reversible and adiabatic. The blade exit angle is 20° and the blade-speed ratio is 0.9.

(*a*) What is the pressure at the exit of the fixed blades?

(*b*) Draw the velocity vector diagram to scale.

(*c*) What is the net work per lbm of steam flowing through the turbine?

CHAPTER 14

THERMODYNAMIC RELATIONS, EQUATIONS OF STATE, AND GENERALIZED CHARTS

In this chapter consideration is given to a number of relations between thermodynamic properties, some general principles involved in developing tables of thermodynamic properties, equations of state, and generalized charts of thermodynamic properties. One danger in a chapter such as this is that a large number of relations will be given, and in so doing the principles and procedures might not be made clear. It should be emphasized, therefore, that the principles and procedures are most important.

14.1 Two Important Relations

This chapter involves partial derivatives, and two important relations are reviewed here. Consider a variable z which is a continuous function of x and y.

$$z = f(x, y)$$

$$dz = \left(\frac{\partial z}{\partial x}\right)_y dx + \left(\frac{\partial z}{\partial y}\right)_x dy$$

It is convenient to write this in the form:

$$dz = M\,dx + N\,dy \tag{14.1}$$

where
$$M = \left(\frac{\partial z}{\partial x}\right)_y$$
$$N = \left(\frac{\partial z}{\partial y}\right)_x$$

The definitions of the terms in this equation are:

dz = total differential of z

$M\,dx$ = partial differential of z with respect to x

$N\,dy$ = partial differential of z with respect to y

$M = \left(\frac{\partial z}{\partial x}\right)_y$ = partial derivative of z with respect to x (the variable y being constant)

$N = \left(\frac{\partial z}{\partial y}\right)_x$ = partial derivative of z with respect to y (the variable x being held constant)

The physical significance of partial derivatives as they relate to the properties of a pure substance can be explained by referring to Fig. 14.1, which shows a P-v-T surface of the superheated vapor region of a pure substance. It shows a constant-temperature, constant-pressure, and a constant–specific-volume plane which intersect at point b on the surface. Thus, the partial derivative $(\partial P/\partial v)_T$ is the slope of curve abc at point b. Line de represents the tangent to curve abc at point b. A similar interpretation can be made of the partial derivative $(\partial P/\partial T)_v$ and $(\partial v/\partial T)_P$.

It should also be noted that if we wish to evaluate the partial derivative along a constant-temperature line, the rules for ordinary derivatives can be applied. Thus, we can write for a constant-temperature process:

$$\left(\frac{\partial P}{\partial v}\right)_T = \frac{dP_T}{dv_T}$$

and the integration can be performed as usual. This point will be demonstrated later in a number of examples.

Let us return to the consideration of the relation

$$dz = M\,dx + N\,dy$$

If x, y, and z are all point functions (that is, quantities that depend only on the state and are independent of the path), the differentials are

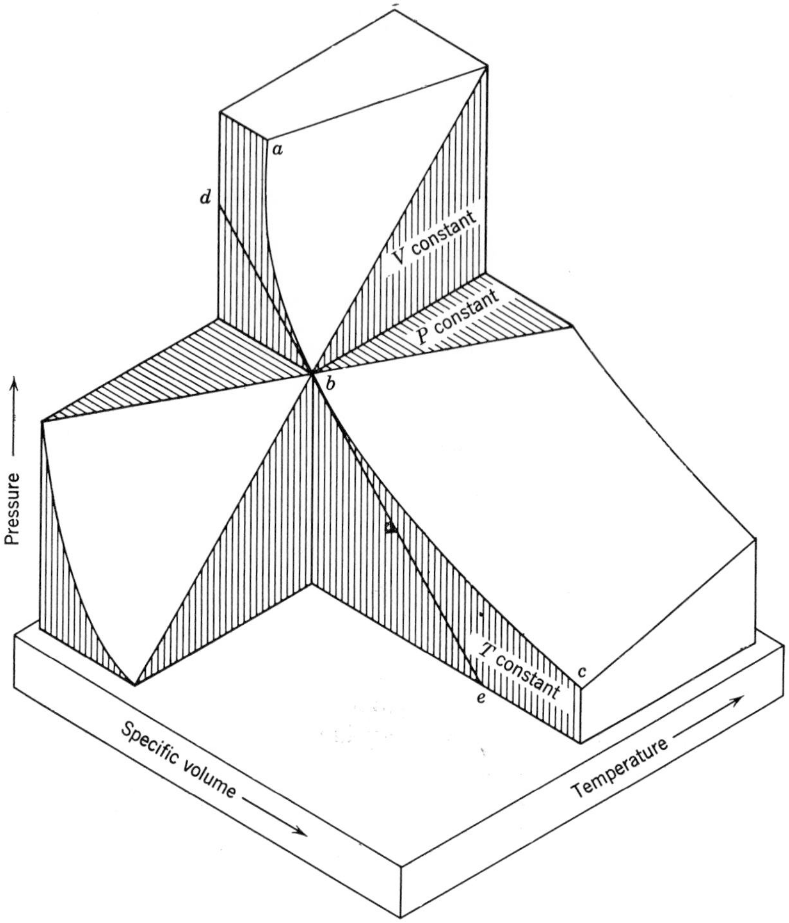

Fig. 14.1 Schematic representation of partial derivatives.

exact differentials. If this is the case, the following important relation holds:

$$\left(\frac{\partial M}{\partial y}\right)_x = \left(\frac{\partial N}{\partial x}\right)_y$$

The proof of this is as follows:

$$\left(\frac{\partial M}{\partial y}\right)_x = \frac{\partial^2 z}{\partial x\, \partial y}$$

$$\left(\frac{\partial N}{\partial x}\right)_y = \frac{\partial^2 z}{\partial y\, \partial x}$$

Since the order of differentiation makes no difference when point functions are involved, it follows that

$$\frac{\partial^2 z}{\partial x\, \partial y} = \frac{\partial^2 z}{\partial y\, \partial x}$$

$$\left(\frac{\partial M}{\partial y}\right)_x = \left(\frac{\partial N}{\partial x}\right)_y$$

The second important mathematical relation is

$$\left(\frac{\partial x}{\partial y}\right)_z \left(\frac{\partial y}{\partial z}\right)_x \left(\frac{\partial z}{\partial x}\right)_y = -1 \tag{14.2}$$

The proof of this relation is as follows: Consider three variables x, y, and z. Suppose there exists a relation between the variables of the form

$$x = f(y, z)$$

then

$$dx = \left(\frac{\partial x}{\partial y}\right)_z dy + \left(\frac{\partial x}{\partial z}\right)_y dz \tag{14.3}$$

Suppose there exists another relation between the three variables of the form

$$y = f(x, z)$$

then

$$dy = \left(\frac{\partial y}{\partial x}\right)_z dx + \left(\frac{\partial y}{\partial z}\right)_x dz \tag{14.4}$$

Substituting Eq. 14.4 in Eq. 14.3 we have

$$dx = \left(\frac{\partial x}{\partial y}\right)_z \left[\left(\frac{\partial y}{\partial x}\right)_z dx + \left(\frac{\partial y}{\partial z}\right)_x dz\right] + \left(\frac{\partial x}{\partial z}\right)_y dz$$

$$= \left(\frac{\partial x}{\partial y}\right)_z \left(\frac{\partial y}{\partial x}\right)_z dx + \left[\left(\frac{\partial x}{\partial y}\right)_z \left(\frac{\partial y}{\partial z}\right)_x + \left(\frac{\partial x}{\partial z}\right)_y\right] dz$$

Since there are only two independent variables, we may select x and z. Suppose that $dz = 0$ and $dx \neq 0$. It then follows that

$$\left(\frac{\partial x}{\partial y}\right)_z \left(\frac{\partial y}{\partial x}\right)_z = 1 \tag{14.5}$$

Similarly, suppose that $dx = 0$ and $dz \neq 0$. It then follows that

$$\left(\frac{\partial x}{\partial y}\right)_z \left(\frac{\partial y}{\partial z}\right)_x + \left(\frac{\partial x}{\partial z}\right)_y = 0$$

$$\left(\frac{\partial x}{\partial y}\right)_z \left(\frac{\partial y}{\partial z}\right)_x = -\left(\frac{\partial x}{\partial z}\right)_y$$

$$\left(\frac{\partial x}{\partial y}\right)_z \left(\frac{\partial y}{\partial z}\right)_x \left(\frac{\partial z}{\partial x}\right)_y = -1$$

This is Eq. 14.2 which we set out to derive.

14.2 The Maxwell Relations

Consider a closed system of fixed chemical composition in the absence of surface, electrical, and magnetic effects. The Maxwell relations, which can be written for such a system, are four equations relating the properties P, v, T, and s.

The Maxwell relations are most easily derived by considering four relations involving thermodynamic properties. Two of these relations have already been derived and are:

$$du = T\,ds - P\,dv \tag{14.6}$$

$$dh = T\,ds + v\,dP \tag{14.7}$$

The other two are derived from the definition of the Helmholtz function, a, and the Gibbs function, g.

$$a = u - Ts$$

$$da = du - T\,ds - s\,dT$$

Substituting Eq. 14.6 into this relation gives the third relation.

$$da = -P\,dv - s\,dT \tag{14.8}$$

Similarly,

$$g = h - Ts$$

$$dg = dh - T\,ds - s\,dT$$

Substituting Eq. 14.7 yields the fourth relation.

$$dg = v\,dP - s\,dT \tag{14.9}$$

Since Eqs. 14.6, 14.7, 14.8, and 14.9 are relations involving properties, we conclude that these are exact differentials, and, therefore, are of the general form

$$dz = M\,dx + N\,dy$$

Since

$$\left(\frac{\partial M}{\partial y}\right)_x = \left(\frac{\partial N}{\partial x}\right)_y$$

it follows from Eq. 14.6 that

$$\left(\frac{\partial T}{\partial v}\right)_s = -\left(\frac{\partial P}{\partial s}\right)_v \tag{14.10}$$

Similarly, from Eqs. 14.7, 14.8, and 14.9 we can write

$$\left(\frac{\partial T}{\partial P}\right)_s = \left(\frac{\partial v}{\partial s}\right)_P \tag{14.11}$$

$$\left(\frac{\partial P}{\partial T}\right)_v = \left(\frac{\partial s}{\partial v}\right)_T \tag{14.12}$$

$$\left(\frac{\partial v}{\partial T}\right)_P = -\left(\frac{\partial s}{\partial P}\right)_T \tag{14.13}$$

These four equations are known as the Maxwell relations, and the great utility of these equations will be demonstrated in later sections of this chapter. In particular, it should be noted that pressure, temperature, and specific volume can be measured by experimental methods, whereas entropy cannot be determined experimentally. By using the Maxwell relations changes in entropy can be determined from quantities that can be measured, i.e., pressure, temperature, and specific volume.

There are a number of other very useful relations that can be derived from Eqs. 14.6 through 14.9. For example, from Eq. 14.6 we can write the relations

$$\left(\frac{\partial u}{\partial s}\right)_v = T \qquad \left(\frac{\partial u}{\partial v}\right)_s = -P \tag{14.14}$$

Similarly, from these other equations we have the following:

$$\left(\frac{\partial h}{\partial s}\right)_P = T \qquad \left(\frac{\partial h}{\partial P}\right)_s = v \tag{14.15}$$

$$\left(\frac{\partial a}{\partial v}\right)_T = -P \qquad \left(\frac{\partial a}{\partial T}\right)_v = -s \tag{14.16}$$

$$\left(\frac{\partial g}{\partial P}\right)_T = v \qquad \left(\frac{\partial g}{\partial T}\right)_P = -s \tag{14.17}$$

Example 14.1 Explain why the entropy correction $(s - s_f)$ in Table 4 of Keenan and Keyes' steam tables is positive at 32 F and negative at all the other temperatures listed.

Table 4 gives the changes in volume, enthalpy, and entropy of liquid water as the pressure increases while the temperature remains constant. Thus, the entropy correction in Table 4 is really $(\partial s/\partial P)_T$. From Eq. 14.13

$$\left(\frac{\partial s}{\partial P}\right)_T = -\left(\frac{\partial v}{\partial T}\right)_P$$

Therefore, the sign of the entropy correction in Table 4 depends on the sign of the term $(\partial v/\partial T)_P$. The physical significance of this term is that it involves the change in specific volume of water as the temperature changes while the pressure remains constant. Now, as water at moderate pressures and 32 F is heated in a constant-pressure process, the specific volume decreases until the point of maximum density is reached at approximately 39 F, after which it increases. This is shown on a v-T diagram in Fig. 14.2. Thus, the quantity $(\partial v/\partial T)_P$

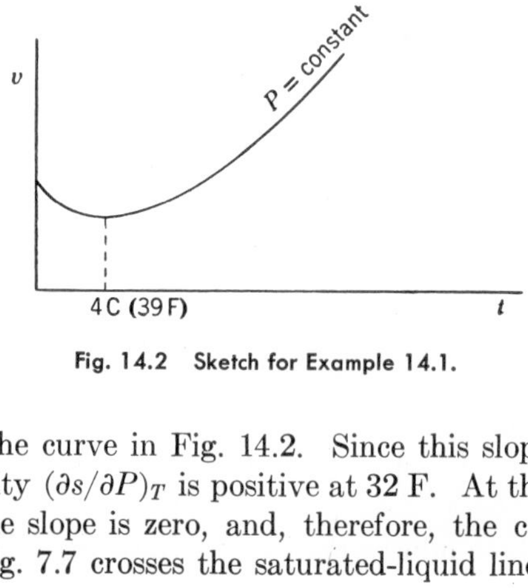

Fig. 14.2 Sketch for Example 14.1.

is the slope of the curve in Fig. 14.2. Since this slope is negative at 32 F, the quantity $(\partial s/\partial P)_T$ is positive at 32 F. At the point of maximum density the slope is zero, and, therefore, the constant-pressure line shown in Fig. 7.7 crosses the saturated-liquid line at the point of maximum density.

14.3 Clapeyron Equation

The Clapeyron equation is an important relation involving the saturation pressure and temperature, the enthalpy of evaporation, and the specific volume of the two phases involved. It can be derived in a number of ways. Here we proceed by considering one of the Maxwell relations, Eq. 14.12.

$$\left(\frac{\partial P}{\partial T}\right)_v = \left(\frac{\partial s}{\partial v}\right)_T$$

Consider, for example, the change of state from saturated liquid to saturated vapor of a pure substance. This is a constant-temperature process, and, therefore, Eq. 14.12 can be integrated between the saturated-liquid and saturated-vapor state. We also note that when saturated states are involved that pressure and temperature are independent of volume. Therefore,

$$\left(\frac{\partial P}{\partial T}\right)_v = \frac{dP}{dT} = \frac{s_g - s_f}{v_g - v_f} = \frac{s_{fg}}{v_{fg}} \tag{14.18}$$

But

$$s_{fg} = \frac{h_{fg}}{T}$$

Therefore,

$$\frac{dP}{dT} = \frac{h_{fg}}{Tv_{fg}} \tag{14.19}$$

The significance of this equation is that dP/dT is the slope of the vapor-pressure curve. Thus, h_{fg} at a given temperature can be determined from the vapor-pressure curve and the specific volume of saturated liquid and saturated vapor at the given temperature. Since enthalpy of evaporation is relatively difficult to measure accurately it is usually determined in this way. At low pressure the equation is usually simplified by assuming that $v_f \ll v_g$ and assuming also that $v_g = RT/P$. The relation then becomes

$$\frac{dP}{dT} = \frac{h_{fg}}{T(RT/P)}$$

$$\frac{dP}{P} = \frac{h_{fg}\,dT}{RT^2} \tag{14.20}$$

Example 14.2 Determine the humidity ratio of a saturated air-water vapor mixture at -100 F and 1 atm pressure.

From Chapter 9

$$\omega = \frac{R_{air}}{R_{H_2O}} \times \frac{p_{H_2O}}{p_{air}} = 0.622\,\frac{p_{H_2O}}{P_T - p_{H_2O}}$$

Thus, if we know the saturation pressure of water at -100 F, we can determine the humidity ratio. However, Table 5 of the steam tables does not give saturation pressures for temperatures less than −40 F. However, we do notice that h_{ig} is relatively constant in this range, and, therefore, we proceed to use Eq. 14.20 (Eqs. 14.19 and 14.20 also apply to equilibrium between a solid-vapor phase), and integrate between the limits −40 F and −100 F.

$$\int_1^2 \frac{dP}{P} = \int_1^2 \frac{h_{ig}}{R} \frac{dT}{T^2} = \frac{h_{ig}}{R} \int_1^2 \frac{dT}{T^2}$$

$$\ln \frac{P_2}{P_1} = \frac{h_{ig}}{R} \left(\frac{T_2 - T_1}{T_1 T_2} \right)$$

Let

$$P_2 = 0.0019 \text{ lbf/in.}^2 \qquad T_2 = 420 \text{ R}$$

$$P_1 = ? \qquad T_1 = 360 \text{ R}$$

Then

$$\ln \frac{P_2}{P_1} = \frac{1221 \times 778}{85.7} \left(\frac{420 - 360}{420 \times 360} \right)$$

$$\ln \frac{P_2}{P_1} = 4.40 \qquad \frac{P_2}{P_1} = 81.5 \qquad P_1 = \frac{0.0019}{81.5} = 2.33 \times 10^{-5} \text{ lbf/in.}^2$$

$$\omega = 0.622 \times \frac{2.33 \times 10^{-5}}{14.7} = 9.86 \times 10^{-7} \text{ lbm } H_2O/\text{lbm air}$$

14.4 Some Important and Useful Thermodynamic Relations

Many equations involving thermodynamic properties can be derived and tables of equations are available. In this chapter only a few equations will be derived, but their physical significance and utility will also be considered.

Let us consider first two relations for the change of enthalpy in a constant-temperature process. Both are derived from the relation

$$dh = T\, ds + v\, dP$$

The first of these relations is derived as follows:

$$\left(\frac{\partial h}{\partial P} \right)_T = T \left(\frac{\partial s}{\partial P} \right)_T + v$$

Substituting the Maxwell relation, Eq. 14.13, we have

$$\left(\frac{\partial h}{\partial P} \right)_T = v - T \left(\frac{\partial v}{\partial T} \right)_P$$

In the integration of an equation such as this it is helpful to recall that a pure substance has, in the absence of gravity, motion, surface effects, electricity, and magnetism, only two independent properties. For example, we can write

$$v = f(P, T)$$

If one of the variables is held constant, there is, in effect, only one independent variable. For example, as long as the pressure is held constant, volume can be expressed as a function of temperature only, and a differential equation involving volume and temperature (while the pressure is held constant) can be solved as an ordinary differential equation. (This actually applies to any number of variables as long as all the variables except one are held constant.) In order to emphasize this we will use a notation such as dv_P to indicate a differential change of volume while the pressure is held constant. Thus, in the solution of the equation we have been considering, we would write

$$\left(\frac{\partial h}{\partial P}\right)_T = v - T\left(\frac{\partial v}{\partial T}\right)_P$$

$$dh_T = \left[v - T\left(\frac{\partial v}{\partial T}\right)_P\right] dP_T$$

On integrating we have

$$(h_2 - h_1)_T = \int_1^2 \left[v - T\left(\frac{\partial v}{\partial T}\right)_P\right] dP_T \qquad (14.21)$$

The second of these relations is derived in a similar manner.

$$\left(\frac{\partial h}{\partial v}\right)_T = T\left(\frac{\partial s}{\partial v}\right)_T + v\left(\frac{\partial P}{\partial v}\right)_T$$

Substituting the Maxwell relation, Eq. 14.12, we have

$$\left(\frac{\partial h}{\partial v}\right)_T = T\left(\frac{\partial P}{\partial T}\right)_v + v\left(\frac{\partial P}{\partial v}\right)_T$$

$$dh_T = \left[T\left(\frac{\partial P}{\partial T}\right)_v + v\left(\frac{\partial P}{\partial v}\right)_T\right] dv_T$$

Integrating along a constant-temperature line, we have

$$(h_2 - h_1)_T = \int_1^2 \left[T\left(\frac{\partial P}{\partial T}\right)_v + v\left(\frac{\partial P}{\partial v}\right)_T\right] dv_T \qquad (14.22)$$

The significance of these two equations (Eqs. 14.21 and 14.22) is that the change in enthalpy along a constant-temperature line can be determined if the equation of state is known, because the quantities $(\partial v/\partial T)_P$, $(\partial P/\partial T)_v$, and $(\partial P/\partial v)_T$ can all be determined from the equation of state. It should also be noted that if the equation of state is explicit in P, i.e., $P = f(v, T)$, Eq. 14.22 is the most convenient form, whereas Eq. 14.21 is the most convenient if the equation of state is explicit in v, i.e., $v = f(P, T)$.

Let us illustrate the use of this equation to demonstrate the fact that the enthalpy of an ideal gas is a function of the temperature only. This statement was made but not proved in Chapter 8.

For an ideal gas

$$Pv = RT$$

$$\left(\frac{\partial v}{\partial T}\right)_P = \frac{R}{P}$$

Therefore, for an ideal gas, we conclude from Eq. 14.21

$$\left(\frac{\partial h}{\partial P}\right)_T = v - \frac{TR}{P} = v - v = 0$$

In words the conclusion is that when the temperature of an ideal gas is held constant while the pressure varies, there is no change in enthalpy. We then conclude that the enthalpy of an ideal gas is a function of the temperature only.

Another relation we will consider is the change of entropy in a constant-temperature process, and again we will consider two relations. The first of these is derived from the Maxwell relation, Eq. 14.13

$$\left(\frac{\partial s}{\partial P}\right)_T = -\left(\frac{\partial v}{\partial T}\right)_P$$

$$ds_T = -\left(\frac{\partial v}{\partial T}\right)_P dP_T$$

Integrating along a constant-temperature line, we have

$$(s_2 - s_1)_T = -\int_1^2 \left(\frac{\partial v}{\partial T}\right)_P dP_T \tag{14.23}$$

The second relation is derived from the Maxwell relation, Eq. 14.12

$$\left(\frac{\partial s}{\partial v}\right)_T = \left(\frac{\partial P}{\partial T}\right)_v$$

$$ds_T = \left(\frac{\partial P}{\partial T}\right)_v dv_T$$

Integrating along a constant-temperature line, we have

$$(s_2 - s_1)_T = \int_1^2 \left(\frac{\partial P}{\partial T}\right)_v dv_T \tag{14.24}$$

Again, we note that the change in entropy at constant temperature can be determined directly from the equation of state. The remarks made regarding the form of the equation of state apply to the calculation of entropy changes as well as to enthalpy changes.

Example 14.3 Over a certain range of pressures and temperatures the equation of state of a certain substance is given by the relation

$$\frac{Pv}{RT} = 1 - C'\frac{P}{T^4}$$

or

$$v = \frac{RT}{P} - \frac{C}{T^3}$$

where C' and C are constants.

Derive an expression for the change of enthalpy and entropy for this substance in a constant-temperature process. Since the equation of state is explicit in v, the most convenient equation to use is Eq. 14.21.

$$(h_2 - h_1)_T = \int_1^2 \left[v - T\left(\frac{\partial v}{\partial T}\right)_P\right] dP_T$$

From the equation of state

$$\left(\frac{\partial v}{\partial T}\right)_P = \frac{R}{P} + \frac{3C}{T^4}$$

$$(h_2 - h_1)_T = \int_1^2 \left[v - T\left(\frac{R}{P} + \frac{3C}{T^4}\right)\right] dP_T$$

$$(h_2 - h_1)_T = \int_1^2 \left[\frac{RT}{P} - \frac{C}{T^3} - \frac{RT}{P} - \frac{3C}{T^3}\right] dP_T$$

$$(h_2 - h_1)_T = -\int_1^2 \frac{4C}{T^3} dP_T = \frac{-4C}{T^3}(P_2 - P_1)_T$$

Similarly, for entropy

$$(s_2 - s_1)_T = -\int_1^2 \left(\frac{\partial v}{\partial T}\right)_P dP_T = -\int_1^2 \left(\frac{R}{P} + \frac{3C}{T^4}\right) dP_T$$

$$(s_2 - s_1)_T = -R \ln\left(\frac{P_2}{P_1}\right)_T - \frac{3C}{T^4}(P_2 - P_1)_T$$

Some important relations involving specific heats can also be developed. The constant-volume and constant-pressure specific heats were defined in Chapter 5 as

$$C_v = \left(\frac{\partial u}{\partial T}\right)_v \qquad C_P = \left(\frac{\partial h}{\partial T}\right)_P$$

Two other relations involving these specific heats can be derived.

$$du = T\,ds - P\,dv$$

$$C_v = \left(\frac{\partial u}{\partial T}\right)_v = T\left(\frac{\partial s}{\partial T}\right)_v \tag{14.25}$$

$$dh = T\,ds + v\,dP$$

$$C_P = \left(\frac{\partial h}{\partial T}\right)_P = T\left(\frac{\partial s}{\partial T}\right)_P \tag{14.26}$$

In Chapter 8 we noted that the specific heat of an ideal gas is a function of the temperature only. For actual gases we frequently are interested in the variation of specific heat with pressure or volume. These relations can be derived as follows:

$$s = f(T, P)$$

$$ds = \left(\frac{\partial s}{\partial T}\right)_P dT + \left(\frac{\partial s}{\partial P}\right)_T dP$$

Substituting Eqs. 14.26 and 14.13 we have

$$ds = \frac{C_P}{T}\,dT - \left(\frac{\partial v}{\partial T}\right)_P dP \tag{14.27}$$

Since this equation is also of the general form $dz = M\,dx + N\,dy$, we can proceed as follows to find a relation that gives the variation of the constant-pressure specific heat with pressure at constant temperature.

$$\left(\frac{\partial(C_P/T)}{\partial P}\right)_T = -\left[\frac{\partial}{\partial T}\left(\frac{\partial v}{\partial T}\right)_P\right]_P$$

$$\left(\frac{\partial C_P}{\partial P}\right)_T = -T\left(\frac{\partial^2 v}{\partial T^2}\right)_P \tag{14.28}$$

The variation of the constant-volume specific heat with pressure as the temperature remains constant can be found in a similar manner.

$$s = f(T, v)$$

$$ds = \left(\frac{\partial s}{\partial T}\right)_v dT + \left(\frac{\partial s}{\partial v}\right)_T dv$$

Substituting Eqs. 14.25 and 14.11 we have

$$ds = \left(\frac{C_v}{T}\right) dT + \left(\frac{\partial P}{\partial T}\right)_v dv \tag{14.29}$$

Since this is of the form $dz = M\,dx + N\,dy$,

$$\left(\frac{\partial C_v/T}{\partial v}\right)_T = \left[\frac{\partial}{\partial T}\left(\frac{\partial P}{\partial T}\right)_v\right]_v$$

$$\left(\frac{\partial C_v}{\partial v}\right)_T = T\left(\frac{\partial^2 P}{\partial T^2}\right)_v \tag{14.30}$$

The important thing to note about Eqs. 14.28 and 14.30 is that the variation of the constant-volume and constant-pressure specific heats in a constant-temperature process can be found from the equation of state.

Example 14.4 Determine the variation of C_p with pressure at constant temperature for a substance such as the one in Example 14.3 over the range where the equation of state is given by the relation

$$v = \frac{RT}{P} - \frac{C}{T^3}$$

Using Eq. 14.28

$$\left(\frac{\partial C_P}{\partial P}\right)_T = -T\left(\frac{\partial^2 v}{\partial T^2}\right)_P$$

$$\left(\frac{\partial v}{\partial T}\right)_P = \frac{R}{P} + \frac{3C}{T^4}$$

$$\left(\frac{\partial^2 v}{\partial T^2}\right)_P = -\frac{12C}{T^5}$$

$$\left(\frac{\partial C_P}{\partial P}\right)_T = -T\left(-\frac{12C}{T^5}\right) = \frac{12C}{T^4}$$

A final interesting and useful relation can be derived by equating Eqs. 14.27 and 14.29.

$$C_P \frac{dT}{T} - \left(\frac{\partial v}{\partial T}\right)_P dP = C_v \frac{dT}{T} + \left(\frac{\partial P}{\partial T}\right)_v dv$$

$$dT = \frac{T(\partial P/\partial T)_v}{C_p - C_v} dv + \frac{T(\partial v/\partial T)_P}{C_p - C_v} dP$$

But

$$T = f(v, P)$$

$$dT = \left(\frac{\partial T}{\partial v}\right)_P dv + \left(\frac{\partial T}{\partial P}\right)_v dP$$

Therefore,

$$\left(\frac{\partial T}{\partial v}\right)_P = \frac{T(\partial P/\partial T)_v}{C_P - C_v}$$

$$\left(\frac{\partial T}{\partial P}\right)_v = \frac{T(\partial v/\partial T)_P}{C_P - C_v}$$

When these equations are solved for $C_p - C_v$, they yield the same result.

$$C_p - C_v = T \left(\frac{\partial v}{\partial T}\right)_P \left(\frac{\partial P}{\partial T}\right)_v \tag{14.31}$$

But from Eq. 14.2

$$\left(\frac{\partial P}{\partial T}\right)_v = - \left(\frac{\partial v}{\partial T}\right)_P \left(\frac{\partial P}{\partial v}\right)_T$$

Therefore,

$$C_p - C_v = -T \left(\frac{\partial v}{\partial T}\right)_P^2 \left(\frac{\partial P}{\partial v}\right)_T \tag{14.32}$$

From this equation we draw several conclusions:

1. For a liquid and solid $(\partial v/\partial T)_P$ is usually relatively small, and, therefore, the difference between the constant-pressure and constant-volume specific heats of a liquid is small. For this reason many tables simply give the specific heat of a solid or a liquid without designating that it is at constant volume or pressure. Further, $C_p = C_v$ exactly when $(\partial v/\partial T)_P = 0$, as is true at the point of maximum density of water.

2. $C_p \rightarrow C_v$ as $T \rightarrow 0$, and, therefore, we conclude that the constant-pressure and constant-volume specific heats are equal at absolute zero.

3. The difference between C_p and C_v is always positive because $(\partial v/\partial T)_P{}^2$ is always positive and $(\partial P/\partial v)_T$ is negative for all known substances.

14.5 Volume Expansivity and Isothermal and Adiabatic Compressibility

The student has most likely encountered the coefficient of linear expansion in his studies of strength of materials. This coefficient indicates how the length of a solid body is influenced by a change in temperature while the pressure remains constant. In terms of the notation of partial derivatives the *coefficient of linear expansion*, δ_t, is defined as follows:

$$\delta_t = \frac{1}{L}\left(\frac{\partial L}{\partial T}\right)_P \tag{14.33}$$

A similar coefficient can be defined for changes in volume, and such a coefficient is applicable to liquids and gases as well as to solids. This coefficient of volume expansion α, also called the volume expansivity, is an indication of the change in volume that results from a change in temperature while the pressure remains constant. The definition of *volume expansivity* is

$$\alpha \equiv \frac{1}{V}\left(\frac{\partial V}{\partial T}\right)_P \equiv \frac{1}{v}\left(\frac{\partial v}{\partial T}\right)_P \tag{14.34}$$

The isothermal compressibility β_T is an indication of the change in volume that results from a change in pressure while the temperature remains constant. The definition of the *isothermal compressibility* is

$$\beta_T \equiv -\frac{1}{V}\left(\frac{\partial V}{\partial P}\right)_T \equiv -\frac{1}{v}\left(\frac{\partial v}{\partial P}\right)_T \tag{14.35}$$

The reciprocal of the isothermal compressibility is called the *isothermal bulk modulus* B_T

$$B_T \equiv -v\left(\frac{\partial P}{\partial v}\right)_T \tag{14.36}$$

The *adiabatic compressibility* β_s is an indication of the change in volume that results from a change in pressure while the entropy remains constant, and is defined as follows:

$$\beta_s \equiv -\frac{1}{v}\left(\frac{\partial v}{\partial P}\right)_s \tag{14.37}$$

The *adiabatic bulk modulus* B_s is the reciprocal of the adiabatic compressibility.

$$B_s \equiv -v\left(\frac{\partial P}{\partial v}\right)_s \tag{14.38}$$

Both the volume expansivity and isothermal compressibility are thermodynamic properties of a substance, and as such are functions of two independent properties. Values of these properties are found in the standard handbooks of physical properties. One example is included which gives an indication of the use and significance of the volume expansivity and isothermal compressibility.

Example 14.5 The pressure on a block of copper having a mass of 1 lbm is increased in a reversible process from 1 atm to 1000 atm while the temperature is held constant at 60 F. Determine the work done on the copper during this process, the change in entropy per pound of copper, the heat transfer, and the change of internal energy per pound.

Over the range of pressures and temperatures involved in this problem the following data can be used:

$$\text{Volume expansivity} = \alpha = 2.8 \times 10^{-5} R^{-1}$$

$$\text{Isothermal compressibility} = \beta_T = 5.9 \times 10^{-8} \text{ in.}^2/\text{lbf}$$

$$\text{Specific volume} = 1.82 \times 10^{-3} \text{ ft}^3/\text{lbm}$$

The work done during the isothermal expansion is

$$w = \int P \, dv_T$$

The isothermal compressibility has been defined:

$$\beta_T = -\frac{1}{v}\left(\frac{\partial v}{\partial p}\right)_T$$

$$v\beta_T \, dP_T = -dv_T$$

Therefore, for this isothermal process

$$w = -\int_1^2 v\beta_T P \, dP_T$$

Since v and β_T remain essentially constant, this is readily integrated:

$$w = -\frac{v\beta_T}{2}(P_2^2 - P_1^2)$$

$$= -1.82 \times 10^{-3} \text{ ft}^3/\text{lbm} \times 5.9 \times 10^{-8} \text{ in.}^2/\text{lbf} \times \frac{1000^2 - 1^2}{2}$$

$$\times (14.7)^2 \text{ lbf}^2/\text{in.}^4 \times 144 \text{ in.}^2/\text{ft}^2 = -1.68 \text{ ft-lbf}$$

The change of entropy can be found by considering the Maxwell relation, Eq. 14.13, and the definition of volume expansivity.

$$\left(\frac{\partial s}{\partial P}\right)_T = -\left(\frac{\partial v}{\partial T}\right)_P = -\frac{v}{v}\left(\frac{\partial v}{\partial T}\right)_P = -v\alpha$$

$$ds_T = -v\alpha\, dP_T$$

This can be readily integrated as follows if we assume that v and α remain constant:

$$(s_2 - s_1)_T = -v\alpha(P_2 - P_1)_T$$

$$= -1.82 \times 10^{-3}\ \text{ft}^3/\text{lbm} \times \frac{2.8 \times 10^{-5}}{R} \times (1000 - 1)$$

$$\times\ 14.7 \times 144\ \text{lbf/ft}^2 = -0.108\ \text{ft-lbf/lbm R}$$

The heat transfer for this reversible isothermal process is $T(s_2 - s_1)$.

$$q = T(s_2 - s_1) = 520(-0.108) = -56.1\ \text{ft-lbf}$$

The change in internal energy follows directly from the first law.

$$(u_2 - u_1) = q - w = -56.1 - (-1.7) = -54.4\ \text{ft-lbf}$$

14.6 Developing Tables of Thermodynamic Properties from Experimental Data

There are many ways in which tables of thermodynamic properties can be developed from experimental data. The purpose of this section is to convey some general principles and concepts by considering only the liquid and vapor region.

Let us assume that the following data for a pure substance have been obtained in the laboratory.

(*a*) Vapor-pressure data: That is, saturation pressures and temperatures have been measured over a wide range.

(*b*) Pressure, specific-volume, temperature data in the vapor region: These data are usually obtained by determining the mass of the substance in a closed vessel (which means a fixed specific volume) and then measuring the pressure as the temperature is varied. This is done for a large number of specific volumes.

(*c*) Density of the saturated liquid and the critical pressure and temperature.

(*d*) Zero-pressure specific heat for the vapor. This might be obtained either calorimetrically or from spectroscopic data.

From these data a complete set of thermodynamic tables for the saturated liquid, saturated vapor, and superheated vapor can be calculated. The first step is to determine an equation for the vapor-pressure curve that accurately fits the data. It may be necessary to use one equation for one portion of the vapor-pressure curve and a different equation for another portion of the curve.

One form of equation that has been used is

$$\ln P = A + \frac{B}{T} + C \ln T + DT$$

Once an equation has been found that accurately represents the data, the saturation pressure for any given temperature can be found by solving this equation. Thus, the saturation pressures in Table 1 of *Thermodynamic Properties of Steam*, would be determined for the given temperatures. The second step is to determine an equation of state for the vapor region that accurately represents the P-v-T data. (An equation of state gives the relation between pressure, specific volume, and temperature.) There are many possible forms of the equation of state which may be selected, and this matter will be considered in detail in the next section. The important considerations are that the equation of state accurately represents the data, and that it be of such a form that the differentiations required can be performed (i.e., in some cases it may be desirable to have an equation of state explicit in v, whereas on other occasions an equation of state that is explicit in P may be more desirable).

Once an equation of state has been determined, the specific volume of superheated vapor at given pressures and temperatures can be determined by solving the equation and tabulating the results as in the superheat tables for steam, ammonia, and Freon-12. The specific volume of saturated vapor at a given temperature may be found by finding the saturation pressure from the vapor-pressure curve and substituting this saturation pressure and temperature into the equation of state.

The procedure followed in determining enthalpy and entropy is best explained with the aid of Fig. 14.3. Let us assume that the enthalpy of saturated liquid in state 1 is zero. The enthalpy of saturated vapor in state 2 can be found from the Clapeyron equation.

$$\frac{dP}{dT} = \frac{h_{fg}}{T(v_g - v_f)}$$

The left side of this equation is found by differentiating the vapor-pressure curve. The specific volume of the saturated vapor is found by the procedure outlined in the last paragraph, and it is assumed the specific volume of the saturated liquid has been measured. Thus, the enthalpy of evaporation, h_{fg}, can be found for this particular temperature, and the enthalpy at state 2 is equal to the enthalpy of evapora-

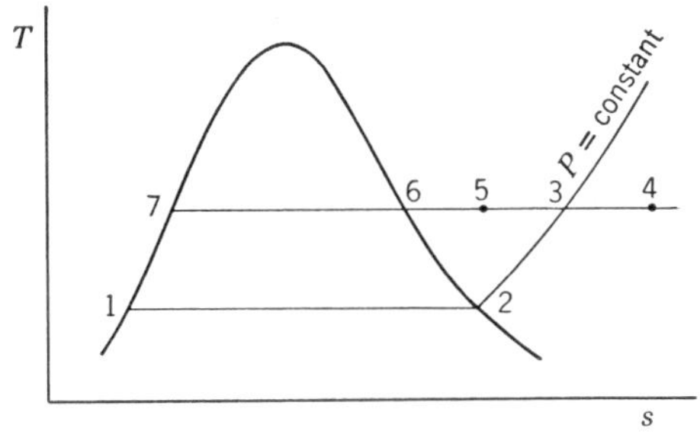

Fig. 14.3 Sketch showing procedure for developing a table of thermodynamic properties from experimental data.

tion (since the enthalpy in state 1 is assumed to be zero). The entropy at state 2 is readily found, since

$$s_{fg} = \frac{h_{fg}}{T}$$

The change in enthalpy between states 2 and 3 is readily found if the specific heat at this pressure is known. We have assumed that the zero-pressure specific heat, C_{Po}, is known. Therefore, at a given pressure the specific heat, C_P, can be found by using Eq. 14.28.

$$\left(\frac{\partial C_P}{\partial P}\right)_T = -T\left(\frac{\partial^2 v}{\partial T^2}\right)_P$$

$$(dC_P)_T = -T\left(\frac{\partial^2 v}{\partial T^2}\right)_P dP_T$$

$$C_P - C_{Po} = -\int_{P=0}^{P} T\left(\frac{\partial^2 v}{\partial T^2}\right)_P dP_T$$

The quantity $(\partial^2 v/\partial T^2)_P$ is found from the equation of state. If this equation of state is of the form $v = f(P, T)$ this differentiation can be readily obtained, and this is the reason for the previous statement regarding the form of the equation of state.

With the specific-heat equation for the given pressure, the change of enthalpy and entropy along the constant-pressure line is readily found.

$$(h_3 - h_2)_P = \int_2^3 C_p \, dT_P$$

$$(s_3 - s_2)_P = \int_2^3 C_p \frac{dT_P}{T}$$

From state 3 changes in enthalpy and entropy at constant temperature are readily found by using Eqs. 14.21 and 14.23 which were developed in the last section. For example:

$$h_3 - h_5 = \int_5^3 \left[v - T \left(\frac{\partial v}{\partial T} \right)_P \right] dP_T$$

$$s_3 - s_5 = \int_5^3 - \left(\frac{\partial v}{\partial T} \right)_P dP_T$$

The enthalpy and entropy of the saturated vapor in state 6 is found by this same procedure. Finally, the enthalpy and entropy of the saturated liquid in state 7 is found by applying the Clapeyron equation between states 6 and 7.

Thus, values for the pressure, temperature, specific volume, enthalpy, entropy, and internal energy of saturated liquid, saturated vapor, and superheated vapor can be tabulated for the entire region for which experimental data were obtained. It is evident that the accuracy of such a table depends both on the accuracy of the experimental data and the degree to which the equation for the vapor pressure and the equation of state represent the experimental data.

14.7 Equations of State

An equation of state is an equation that relates pressure, temperature, and specific volume. Many different equations of state have been developed over the years. Some have been purely empirical and others have been based on kinetic theory and/or thermodynamic considerations. However, all are empirical in the sense that the constants must be determined from experimental observations, and the equations are valid only for interpolation in the region where experimental data have been obtained.

Some general forms of equations of state are:

$$v = f(P, T)$$

$$P = f(v, T)$$

$$Pv = f(P, T)$$

$$\frac{Pv}{RT} = f(P, T)$$

Every equation of state has a number of experimentally determined constants. For example, two forms of equations of state are:

$$P = A + BT + \frac{C}{T^2}$$

where A, B, and C are functions of v, and each involves a number of constants.

$$Pv = A + \frac{B}{v} + \frac{C}{v^2} + \cdots$$

where A, B, and C are functions of temperature and are called virial coefficients.

One of the earliest equations of state to be proposed was Van der Waals', which is

$$P = \frac{RT}{v - b} - \frac{a}{v^2} \tag{14.39}$$

This equation was based on kinetic considerations and the fact that on a P-v plot the critical isotherm has a zero slope and passes through a point of inflection at the critical point (i.e., $(\partial P/\partial v)_T = 0$ and $(\partial^2 P/\partial v^2)_T = 0$ at the critical point). Regarding the kinetic considerations, the constant b is intended to correct for the volume occupied by the molecules and the term a/v^2 is intended to account for the intermolecular attraction.

Applying the criterion that the critical isotherm passes through a point of inflection at the critical point, we have

$$\left(\frac{\partial P}{\partial v}\right)_T = 0 = \frac{-RT_c}{(v_c - b)^2} + \frac{2a}{v_c^3}$$

where subscript c indicates the property at the critical state.

$$\left(\frac{\partial^2 P}{\partial v^2}\right)_T = 0 = \frac{2RT_c}{(v_c - b)^3} - \frac{6a}{v_c^4}$$

From these two equations the following relations can be obtained:

$$a = \frac{27}{64} \frac{R^2 T_c^{\,2}}{P_c}$$

$$b = \frac{RT_c}{8P_c}$$

$$\frac{P_c v_c}{RT_c} = \frac{3}{8}$$

Thus, Van der Waals' equation predicts that the constants a and b are determined by critical pressure and temperature, and the quantity $P_c v_c / RT_c$ is a constant for all substances and is equal to $\frac{3}{8}$. The fact that $P_c v_c / RT_c$ is not the same for all substances and is significantly different from $\frac{3}{8}$ is evidence that Van der Waals' equation is of limited accuracy and utility. It is of primary interest because of its historical significance and its relation to the law of corresponding states, which will be discussed in the next section.

Two other two-constant equations of state have been developed.

Dieterici equation: $$P = \frac{RT}{v - b} e^{-a/RTv} \tag{14.40}$$

Berthelot equation: $$P = \frac{RT}{v - b} - \frac{a}{Tv^2} \tag{14.41}$$

In general these are not significantly better than the Van der Waals' equation.

One of the best known and most useful equations of state is the Beattie-Bridgeman equation of state. This was introduced in Chapter 3 and is presented here for the sake of completeness.

$$P = \frac{RT(1 - \epsilon)}{v^2} (v + B) - \frac{A}{v^2} \tag{14.42}$$

where $A = A_0(1 - a/v)$

$B = B_0(1 - b/v)$

$$\epsilon = \frac{c}{vT^3}$$

A_o, a, B_o, b, and c are constants that must be determined experimentally for each substance. Table 14.1 gives a list of these constants for a number of substances.

A less accurate form of this equation has been developed, which is explicit in specific volume, and is frequently very useful:

$$v = (\pi + B)(1 - \epsilon) - \frac{A}{RT}$$

where $A = A_o(1 - a/\pi)$

$$B = B_o(1 - b/\pi)$$

$$\epsilon = \frac{c}{\pi T^3}$$

$$\pi = \frac{RT}{P}$$

TABLE 14.1

Constants of the Beattie-Bridgeman Equation of State *

Pressure in atmospheres, volume in liters/gm mole, temperature in deg K
R = 0.08206 atm liters/gm mole K

Gas	A_o	a	B_o	b	$10^{-4}c$
Helium	0.0216	0.05984	0.01400	0.0	0.0040
Neon	0.2125	0.02196	0.02060	0.0	0.101
Argon	1.2907	0.02328	0.03931	0.0	5.99
Hydrogen	0.1975	−0.00506	0.02096	−0.04359	0.0504
Nitrogen	1.3445	0.02617	0.05046	−0.00691	4.20
Oxygen	1.4911	0.02562	0.04624	0.004208	4.80
Air	1.3012	0.01931	0.04611	−0.001101	4.34
CO_2	5.0065	0.07132	0.10476	0.07235	66.00
$(C_2H_5)_2O$	31.278	0.12426	0.45446	0.11954	33.33
C_2H_4	6.152	0.04964	0.12156	0.03597	22.68
Ammonia	2.3930	0.17031	0.03415	0.19112	476.87
CO	1.3445	0.02617	0.05046	−0.00691	4.20
N_2O	5.0065	0.07132	0.10476	0.07235	66.0
CH_4	2.2769	0.01855	0.05587	−0.01587	12.83
C_2H_6	5.8800	0.05861	0.09400	0.01915	90.00
C_3H_8	11.9200	0.07321	0.18100	0.04293	120
n-C_4H_{10}	17.794	0.12161	0.24620	0.09423	350
n-C_7H_{16}	54.520	0.20066	0.70816	0.19179	400

* *Proc. Am. Acad. Arts Sci.*, Vol. 63 (1928), pp. 229–308; *Z. Physik*, Vol. 62 (1930), pp. 95–101; *J. Chem. Phys.*, Vol. 3 (1935), pp. 93–96; *J. Am. Chem. Soc.*, Vol. 59 (1937), pp. 1587–1589, 1589–1590; Vol. 61 (1939), pp. 26–27.

14.8 Generalized Compressibility Charts

In dealing with actual fluids, particularly gases, it is advantageous to express the deviation from the ideal-gas equation of state by the *compressibility factor* Z, which is defined by the relation

$$Z = \frac{Pv}{RT} \tag{14.43}$$

It is advantageous of course only if we know Z as a function, say, of P and T, and in this section we will consider how this relation is obtained with a fair degree of accuracy by use of the generalized compressibility charts (Fig. 14.4). In the next section we will consider how information concerning other thermodynamic properties can be developed from this compressibility chart.

In order to understand the basis for the generalized compressibility charts we must first consider the law of corresponding states. It has been stated that Van der Waals' equation of state, which contains two arbitrary constants, depends upon the fact that the slope of the critical isotherm is zero at the critical point and that it also goes through a point of inflection at the critical point. Further, these two constants can be evaluated simply from data for the critical pressure and temperature, according to the relations:

$$P = \frac{RT}{v - b} - \frac{a}{v^2} \quad \text{(Van der Waals' equation of state)}$$

$$a = \frac{27R^2T_c^2}{64P_c}$$

$$b = \frac{RT_c}{8P_c}$$

We now introduce the concept of reduced pressures, temperatures, and specific volumes. These are defined as follows:

$$\text{Reduced pressure} = P_r = \frac{P}{P_c} \qquad P_c = \text{Critical pressure}$$

$$\text{Reduced temperature} = T_r = \frac{T}{T_c} \qquad T_c = \text{Critical temperature}$$

$$\text{Reduced specific volume} = v_r = \frac{v}{v_c} \qquad v_c = \text{Critical specific volume}$$

In words these equations state that the reduced property for a given state is the value of that property divided by the value of that same property in the critical state.

Let us now rewrite Van der Waals' equation of state in terms of the reduced pressure and the compressibility factor Z. The equation then becomes

$$Z^3 - \left(\frac{P_r}{8T_r} + 1\right) Z^2 + \left(\frac{27P_r}{64T_r^2}\right) Z - \frac{27P_r^2}{512T_r^3} = 0 \qquad (14.44)$$

The important thing to note is that according to this equation $Z = f(P_r, T_r)$. Even though Van der Waals' equation of state is of limited accuracy, the relation $Z = f(P_r, T_r)$ has proved to be quite accurate, and is known as the law of corresponding states. In words this law states that when different substances have the same reduced pressure and temperature they will have the same compressibility factor.

This means that if lines of constant reduced temperature are plotted on a Z vs. P_r chart, the chart would be the same (within the accuracy of the law of corresponding states) for all substances; it is known as the generalized compressibility chart, and is made by plotting lines of constant T_r on a Z vs. P_r chart for a large number of substances, and drawing the chart that most accurately represents all the data. The fact that such charts for different substances do essentially coincide is the justification for the law of corresponding states.

The generalized compressibility chart is shown in Fig. 14.4 and its use is demonstrated in the example that follows. Table A.5 in the Appendix gives the critical constants for a number of substances.

Example 14.6 (*a*) Volume unknown. Calculate the specific volume of propane at a pressure of 1000 lbf/in.^{2} and a temperature of 300 F, and compare this with the specific volume given by the ideal-gas equation of state.

For propane

$$T_c = 666 \text{ R}$$

$$P_c = 617 \text{ lbf/in.}^2$$

$$R = 35.1 \text{ ft-lbf/lbm R}$$

$$T_r = \tfrac{760}{666} = 1.141 \qquad P_r = \tfrac{1000}{617} = 1.62$$

From the compressibility chart

$$Z = 0.55$$

$$Pv = ZRT$$

$$v = \frac{0.55 \times 35.1 \times 760}{1000 \times 144} = 0.102 \text{ ft}^3\text{/lbm}$$

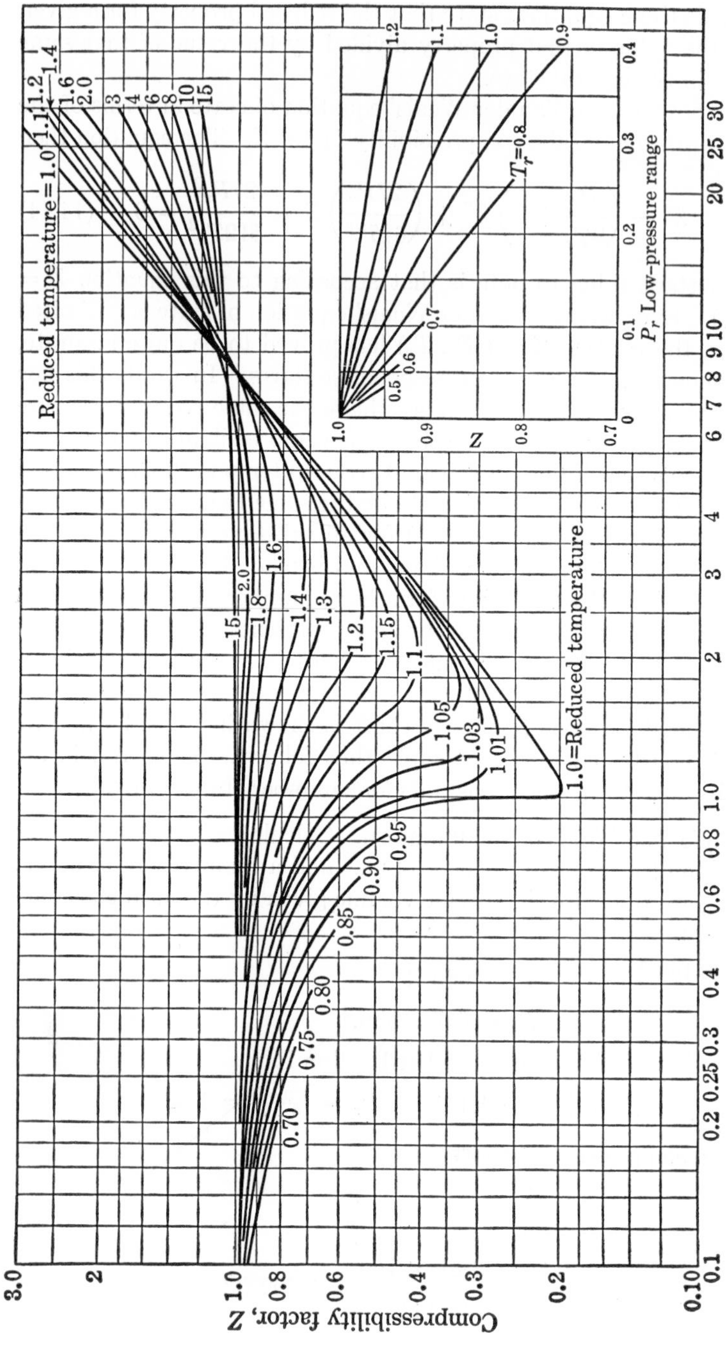

Fig. 14.4 Compressibility factors of gases and vapors. (Reprinted by permission from Hougen and Watson, *Chemical Process Principles*, Part II, Wiley, New York.)

The ideal-gas equation would give the value

$$v = \frac{35.1 \times 760}{1000 \times 144} = 0.185 \text{ ft}^3/\text{lbm}$$

(*b*) Pressure unknown. What pressure is required in order that propane have a specific volume of 0.102 ft^3/lbm at a temperature of 300 F?

$$Pv = ZRT \qquad T_r = \tfrac{760}{666} = 1.141$$

$$P = P_c P_r$$

Therefore,

$$P_r = \frac{ZRT}{vP_c} = \frac{Z \times 35.1 \times 760}{0.102 \times 617 \times 144} = 2.94Z$$

$$Z = \frac{P_r}{2.94} = 0.34P_r$$

The slope of this line on a log log plot is 45°. Therefore, if a 45° line is drawn from the point where $Z = 1$, $P_r = 2.94$, this line will intersect the desired T_r line, which in this case is $T_r = 1.141$. At this point P_r is found to be 1.62. Therefore,

$$P = P_r P_c = 1.62(617) = 1000 \text{ lbf/in.}^2$$

(*c*) Temperature unknown. What will be the temperature of propane when it has a specific volume of 0.102 ft^3/lbm and a pressure of 1000 lbf/in.^{2}?

$$Pv = ZRT \qquad T = T_c T_r$$

$$Pv = ZRT_c T_r$$

$$T_r = \frac{Pv}{ZRT_c} = \frac{1000 \times 144 \times 0.102}{Z \times 35.1 \times 666} = \frac{0.628}{Z}$$

This line must be plotted on the compressibility chart (it is not a straight line in this case) and the state point is given by the intersection of this line with the known reduced-pressure line ($P_r = 1.62$ in this case).

The reduced temperature is thus found to be

$$T_r = 1.14$$

$$T = T_r \times T_c = 1.14 \times 666 = 760 \text{ R}$$

The chief value of the generalized compressibility charts is that if P-v-T information is needed for a substance for which limited data are available, the generalized charts can be used with fair accuracy. It is only necessary to know the critical pressure and temperature of the particular substance concerned. Whenever accurate thermodynamic data are available one would, of course, use these in preference to the generalized charts.

14.9 Generalized Chart for Change of Enthalpy at Constant Temperature

One might frequently be interested not only in P-v-T information for a substance for which no data are available, but also in changes in enthalpy and internal energy. This section is concerned with a generalized chart that enables one to find changes in enthalpy in a constant-temperature process. Again, it is necessary to know only the critical pressure and temperature of the particular substance. This chart can be developed from the generalized compressibility chart as shown in the following derivation.

From the definition of compressibility factor we can write

$$v = \frac{ZRT}{P}$$

Therefore,

$$\left(\frac{\partial v}{\partial T}\right)_P = \frac{ZR}{P} + \frac{RT}{P}\left(\frac{\partial Z}{\partial T}\right)_P$$

It has been shown that

$$\left(\frac{\partial h}{\partial P}\right)_T = v - T\left(\frac{\partial v}{\partial T}\right)_P$$

$$= \frac{ZRT}{P} - \frac{ZRT}{P} - \frac{RT^2}{P}\left(\frac{\partial Z}{\partial T}\right)_P = -\frac{RT^2}{P}\left(\frac{\partial Z}{\partial T}\right)_P$$

$$dh_T = -\frac{RT^2}{P}\left(\frac{\partial Z}{\partial T}\right)_P dP_T$$

But $T = T_cT_r$ and $P = P_cP_r$. Therefore,

$$\frac{dT}{T} = \frac{dT_r}{T_r} \qquad \frac{dP}{P} = \frac{dP_r}{P_r}$$

$$dT = T_c\,dT_r \qquad dP = P_c\,dP_r$$

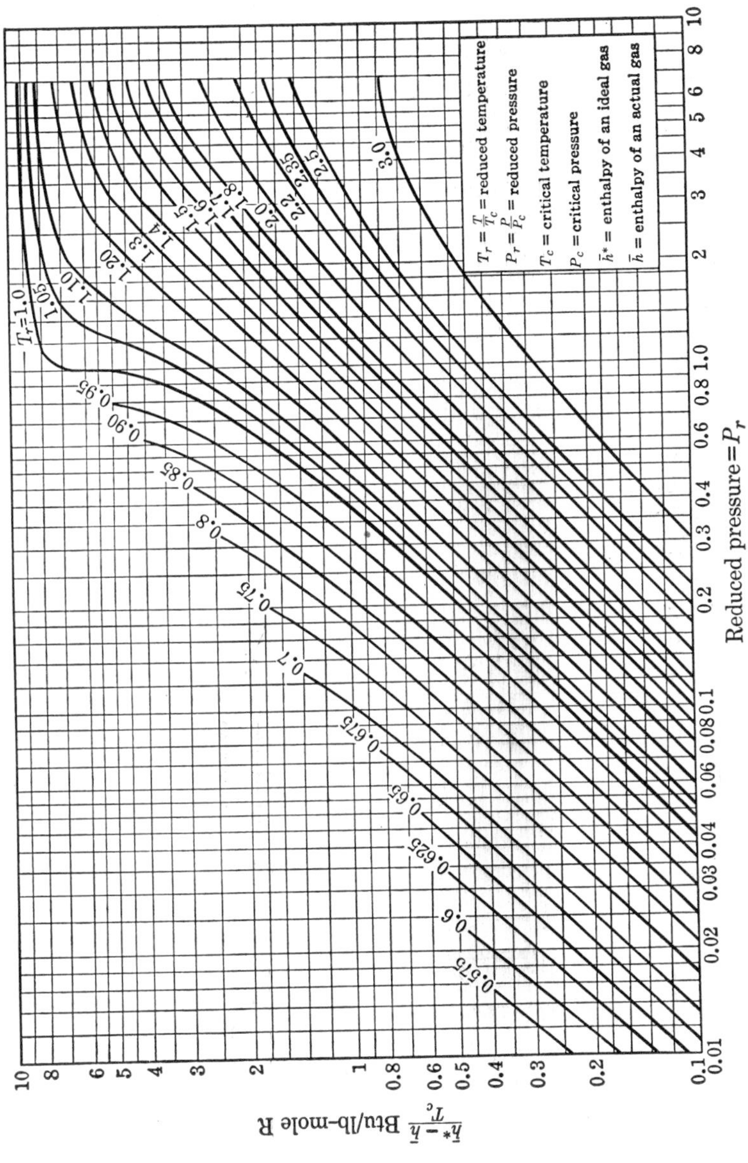

Fig. 14.5 Enthalpy correction for gases and vapors. (Reprinted by permission from Hougen and Watson, *Chemical Process Principles,* Part II, Wiley, New York.)

Substituting these values we have

$$dh_T = -\frac{RT_c^2T_r^2}{P_cP_r}\left(\frac{\partial Z}{T_c\,\partial T_r}\right)_{P_r} P_c(dP_r)_T$$

$$= -RT_cT_r^2\left(\frac{\partial Z}{\partial T_r}\right)_{P_r}(d\ln P_r)_T$$

Integrating at constant temperature we have

$$\frac{\Delta h_T}{T_c} = -R\int_1^2 T_r^2\left(\frac{dZ}{dT_r}\right)_{P_r} d\ln P_r \qquad (14.45)$$

Let us designate by an asterisk the enthalpy at a given temperature and a very low pressure, such that $P \to 0$, and let us designate without superscript the enthalpy at any state at the same temperature and a given pressure P_r. Then the integration is as follows:

$$\frac{h^* - h}{T_c} = R\int_{P=0}^{P} T_r^2\left(\frac{dZ}{dT_r}\right)_{P_r} d\ln P_r \qquad (14.46)$$

The right side of this equation can be obtained by graphical integration of the compressibility chart, and is a function of only the reduced pressure and temperature. The left side represents the change in enthalpy as the pressure is increased from zero to the given pressure while the temperature is held constant. Sometimes this quantity is called the residual enthalpy. Figure 14.5 shows a plot of this quantity, $(\bar{h}^* - \bar{h})/T_c$, as a function of P_r for various lines of constant T_r.

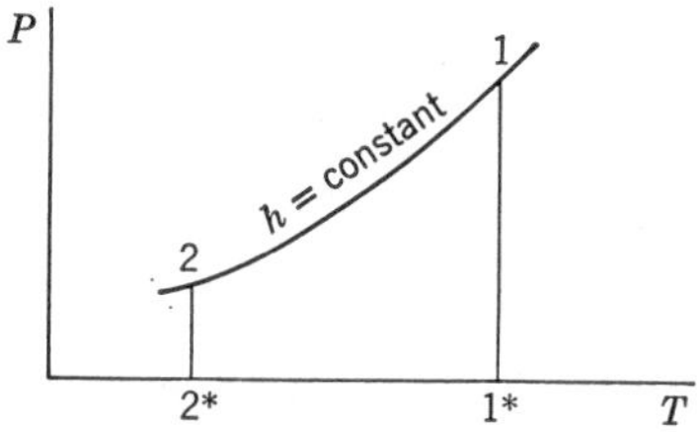

Fig. 14.6 Sketch for Example 14.7.

Example 14.7 Nitrogen is throttled from 3000 lbf/in.^{2}, -100 F to 200 lbf/in.^{2} Determine the final temperature of the nitrogen. The zero-pressure specific heat of nitrogen in this temperature range is 0.248 Btu/lbm R.

This is a constant-enthalpy process, and in solving such a problem it is usually helpful to use a pressure-temperature plot such as the one shown in Fig. 14.6.

$$P_1 = 3000 \text{ lbf/in.}^2 \qquad P_2 = 200 \text{ lbf/in.}^2$$

$$P_{r1} = \tfrac{3000}{492} = 6.1 \qquad P_{r2} = \tfrac{200}{492} = 0.406$$

$$T_1 = 360 \text{ R}$$

$$T_{r1} = \tfrac{360}{227} = 1.586$$

From the generalized chart for enthalpy change at constant temperature

$$\frac{\bar{h}_1{}^* - \bar{h}_1}{T_c} = 4.4 \text{ Btu/lb-mole R}$$

$$h_1{}^* - h_1 = \frac{4.4 \times 227}{28} = 35.7 \text{ Btu/lbm}$$

It is now necessary to assume a final temperature and check to see if the net change in enthalpy for the process is zero. Let us assume that $T_2 = 244$ R. Then the change in enthalpy between 1* and 2* can be found from the zero-pressure specific-heat data

$$h_1{}^* - h_2{}^* = C_{p0}(T_1{}^* - T_2{}^*) = 0.248(360 - 244) = 28.8 \text{ Btu/lbm}$$

(The *Gas Tables* can be used if one wishes to take into account the variation in C_{p0} with temperature.)

We now find the enthalpy between 2* and 2

$$T_{r2} = \tfrac{244}{227} = 1.075 \qquad P_{r2} = 0.406$$

Therefore, from the charts

$$\frac{\bar{h}_2{}^* - \bar{h}_2}{T_c} = 0.87 \text{ Btu/lb-mole R}$$

$$h_2{}^* - h_2 = \frac{0.87(227)}{28} = 7.06 \text{ Btu/lbm}$$

We now check to see if the net change in enthalpy for the process is zero.

$$h_1 - h_2 = 0 = -(h_1{}^* - h_1) + (h_1{}^* - h_2{}^*) + (h_2{}^* - h_2)$$

$$= -35.7 + 28.8 + 7.0 = 0.1 \text{ Btu/lbm}$$

This essentially checks, and we conclude that the final temperature is 244 R.

14.10 Work and Heat Transfer in a Reversible Isothermal Process

A similar development to that made in the last section for enthalpy can be made to enable one to determine the work and heat transfer in a reversible isothermal process. This is done by introducing the fugacity.

We begin by deriving an expression for the work done in a steady-flow reversible isothermal compression process (Fig. 14.7).

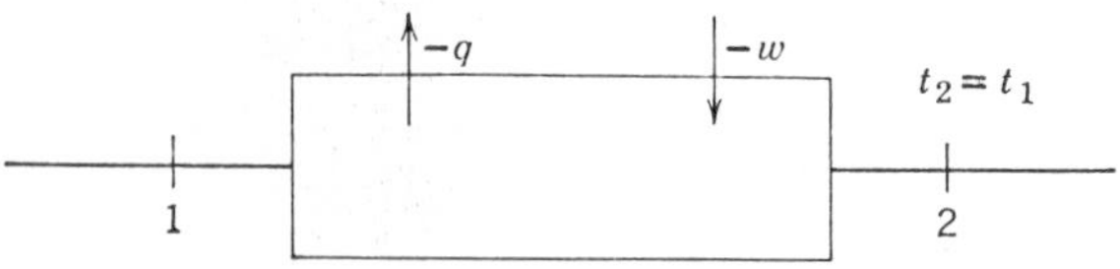

Fig. 14.7 A reversible isothermal steady-flow process.

From the steady-flow energy equation

$$q + h_1 + KE_1 + PE_1 = h_2 + KE_2 + PE_2 + w$$

Since the process is reversible and isothermal

$$q = T(s_2 - s_1) = T_2 s_2 - T_1 s_1$$

Therefore,

$$-w = h_2 - h_1 - (T_2 s_2 - T_1 s_1) + \Delta KE + \Delta PE$$

$$-w = g_2 - g_1 + \Delta KE + \Delta PE$$

where $g = h - Ts$ = Gibbs function

The fugacity f is defined by the relation

$$g = RT \ln f + y \tag{14.47}$$

where y is a function of temperature only.
Therefore,

$$dg_T = RT(d \ln f)_T$$

But

$$dg = v\,dP - s\,dT$$

Therefore,

$$dg_T = v\,dP_T = RT(d \ln f)_T$$

From the definition of compressibility factor,

$$v = \frac{ZRT}{P}$$

Therefore,

$$dg_T = ZRT(d \ln P)_T = RT(d \ln f)_T$$

or

$$Z(d \ln P)_T = (d \ln f)_T$$

We have noted previously that

$$\frac{dP}{P} = \frac{dP_r}{P_r};\ d \ln P = d \ln P_r$$

Therefore,

$$Z(d \ln P_r)_T = (d \ln f)_T$$

$$Z(d \ln P_r)_T - (d \ln P_r)_T = (d \ln f)_T - (d \ln P_r)_T$$

$$(Z - 1)(d \ln P_r)_T = (d \ln f)_T - (d \ln P)_T = (d \ln f/P)_T$$

$$(d \ln f/P)_T = (Z - 1)(d \ln P_r)_T$$

$$\ln f/P = \int (Z - 1)(d \ln P_r)_T + \text{constant} \qquad (14.48)$$

The characteristic of the fugacity is such that as P approaches zero, f/P approaches 1. In other words, at very low pressures the fugacity is equal to the pressure. Therefore, if we integrate the above equation between $P = 0$ and some finite pressure, we have the value of $\ln f/P$ at that pressure. Thus, we can write

$$\ln f/P = \int_{P_r=0}^{P_r} (Z - 1)(d \ln P_r)_T \qquad (14.49)$$

The right side can be obtained by graphical integration from the compressibility chart. The quantity f/P is called the fugacity coefficient and is given the symbol ν, and Fig. 14.8 gives values of ν as a function of P_r and T_r.

Example 14.8 Calculate the work of compression and the heat transfer per pound when ethane is compressed reversibly and isothermally from 14.7 lbf/in.² to 1000 lbf/in.² at a temperature of 110 F.

For ethane,

$$P_c = 708 \text{ lbf/in.}^2 \qquad T_c = 549.8 \text{ R}$$

$$P_{r1} = \frac{14.7}{708} = 0.021 \qquad T_{r1} = T_{r2} = \frac{570}{549.8} = 1.037$$

$$P_{r2} = \frac{1000}{708} = 1.412$$

From the enthalpy chart

$$\frac{\bar{h}_2{}^* - \bar{h}_2}{T_c} = 7.1 \qquad \frac{\bar{h}_1{}^* - \bar{h}_1}{T_c} < 0.1$$

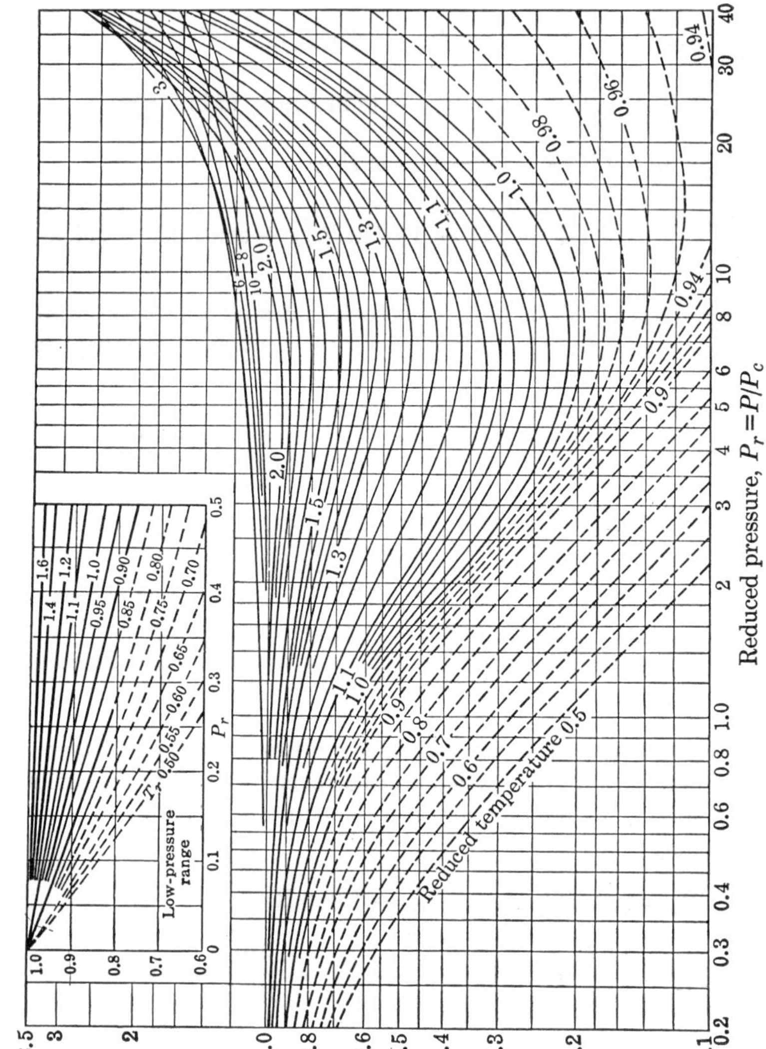

Fig. 14.8 Fugacity coefficients of gases and vapors. (Reprinted by permission from Hougen and Watson, *Chemical Process Principles*, Part II, Wiley, New York.)

Since $h_1{}^* = h_2{}^* = h_1$

$$h_2{}^* - h_2 = h_1 - h_2 = \frac{7.1(549.8)}{30} = 130 \text{ Btu/lbm}$$

$$\nu_2 = f_2/P_2 = 0.56 \qquad f_2 = 1000(0.56) = 560 \text{ lbf/in.}^2$$

$$\nu_1 = f_1/P_1 = 1.0 \qquad f_1 = 14.7(1.0) = 14.7 \text{ lbf/in.}^2$$

Assuming negligible changes in kinetic energy during the compression process we have

$$-w = g_2 - g_1 = RT \ln f_2/f_1 = \frac{1.986}{30} \times 570 \ln \frac{560}{14.7}$$
$$= 137.5 \text{ Btu/lbm}$$

$$-w = g_2 - g_1 = (h_2 - h_1) - T_2(s_2 - s_1) = (h_2 - h_1) - q$$
$$= 137.5 \text{ Btu/lbm} = -130 - T_2(s_2 - s_1) = -130 - q$$
$$q = T_2(s_2 - s_1) = -130 - 137.5 = -268 \text{ Btu/lbm}$$

$$s_2 - s_1 = \frac{-268}{570} = -0.470 \text{ Btu/lbm R}$$

Occasionally one may find it necessary to find the work done during an isentropic process by use of the compressibility charts. The solution to a problem of this type is illustrated in the example that follows.

Example 14.9 Carbon dioxide at 500 lbf/in.^{2}, 100 F is compressed in a reversible adiabatic steady-flow process to 1000 lbf/in.^{2} Determine the work of compression per pound of CO_2 by use of compressibility charts and the zero-pressure specific-heat data.

The approach to this problem is indicated in Fig. 14.9. The actual compression process is represented by line 1–2 on the T-s diagram.

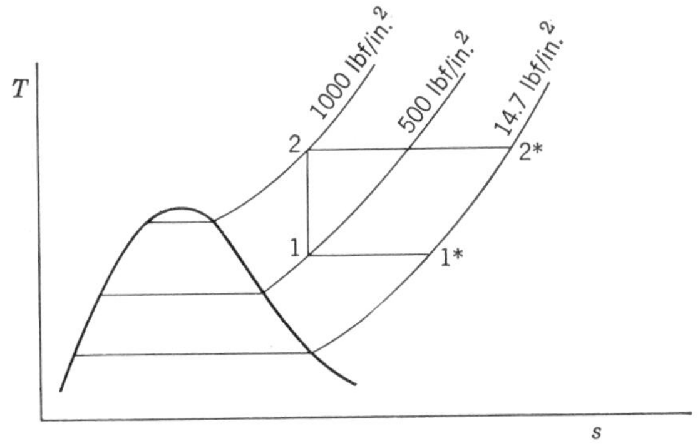

Fig. 14.9 Sketch for Example 14.9.

However, since we have only zero-pressure specific-heat data, it is necessary to follow the path 1–1*–2*–2 in evaluating the change of enthalpy and entropy. The line of one standard atmosphere is selected to evaluate the zero-pressure change in enthalpy and entropy. Actually, any low-pressure line could be selected, but a finite pressure must be used, and the one-atmosphere line is arbitrarily selected.

In essence the problem is a trial-and-error solution and involves selecting a final temperature and then checking to see if the entropy at this temperature and the given final pressure is equal to the initial entropy. When the correct final temperature has been determined the work is found from the steady-flow energy equation, $w = h_1 - h_2$.

For the given problem let us assume that $T_2 = 660$ R.

$$P_{r1} = \tfrac{500}{1072} = 0.467 \qquad T_{r1} = \tfrac{560}{547} = 1.023$$

$$P_{r2} = \tfrac{1000}{1072} = 0.93 \qquad T_{r2} = \tfrac{660}{547} = 1.21$$

From the respective charts

$$\nu_1 = 0.85 \qquad \nu_2 = 0.82$$

$$f_1 = 0.85(500) = 425 \text{ lbf/in.}^2 \qquad f_2 = 0.82(1000) = 820 \text{ lbf/in.}^2$$

$$\frac{\bar{h}_1{}^* - \bar{h}_1}{T_c} = 1.2 \qquad \frac{\bar{h}_2{}^* - \bar{h}_2}{T_c} = 1.7$$

We now proceed to find the net change of entropy, $s_2 - s_1$, for the assumed final temperature. To do so we follow path 1–1*–2*–2 and find the change of entropy for processes 1–1*, 1*–2*, and 2*–2.

$$h_1{}^* - h_1 = \frac{1.2 \times 547}{44} = 14.9 \text{ Btu/lbm}$$

$$h_2{}^* - h_2 = \frac{1.7 \times 547}{44} = 21.1 \text{ Btu/lbm}$$

$$g_1 - g_1{}^* = RT \ln f_1/f_1{}^* = \frac{35.1 \times 560}{778} \ln \frac{425}{14.7} = 85.1 \text{ Btu/lbm}$$

$$g_2 - g_2{}^* = RT \ln f_2/f_2{}^* = \frac{35.1 \times 660}{778} \ln \frac{820}{14.7} = 119.8 \text{ Btu/lbm}$$

$$g_1 - g_1{}^* = (h_1 - h_1{}^*) - T_1(s_1 - s_1{}^*)$$

$$s_1{}^* - s_1 = \frac{85.1 + 14.9}{560} = 0.1786 \text{ Btu/lbm R}$$

$$g_2 - g_2^* = (h_2 - h_2^*) - T_2(s_2 - s_2^*)$$

$$s_2^* - s_2 = \frac{119.8 + 21.1}{660} = 0.2135 \text{ Btu/lbm R}$$

The change in entropy between s_2^* and s_1^* can be found from the *Gas Tables*, Table 17 (or Appendix Table A.9).

$$s_2^* - s_1^* = \frac{\bar{\phi}_2 - \bar{\phi}_1}{M} = \frac{52.934 - 51.408}{44} = 0.0347 \text{ Btu/lbm R}$$

$$s_2 - s_1 = -(s_2^* - s_2) + (s_2^* - s_1^*) + (s_1^* - s_1)$$

$$= -0.2135 + 0.0347 + 0.1786 = 0.0002 \text{ Btu/lbm R}$$

Thus, the entropy change is essentially zero and we can assume that the final temperature is 660 R.

The net work is now found from the first law

$$w = h_1 - h_2$$

$$= -(h_1^* - h_1) - (h_2^* - h_1^*) + (h_2^* - h_2)$$

From the *Gas Tables*, Table 17

$$(h_2^* - h_1^*) = \frac{\bar{h}_2^* - \bar{h}_1^*}{M} = \frac{5165 - 4236}{44} = 21.1 \text{ Btu/lbm}$$

Then

$$w = -14.9 - 21.1 + 21.1 = -14.9 \text{ Btu/lbm}$$

PROBLEMS

14.1 Show that on a Mollier diagram (h-s diagram) the slope of a constant-pressure line increases with temperature in the superheat region and that the constant-pressure line is straight in the two-phase region.

14.2 From the first Maxwell relation derived and the relation

$$\left(\frac{\partial x}{\partial y}\right)_z \left(\frac{\partial y}{\partial z}\right)_x \left(\frac{\partial z}{\partial x}\right)_y = -1$$

derive the other three Maxwell relations.

14.3 The following data are available at the triple point of water:

Pressure:	0.0888 lbf/in.^2	
Temperature:	32.02 F	
Enthalpy:	Liquid:	0.02 Btu/lbm
	Solid:	−143.3 Btu/lbm
Specific Volume:	Liquid:	$0.01603 \text{ ft}^3\text{/lbm}$
	Solid:	$0.01747 \text{ ft}^3\text{/lbm}$

A man weighing 200 lbm is ice skating on "hollow ground" blades which have a total area in contact with the ice of 0.033 in.^2 The temperature of the ice is 28 F. Will the ice melt under the blades?

14.4 At 1 atm pressure helium boils at 4.22 K, and the latent heat of vaporization is 19.7 cal/gm mole. Suppose one wishes to determine how low a temperature could be obtained by producing a vacuum over liquid helium, thus causing it to boil at a lower pressure and temperature. Determine what pressure would be necessary to produce a temperature of 1 K and 0.1 K.

What do you think is the lowest temperature that could be obtained by pumping a vacuum over liquid helium?

14.5 The equation for the vapor-pressure curve used in the Ammonia Tables (a summary of which appears in the Appendix) is as follows:

$$\log_{10} P = A - \frac{B}{T} - C \log_{10} T - DT + ET^2$$

where $A = 25.5743247$
$B = 3.2951254 \times 10^3$
$C = 6.4012471$
$D = 4.148279 \times 10^{-4}$
$E = 1.4759945 \times 10^{-6}$
$P =$ pressure in lbf/in.^2
$T =$ absolute temperature in °R

(*a*) Derive an expression for h_{fg} in terms of the constants in the vapor-pressure curve, the absolute temperature, and the specific volume of the saturated liquid and saturated vapor.

(*b*) Calculate the enthalpy of evaporation at 40 F and compare the value thus calculated with the value given in the Ammonia Tables.

14.6 Derive expressions for $(\partial u/\partial P)_T$ and $(\partial u/\partial v)_T$ that involve only pressure, temperature, and specific volume.

14.7 Show that for Van der Waals' equation the following relations apply.

(a) $(h_2 - h_1)_T = (P_2 v_2 - P_1 v_1) + a\left(\frac{1}{v_1} - \frac{1}{v_2}\right)$

(b) $(s_2 - s_1)_T = R \ln \left(\frac{v_2 - b}{v_1 - b}\right)$

(c) $\left(\frac{\partial C_v}{\partial v}\right)_T = 0$

(d) $C_P - C_v = \frac{R}{1 - 2a(v - b)^2/RTv^3}$

14.8 Derive an expression for the change of enthalpy and entropy during a constant-temperature process for a substance that follows the Berthelot equation of state.

14.9 Show that the Joule-Thomson coefficient μ_J is given by the relation

$$\mu_J = \frac{1}{C_p}\left[T\left(\frac{\partial v}{\partial T}\right)_P - v\right]$$

14.10 (a) Show that the difference between the constant-pressure and constant-volume specific heats is given by the relation

$$C_P - C_v = \frac{Tv\alpha^2}{\beta_T}$$

(b) Show that if the volume expansivity is constant, the following relation is valid:

$$\left(\frac{\partial C_P}{\partial P}\right)_T = -vT\alpha^2$$

14.11 From data given in Table 4 of the steam tables determine the average value of the isothermal compressibility at 32 F, 100 F, and 400 F. Does the isothermal compressibility vary significantly with pressure at a given temperature?

14.12 (a) Show that the velocity of sound is given by the relation

$$c = \sqrt{\frac{g_c}{\beta_s \rho}}$$

where β_s is the adiabatic compressibility.

(b) At a temperature of 100 F the velocity of sound in water is 5040 ft/sec. Verify this by determining the adiabatic compressibility of water at 100 F by using Tables 1 and 4 of the steam tables. (Note that the state of water after an isentropic process can be determined by using both Tables 1 and 4.)

14.13 The following data are available for ethyl alcohol at 20 C.

$$\rho = 49.2 \text{ lbm/ft}^3$$

$$\beta_s = 6.5 \times 10^{-6} \text{ in.}^2\text{/lbf}$$

$$\beta_T = 7.7 \times 10^{-6} \text{ in.}^2\text{/lbf}$$

(*a*) What is the velocity of sound in ethyl alcohol?

(*b*) The pressure on a cylinder having a volume of 1 ft^3 containing ethyl alcohol at 20 C is increased from 1 atm to 200 atm while the temperature remains constant. Determine the work done on the ethyl alcohol during this process.

14.14 Suppose that the following data are available for a given pure substance:

A vapor-pressure curve.
An equation of state of the form $P = f(v, T)$.
Specific volume of saturated liquid.
Critical pressure and temperature.
Constant-volume specific heat of the vapor at one specific volume.

Outline the steps you would follow in order to develop a table of thermodynamic properties comparable to Tables 1, 2, and 3 of the steam tables.

14.15 Propane is compressed in a reversible isothermal steady-flow process from 1 atm pressure, 700 R to 40 atm pressure. Determine the work of compression and heat transfer per lbm of propane.

14.16 Propane is compressed in a reversible adiabatic steady-flow process from 1 atm pressure, 700 R to 40 atm pressure. Determine the work of compression per lbm of propane.

14.17 Freon-12 is compressed in a reversible isothermal steady-flow process from 100 lbf/in.^{2}, 300 F to 600 lbf/in.^{2} Determine the work of compression and the heat transfer by using the generalized charts, and compare these results with those obtained by using the Freon-12 tables in the Appendix.

14.18 Repeat Problem 14.17 but let the Freon-12 be contained in a cylinder fitted with a piston.

14.19 Find the change of enthalpy, internal energy, and entropy as propane undergoes a change of state from 60 F, 14.7 lbf/in.^{2} pressure to 300 F, 800 lbf/in.^{2} pressure. $C_{po} = 0.455$ Btu/lbm R at 1 atm pressure.

14.20 Propane (C_3H_8) flows through a pipe of constant cross-sectional area. At the inlet section of this pipe the velocity is 60 ft/sec, the pressure is 400 lbf/in.^{2}, and the temperature is 200 F. At the exit section the pressure is 100 lbf/in.^{2} There is no heat transfer to or from the gas as it flows in the pipe. Assume C_{po} = 0.4 Btu/lbm R. Determine the temperature and velocity of the propane leaving the pipe.

14.21 Nitrogen at a pressure of 1500 lbf/in.^{2}, -100 F is contained in a tank of 10 ft^3 volume. Heat is transferred to the nitrogen until the temperature is 200 F. Determine the heat transfer and the final pressure of the nitrogen.

14.22 Methane is cooled in a constant-pressure process at 800 lbf/in.^{2} from 200 F to 100 F. The constant-pressure specific heat at zero pressure is

$$\bar{C}_{po} = 4.52 + 0.00737T \text{ Btu/lb-mole R}$$

Calculate the heat transfer per lbm of methane.

14.23 Methane at a pressure of 1000 lbf/in.^{2}, 200 F is throttled in a steady-flow process to 100 lbf/in.^{2} Determine the final temperature of the methane.

14.24 Ethane (C_2H_6) at 800 lbf/in.^{2}, 300 F is cooled in a heat exchanger to 120 F. The volume rate of flow entering the heat exchanger is 100 ft^3/min, and the pressure drop through the heat exchanger is small. Determine the heat transfer per min.

CHAPTER **15**

CHEMICAL REACTIONS; COMBUSTION

Many thermodynamic problems involve chemical reactions. Perhaps the most familiar of these is the combustion process, because most power is generated by utilization of the combustion of hydrocarbon fuels. However, we can all think of a host of other processes involving chemical reactions, some of the most important being those of the human body.

It is our purpose in this chapter to consider a first- and second-law analysis of a system undergoing a chemical reaction. In this sense, it is simply an extension of the first and second laws, which we considered previously. However, a number of new terms are introduced, and it will also be necessary to introduce the third law of thermodynamics.

Because so many thermodynamic problems involve combustion, fuels and combustion are considered. However, the student should focus his attention on the basic principles and definitions, which apply to all chemical reactions, and thereby be prepared to handle not only combustion, but chemical reactions in general.

Chemical equilibrium is considered in Chapter 16. Therefore, the matter of dissociation is not considered in this chapter.

15.1 Fuels

A thermodynamics textbook is not the place for a detailed treatment of fuels. However, some knowledge of them is a prerequisite to consideration of combustion, and this section is therefore devoted to a brief

TABLE 15.1

Analyses of Delivery Samples of Some Representative Coals from Various Regions of the United States

		Proximate Analysis, %			Prox. & Ult. Anal., %	Ultimate Analysis, %					Higher Heating-Value, Btu/lbm
State	Locality	Moisture	Volatile Matter	Fixed Carbon	Ash	Sulfur	Hydrogen	Carbon	Oxygen	Nitrogen	
Ill.	Eldorado	7.9	33.2	50.4	8.5	2.5	5.3	68.2	13.9	1.6	12,190
Ky.	Jenkin	3.1	35.0	58.9	3.0	.6	5.7	79.2	10.0	1.5	14,290
Ohio	Piney Fork	4.9	36.6	51.2	7.3	2.6	5.4	71.9	11.4	1.4	12,990
Okla.	Williams	2.6	16.5	72.2	8.7	1.0	4.3	80.1	4.2	1.7	13,800
Pa.	Nanty-Glo	3.3	20.5	70.0	6.2	1.8	4.9	80.7	5.3	1.1	14,310
Tenn.	Cotula	4.3	37.1	55.1	3.5	1.2	5.6	76.7	11.2	1.8	13,660
Va.	Patterson	3.1	21.8	67.9	7.2	1.0	5.0	80.1	5.2	1.5	14,030
Wash.	Ronald	4.3	37.7	47.1	10.9	0.5	5.9	68.9	12.3	1.5	12,610
W. Va.	Farmington	2.2	35.7	56.2	5.9	0.8	5.2	78.4	8.2	1.5	13,940
Wyo.	Diamondville	5.1	40.5	49.8	4.6	0.5	5.6	73.0	15.1	1.2	12,960

consideration of some important fuels. Most fuels fall into one of three categories, namely, coal, liquid hydrocarbons, and gaseous hydrocarbons.

Coal Coal consists of the remains of vegetation deposists of past geologic ages, after subjection to biochemical actions, pressure, temperature, and submersion. The characteristics of coal vary considerably with location, and even within a given mine there is some variation in composition.

The proximate analysis of coal gives, on a mass basis, the relative amounts of moisture, volatile matter, fixed carbon, and ash. This is the most frequently made analysis of coal, and the laboratory methods used in making it are given in the ASTM Bulletin, *Standards on Coal and Coke*.

The ultimate analysis of coal gives, on a mass basis, the relative amounts of carbon, sulfur, hydrogen, nitrogen, oxygen, and ash. It may be given on an "as received" basis or on a dry basis. In the latter case the ultimate analysis does not include the moisture as determined by the proximate analysis.

Table 15.1 gives the proximate analysis, the ultimate analysis, and the higher heating value of a number of coals. This last term will be defined in a subsequent section.

There are a number of other properties of coal that are important in evaluating a coal for a given use. Some of these properties are the fusibility of the ash, the grindability or ease of pulverization, the weathering or slacking characteristics, and size.

Most liquid and gaseous hydrocarbon fuels are a mixture of many different hydrocarbons. For example, gasoline is primarily a mixture of about forty, but many others are found in very small quantities. In discussing hydrocarbon fuels, therefore, brief consideration should be given to the most important families of hydrocarbons, a summary of which is given in Table 15.2.

TABLE 15.2

Characteristics of Some of the Hydrocarbon Families

Family	Formula	Structure	Saturated
Paraffin	C_nH_{2n+2}	Chain	Yes
Olefin	C_nH_{2n}	Chain	No
Diolefin	C_nH_{2n-2}	Chain	No
Naphthene	C_nH_{2n}	Ring	Yes
Aromatic			
Benzene	C_nH_{2n-6}	Ring	No
Naphthalene	C_nH_{2n-12}	Ring	No

Three terms should be defined. The first pertains to the structure of the molecule. The important types are the ring and chain structures; the difference between the two is illustrated in Fig. 15.1. The

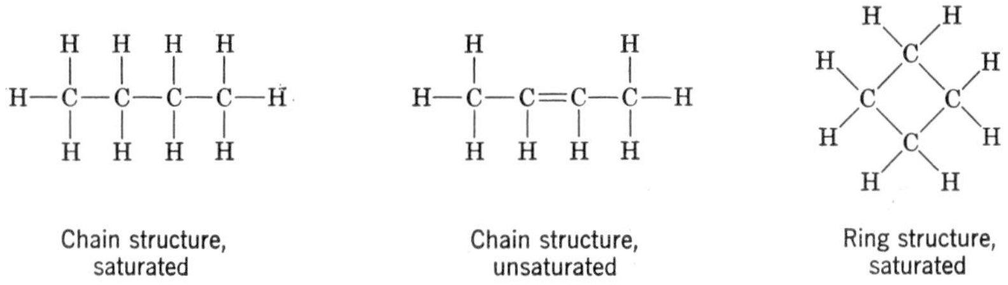

Fig. 15.1 Molecular structure of some hydrocarbon fuels.

same figure illustrates the definition of saturated and unsaturated hydrocarbons. An unsaturated hydrocarbon has two or more adjacent carbon atoms joined by a double or triple bond, whereas in a saturated hydrocarbon all the carbon atoms are joined by a single bond. The third term to be defined is an isomer. Two hydrocarbons that have the same number of carbon and hydrogen atoms but different structures are called isomers. Thus, there are several different octanes (C_8H_{18}), each having 8 carbon atoms and 18 hydrogen atoms, but each has a different structure.

The various hydrocarbon families are identified by a common suffix. The compounds comprising the paraffin family all end in "ane" (as propane and octane). Similarly, the compounds comprising the olefin family end in "ylene" or "ene" (as propene and octene), and the diolefin family ends in "diene" (as butadiene). The naphthene family has the same chemical formula as the olefin family, but has a ring structure rather than chain. The hydrocarbons in the naphthene family are named by adding the prefix "cyclo" (as cyclopentane).

The aromatic family includes the benzene series (C_nH_{2n-6}) and the naphthalene series (C_nH_{2n-12}). The benzene series has a ring structure and is unsaturated.

Alcohols are sometimes used as a fuel in internal-combustion engines. The characteristic feature of the alcohol family is that one of the hydrogen atoms has been replaced by an OH radical. Thus methyl alcohol, also called methanol, is CH_3OH.

Most liquid hydrocarbon fuels are mixtures of hydrocarbons that are derived from crude oil through distillation and cracking processes. Thus, from a given crude oil a variety of different fuels can be produced, some of the common ones being gasoline, kerosene, Diesel fuel, and fuel oil. Within each of these classifications there is a wide variety of grades, and each is made up of a large number of different hydrocarbons. The important distinction between these fuels is the distillation curve; Fig. 15.2 gives typical ones. The distillation curve is obtained by slowly heating a sample of fuel so that it vaporizes. The vapor is then condensed and the amount measured. The more volatile hydrocarbons are vaporized first, and thus the temperature of the nonvaporized fraction increases during the process. The distillation curve, which is a plot of the temperature of the nonvaporized fraction vs. the amount of vapor condensed, is an indication of the volatility of the fuel.

In dealing with combustion of liquid fuels it is convenient to express the composition in terms of a single hydrocarbon, even though it is a mixture of many hydrocarbons. Thus gasoline is usually considered to be octane, C_8H_{18}, and Diesel fuel is considered to be dodecane, $C_{12}H_{26}$. The composition of a hydrocarbon fuel may also be given in terms of percentage of carbon and hydrogen.

The two primary sources of gaseous hydrocarbon fuels are natural gas wells and manufacturing processes. Table 15.3 gives the composi-

TABLE 15.3

Volumetric Analyses of Some Typical Gaseous Fuels

Constituent	Various Natural Gases A	B	C	D	Producer Gas from Bituminous Coal	Carburetted Water Gas	Coke-Oven Gas
Methane	93.9	60.1	67.4	54.3	3.0	10.2	32.1
Ethane	3.6	14.8	16.8	16.3			
Propane	1.2	13.4	15.8	16.2			
Butanes plus *	1.3	4.2		7.4			
Ethene						6.1	3.5
Benzene						2.8	0.5
Hydrogen					14.0	40.5	46.5
Nitrogen		7.5		5.8	50.9	2.9	8.1
Oxygen					0.6	0.5	0.8
Carbon monoxide					27.0	34.0	6.3
Carbon dioxide					4.5	3.0	2.2

* This includes butane and all heavier hydrocarbons.

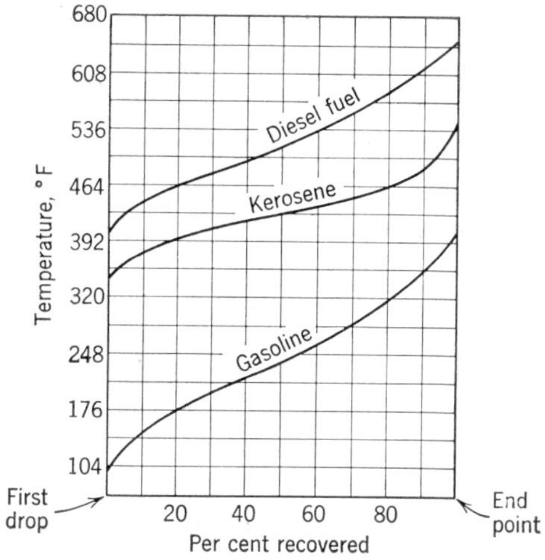

Fig. 15.2 Typical distillation curves of some hydrocarbon fuels.

tion of a number of gaseous fuels. The major constituent of natural gas is methane, which distinguishes it from manufactured gas.

15.2 The Combustion Process

The combustion process essentially involves the oxidation of those constituents in the fuel that are capable of being oxidized, and can therefore be represented by a chemical equation. During a combustion process the mass of each element remains the same. Thus, writing chemical equations and solving problems involving quantities of the various constituents basically involves the conservation of mass of each element.

Consider first the reaction of carbon with oxygen.

$$\begin{array}{ccc} \text{Reactants} & & \text{Products} \\ C + O_2 & \rightarrow & CO_2 \end{array}$$

In words this equation states that one mole of carbon reacts with one mole of oxygen to form one mole of carbon dioxide. This also means that 12 lbm of carbon react with 32 lbm of oxygen to form 44 lbm of carbon dioxide. All the initial substances that undergo the combustion process are called the reactants, and the substances that result from the combustion process are called the products.

When a hydrocarbon fuel is burned both the carbon and the hydrogen are oxidized. Consider the combustion of methane as an example.

$$CH_4 + 2O_2 \rightarrow CO_2 + 2H_2O \tag{15.1}$$

In this case the products of combustion include both carbon dioxide and water. The water may be in the vapor, liquid, or solid phase, depending on the temperature and pressure of the products of combustion.

It should be pointed out that in the combustion process there are many intermediate products formed during the chemical reaction. In this course we are concerned with the initial and final products and not with the intermediate products, but this aspect should not be ignored in a detailed consideration of combustion.

In most combustion processes the oxygen is supplied as air rather than as pure oxygen. The composition of air on a molal basis is approximately 21 per cent oxygen, 78 per cent nitrogen, and 1 per cent argon. The nitrogen and the argon do not undergo chemical reaction (except for dissociation which will be considered in Chapter 16). However, they do leave at the same temperature as the other products, and therefore undergo a change of state if the products are at a temperature other than the original air temperature.

In combustion calculations involving air, the argon is usually neglected, and the air is considered to be composed of 21 per cent oxygen and 79 per cent nitrogen by volume. When this assumption is made the nitrogen is sometimes referred to as "atmospheric nitrogen." Atmospheric nitrogen has a molecular weight of 28.16 (which takes the argon into account) as compared to 28.016 for pure nitrogen. This distinction will not be made in this text, and we will consider the 79 per cent nitrogen to be pure nitrogen.

The assumption that air is 21.0 per cent oxygen and 79.0 per cent nitrogen by volume leads to the conclusion that for each mole of oxygen, $79.0/21.0 = 3.76$ moles of nitrogen are involved. Therefore, when the oxygen for the combustion of methane is supplied as air, the reaction can be written

$$CH_4 + 2O_2 + 2(3.76)N_2 \rightarrow CO_2 + 2H_2O + 7.52N_2 \tag{15.2}$$

The minimum amount of air that supplies sufficient oxygen for the complete combustion of all the carbon, hydrogen, and any other elements in the fuel that may oxidize, is called the "theoretical air." When complete combustion is achieved with theoretical air the products contain no oxygen. In practice it is found that complete combustion is not likely to be achieved unless the amount of air supplied is somewhat greater than the theoretical amount. The amount of air actually supplied is expressed in terms of per cent theoretical air. Thus 150 per cent theoretical air means that the air actually supplied is 1.5 times the theoretical air. The complete combustion of methane with 150 per cent theoretical air is written

$$CH_4 + 2(1.5)O_2 + 2(3.76)(1.5)N_2 \rightarrow CO_2 + 2H_2O + O_2 + 11.28N_2 \quad (15.3)$$

The amount of air actually supplied may also be expressed in terms of per cent excess air. The excess air is the amount of air supplied over and above the theoretical air. Thus 150 per cent theoretical air is equivalent to 50 per cent excess air. The terms theoretical air and excess air are both in current usage.

Two important parameters applied to combustion processes are the air-fuel ratio (abbreviated AF) and its reciprocal, the fuel-air ratio (abbreviated FA). The air-fuel ratio is usually expressed on a mass basis, but a mole basis is also used at times. The theoretical air-fuel ratio is the ratio of the mass (or moles) of theoretical air to the mass (or moles) of fuel.

When the amount of air supplied is less than the theoretical air required, the combustion is incomplete. If there is only a slight deficiency of air, the usual result is that some of the carbon unites with the oxygen to form carbon monoxide (CO) instead of carbon dioxide (CO_2). If the air supplied is considerably less than the theoretical air, there may also be some hydrocarbons in the products of combustion.

Even when some excess air is supplied there may be small amounts of carbon monoxide present, the exact amount depending on a number of factors including the mixing and turbulence during combustion. Thus, the combustion of methane with 110 per cent theoretical air might be as follows:

$$CH_4 + 2(1.1)O_2 + 2(1.1)3.76N_2 \rightarrow$$
$$0.95CO_2 + 0.05CO + 2H_2O + 0.225O_2 + 8.27N_2 \quad (15.4)$$

The material covered so far in this section is illustrated by the following examples.

Example 15.1 Calculate the theoretical air-fuel ratio for the combustion of octane, C_8H_{18}.

The combustion equation is

$$C_8H_{18} + 12.5O_2 + 12.5(3.76)N_2 \rightarrow 8CO_2 + 9H_2O + 47.0N_2$$

The air-fuel ratio on a mole basis is

$$AF = \frac{12.5 + 47.0}{1} = 59.5 \text{ moles air/mole fuel}$$

The theoretical air-fuel ratio on a mass basis is found by introducing the molecular weight of the air and fuel.

$$AF = \frac{59.5(28.96)}{114} = 15.0 \text{ lbm air/lbm fuel}$$

Example 15.2 Calculate the molal analysis of the products of combustion when octane, C_8H_{18}, is burned with 200 per cent theoretical air, and determine the dew point of the products if the pressure is 14.7 lbf/in.^{2}

The equation for the combustion of octane with 200 per cent theoretical air is

$$C_8H_{18} + 12.5(2)O_2 + 12.5(2)(3.76)N_2 \rightarrow$$
$$8CO_2 + 9H_2O + 12.5O_2 + 94.0N_2$$

$$\text{Total moles of product} = 8 + 9 + 12.5 + 94.0 = 123.5$$

Molal analysis of products:

$$\begin{aligned} CO_2 &= 8/123.5 &&= 6.47\% \\ H_2O &= 9/123.5 &&= 7.29 \\ O_2 &= 12.5/123.5 &&= 10.12 \\ N_2 &= 94/123.5 &&= \underline{76.12} \\ & && 100.00 \end{aligned}$$

The partial pressure of the H_2O is 14.7(0.0729) = 1.072 lbf/in.^{2}

The saturation temperature corresponding to this pressure is 104 F, which is also the dew-point temperature.

The water condensed from the products of combustion usually contains some dissolved gases and therefore may be quite corrosive. For this reason the products of combustion are often kept above the dew point until discharged to the atmosphere.

Example 15.3 Producer gas from bituminous coal (see Table 15.3) is burned with 20 per cent excess air. Calculate the air-fuel ratio on a volumetric basis and on a mass basis.

To calculate the theoretical air requirement, let us write the combustion equation for the combustible substances in 1 mole of fuel.

$$
\begin{aligned}
&0.14H_2 \quad + 0.070O_2 \rightarrow 0.14H_2O \\
&0.27CO \quad + 0.135O_2 \rightarrow 0.27CO_2 \\
&0.03CH_4 + \underline{0.060_2} \rightarrow 0.03CO_2 + 0.06H_2O \\
&\qquad\qquad 0.265 = \text{moles oxygen required/mole fuel} \\
&\qquad\quad - \underline{0.006} = \text{oxygen in fuel/mole fuel} \\
&\qquad\qquad 0.259 = \text{moles oxygen required from air/mole fuel}
\end{aligned}
$$

Therefore, the complete combustion equation for 1 mole of fuel is

$$
\overbrace{0.14H_2 + 0.27CO + 0.03CH_4 + 0.006O_2 + 0.509N_2 + 0.045CO_2}^{\text{fuel}} +
$$

$$
\overbrace{0.259O_2 + 0.259(3.76)N_2}^{\text{air}} \rightarrow 0.20H_2O + 0.345CO_2 + 1.482N_2
$$

$$
\left(\frac{\text{moles air}}{\text{mole fuel}}\right)_{\text{theo}} = 0.259 \times \frac{1}{0.21} = 1.233 \frac{\text{moles air}}{\text{mole fuel}} = \frac{\text{volume air}}{\text{volume fuel}}
$$

For 20 per cent excess air, $\dfrac{\text{moles air}}{\text{mole fuel}} = 1.233 \times 1.200 = 1.48 \dfrac{\text{moles air}}{\text{mole fuel}}$

On a mass basis the air-fuel ratio is

$$
AF = \frac{1.48(28.96)}{0.14(2)+0.27(28)+0.03(16)+0.006(32)+0.509(28)+0.045(44)}
$$

$$
= \frac{1.48(28.96)}{24.74} = 1.73 \text{ lbm air/lbm fuel}
$$

Example 15.4 Coal from Jenkin, Kentucky (see Table 15.1) is to be burned with 30 per cent excess air. Calculate the air-fuel ratio on a mass basis.

One approach to this problem is to write the combustion equation for each of the combustible elements per 100 lbm of fuel. The molal composition per 100 lbm of fuel is found first.

$$
\text{moles S/100 lbm fuel} = \frac{0.6}{32} = 0.02
$$

$$
\text{moles } H_2\text{/100 lbm fuel} = \frac{5.7}{2} = 2.85
$$

$$
\text{moles C/100 lbm fuel} = \frac{79.2}{12} = 6.60
$$

$$\text{moles } O_2/100 \text{ lbm fuel} = \frac{10}{32} = 0.31$$

$$\text{moles } N_2/100 \text{ lbm fuel} = \frac{1.5}{28} = 0.05$$

The combustion equations for the combustible elements are now written, which enables us to find the theoretical oxygen required.

$$\begin{array}{lll} 0.02S & + 0.02O_2 & \rightarrow 0.02SO_2 \\ 2.85H_2 & + 1.42O_2 & \rightarrow 2.85H_2O \\ 6.60C & + \underline{6.60O_2} & \rightarrow 6.60CO_2 \\ & 8.04 \text{ moles } O_2 \text{ required}/100 \text{ lbm fuel} & \\ & -\underline{0.31} \text{ mole } O_2 \text{ in fuel}/100 \text{ lbm fuel} & \\ & 7.73 \text{ moles } O_2 \text{ from air}/100 \text{ lbm fuel} & \end{array}$$

$$AF_{\text{theo}} = \frac{[7.73 + 7.73(3.76)]28.96}{100} = 10.63 \text{ lbm air/lbm fuel}$$

For 30 per cent excess air the air-fuel ratio is

$$AF = 1.3 \times 10.63 = 13.82 \text{ lbm air/lbm fuel}$$

15.3 Analysis of the Products of Combustion

An analysis of the products of combustion affords a very simple method for calculating the actual amount of air supplied in a combustion process. In some cases the analysis of the fuel may also be calculated from an analysis of the products. This analysis is usually made with an Orsat apparatus, a schematic sketch of which is shown in Fig. 15.3.

In this apparatus a measured volume of combustion gases is successively passed through a number of chemical solutions, each of which absorbs one of the components of the products of combustion. The volume is measured before and after the gas is passed through each solution, and the decrease in volume is the volume of that component absorbed by the given solution. From these data a volumetric analysis of the combustion gases can be determined. The gas is moved into and out of the apparatus by lowering or raising the leveling bottle, which contains water (occasionally brine or mercury are also used).

In detail the operation is as follows. The measuring burette is filled with the combustion products and the volume is accurately measured. The gas is then passed into pipette A, containing potassium hydroxide, which absorbs carbon dioxide. The remaining gas is brought back into the measuring burette and the volume measured. The difference between the original volume and this volume is the volume of carbon dioxide. In a similar manner the gas is successively passed into pipettes *B* and *C*. Pipette *B* contains a solution of pyrogallic acid which absorbs oxygen, and pipette *C* contains cuprous chloride, which absorbs carbon monoxide. When it is necessary to analyze products

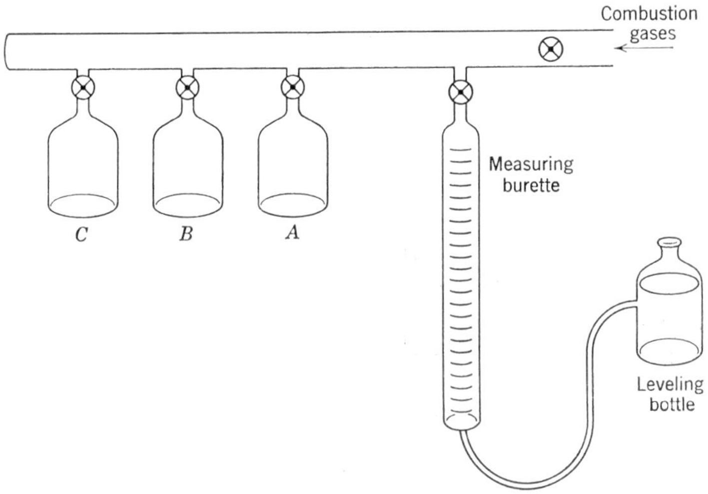

Fig. 15.3 Schematic arrangement of Orsat apparatus.

of combustion containing other constituents, the gas can be passed through additional solutions which absorb these constituents. The gas that is not absorbed is assumed to be nitrogen.

The volumetric analysis obtained when using the Orsat apparatus is on a dry basis; that is, the water formed during combustion does not appear in the analysis. This is true because the analysis is made at room temperature, which is considerably below the dew point of the products of combustion of most hydrocarbon fuels, because the gas remains saturated during the analysis, and because the analysis takes place at constant pressure and temperature.

In using the analysis of the products of combustion to obtain the actual fuel-air ratio, the basic principle is the conservation of the mass of each of the elements. Thus, in changing from reactants to products we can make a carbon balance, hydrogen balance, oxygen balance, and nitrogen balance (plus any other elements that may be involved). Further, we recognize that there is a definite ratio between the amounts of some of these elements. Thus, the ratio between the nitrogen and oxygen supplied in the air is fixed, as well as the ratio between carbon and hydrogen if the composition of a hydrocarbon fuel is known.

These principles involving the analysis of the products of combustion are illustrated in the following examples.

Example 15.5 Methane (CH_4) is burned with atmospheric air. The analysis of the products as determined by an Orsat apparatus is as follows:

CO_2	10.00%
O_2	2.37
CO	0.53
N_2	87.10
	100.00%

Calculate the air-fuel ratio, the per cent theoretical air, and determine the combustion equation.

Two solutions are given to this problem. The first is somewhat longer, but illustrates the basic principle quite clearly. The second uses the same principle, but is somewhat streamlined.

By using a nitrogen balance and the known composition of air, the moles of air per mole dry product can be found as follows: (In this example all the nitrogen in the products comes from the air.)

$$\text{moles } N_2/\text{mole dry product (from Orsat analysis)} = 0.8710$$

$$\text{moles } N_2/\text{mole air (from known composition of air)} = 0.790$$

Therefore

$$\frac{\text{moles air}}{\text{mole dry product}} = \frac{\text{moles } N_2/\text{mole dry product}}{\text{moles } N_2/\text{mole air}} = \frac{0.8710}{0.790} = 1.102 \qquad \text{(a)}$$

By using a carbon balance and the composition of the fuel, the moles of fuel per mole of dry product can be found.

$$\text{moles C/mole dry product (from Orsat analysis)} = 0.1053$$

$$\text{moles C/mole fuel (from composition of the fuel)} = 1$$

Therefore

$$\frac{\text{moles fuel}}{\text{mole dry product}} = \frac{\text{moles C/mole dry product}}{\text{moles C/mole fuel}} = \frac{0.1053}{1} = 0.1053 \qquad \text{(b)}$$

The air-fuel ratio on a mole basis can be found by dividing Eq. a by Eq. b

$$\frac{\text{moles air}}{\text{mole fuel}} = \frac{\text{moles air/mole dry product}}{\text{moles fuel/mole dry product}} = \frac{1.102}{0.1053} = 10.47$$

The air-fuel ratio on a mass basis is found by introducing the molecular weight.

$$AF = \frac{10.47 \times 28.95}{16.0} = 18.97 \text{ lbm air/lbm fuel}$$

The theoretical air-fuel ratio is found by writing the combustion equation for theoretical air.

$$CH_4 + 2O_2 + 2(3.76)N_2 \rightarrow CO_2 + 2H_2O + 7.52N_2$$

$$AF_{theo} = \frac{(2 + 7.52)28.95}{16.0} = 17.23 \text{ lbm air/lbm fuel}$$

$$\text{The per cent theoretical air is } \frac{18.97}{17.23} = 110\%$$

The actual combustion equation can be found from Eq. b, the Orsat analysis, and a hydrogen balance. From Eq. b

$$\text{moles dry product/mole fuel} = \frac{1}{0.1053} = 9.50$$

Then

$$\frac{\text{moles } CO_2}{\text{mole fuel}} = \frac{\text{moles dry product}}{\text{mole fuel}} \times \frac{\text{moles } CO_2}{\text{mole dry product}} =$$

$$9.50 \times 0.100 = 0.950$$

Similarly

$$\frac{\text{moles CO}}{\text{mole fuel}} = 9.50 \times 0.0053 = 0.050$$

$$\frac{\text{mole } O_2}{\text{mole fuel}} = 9.50 \times 0.0237 = 0.225$$

$$\frac{\text{mole } N_2}{\text{mole fuel}} = 9.50 \times 0.8710 = 8.27$$

The combustion equation can now be written, the water in the products being determined from a hydrogen balance

$$CH_4 + 2(1.1)O_2 + 2(1.1)(3.76)N_2 \rightarrow$$

$$0.95CO_2 + 0.05CO + 2H_2O + 0.225O_2 + 8.27N_2$$

It will be observed that this is the same equation as Eq. 15.4, which is the combustion equation used to develop this problem.

The second method of solution involves writing the combustion equation for 100 moles of dry products, and introducing letter coefficients for the unknown quantities, and then solving for them.

From the Orsat analysis the following equation can be written, keeping in mind that the Orsat analysis is on a dry basis.

$$a\,CH_4 + b\,O_2 + c\,N_2 \rightarrow$$
$$10.0CO_2 + 0.53CO + 2.37O_2 + d\,H_2O + 87.1N_2$$

A balance for each of the elements involved will enable us to solve for all the unknown coefficients:

Nitrogen balance: $c = 87.1$

Since all the nitrogen comes from the air,

$$\frac{c}{b} = 3.76 \qquad b = \frac{87.1}{3.76} = 23.16$$

Carbon balance: $a = 10.00 + 0.53 = 10.53$
Hydrogen balance: $d = 2a = 21.06$
Oxygen balance: All the unknown coefficients have been solved for, and in this case the oxygen balance provides a check on the accuracy. Thus, b can also be determined by an oxygen balance.

$$b = 10.00 + \frac{0.53}{2} + 2.37 + \frac{21.06}{2} = 23.16$$

The balance of the solution is the same as has already been given.

Example 15.6 The products of combustion of a hydrocarbon fuel of unknown composition have the following composition as measured by an Orsat apparatus:

CO_2	8.0%
CO	0.9
O_2	8.8
N_2	82.3
	100.0%

Calculate

(*a*) The air-fuel ratio.
(*b*) The composition of the fuel on a mass basis.
(*c*) The per cent theoretical air on a mass basis.

Let us first write the combustion equation for 100 moles of dry product.

$$C_aH_b + d\,O_2 + c\,N_2 \rightarrow$$
$$8.0CO_2 + 0.9CO + 8.8O_2 + e\,H_2O + 82.3N_2$$

Nitrogen balance: $c = 82.3$

From composition of air, $\frac{c}{d} = 3.76$; $d = \frac{82.3}{3.76} = 21.9$

Oxygen balance: $21.9 = 8.0 + \frac{0.9}{2} + 8.8 + \frac{e}{2}$; $e = 9.3$

Carbon balance: $a = 8.0 + 0.9 = 8.9$

Hydrogen balance: $b = 2e = 2 \times 9.3 = 18.6$

The air-fuel ratio can now be found, using the molecular weights.

$$\text{AF} = \frac{(21.9 + 82.3)28.95}{8.9(12) + 18.6(1)} = 24.1 \text{ lbm air/lbm fuel}$$

On a mass basis the composition of fuel is:

$$\text{Carbon} = \frac{8.9(12)}{8.9(12) + 18.6(1)} = 85.2\%$$

$$\text{Hydrogen} = \frac{18.6(1)}{8.9(12) + 18.6(1)} = 14.8\%$$

The theoretical air requirement can be found by writing the theoretical combustion equation.

$$C_{8.9}H_{18.6} + 13.5O_2 + 13.5(3.76)N_2 \rightarrow 8.9CO_2 + 9.3H_2O + 50.8N_2$$

$$\text{AF}_{\text{theo}} = \frac{(13.5 + 50.8)28.95}{8.9(12) + 18.6(1)} = 14.9 \text{ lbm air/lbm fuel}$$

$$\% \text{ theoretical air} = \frac{24.1}{14.9} = 162\%$$

Example 15.7 Producer gas from bituminous coal (Table 15.3) is burned with air. The composition of the products of combustion are analyzed with an Orsat apparatus and are as follows:

CO_2	11.9%
CO	1.8
O_2	6.5
N_2	79.8

Calculate the actual air-fuel ratio on a mole basis and the per cent theoretical air.

The combustion equation for 100 moles of dry product is first written, using letter coefficients for the unknown constituents.

$$a\,(0.14H_2 + 0.27CO + 0.03CH_4 + 0.006O_2 + 0.509N_2 + 0.045CO_2) + b\,O_2 + 3.76(b)N_2 \rightarrow 11.9CO_2 + 1.8CO + 6.5O_2 + c\,H_2O + 79.8N_2$$

Carbon balance: $a(0.27 + 0.03 + 0.045) = 11.9 + 1.8$

$$a = \frac{13.7}{0.345} = 39.7$$

Nitrogen balance: $0.509a + 3.76b = 79.8$

$$0.509(39.7) + 3.76b = 79.8$$
$$b = 15.85$$

Hydrogen balance: $a(0.14 + 0.06) = c$

$$c = 39.7(0.2) = 7.94$$

The oxygen balance provides a check on the accuracy of our work.

$$a(0.135 + 0.006 + 0.045) + b = 11.9 + \frac{1.8}{2} + 6.5 + \frac{c}{2}$$

$$39.7(0.186) + 15.85 = 11.9 + 0.9 + 6.5 + 3.97$$
$$23.24 = 23.27$$

$$AF_{\text{actual}} = \frac{b + 3.76(b)}{a} = \frac{15.85 + 3.76(15.85)}{39.7}$$
$$= 1.903 \text{ moles air/mole fuel}$$

$AF_{\text{theoretical}}$ (From Example 15.3) $= 1.233$ moles air/mole fuel

$$\text{Per cent theoretical air} = \frac{1.903}{1.233} = 155\%$$

Example 15.8 Coal from Jenkin, Kentucky is burned in a furnace. It is determined that the refuse removed from the furnace contains 33 per cent carbon by mass. The Orsat analysis of the products of combustion is as follows:

CO_2	12.8%
CO	1.1
O_2	5.7
N_2	80.4

Calculate the actual air-fuel ratio and the per cent excess air.

The refuse contains the 3.0 per cent ash this coal contains, plus the unburned carbon.

$$\frac{\text{unburned carbon}}{\text{ash} + \text{unburned carbon}} = 0.33 \qquad \text{Ash} = 0.03$$

Therefore, the unburned carbon = 0.015
The carbon burned per lbm coal = 0.792 − 0.015 = 0.777
Let us now make a balance of the carbon burned.

$$\text{moles carbon burned/100 lbm fuel} = \frac{77.7}{12} = 6.48$$

$$\text{moles carbon burned/mole dry product} = 0.128 + 0.011 = 0.139$$

Therefore,

$$\frac{\text{moles dry product}}{\text{100 lbm fuel}} = \frac{\text{moles carbon burned/100 lbm fuel}}{\text{moles carbon burned/mole dry product}}$$

$$= \frac{6.48}{0.139} = 46.6$$

The nitrogen in the products is essentially all from the air and very little error would result from ignoring the small amount of nitrogen in the coal. However, to make this example complete, let us consider the nitrogen in the coal.

$$\frac{\text{moles } N_2 \text{ in fuel}}{\text{100 lbm fuel}} = 0.05 \text{ (see Example 15.4)}$$

$$\frac{\text{moles } N_2 \text{ in products}}{\text{100 lbm fuel}} = 46.6\,(0.804) = 37.47$$

$$\frac{\text{moles } N_2 \text{ from air}}{\text{100 lbm fuel}} = 37.47 - 0.05 = 37.42$$

$$\frac{\text{moles } O_2 \text{ from air}}{\text{100 lbm fuel}} = \frac{37.42}{3.76} = 9.95$$

$$\frac{\text{moles air}}{\text{100 lbm fuel}} = 9.95 + 37.42 = 47.37$$

The air-fuel ratio is found by introducing the molecular weight of air.

$$AF_{\text{actual}} = \frac{47.37 \times 28.95}{100} = 13.7 \text{ lbm air/lbm fuel}$$

In Example 15.4 it was found that the theoretical air-fuel ratio was 10.63 lbm air/lbm fuel.

Therefore,

$$\% \text{ excess air} = \frac{13.7 - 10.63}{10.63} = \frac{3.07}{10.63} = 28.9\%$$

15.4 Enthalpy of Formation

In the first fourteen chapters of this book, the problems considered always involved a fixed composition, never a change of composition due to chemical reaction. Therefore, in dealing with a thermodynamic property we made use of tables of thermodynamic properties for the given substance, and in each of these tables the thermodynamic properties were given relative to some arbitrarily assumed base. In the steam tables, for example, the enthalpy of saturated liquid at 32 F is assumed to be zero. This procedure is quite adequate when no change in composition is involved, because we are concerned with the changes in the properties of a given substance. However, it is quite evident that this procedure is quite inadequate when dealing with a chemical reaction, because the composition changes during the process.

The technique used for chemical reactions is to assume that the enthalpy of all the elements is zero at the reference state of 25 C (77 F) and 1 atm pressure. The *enthalpy of formation* of a compound is its enthalpy at this same temperature and pressure (25 C, 1 atm pressure) with reference to this base in which the enthalpy of the elements is assumed to be zero.

Consider carbon dioxide as an example. Let 1 mole of carbon at 25 C and 1 atm pressure react with 1 mole of oxygen at 25 C and 1 atm pressure in a steady-flow process, to form 1 mole of CO_2. Let sufficient heat be transferred so that the CO_2 formed exists at 25 C and let the pressure of the CO_2 formed be 1 atm. If heat transfer is carefully determined it is found to be $-169{,}293$ Btu. This reaction can be written

$$C + O_2 \rightarrow CO_2$$

Applying the first law to this process we have

$$H_R + Q = H_P$$

where the subscripts R and P refer to the reactants and products respectively. However, since the reactants consist of carbon and oxygen, both of which are elements and both are at 25 C and 1 atm pressure, the enthalpy of the reactants is zero. Therefore,

$$H_P = Q = -169{,}293 \text{ Btu/mole fuel}$$

The enthalpy of CO_2 at 25 C and 1 atm pressure (with reference to this base at which the enthalpy of elements is assumed to be zero) is called the enthalpy of formation. We will designate this with the symbol $h°$. Thus,

$$\bar{h}^{\circ}_{CO_2} = -169{,}293 \text{ Btu/mole fuel}$$

Frequently students are bothered by the minus sign when the enthalpy of formation is negative. The significance of this is simply that the enthalpy of the CO_2 is less than the enthalpy of the carbon and the oxygen. This is quite evident because the heat transfer was negative during the steady-flow chemical reaction, and the energy of the CO_2 must be less than the sum of energy of the carbon and oxygen initially. This is quite analogous to the situation we would have in the steam tables if we let the enthalpy of saturated vapor be zero at 1 atm pressure, for in this case the enthalpy of the liquid would be negative, and we would simply use the negative value for the enthalpy of the liquid when solving problems.

Table 15.4 gives values of the enthalpy of formation for a number of substances. The significance of using the enthalpy of formation is

TABLE 15.4

Enthalpy and Gibbs Function of Formation and Absolute Entropy of Various Substances at 77 F (25 C) and 1 Atmosphere Pressure

Substance	Formula	State	$\bar{h}_f°$, Btu/lb mole	$\bar{g}_f°$, Btu/lb mole	$\bar{s}_f°$, Btu/lb mole R
Carbon Monoxide	CO	gas	−47,548	−59,054	47.300
Carbon Dioxide	CO_2	gas	−169,293	−169,667	51.061
Water	H_2O	gas	−104,071	−98,344	45.106
Water	H_2O	liquid	−122,971	−102,042	16.716
Methane	CH_4	gas	−32,200	−21,852	44.50
Ethene	C_2H_4	gas	22,493	29,308	52.45
Ethane	C_2H_6	gas	−36,425	−14,148	54.85
Propane	C_3H_8	gas	−44,676	−10,105	64.51
Butane	C_4H_{10}	gas	−54,270	−7,380	74.12
Octane	C_8H_{18}	liquid	−107,530	2,840	86.23
Oxygen	O_2	gas	0	0	49.003
Nitrogen	N_2	gas	0	0	45.767
Carbon (graphite)	C	solid	0	0	1.361

that the enthalpies of different substances can be added and subtracted from one another. This is best demonstrated by an example.

Example 15.9 Consider the following reaction, which occurs in a steady-flow process.

$$CH_4 + 2O_2 \rightarrow CO_2 + 2H_2O\ (l)$$

The reactants and products are each at a total pressure of 1 atm and 25 C. Determine the heat transfer.

Applying the first law,

$$H_R + Q = H_P$$

From Table 15.4

$$H_R = \bar{h}^\circ_{f\,CH_4} = -32{,}200 \text{ Btu/mole fuel}$$

$$H_P = \bar{h}^\circ_{f\,CO_2} + 2\bar{h}^\circ_{f\,H_2O(l)} = -169{,}293 + 2(-122{,}971)$$

$$= -415{,}235 \text{ Btu/mole fuel}$$

$$Q = -415{,}235 - (-32{,}200) = -383{,}035 \text{ Btu/mole fuel}$$

15.5 Enthalpy and Internal Energy of Combustion

The concept of enthalpy of combustion can be understood with the aid of Fig. 15.4, which shows a steady-flow calorimeter in schematic

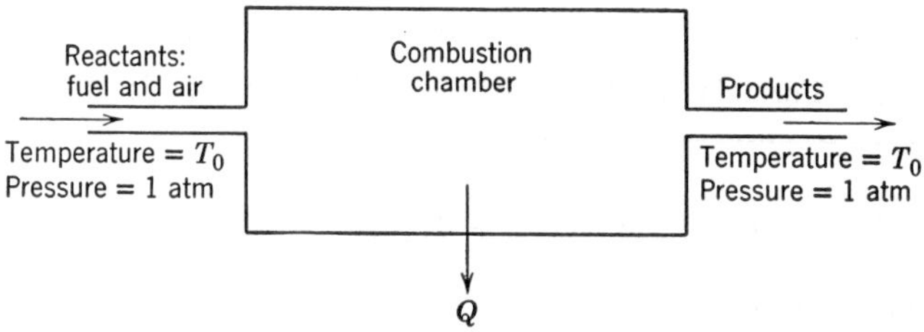

Fig. 15.4 Schematic arrangement of a steady-flow calorimeter.

form. An air-fuel mixture at temperature T_o enters the calorimeter, where combustion takes place. Heat is transferred from the products of combustion at such a rate that they leave at temperature T_o. Thus, the products leave at the same temperature as the reactants enter and the entire process is a steady-flow process. If the changes in kinetic and potential energies are negligible, as is usually the case, the steady-flow energy equation is

$$h_{Ro}' + q_o = h_{Po}' \qquad (15.5)$$

where h_{Ro}' = enthalpy of the reactants per unit mass of fuel at temperature T_o, $h_{Po}' =$ enthalpy of the products per unit mass of fuel at temperature T_o, and $q_o =$ heat transfer per unit mass of fuel.

In this chapter the lower-case letter with a prime will designate the value of the property per unit mass of fuel.

Equation 15.5 can be written

$$q_o = h_{Po}' - h_{Ro}' = h_{RPo} \tag{15.6}$$

The quantity h_{RPo} is called the *enthalpy of combustion* and is the change in enthalpy during the combustion process. We note that it is equal to the heat transfer per unit mass of fuel. In this text the enthalpy of combustion will usually be given in Btu per pound of fuel or Btu per mole of fuel. The enthalpy of combustion is negative since the heat transfer is always negative when this experiment is performed.

The term heating value is defined (in this same experiment) as the heat transferred *from* the products per pound of fuel. Thus the heating value and enthalpy of combustion have the same magnitude but opposite sign; heating value is always positive. Since the heat transfer in a constant-pressure process is equal to the change of enthalpy (see Section 5.10), the heating value associated with the enthalpy of combustion is usually called the constant-pressure heating value.

The term heat of combustion is also in use. This term is essentially equivalent to the term heating value. A general term which applies to all chemical reactions is heat of reaction, and this is defined in the same way as heat of combustion.

A similar experiment might have been carried out in a constant-volume process, as shown in Fig. 15.5. The constant-volume process

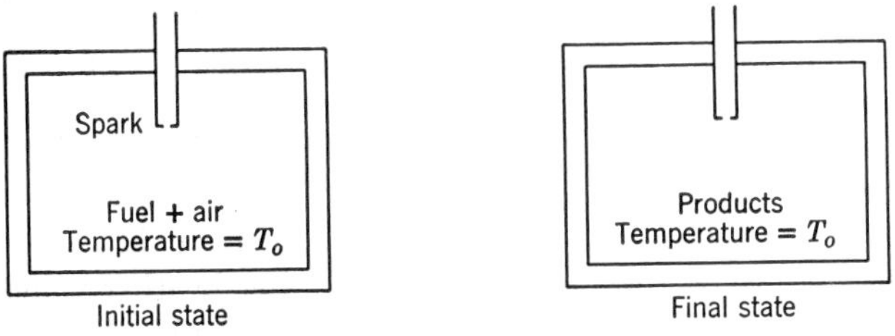

Fig. 15.5 Schematic arrangement of a constant-volume calorimeter.

is used in particular for solid fuels. The fuel and air are initially at temperature T_o. Ignition is achieved with a spark or hot wire, and the products of combustion are cooled to temperature T_o. This is a closed system, and the first law for this process is

$$q_o = u_{Po}' - u_{Ro}' = u_{RPo} \tag{15.7}$$

where $u_{Ro}' =$ internal energy of the reactants per unit mass of fuel at temperature T_o, $u_{Po}{}'$ = internal energy of the products per unit mass of fuel at temperature T_o, and q_o = heat transfer per unit mass of fuel.

The quantity u_{RPo} is the internal energy of combustion and is equal to the heat transfer per unit mass of fuel. The constant-volume heating value is defined as the heat transfer *from* the products in this process. Thus, the internal energy of combustion and the constant-volume heating value have the same magnitude, but opposite sign.

The difference between the enthalpy and internal energy of combustion (i.e., the difference between the constant-pressure and constant-volume heating value) can be found by introducing the definition of enthalpy.

$$h = u + Pv$$

$$h_{Po}{}' - h_{Ro}{}' = (u_{Po}{}' - u_{Ro}{}') + (Pv_{Po}{}' - Pv_{Ro}{}') \qquad (15.8)$$

where v' = volume per unit mass of fuel

Thus, the difference between the enthalpy and internal energy of combustion depends on the difference between the Pv of the products and the reactants. Since the Pv of liquids and solids is usually much less than the Pv of the gases, let us consider only the constituents that are gases, and assume that these are ideal gases. Then

$$Pv_{Po}{}' = n_P{}'\bar{R}T_o \qquad Pv_{Ro}{}' = n_R{}'\bar{R}T_o$$

where n' = number of moles of gaseous constituents per unit mass of fuel. Then

$$h_{Po}{}' - h_{Ro}{}' = (u_{Po}{}' - u_{Ro}{}') + \bar{R}T_o(n_P{}' - n_R{}') \qquad (15.9)$$

Therefore, the difference between the enthalpy and internal energy of combustion depends on the change in the number of moles of gaseous constituents during the combustion process.

Another important factor must be pointed out in regard to the enthalpy and internal energy of combustion. This concerns the phase of the water formed when the hydrogen in the fuel is oxidized, and leads to the distinction between the higher and lower heating value. If all the water formed during combustion is in the liquid phase when the products are at temperature T_o, the heating value measured is the higher heating value. However, if this water is in the vapor phase, the heat transferred from the products will be less, and the heating value measured is the lower heating value. Thus, when stating or specifying heating value—or to use the preferred term, enthalpy or internal energy of combustion—it is necessary to state clearly whether the water formed during combustion is in the liquid or vapor phase. This may be done by stating the phase of the water [when writing

TABLE 15.5

Enthalpy of Combustion of Some Hydrocarbons at 25 C (77 F)

Hydrocarbon	Formula	Liquid H_2O in Products (Negative of Higher Heating Value)		Vapor H_2O in Products (Negative of Lower Heating Value)	
		Liquid Hydrocarbon, Btu/lbm fuel	Gaseous Hydrocarbon, Btu/lbm fuel	Liquid Hydrocarbon, Btu/lbm fuel	Gaseous Hydrocarbon, Btu/lbm fuel
Paraffin Family:					
Methane	CH_4		−23,861		−21,502
Ethane	C_2H_6		−22,304		−20,416
Propane	C_3H_8	−21,490	−21,649	−19,773	−19,929
Butane	C_4H_{10}	−21,134	−21,293	−19,506	−19,665
Pentane	C_5H_{12}	−20,914	−21,072	−19,340	−19,499
Hexane	C_6H_{14}	−20,772	−20,930	−19,233	−19,391
Heptane	C_7H_{16}	−20,668	−20,825	−19,157	−19,314
Octane	C_8H_{18}	−20,591	−20,747	−19,100	−19,256
Decane	$C_{10}H_{22}$	−20,484	−20,638	−19,020	−19,175
Dodecane	$C_{12}H_{26}$	−20,410	−20,564	−18,964	−19,118
Olefin Family:					
Ethene	C_2H_4		−21,626		−20,276
Propene	C_3H_6		−21,033		−19,683
Butene	C_4H_8		−20,833		−19,483
Pentene	C_5H_{10}		−20,696		−19,346
Hexene	C_6H_{12}		−20,612		−19,262
Heptene	C_7H_{14}		−20,552		−19,202
Octene	C_8H_{16}		−20,507		−19,157
Nonene	C_9H_{18}		−20,472		−19,122
Decene	$C_{10}H_{20}$		−20,444		−19,094
Alkylbenzene Family:					
Benzene	C_6H_6	−17,985	−18,172	−17,259	−17,446
Methylbenzene	C_7H_8	−18,247	−18,423	−17,424	−17,601
Ethylbenzene	C_8H_{10}	−18,488	−18,659	−17,596	−17,767
Propylbenzene	C_9H_{12}	−18,667	−18,832	−17,722	−17,887
Butylbenzene	$C_{10}H_{14}$	−18,809	−18,970	−17,823	−17,984

chemical equations this is usually done by using a lower-case g for gas or vapor and l for liquid, as H_2O (g) or H_2O(l)] or by implication in the use of the term lower or higher heating value.

Before illustrating this with some examples, two other points should be made. If the phase of the fuel entering the calorimeter (Fig. 15.4) may vary (gasoline, for example, might be either liquid or vapor), the enthalpy and internal energy of combustion will depend on the phase of the fuel. Therefore it is important to designate the phase of the fuel in stating a heating value or enthalpy or internal energy of combustion.

The enthalpy and internal energy of combustion also depend on the temperature at which the measurement is made, although the variation with temperature is often not large, as is demonstrated by an example below. To insure accuracy however, the temperature at which the enthalpy and internal energy of combustion have been measured should be specified.

Table 15.5 gives the enthalpy of combustion for a number of hydrocarbons from the paraffin, olefin, and alkylbenzene families. Table 10 of Keenan and Kaye's *Gas Tables* is similar to this.

In the examples that follow the following scheme is used to designate the phase of the fuel in the reactants and the phase of the water in the products: If the fuel in the reactants is in the vapor phase, the symbols $h_{R(\mathrm{g})}$ and $u_{R(\mathrm{g})}$ are used, and if the fuel is in the liquid phase the symbols $h_{R(\mathrm{l})}$ and $u_{R(\mathrm{l})}$ are used. Similarly, $h_{P(\mathrm{l})}$ and $u_{P(\mathrm{l})}$ indicate that the water is in the liquid phase, and $h_{P(\mathrm{g})}$ and $u_{P(\mathrm{g})}$ indicate that the water in the products is in the vapor phase.

Example 15.10 The enthalpy of combustion of gaseous propane (C_3H_8) at 77 F with the water in the products in the vapor phase, is −19,929 Btu/lbm fuel. (This corresponds to the lower heating value at constant pressure of gaseous propane at 77 F). Calculate the enthalpy of combustion of this fuel when the water formed during combustion is in the liquid phase.

$$h_{RPo}\begin{Bmatrix}\mathrm{C_3H_8\ (g)}\\ \mathrm{H_2O\ (g)}\end{Bmatrix} = h'_{Po(\mathrm{g})} - h'_{Ro(\mathrm{g})} = -19{,}929 \text{ Btu/lbm fuel}$$

The enthalpy of the gaseous products is equal to the enthalpy of the products with liquid water plus the enthalpy of evaporation. Therefore

$$h'_{Po(\mathrm{g})} = h'_{Po(\mathrm{l})} + \frac{\text{lbm H}_2\text{O}}{\text{lbm fuel}}\, h_{fgo}$$

The pounds of water vapor formed per pound of fuel can be calculated from the combustion equation.

$$C_3H_8 + 5O_2 \rightarrow 3CO_2 + 4H_2O$$

$$\frac{\text{lbm } H_2O}{\text{lbm fuel}} = \frac{4(18)}{3(12) + 8(1)} = 1.638$$

therefore

$$h_{RPo}\begin{Bmatrix} C_3H_8\ (g) \\ H_2O\ (l) \end{Bmatrix} = h'_{Po(l)} - h'_{Ro(g)}$$

$$= h'_{Po(g)} - \frac{\text{lbm } H_2O}{\text{lbm fuel}} h_{fgo} - h'_{Ro(g)}$$

$$= -19{,}929 - 1.638(1050.4) = -21{,}649 \text{ Btu/lbm fuel}$$

Thus, the higher heating value at constant pressure of gaseous propane at 77 F is 21,649 Btu/lbm fuel.

Example 15.11 From the given data and the solution of the previous example, calculate the enthalpy of combustion at 77 F of liquid propane (C_3H_8) with liquid H_2O in the products. The latent heat of vaporization of propane is 159 Btu/lbm.

For the reactants

$$h'_{Ro(g)} - h'_{Ro(l)} = h_{fg(\text{fuel})} = 159 \text{ Btu/lbm fuel}$$

Therefore,

$$h_{RPo}\begin{Bmatrix} C_3H_8\ (l) \\ H_2O\ (l) \end{Bmatrix} = h'_{Po(l)} - h'_{Ro(l)}$$

$$= h'_{Po(l)} - h'_{Ro(g)} + 159$$

Using the results of Example 15.10

$$h_{RPo}\begin{Bmatrix} C_3H_8\ (l) \\ H_2O\ (l) \end{Bmatrix} = -21{,}649 + 159 = -21{,}490 \text{ Btu/lbm fuel}$$

In general the phase of the fuel influences the heating value of a hydrocarbon fuel less than 1 per cent. The phase of the water in the products is a much more significant factor.

Example 15.12 Calculate the enthalpy of combustion of gaseous propane at 400 F. (At this temperature all the water formed during combustion will be in the vapor phase.) This example will enable us to evaluate how much the enthalpy of combustion of propane varies with temperature.

The average constant-pressure specific heat of propane is 0.4 Btu/lbm R over this range of temperature. Other values of enthalpy may be taken from the *Gas Tables*.

In solving this problem it may be helpful to refer to Fig. 15.6. Figure 15.6*a* shows that the reaction might proceed in two ways. It might simply take place at 400 F, with the reactants entering and the products leaving at 400 F. It might also take place by first cooling the reactants to 77 F, followed by a reaction at 77 F and subsequent heating of the products to 400 F. Since enthalpy is a property, the net

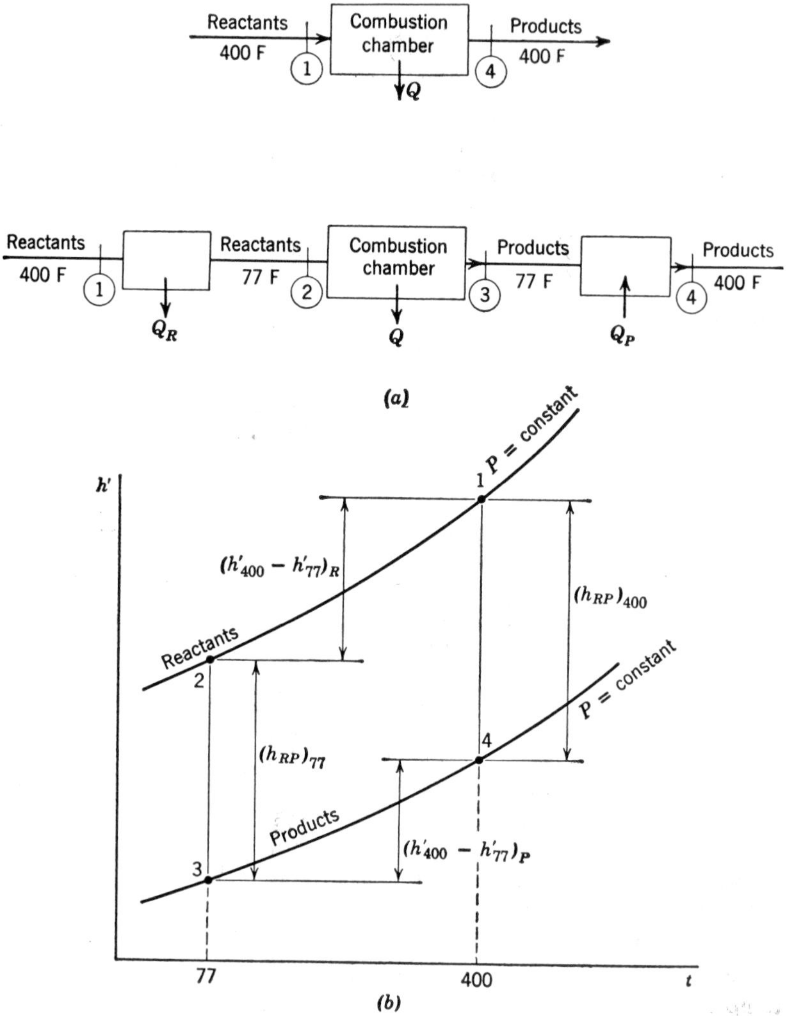

Fig. 15.6 Sketch for Example 15.12.

change of enthalpy is independent of the path and depends only on the initial and final states. Therefore, the net change of enthalpy is the same in both cases.

Figure 15.6*b* shows this process on an enthalpy-temperature diagram. The two lines marked reactants and products are constant-pressure lines for the reactants and products respectively, and the slope of this line at any point is equal to the constant-pressure specific heat at that temperature: $C_P = (\partial h/\partial T)_P$. The vertical distance between the two lines at any temperature is the enthalpy of combustion at that temperature. Thus, we can proceed from point 1 to 2 and from 3 to 4 if we have specific-heat data or tables of properties such as the *Gas Tables*. We know the enthalpy of combustion at 77 F from standard tables. Thus, the enthalpy of combustion at 400 F is found by finding the net change of enthalpy along path 1–2–3–4. This procedure is used in the calculation that follows.

The difference between the enthalpy at point 4 and the enthalpy at point 1 is the enthalpy of combustion at 400 F, and the difference in enthalpy between points 3 and 2 is the enthalpy of combustion at 77 F. Therefore we can write

$$h_4 - h_1 = (h_4 - h_3) + (h_3 - h_2) + (h_2 - h_1) = (h_{RP})$$

$$(h_{RP})_{77} = (h'_{400} - h'_{77})_P + (h_{RP})_{77} + (h'_{77} - h'_{400})_R$$

$$(h_{RP})_{400} = (h'_{400} - h'_{77})_P - 19{,}929 + (h'_{400} - h'_{77})_R$$

This equation has been written for one pound of fuel. Therefore, in solving for the quantities $(h'_{400} - h'_{77})_P$ and $(h'_{400} - h'_{77})_R$ the combustion equation must be considered.

$$C_3H_8 + 5O_2 \rightarrow 3CO_2 + 4H_2O$$

$$(h'_{400} - h'_{77})_P = \frac{n_{CO_2}(\bar{h}_{400} - \bar{h}_{77})_{CO_2} + n_{H_2O}(\bar{h}_{400} - \bar{h}_{77})_{H_2O}}{M_{\text{fuel}}}$$

$$= \frac{3(7175 - 4030) + 4(6896 - 4258)}{44}$$

$$= 453 \text{ Btu/lbm fuel}$$

$$(h'_{400} - h'_{77})_R = C_P(400 - 77)_{\text{fuel}} + \frac{n_{O_2}(\bar{h}_{400} - \bar{h}_{77})_{O_2}}{M_{\text{fuel}}}$$

$$= 0.4(323) + \frac{5(6042 - 3725)}{44} = 394 \text{ Btu/lbm fuel}$$

Therefore,

$$(h_{RP})_{400} = 453 - 19{,}929 + 394 = 19{,}870 \text{ Btu/lbm fuel}$$

The enthalpy of combustion of most hydrocarbon fuels varies only slightly with the temperature, as is shown by the example above.

One other observation might be made concerning this example. The combustion equation was written without nitrogen. This is permissible because the increase in enthalpy of the nitrogen between states 1 and 2 is exactly equal to the decrease in enthalpy of nitrogen between states 3 and 4.

Example 15.13 This example demonstrates the difference between the enthalpy of combustion and the internal energy of combustion. This is equivalent to the negative of the difference between the constant-pressure and the constant-volume heating values.

Calculate the internal energy of combustion at 77 F of liquid propane when the water in the products is a vapor. The enthalpy of combustion under these conditions is −19,773 Btu/lbm fuel.

The reaction equation is

$$C_3H_8\ (l) + 5O_2 \rightarrow 3CO_2 + 4H_2O\ (g)$$

$$\text{moles of gaseous product/mole fuel} = 7$$

$$\text{moles of gaseous reactant/mole fuel} = 5$$

Using Eq. 15.9

$$(u_{Po}' - u_{Ro}') = h_{Po}' - h_{Ro}' - \bar{R}T_o(n_P' - n_R')$$

$$= -19{,}773 - 1.986\,\frac{537 \times (7 - 5)}{(44)}$$

$$= -19{,}773 - 48 = -19{,}821 \text{ Btu/lbm fuel}$$

The enthalpy of combustion for a compound or fuel can readily be determined by using the enthalpy of formation. This is illustrated in the two examples that follow.

Example 15.14 Determine the enthalpy of combustion at 77 F of liquid octane with gaseous H_2O in the products by use of the enthalpies of formation given in Table 15.4.

The combustion equation is

$$C_8H_{18}\ (l) + 12.5O_2 \rightarrow 8CO_2 + 9H_2O\ (g)$$

$$H_{RPo} = H_{Po} - H_{Ro}$$

$$H_{Po} = 8\bar{h}^{\circ}_{CO_2} + 9\bar{h}^{\circ}_{H_2O(g)}$$

$$H_{Po} = 8(-169{,}293) + 9(-104{,}071) = -2{,}290{,}983 \text{ Btu/mole fuel}$$

$$H_{Ro} = \bar{h}^\circ_{C_8H_{18}(l)} = -107{,}530 \text{ Btu/mole fuel}$$

$$H_{RPo} = -2{,}290{,}983 - (-107{,}530) = -2{,}183{,}453 \text{ Btu/mole fuel}$$

$$h_{RPo} = \frac{-2{,}183{,}453}{114.23} = -19{,}100 \text{ Btu/lbm fuel}$$

Example 15.15 Determine the lower heating value of the producer gas from bituminous coal listed in Table 15.3. Since the lower heating value involves gaseous H_2O in the products, the combustion equation is

$$0.03CH_4 + 0.14H_2 + 0.27CO + 0.265O_2 \rightarrow 0.30CO_2 + 0.20H_2O\ (g)$$

$$H_R + Q = H_P$$

$$H_P = 0.30\bar{h}^\circ_{CO_2} + 0.20\bar{h}^\circ_{H_2O(g)}$$

$$H_P = 0.30(-169{,}293) + 0.20(-104{,}071)$$

$$= -71{,}602 \text{ Btu/mole fuel}$$

$$H_R = 0.03\bar{h}^\circ_{CH_4} + 0.27\bar{h}^\circ_{CO}$$

$$H_R = 0.03(-32{,}200) + 0.27(-47{,}548) = -13{,}804 \text{ Btu/mole fuel}$$

$$Q = H_P - H_R = -71{,}602 - (-13{,}804) = -57{,}798 \text{ Btu/mole fuel}$$

Thus, the lower heating value = 57,798 Btu/mole fuel.

15.6 First-Law Analysis of Reacting Systems

The most convenient way of performing a first-law analysis of a reacting system is to utilize the enthalpy of formation. When gases that can be treated as ideal gases are involved, the enthalpy is a function of temperature only, and changes in pressure do not cause a change in enthalpy. Thus, at a given temperature T, the enthalpy of a gas relative to the base at which the enthalpy of the elements is zero is

$$\bar{h} = \bar{h}^\circ + \bar{h}_T - \bar{h}_{537}$$

where $h_T - h_{537}$ is the change in enthalpy between 537 R, which is the reference temperature for the enthalpy of formation, and the given temperature T. It is convenient to use Tables 11 through 23 of the *Gas Tables* in this case. These tables are similar to the Air Tables, and take into account the variation of specific heat with temperature. Table A.9 of the Appendix is taken from these tables, and lists only the values of h and ϕ for these gases. As will be noted later, ϕ is essentially equal to the absolute entropy of these gases at a pressure of 1 atm.

Example 15.16 A gasoline engine delivers 200 hp. The fuel used is C_8H_{18} (l), and it enters the engine at 77 F; 150 per cent theoretical air is used and enters at 110 F. The products of combustion leave the engine at 920 F, and the heat transfer from the engine is 700,000 Btu/hr. Determine the fuel consumption per hour if complete combustion is achieved.

The chemical reaction is

$$C_8H_{18}\ (l) + 1.5(12.5)O_2 + 1.5(3.76)(12.5)N_2 \rightarrow$$
$$8CO_2 + 9H_2O\ (g) + 6.25O_2 + 70.5N_2$$

The first law for this process, considering the engine as an open system is

$$\dot{n}_f \bar{h}_R + \dot{Q} = \dot{W} + \dot{n}_f \bar{h}_P$$

where $\dot{n}_f$ = number of moles of fuel required per hour
$\dot{Q}$ = heat transfer per hour
$\dot{W}$ = work per hour
$\bar{h}_R$ = enthalpy of the reactants per mole of fuel
$\bar{h}_P$ = enthalpy of the products per mole of fuel

Solving first for $\bar{h}_R$ and $\bar{h}_P$ we have

$$\bar{h}_R = \bar{h}^\circ_{C_8H_{18}(l)} + 18.75(\bar{h}_{570} - \bar{h}_{537})_{O_2} + 70.5(\bar{h}_{570} - \bar{h}_{537})_{N_2}$$
$$= -107{,}530 + 18.75(3957 - 3725) + 70.5(3959 - 3730)$$
$$= -107{,}530 + 4350 + 16{,}144 = -87{,}036 \text{ Btu/mole fuel}$$

$$\bar{h}_P = 8(\bar{h}^\circ + \bar{h}_{1380} - \bar{h}_{537})_{CO_2} + 9(\bar{h}^\circ + \bar{h}_{1380} - \bar{h}_{537})_{H_2O}$$
$$+ 6.25(\bar{h}_{1380} - \bar{h}_{537})_{O_2} + 70.5(\bar{h}_{1380} - \bar{h}_{537})_{N_2}$$
$$= 8(-169{,}293 + 13{,}101 - 4030) + 9(-104{,}071 + 11{,}441 - 4258)$$
$$+ 6.25(10{,}050 - 3725) + 70.5(9748 - 3730)$$
$$= -1{,}281{,}776 - 880{,}992 + 39{,}531 + 424{,}269 = -1{,}698{,}968 \text{ Btu/mole fuel}$$

$$\dot{n}_f(-87{,}036) - 700{,}000 = 200(2545) + \dot{n}_f(-1{,}698{,}968)$$

$$\dot{n}_f = \frac{1{,}209{,}000}{1{,}611{,}932} = 0.750 \text{ mole fuel/hr}$$

$$\text{lbm fuel/hr} = 0.750(114.23) = 85.7$$

15.7 Adiabatic Flame Temperature

If a given reaction goes to completion (that is, equilibrium is achieved) and takes place adiabatically and with very low kinetic energy changes, the temperature of the products is the highest possible temperature for the given reactants, and this temperature is known as the *adiabatic flame temperature.* There are many situations where it is important to know the adiabatic flame temperature. In a gas turbine for example, the maximum temperature permissible is determined by metallurgical considerations in the turbine, so it is important to know the adiabatic flame temperature for the given fuel and the given fuel-air ratio. (Since there is usually some heat transfer during the combustion process, and since the reaction is usually less than complete to some degree, the actual flame temperature is always somewhat less than the adiabatic flame temperature.)

In the examples that follow the effects of dissociation are not considered. This matter will be considered in Chapter 16.

Example 15.17 Liquid octane at 77 F is burned with 400 per cent theoretical air at 77 F in a steady-flow process. Determine the adiabatic flame temperature.

The reaction is

$$C_8H_{18}\ (l) + 4(12.5)O_2 + 4(12.5)(3.76)N_2 \rightarrow$$
$$8CO_2 + 9H_2O\ (g) + 37.5O_2 + 188.0N_2$$

In this case the first law reduces to

$$H_R = H_P$$

That is, this is a constant-enthalpy process, and we let h_T refer to the enthalpy at the adiabatic flame temperature.

$$H_R = \bar{h}^{\circ}_{C_8H_{18}(l)} = -107{,}530 \text{ Btu/mole fuel}$$

$$\begin{aligned} H_P &= 8(\bar{h}^{\circ} + \bar{h}_T - \bar{h}_{537})_{CO_2} + 9(\bar{h}^{\circ} + \bar{h}_T - \bar{h}_{537})_{H_2O} \\ &\quad + 37.5(\bar{h}_T - \bar{h}_{537})_{O_2} + 188.0(\bar{h}_T - \bar{h}_{537})_{N_2} \\ &= 8(-169{,}293 + \bar{h}_{TCO_2} - 4030) + 9(-104{,}071 + \bar{h}_{TH_2O} - 4268) \\ &\quad + 37.5(\bar{h}_{TO_2} - 3725) + 188.0(\bar{h}_{TN_2} - 3730) \end{aligned}$$

Therefore,

$$\begin{aligned} -107{,}530 &= 8(\bar{h}_{TCO_2} - 173{,}323) + 9(\bar{h}_{TH_2O} - 108{,}329) \\ &\quad + 37.5(\bar{h}_{TO_2} - 3725) + 188.0(\bar{h}_{TN_2} - 3730) \end{aligned}$$

By trial-and-error solution, a temperature of the products is found that satisfies this equation.

Assume $T_2 = 1730$ R

$$-107{,}530 \overset{?}{=} 8(17{,}487 - 173{,}323) + 9(14{,}747 - 108{,}329)$$
$$+ 37.5(12{,}904 - 3725)$$
$$+ 188.0(12{,}411 - 3730) \text{ Btu/mole fuel}$$
$$-107{,}530 \overset{?}{=} -96{,}966 \text{ Btu/mole fuel}$$

Thus we find the adiabatic flame temperature is slightly less than 1730 R. A simplified procedure for finding the adiabatic flame temperature for hydrocarbon fuels will be considered in Example 15.20.

15.8 Use of Enthalpy and Internal Energy of Combustion for Problems Involving Complete Combustion

In problems involving complete combustion, the enthalpy and internal energy of combustion can be utilized in the solution. Consider Fig. 15.7, which shows a steady-flow combustion process, the reactants

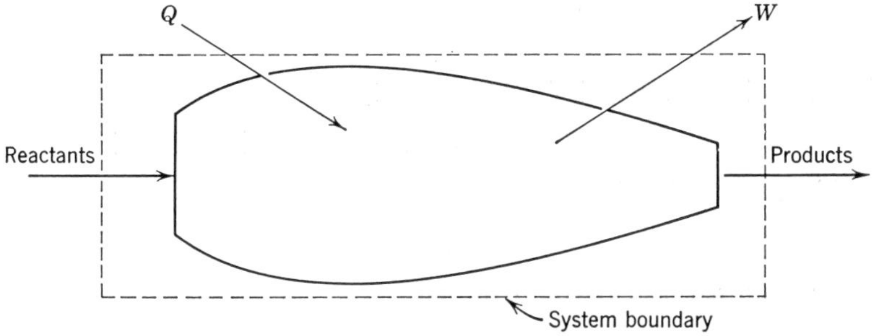

Fig. 15.7 Steady-flow combustion process involving complete combustion.

entering the system with a certain kinetic energy and enthalpy, and the products leaving the system with a certain kinetic energy and enthalpy. Heat and work may also be involved as indicated. In almost every case potential energy changes will not be important, and therefore are neglected in this section. The steady-flow energy equation per pound of fuel for this process is

$$q + h_R' + KE_R' = h_P' + KE_P' + w \tag{15.10}$$

The enthalpy of combustion was defined by Eq. 15.6.

$$h_{RP_0} = h_{P_0}' - h_{R_0}'$$

Subtracting Eq. 15.6 from Eq. 15.10 and rearranging, we have

$$(h_P' + KE_P' - h_{P_o}') - (h_R' + KE_R' - h_{R_o}') = q - w - h_{RP_o} \quad (15.11)$$

Note however that if we assume complete combustion, then the quantities h_P and h_{P_o} involve exactly the same composition (because the measurement of the heating value involved complete combustion), and therefore the quantity $(h_P' - h_{P_o}')$ can be found using any base whatsoever for the enthalpies, because only differences of enthalpy are involved. Thus, the problem can be solved without using enthalpies of formation, and the enthalpy of combustion, which was measured at some temperature T_o, may be used in a combustion process involving any temperature of reactants and products.

Equation 15.10 does assume, however, that the combustion process is complete, and therefore that the composition of the products of combustion of the actual process are identical to the composition of those obtained when determining the enthalpy of combustion.

Equation 15.10 was written for 1 lbm of fuel, as is indicated by the primes in the equation. It may be put into slightly different form by introducing the air-fuel ratio, AF. The mass of air per unit mass of fuel is AF, and the mass of products per unit mass of fuel is (1 + AF). The equation is then written

$$(1 + \mathrm{AF})(h_P + KE_P - h_{P_o}) - \mathrm{AF}(h_a + KE_a - h_{ao}) - (h_f - h_{fo}) = q - w - h_{RP_o} \quad (15.12)$$

The enthalpy of the products in this equation is per unit mass of products, and the enthalpy of the air is per unit mass of air, as is evidenced by the absence of the primes. The kinetic energy of the fuel has been dropped, as it is usually insignificant.

This equation can also be written per unit mass of air by introducing the fuel-air ratio, FA, which is the reciprocal of the air-fuel ratio. The equation then becomes

$$(1 + \mathrm{FA})(h_P + KE_P - h_{P_o}) - (h_a + KE_a - h_{ao}) - \mathrm{FA}(h_f - h_{fo}) = \mathrm{FA}\,(q - w - h_{RP_o}) \quad (15.13)$$

(Note that in Eqs. 15.10, 15.11, and 15.12, q and w are defined as heat transfer and work per pound of fuel.)

Example 15.18 Liquid octane, C_8H_{18}, is burned with 200 per cent theoretical air in a steady-flow process. The initial temperature of the octane is 200 F and the initial temperature of the air is 400 F. The products of combustion leave at 1600 F, and changes in kinetic energy are not significant. Assume complete combustion. Calculate the heat transfer per pound of fuel.

The combustion equation is

$$C_8H_{18} + 25O_2 + 25(3.76)N_2 \rightarrow 8CO_2 + 9H_2O + 12.5O_2 + 94.0N_2$$

From the Table 15.5, $h_{RPo} = -19{,}100$ Btu/lbm fuel at 77 F.

Equation 15.11 is applicable to this process, and we note that the work is zero in this case, and changes in kinetic energy are negligible.

$$(h_P' - h_{Po}') - (h_R' - h_{Ro}') = q - h_{RPo}$$

The quantity $(h_P' - h_{Po}')$ can be calculated as follows, using values of enthalpy from the gas tables.

Constituent	n	$\bar{h}_{2060}$	$\bar{h}_{537}$	$n(\bar{h}_{2060} - \bar{h}_{537})$
CO_2	8.0	21,818	4030	142,300
H_2O	9.0	18,054	4258	124,200
O_2	12.5	15,671	3725	149,300
N_2	94.0	15,013	3730	1,060,600
				1,476,400 Btu/mole fuel

$$h_P' - h_{Po}' = \frac{1{,}476{,}400}{114.2} = 12{,}920 \text{ Btu/lbm fuel}$$

The quantity $(h_R' - h_{Ro}')$ can be calculated by considering the air and the fuel separately. That is,

$$h_R' - h_{Ro}' = (h' - h_o')_{\text{air}} + (h' - h_o')_{\text{fuel}}$$

$$(h' - h_o')_{\text{air}} = n_{\text{air}} \times \frac{M_{\text{air}}}{M_{\text{fuel}}}(h_{860} - h_{537})$$

$$= \frac{(25 + 94)28.95}{114.2}(206.46 - 128.34) = 2360 \text{ Btu/lbm fuel}$$

For the fuel (assume C_p for fuel is 0.5 Btu/lbm F)

$$(h' - h_o')_{\text{fuel}} = C_{p\ \text{fuel}}(t - t_o)_{\text{fuel}} = 0.5(200 - 77) = 62 \text{ Btu/lbm fuel}$$

Therefore

$$q = (h_P' - h_{Po}') - (h_R' - h_{Ro}') + h_{RPo}$$
$$= 12{,}920 - 2356 - 62 - 19{,}100 = -8{,}598 \text{ Btu/lbm fuel}$$

Tables 4 and 7 of the *Gas Tables* have been developed in order to simplify calculations in problems such as the one cited in the example above. These tables have been developed for a hydrocarbon fuel having the composition $(CH_2)_n$. However, they represent with high precision the properties of the products of combustion of hydrocarbon fuels having a wide range of composition. This is evident from Tables 5 and 8 of the *Gas Tables* which give the specific heats of three hydrocarbon fuels representing a wide range of composition. Table 9 gives the molecular weight for the products of combustion as a function of the composition of the fuel and the per cent theoretical fuel. Table 4 applies to the products of combustion when 400 per cent theoretical air is used, and Table 7 is for 200 per cent theoretical air. A portion of Table A.9 in the Appendix has been abstracted from Tables 4 and 7 of the *Gas Tables.*

Example 15.19 To illustrate the use of these tables let us repeat Example 15.18, using data from Table 7 of the *Gas Tables* (Appendix Table A.9).

$$q = (h_P' - h_{P_o}') - (h_R' - h_{R_o}') + h_{RP_o}$$

$$(h_P' - h_{P_o}') = \frac{n_P(\bar{h}_{2060} - \bar{h}_{537})}{M_{\text{fuel}}}$$

$$= \frac{(8 + 9 + 12.5 + 94.0)(15{,}700 - 3775)}{114.2}$$

$$= 12{,}900 \text{ Btu/lbm fuel}$$

From Example 15.18

$$(h_R' - h_{R_o}') = (h' - h_o')_{\text{air}} + (h' - h_o')_{\text{fuel}} = 2356 + 62$$

$$= 2418 \text{ Btu/lbm fuel}$$

$$q = 12{,}900 - 2418 - 19{,}100 = -8618 \text{ Btu/lbm fuel}$$

Example 15.20 Repeat Example 15.17 using the *Gas Tables.*
The reaction is

$$C_8H_{18} + 4(12.5)O_2 + 4(12.5)(3.76)N_2 \rightarrow$$
$$8CO_2 + 9H_2O + 37.5O_2 + 188.0N_2$$

$$(h_P' - h_{P_o}') - (h_R' - h_{R_o}') = q - w - h_{RP_o}$$

For this process, $q = 0$, $w = 0$, and since the fuel and air are both at 77 F, $(h_R' - h_{R_o}') = 0$. Therefore

$$h_P' - h_{P_o}' = \frac{n_P(\bar{h}_P - \bar{h}_{P_o})}{M_{\text{fuel}}} = -h_{RP_o} = 19{,}100 \text{ Btu/lbm fuel}$$

$$\bar{h}_P - \bar{h}_{P_o} = \frac{19{,}100 \times 114.2}{242.5} = 8990 \text{ Btu/mole products}$$

From Table 4 of the *Gas Tables* (Table A.9 of the Appendix)

$$\bar{h}_{P_0} = 3747 \text{ Btu/mole products}$$

$$\bar{h}_P = 8990 + 3747 = 12{,}737 \text{ Btu/mole products}$$

$$T_{\text{products}} = 1734 \text{ R}$$

15.9 Evaluation of Actual Combustion Processes

In evaluating the performance of an actual combustion process a number of different parameters can be defined, depending on the nature of the process and the system considered. In the combustion chamber of a gas turbine, for example, the objective is to raise the temperature of the products to a given temperature (usually the maximum temperature the metals in the turbine can withstand). If we had a combustion process in which complete combustion was achieved and which was adiabatic, the temperature of the products would be the adiabatic flame temperature. Let us designate the fuel-air ratio to reach a given temperature under these conditions as the ideal fuel-air ratio. In the actual combustion chamber the combustion will be incomplete to some extent and there will be some heat transfer to the surroundings. Therefore more fuel will be required to reach the given temperature, and this we designate as the actual fuel-air ratio. In this case, the combustion efficiency, $\eta_{\text{comb.}}$ is defined as

$$\eta_{\text{comb.}} = \frac{\text{FA}_{\text{ideal}}}{\text{FA}_{\text{actual}}} \tag{15.14}$$

On the other hand, in the furnace of a steam generator (boiler) the purpose is to transfer the maximum possible amount of heat to the steam (water). In practice, the efficiency of a steam generator is defined as the ratio of the heat transferred to the steam to the higher heating value of the fuel. For a coal this is the heating value as measured in a bomb calorimeter, which is the constant-volume heating value, and corresponds to the internal energy of combustion. One observes a minor inconsistency, since the boiler is really an open system, and the change in enthalpy is the significant factor. However, in most cases the error thus introduced is less than the experimental error involved in measuring the heating value.

$$\eta_{\text{steam generator}} = \frac{\text{heat transferred to steam/lbm fuel}}{\text{higher heating value of the fuel}} \tag{15.15}$$

In an internal-combustion engine the purpose is to do work. The logical way to evaluate the performance of an internal-combustion engine would be to compare the actual work done to the maximum work which would be done by a reversible change of state from the reactants to the products. This, as we noted in Chapter 10 is equal to the decrease in Gibbs function $(h - Ts)$.

However, in practice the efficiency is defined as the ratio of the actual work to the negative of the enthalpy of combustion of the fuel (i.e., the constant-pressure heating value). This is usually called the thermal efficiency, η_{th}.

$$\eta_{th} = \frac{w}{(-h_{RP_o})} = \frac{w}{HV} \tag{15.16}$$

The over-all efficiency of a gas turbine or steam power plant is defined in the same way. It should be pointed out that in an internal-combustion engine or fuel-burning steam power plant it is impossible to do an amount of work equal to the decrease in Gibbs function because the combustion process is itself an irreversible process. This will be demonstrated later.

One other factor should be pointed out regarding efficiency. We have noted that the enthalpy of combustion of a hydrocarbon fuel varies considerably with the phase of the water in the products (which leads to the concept of higher and lower heating values). Therefore, in considering the thermal efficiency of an engine, the heating value used to determine this efficiency must be borne in mind. Two engines made by different manufacturers might have identical performance, but if one manufacturer bases his efficiency on the higher heating value and the other on the lower heating value, the latter will be able to claim a higher thermal efficiency. This claim is not significant, of course, as the performance is the same, and a consideration of the way in which the efficiency was defined would reveal this.

This whole matter of efficiencies of devices involving combustion processes is treated in detail in textbooks dealing with particular applications, and this discussion has been intended only as an introduction to the subject. However, a few examples will be cited to illustrate these remarks.

Example 15.21 The combustion chamber of a gas turbine uses a liquid hydrocarbon fuel which has an approximate composition of $C_{10}H_{22}$. On test the following data are obtained.

$$t_{air} = 280\text{ F}(740\text{ R}) \qquad t_{products} = 1460\text{ F}(1920\text{ R})$$

$$\overline{V}_{air} = 300\text{ ft/sec} \qquad \overline{V}_{products} = 450\text{ ft/sec}$$

$$t_{fuel} = 120\text{ F} \qquad \text{FA}_{actual} = 0.0184$$

Calculate the efficiency of combustion of this combustion chamber.

The fuel-air ratio for the ideal case, which involves complete combustion and no heat transfer, can be found from Eq. 15.13. It will be necessary to know the per cent theoretical fuel, so we first solve for the theoretical fuel-air ratio.

$$C_{10}H_{22} + 15.5O_2 + 15.5(3.76)N_2 \rightarrow 10CO_2 + 11H_2O + 58.2(N_2)$$

$$FA_{theo} = \frac{10(12) + 22(1)}{(15.5 + 58.2)28.96} = 0.066$$

$$\% \text{ theoretical fuel} = \frac{0.0184}{0.066} = 28\%; \ \% \text{ theoretical air} = 360\%$$

In using Eq. 15.13 the molecular weight of the products is found from Table 9 of the *Gas Tables.*

$$(1 + FA)\left(\frac{(\bar{h}_P - \bar{h}_{P_o})}{M_P} + \frac{\overline{V}_P^2}{2g_c}\right) - \left(h_a + \frac{\overline{V}_a^2}{2g_c} - h_{ao}\right) - FA(h_f - h_{fo})$$
$$= FA(q - w - h_{RP_o})$$

$$(1 + FA)\left(\frac{(14{,}303 - 3752)}{28.89} + \frac{450^2}{64.34 \times 778}\right)$$
$$- \left(177.2 + \frac{300^2}{64.34 \times 778} - 128.3\right)$$
$$- FA(0.5)(120 - 77) = FA(19{,}020)$$

$$(1 + FA)(368.5) - (50.7) - FA(22) = FA(19{,}020)$$

$FA = 0.0170.$ This is the ideal fuel-air ratio.

Therefore, from Eq. 15.13, the combustion efficiency is

$$\eta_c = \frac{0.0170}{0.0184} = 92.5\%$$

Example 15.22 In a certain steam power plant 715,000 lbm of water per hour enter the boiler at a pressure of 1850 lbf/in.^{2} and a temperature of 415 F. Steam leaves the boiler at 1320 lbf/in.^{2}, 925 F. The power output of the turbine is 81,000 kw. Coal is used at the rate of 59,000 lbm/hr, and the coal used is the Pennsylvania coal listed in Table 15.1. Determine the efficiency of boiler and over-all thermal efficiency of the plant.

In power plants the efficiency of both the boiler and the over-all efficiency of the plant are based on the higher heating value of the fuel, which is given in Table 15.1 as 14,310 Btu/lbm.

The efficiency of the boiler is defined by Eq. 15.15 as

$$\eta_{\text{boiler}} = \frac{\text{heat transferred to } H_2O/\text{lbm fuel}}{\text{higher heating value}}$$

Therefore

$$\eta_{\text{boiler}} = \frac{715{,}000}{59{,}000} \times \frac{(1448.0 - 302.8)}{14{,}310} = 89.4\%$$

The thermal efficiency is defined by Eq. 15.16

$$\eta_{\text{thermal}} = \frac{w}{\text{heating value}} = \frac{81{,}000 \times 3412}{59{,}000 \times 14{,}310} = 32.7\%$$

15.10 Second-Law Analysis of Combustion Processes

From our consideration of the second law we would conclude that since the combustion process proceeds so readily, it is probably an irreversible process. This can be verified by determining the entropy change during an adiabatic combustion process. However, to do this we must know the entropy of each of the reactants and products relative to the same base. This would be similar to our procedure in defining the enthalpy of formation.

This leads to a consideration of the third law of thermodynamics even though a detailed discussion of the third law is beyond the scope of this book. In essence the third law states that the entropy of a pure substance is zero at absolute zero. Entropy measured relative to this base is called the *absolute entropy*, and is given the symbol $S°$. It may be determined by either calorimetric or spectrographic means. Table 15.4 gives the absolute entropy at 77 F (25 C) and 1 atm pressure of a number of substances. The quantity ϕ, which is given in the *Gas Tables* for a number of substances, is equal to the absolute entropy at a pressure of 1 atm, and it is so used in a number of the examples that follow.

Another term to be defined is the Gibbs function $(h - Ts)$ of formation (often called the free energy of formation). This term is similar to the enthalpy of formation in that the Gibbs function of the elements at 25 C (77 F) and 1 atm pressure is assumed to be zero. The Gibbs function of any substance relative to this base is found from the enthalpy of formation and the absolute entropy.

Example 15.23 Determine the Gibbs function of formation of CO_2. Consider the reaction

$$C + O_2 \rightarrow CO_2$$

Assume that the carbon and oxygen are initially at 77 F and 1 atm pressure, and that the CO_2 is finally at 77 F and 1 atm pressure.

The change in Gibbs function for this reaction is found first.

$$G_P - G_R = (H_P - H_R) - T_o(S_P - S_R)$$
$$= \bar{h}^\circ_{CO_2} - 536.7(51.061 - 49.003 - 1.361)$$
$$= -169{,}293 - 374 = -169{,}667 \text{ Btu/lb mole}$$

Since the Gibbs function of the reactants, G_R, is zero (in accordance with the assumption that the Gibbs function of the elements is zero at 25 C and 1 atm pressure), it follows that

$$G_P = \bar{g}^\circ_{CO_2} = -169{,}667 \text{ Btu/lb mole}$$

This is the value given in Table 15.4.

In Section 10.7 it was pointed out that the maximum useful work a system could do during a process in which it is in pressure and temperature equilibrium with the environment is equal to the decrease in the Gibbs function, $H - TS$, during the process. This is particularly applicable to a chemical reaction, since in many cases the reactants are in pressure and temperature equilibrium with the environment before the reaction takes place, and the same is true of the products after the reaction takes place. Consider an automobile engine as an example. The air and fuel are initially at the pressure and temperature of the surroundings, and work may be done by the products until they are at the pressure and temperature of the surroundings. For this reaction therefore, the maximum work that could be done is the decrease in Gibbs function for this reaction.

For this reason it would seem logical to rate the efficiency of a device designed to do work by utilization of a combustion process, such as an internal-combustion engine or a steam power plant, as the ratio of actual work to the decrease in free energy for the given reaction, rather than to the heating value, as is current practice. However, as the example that follows demonstrates, the difference between the decrease in Gibbs function and the heating value is small for hydrocarbon fuels; and the efficiency defined in terms of heating value is essentially equal to that defined in terms of decrease in free energy.

In the examples that follow the maximum useful work for a combustion process is calculated by determining the decrease in Gibbs function for the combustion process when it takes place in pressure and temperature equilibrium with the environment. This is then compared with the availability of the products of combustion after an adiabatic combustion process. The fact that the availability is much less than the maximum useful work is indicative of the irreversibility of the combustion process. This can be verified by determining the increase in entropy during the adiabatic combustion process.

Example 15.24 Ethene (g) at 77 F and 1 atm pressure is burned with 400 per cent theoretical air at 77 F and 1 atm pressure. Assume that this reaction takes place reversibly at 77 F and that the products leave at 77 F and 1 atm pressure. To further simplify this problem assume that the oxygen and nitrogen are separated before the reaction takes place (each at 1 atm, 77 F), and that the constituents in the products are separated, and that each is at 77 F and 1 atm pressure. Thus, the reaction takes place as shown in Fig. 15.8.

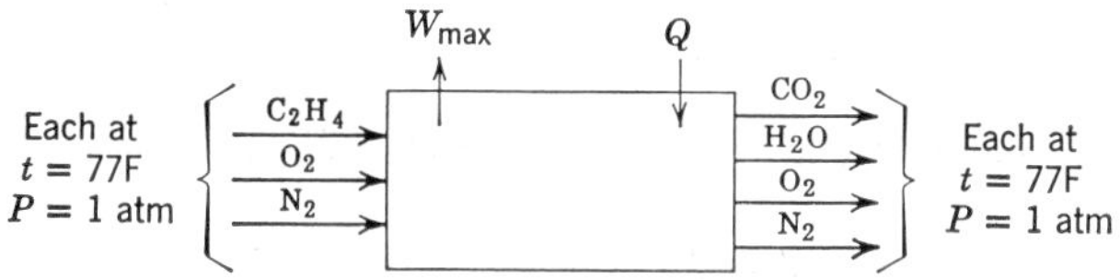

Fig. 15.8 Sketch for Example 15.24.

Determine the maximum work for this process (i.e., the work that would be done if this chemical action takes place reversibly and isothermally). The equation for this chemical reaction is

$$C_2H_4\ (g) + 3(4)O_2 + 3(4)(3.76)N_2 \rightarrow 2CO_2 + 2H_2O\ (g) + 9O_2 + 45.1N_2$$

The maximum work for this process is equal to the decrease in Gibbs function during this reaction (see Section 10.7). The values for the Gibbs function can be taken directly from Table 15.4, since each is at 77 F and 1 atm pressure. From Eq. 10.35 (neglecting kinetic and potential energy changes)

$$W_{\max} = G_R - G_P$$

$$G_R = \mathring{\bar{g}}_{C_2H_4} = 29{,}308 \text{ Btu/mole fuel}$$

$$G_P = n_{CO_2}\mathring{\bar{g}}_{CO_2} + n_{H_2O}\mathring{\bar{g}}_{H_2O}$$

$$= 2(-169{,}667) + 2(-98{,}344) = -536{,}022 \text{ Btu/mole fuel}$$

$$\bar{g}^\circ = 0 \text{ for the elements at 77 F and 1 atm pressure}$$

$$W_{\max} = 29{,}308 - (-536{,}022) = 565{,}330 \text{ Btu/mole fuel}$$

$$w_{\max} = 20{,}140 \text{ Btu/lbm fuel}$$

Example 15.25 Consider the same combustion process as in Example 15.24, but let it take place adiabatically, and assume that each constituent in the products is at 1 atm pressure and at the adiabatic flame temperature. This combustion process is shown schematically in Fig. 15.9. The temperature of the surroundings is 77 F.

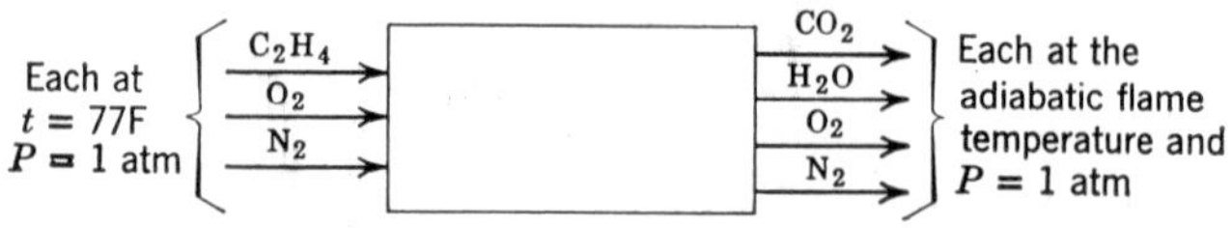

Fig. 15.9 Sketch for Example 15.25.

For this combustion process determine:

(*a*) The increase in entropy during combustion.

(*b*) The availability of the products of combustion.

The adiabatic flame temperature is determined first. The simplest procedure is to use Eq. 15.10.

$$(h_P' - h_{P_o}') - (h_R' - h_{R_o}') = q - w - h_{RP_o}$$

$$(h_P' - h_{P_o}') = \frac{n_P}{M_{\text{fuel}}} (\bar{h}_P - \bar{h}_{P_o}) = 20{,}276$$

$$\bar{h}_P = \frac{20{,}276 \times 28}{58.1} + 3747$$

$$= 13{,}527 \text{ Btu/mole product}$$

From Table 4 of the *Gas Tables* the adiabatic flame temperature is found to be 1832 R.

In determining the change of entropy let us assume, as noted previously, that the value for ϕ given in the *Gas Tables* (and Table A.9 of the Appendix) represents the absolute entropy at the given temperature and 1 atm pressure. That is, $\bar{s}° = \bar{\phi}$.

The change of entropy for the reaction is $S_P - S_R$.

$$S_R = \bar{s}^{\circ}_{C_2H_4} + n_{O_2}\bar{s}^{\circ}_{O_2} + n_{N_2}\bar{s}^{\circ}_{N_2}$$

$$= 52.45 + 12(49.003) + 45.1(45.767)$$

$$= 2704.58 \text{ Btu/mole fuel R}$$

$$S_P = n_{CO_2}\bar{s}^{\circ}_{CO_2} + n_{H_2O}\bar{s}^{\circ}_{H_2O} + n_{O_2}\bar{s}^{\circ}_{O_2} + n_{N_2}\bar{s}^{\circ}_{N_2}$$

$$= 2(64.521) + 2(55.733) + 9(58.302) + 45.1(54.610)$$

$$= 3488.13 \text{ Btu/mole fuel R}$$

Since this is an adiabatic process the increase in entropy indicates that the adiabatic combustion process is irreversible. This will also be reflected by the fact that the availability of the products of combustion is less than the work for the reversible isothermal chemical reaction as calculated in Example 15.24.

The availability of the products of combustion is the work that could be done by reversibly cooling the products of combustion from the adiabatic flame temperature to 77 F. This is found most readily by using Eq. 10.12.

$$\psi = (h - T_o s) - (h_o - T_o s_o)$$

$$= (h - h_o) - T_o(s - s_o)$$

We will take as our parameter 1 lbm of fuel. The quantity $(h - h_o)$ is simply the negative of enthalpy of combustion (since the combustion process was adiabatic) or 20,276 Btu/lbm fuel. The quantity $T_o(s - s_o)$ can be found from the *Gas Tables*, assuming $\bar{s}^\circ = \bar{\phi}$.

$$(s - s_o) = \frac{1}{M_f}[n_{CO_2}(\bar{\phi}_{1832} - \bar{\phi}_{537})_{CO_2} + n_{H_2O}(\bar{\phi}_{1832} - \bar{\phi}_{537})_{H_2O}$$

$$+ n_{O_2}(\bar{\phi}_{1832} - \bar{\phi}_{537})_{O_2} + n_{N_2}(\bar{\phi}_{1832} - \bar{\phi}_{537})_{N_2}]$$

$$= \tfrac{1}{28}[2(64.521 - 51.032) + 2(55.733 - 45.079)$$

$$+ 9(58.302 - 48.986) + 45.1(54.610 - 45.755)]$$

$$= \frac{531.49}{28} = 18.98 \text{ Btu/lbm fuel R}$$

$$\psi = 20{,}276 - 537(18.98) = 20{,}276 - 10{,}192$$

$$= 10{,}084 \text{ Btu/lbm fuel}$$

That is, if every process after the adiabatic combustion process were reversible, the maximum amount of work that could be done is 10,084 Btu/lbm fuel. This compares to a value of 20,140 Btu/lbm for the reversible isothermal reaction. This means that if we had an engine that had the indicated adiabatic combustion process, and if all other processes were completely reversible, the efficiency would be about 50 per cent.

Example 15.26 Consider the same combustion process as in the two previous examples but assume that the reactants consist of a mixture at 1 atm pressure and 77 F and that the products also consist of a mixture at 1 atm and 77 F. Thus, the combustion process is as shown in Fig. 15.10. Assume each constituent to be an ideal gas.

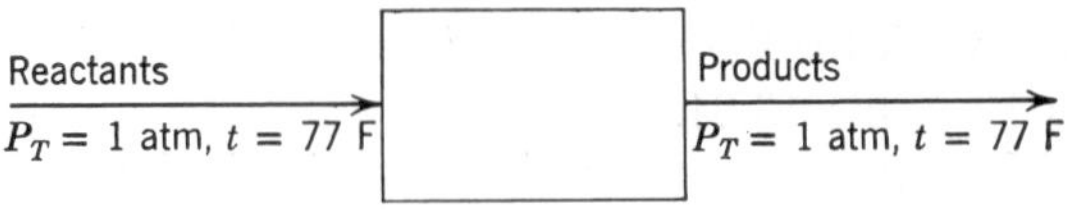

Fig. 15.10 Sketch for Example 15.26.

Determine the work that would be done if this combustion process took place reversibly and in pressure and temperature equilibrium with the surroundings.

The combustion equation, as noted previously, is

$$C_2H_4\ (g) + 3(4)O_2 + 3(4)(3.76)N_2 \rightarrow$$
$$2CO_2 + 2H_2O\ (g) + 9O_2 + 45.1N_2$$

In this case we must find the entropy of each substance as it exists in the mixture; i.e., at its partial pressure and the given temperature of 77 F. Since the absolute entropies given in Table 15.4 are at 1 atm pressure and 77 F, the entropy of each constituent in the mixture can be found from the relation (derived from Eq. 8.26)

$$\bar{s} - \bar{s}^\circ = -\bar{R} \ln \frac{p}{P^\circ}$$

where $\bar{s}$ = entropy of the constituent in the mixture
$\bar{s}^\circ$ = absolute entropy at the same temperature and 1 atm pressure
p = partial pressure of the constituent in the mixture
P° = 1 atm pressure.

Since P° and the pressure of the mixture are both 1 atm pressure, the entropy of each constituent can be found by the relation

$$\bar{s} = \bar{s}^\circ - \bar{R} \ln x = \bar{s}^\circ + \bar{R} \ln \frac{1}{x}$$

For the reactants:

	n	$1/x$	$\bar{R} \ln 1/x$	$\bar{s}^\circ$	$\bar{s}$
C_2H_4	1	58.100	8.060	52.450	60.510
O_2	12	4.850	3.135	49.003	52.138
N_2	45.1	1.285	0.498	45.767	46.265
	58.1				

For the products:

	n	$1/x$	$\bar{R} \ln 1/x$	$\bar{s}^\circ$	$\bar{s}$
CO_2	2	29.050	6.690	51.061	57.751
H_2O	2	29.050	6.690	45.106	51.796
O_2	9	6.460	3.700	49.003	52.703
N_2	45.1	1.285	0.498	45.767	46.265
	58.1				

$$w_{\max} = g_R - g_P = (h_R - h_P) - T_0(s_R - s_P)$$

$$s_R = \frac{1}{M_f}(\bar{s}_{C_2H_4} + n_{O_2}\bar{s}_{O_2} + n_{N_2}\bar{s}_{N_2}) = \tfrac{1}{28}[60.510 + 12(52.138)$$

$$+ 45.1(46.265)] = \frac{2772.72}{28} = 99.03 \text{ Btu/lbm fuel R}$$

$$s_P = \frac{1}{M_f}(n_{CO_2}\bar{s}_{CO_2} + n_{H_2O}\bar{s}_{H_2O} + n_{O_2}\bar{s}_{O_2} + n_{N_2}\bar{s}_{N_2})$$

$$= \tfrac{1}{28}[2(57.751) + 2(51.796) + 9(52.703) + 45.1(46.265)]$$

$$= \frac{2779.97}{28} = 99.29 \text{ Btu/lbm fuel R}$$

$$w_{\max} = 20{,}276 - 537(99.03 - 99.29) = 20{,}416 \text{ Btu/lbm fuel}$$

This raises the question of the possibility of a reversible chemical reaction. Some reactions can be made to approach reversibility by having them take place in an electrolytic cell. When a potential just equal to the electromotive force of the cell is applied, no reaction takes place. When the applied potential is increased slightly, the reaction proceeds in one direction, and if the applied potential is decreased slightly, the reaction proceeds in the opposite direction. The work involved is the electrical energy supplied or delivered.

Much effort is being directed toward the development of fuel cells in which carbon, hydrogen, or hydrocarbons react with oxygen and produce electricity directly. If a fuel cell can be developed that utilizes a hydrocarbon fuel and has a sufficiently high efficiency, drastic changes will be made in our techniques of generating electricity on a commercial scale. At the present time, however, fuel cells are not competitive with conventional power plants for the production of electricity on a large scale.

PROBLEMS

15.1 Propane is burned with air and an Orsat analysis of the products of combustion is as follows:

CO_2	8.2%
CO	1.3
O_2	7.0
N_2	83.5

Determine the per cent of theoretical air used for this combustion process.

15.2 A hydrocarbon fuel is burned with air and the following Orsat analysis is obtained from the products of combustion.

CO_2	10.5%
O_2	5.3
N_2	84.2

Determine the composition of the fuel on a mass basis and the per cent of theoretical air utilized in the combustion process.

15.3 The Pennsylvania coal listed in Table 15.1 is used in a steam power plant. An Orsat analysis of the products of combustion yields the following results:

CO_2	13.1%
CO	1.3
O_2	3.3
N_2	82.3

It is determined that the ash contains 40% combustible material, which is assumed to be carbon. Determine the air-fuel ratio and the per cent of excess air.

15.4 Natural gas *B* in Table 15.3 is burned with 30% excess air.

(*a*) What is the dew point of the products if the total pressure is 1 atm?

(*b*) How many moles of water will be condensed per mole of fuel if the products are cooled to 100 F while the pressure remains 1 atm?

15.5 Octane is burned with the theoretical air in a constant pressure process ($P = 14.7$ lbf/in.^{2}) and the products are cooled to 80 F.

(*a*) How many lbm of water are condensed per lbm of fuel?

(*b*) Suppose the air used for combustion has a relative humidity of 90% and is at a temperature of 80 F and 14.7 lbf/in.^{2} pressure. How many lbm of water will be condensed per lbm of fuel when the products are cooled to 80 F?

15.6 Determine the enthalpy of combustion at 77 F and 1 atm. pressure of natural gas *B* in Table 15.3.

15.7 (*a*) Determine the enthalpy of formation of liquid benzene at 77 F.

(*b*) Benzene (l) at 77 F is burned with air at 400 F in a steady flow process. The products of combustion are cooled to 2060 F and have the following Orsat analysis:

	% by volume
CO_2	10.7
CO	3.6
O_2	5.3
N_2	80.4

Calculate the heat transfer per mole of fuel during the combustion process.

15.8 A small air-cooled gasoline engine is tested, and the output is found to be 1.34 hp. The temperature of the products is measured and found to be 730 F. The products are analyzed with an Orsat, with the following results:

CO_2	11.4%
O_2	1.6
CO	2.9
N_2	84.1

The fuel used may be considered to be C_8H_{18}, and the air and fuel enter the engine at 77 F. The rate at which fuel is used is 1.20 lbm/hr.

(*a*) Determine the heat transfer from the engine.

(*b*) What is the efficiency of the engine?

15.9 The boron hydrides have been considered as a "superfuel." Determine the enthalpy of combustion of pentaborane, B_5H_9 (g). The oxide is B_2O_3 which is a solid at room temperature. The enthalpies of formation at 77 F are as follows:

B_5H_9 (g)	27,000 Btu/lb mole
B_2O_3 (s)	−544,000 Btu/lb mole

Determine the enthalpy of combustion per lbm of fuel. How does this compare with a hydrocarbon fuel?

15.10 From data given for the enthalpy of combustion at 77 F of gaseous octane (C_8H_{18}) with vapor H_2O in the products, calculate the enthalpy of combustion of liquid octane at 77 F with liquid H_2O in the products. The latent heat of vaporization of octane is 156 Btu/lbm.

15.11 Calculate the internal energy of combustion of gaseous octane (C_8H_{18}) with liquid H_2O in the products.

15.12 Gaseous propane at 77 F is mixed with air at 300 F and burned; 300% theoretical air is used. What is the adiabatic flame temperature?

15.13 Gaseous propane at 77 F is mixed with air at 300 F and burned. What percentage of theoretical air must be used if the temperature of the products is to be 1600 F? Assume an adiabatic process and complete combustion.

15.14 Repeat Problem 15.13 assuming a heat transfer to the surroundings of 100,000 Btu/mol fuel and that 5% of the carbon in the fuel burns to form CO and 95% burns to form CO_2.

15.15 A bomb is charged with 1 mole of CO and 1 mole of O_2. The total pressure is 1 atm and the temperature is 537 R before combustion. Combustion then occurs and the products are cooled to 2500 R. Assume complete combustion. Determine:

(*a*) The final pressure. (Note that the number of moles changes during combustion.)

(*b*) The heat transfer.

15.16 A stoichiometric mixture of CO and air, initially at 77 F, reacts in a steady-flow process. What is the adiabatic flame temperature? Assume complete combustion.

15.17 Nitric acid is sometimes used as the oxidizer in liquid-fuel rockets. As an initial approach to the problem consider the following reactions involving the combustion of C_8H_{18} (l):

$$10HNO_3 + C_8H_{18}\ (l) \rightarrow 8CO_2 + 14H_2O\ (g) + 5N_2$$

$$C_8H_{18} + 12.5O_2 \rightarrow 8CO_2 + 9H_2O\ (g)$$

Determine the enthalpy of combustion for these two reactions per lbm of *reactants*. The enthalpy of formation of HNO_3 is $-88{,}600$ Btu/lb mole.

15.18 The natural gas *A* from Table 15.3 is burned with 150% theoretical air. Calculate the adiabatic flame temperature for complete combustion. Initial temperature of reactants = 77 F.

15.19 The Pennsylvania coal listed in Table 15.1 is burned with 30% excess air. If complete combustion is achieved, estimate the adiabatic flame temperature. Initial temperature of reactants = 77 F.

15.20 Consider the combustion chamber of a gas turbine. Liquid octane enters at 77 F and the air enters at 300 F. Of the carbon in the fuel 95% burns to CO_2 and 5% burns to CO. What amount of excess air will be required if the temperature of the products is to be 1600 F?

15.21 Sulfur at 77 F is burned with 30% excess air, the air being at a temperature of 300 F. Assuming all the sulfur is burned to SO_2, calculate the adiabatic flame temperature. The enthalpy of formation of SO_2 is −127,700 Btu/lb mole. The constant-pressure specific heat of SO_2 is given by the relation

$$\bar{C}_p = 7.70 + .0029T - 0.26 \times 10^{-6}T^2$$

where $\bar{C}_p$ = Btu/lb mole F

T = °R

15.22 The enthalpy of formation of magnesium oxide, MgO (s) is −143.84 k cal/gm mole at 25 C. The melting point of magnesium oxide is approximately 3000 K, and the increase in enthalpy between 298 K and 3000 K is 30.7 k cal/gm mole. The enthalpy of sublimation at 3000 K is estimated at 100 k cal/gm mole, and the specific heat of magnesium-oxide vapor above 3000 K is estimated at 8.9 cal/gm mole K.

(*a*) Determine the enthalpy of combustion per lbm of magnesium.

(*b*) Estimate the adiabatic flame temperature when magnesium is burned with theoretical oxygen.

15.23 Hydrogen peroxide is sometimes used as the oxidizer in special power plants such as torpedoes and rockets. Determine the enthalpy of combustion at 77 F per lbm of reactants for the following combustion process:

$$4H_2O_2\ (l) + CH_4 \rightarrow 6H_2O + CO_2$$

The enthalpy of formation H_2O_2 (l) is −80,700 Btu/lb mole.

15.24 One mole of carbon at 1 atm, 77 F and 1 mole of O_2 at 1 atm, 400 F react to form CO_2 in a steady-flow process. The CO_2 leaves at 2000 F, 1 atm pressure. Determine:

(*a*) The heat transfer per mole of fuel.

(*b*) The change of entropy per mole of fuel.

(*c*) The total change of entropy (system plus surroundings) if the heat is transferred to the surroundings.

15.25 Consider the combustion of gaseous propane at 14.7 lbf/in.^{2}, 77 F, with theoretical air at 14.7 lbf/in.^{2}, 77 F in a steady-flow process. Assume that combustion is complete and that the products leave at 77 F. Determine the decrease in Gibbs function for the two following cases:

(*a*) The reactants and products are both separated into their various constituents, and each constituent is at 1 atm pressure, 77 F. This is shown schematically in Fig. 15.11.

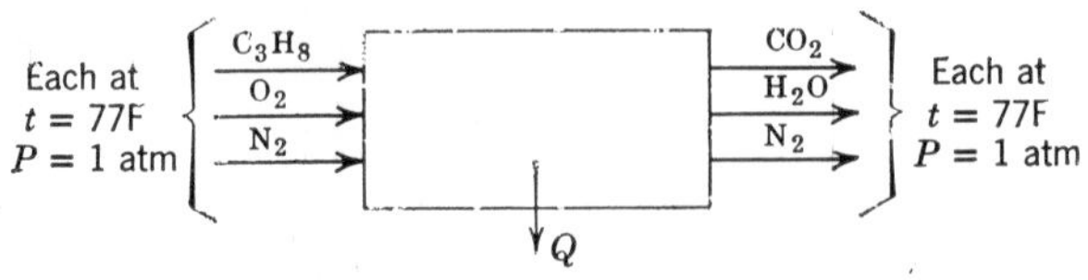

Fig. 15.11 Sketch for Problem 15.25*a*

(*b*) The reactants and products each consist of a mixture at a total pressure of 1 atm and a temperature of 77 F. This is shown schematically in Fig. 15.12.

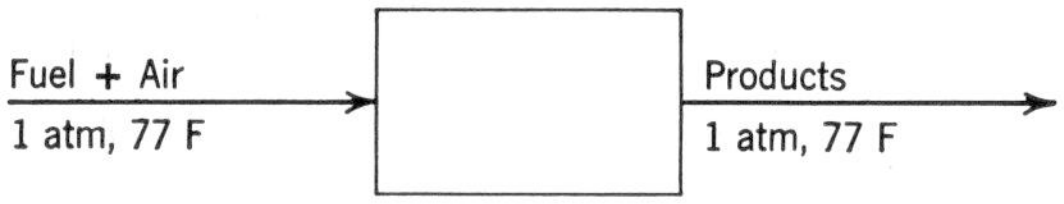

Fig. 15.12 Sketch for Problem 15.25*b*.

15.26 Determine the increase of entropy that takes place during the combustion process of Problem 15.16. What is the irreversibility of this process per mole of fuel? The total pressure during this process is 1 atm.

15.27 Consider one cylinder of a spark-ignition internal-combustion engine. Before the compression stroke the cylinder is filled with a mixture of air and ethane. Assume that 150% theoretical air has been used, and that the pressure is 1 atm and the temperature 77 F before compression. The compression ratio of the engine is 9 to 1.

(*a*) Determine the pressure and temperature after compression assuming a reversible adiabatic compression. Assume that the ethane behaves as an ideal gas.

(*b*) Assume further that complete combustion is achieved while the piston remains at head end dead center (i.e., after the reversible adiabatic compression), and that the combustion process is adiabatic. Determine the temperature and pressure after combustion, and the increase in entropy during the combustion process.

(*c*) What is the irreversibility for this process?

CHAPTER 16

EQUILIBRIUM

The matter of equilibrium is of great importance in many thermodynamic problems. The equilibrium between the constituents of a chemically reacting system; the equilibrium between various solutes in a solvent, and the equilibrium of a vapor bubble in a liquid are examples. The objectives of this text preclude a detailed consideration of this matter, but an introduction to the subject is considered an expedient topic with which to conclude. Particular attention will be focused on the equilibrium of chemical reactions that involve ideal gases, as this is a frequently encountered problem.

16.1 General Requirements for Equilibrium

As a general requirement for equilibrium we postulate that two systems are in equilibrium when there is no possibility of doing work by any interaction between the two systems. In this regard consider

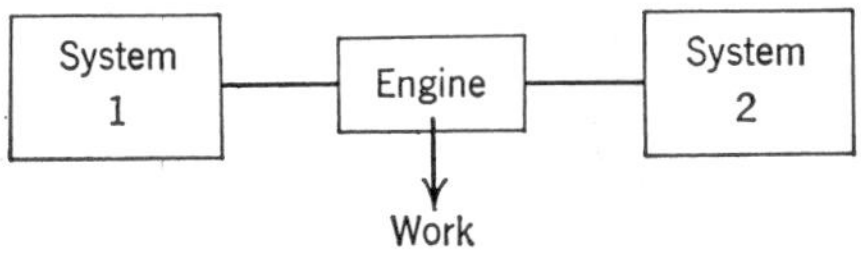

Fig. 16.1 Two systems, which communicate through an engine.

Fig. 16.1, which shows two systems connected by an engine. Now we must consider all possible reversible interactions between these two systems, and if for all these possible interactions the maximum useful work (as defined in Chapter 10, the maximum useful work is the total work done during a reversible process minus the work done against the surroundings) is zero, we conclude that the two systems are in equilibrium. We should also note that one of the systems could be the surroundings, and therefore we are also considering the matter of equilibrium between a system and its surroundings.

Now the first requirement for equilibrium is that these two systems have the same temperature, for otherwise we could operate a heat engine between the two systems and do work. It is also evident that these two systems must have the same pressure, for otherwise we could operate a turbine or piston type engine and do work. Also these two systems must have the same kinetic and potential energy, or else engines could be devised which would permit work to be done by an interaction between the two systems. Similar requirements could be set forth for other effects, such as surface, electrical or magnetic effects.

Now a general requirement for equilibrium may be developed by a consideration of maximum useful work, as developed in Chapter 10. We noted in Eq. 10.35 that the maximum useful work, W_{max}, that a system in temperature and pressure equilibrium with its surroundings (and in the absence of the effects as noted above) could produce is given by the relation

$$-W_{max} = \Delta G + \Delta KE + \Delta PE$$

Since the requirement for equilibrium is that $W_{max} = 0$, it follows that, if we assume that $\Delta KE = 0$ and $\Delta PE = 0$, the requirement for equilibrium is given by the relation

$$\Delta G = 0 \tag{16.1}$$

This will be the general requirement for equilibrium that we will use in this chapter, namely, that two systems (or a system and its surroundings) which have the same pressure and temperature are in equilibrium when for all possible changes of state $\Delta G = 0$.

16.2 Equilibrium Between Two Phases of a Pure Substance

As a simple application of this requirement for equilibrium, let us consider the equilibrium between two phases of a pure substance. If the liquid and vapor phases are at the same pressure and temperature we know that they are in equilibrium, and we would expect the Gibbs function for the liquid and vapor to be equal. Let us check this by determining the Gibbs function of saturated liquid (water) and saturated vapor (steam) at 14.7 lbf/in.[2] From the steam tables:

For the liquid:

$$g_f = h_f - Ts_f = 180.07 - 671.7 \times 0.3120 = -29.5 \text{ Btu/lbm}$$

For the vapor:

$$g_g = h_g - Ts_g = 1150.4 - 671.7 \times 1.7566 = -29.5 \text{ Btu/lbm}$$

This is exactly what we would expect from the relation

$$T\,ds = dh - v\,dP$$

Since the change of phase takes place at constant pressure and temperature this relation can be integrated as follows:

$$\int_f^g T\,ds = \int_f^g dh$$

$$T(s_g - s_f) = (h_g - h_f)$$

$$h_f - Ts_f = h_g - Ts_g$$

$$g_f = g_g$$

The Clapeyron equation, which was derived in Section 14.3 can be derived by an alternate method by considering the fact that the Gibbs functions of two phases in equilibrium are equal. In Chapter 14 we considered the relation (Eq. 14.9)

$$dg = v\,dP - s\,dT$$

Consider a system that consists of a saturated liquid and a saturated vapor in equilibrium, and let this system undergo a change of pressure dP. The corresponding change in temperature, as determined from the vapor-pressure curve, is dT. Both phases will undergo the change in Gibbs function, dg, but since the phases always have the same value of the Gibbs function when they are in equilibrium, it follows that

$$dg_f = dg_g$$

But, from Eq. 14.9,

$$dg = v\,dP - s\,dT$$

it follows that

$$dg_f = v_f\,dP - s_f\,dT$$

$$dg_g = v_g\,dP - s_g\,dT$$

Since

$$dg_f = dg_g$$

it follows that

$$v_f\,dP - s_f\,dT = v_g\,dP - s_g\,dT$$

$$dP(v_g - v_f) = dT(s_g - s_f)$$

$$\frac{dP}{dT} = \frac{s_{fg}}{v_{fg}} = \frac{h_{fg}}{Tv_{fg}} \tag{16.2}$$

In summary, when different phases of a pure substance are in equilibrium each phase has the same value of the Gibbs function per unit mass. This fact is relevant to different solid phases of a pure substance and is important in metallurgical applications of thermodynamics. The example that follows, though slightly different from most of our examples, illustrates this principle.

Example 16.1 What pressure is required to make diamonds from graphite at a temperature of 25 C? The following data are given for a temperature of 25 C and a pressure of 1 atm.

	Graphite	*Diamond*
$\bar{g}$	0	1,233 Btu/lb mole R
v	0.00712 ft³/lbm	0.00456 ft³/lbm
β_T	3.0×10^{-6} atm^{-1}	0.16×10^{-6} atm^{-1}

The basic principle in the solution is that graphite and diamond can exist in equilibrium when they have the same value of the Gibbs function. At 1 atm pressure the Gibbs function of the diamond is greater than that of the graphite. However, the rate of increase in Gibbs function with pressure is greater for the graphite than the diamond, and therefore, at some pressure they can exist in equilibrium, and our problem is to find this pressure.

We have already considered the relation

$$dg = v\,dP - s\,dT$$

Since we are considering a process that takes place at constant temperature this reduces to

$$dg_T = v\,dP_T \tag{a}$$

Now at any pressure P and the given temperature the specific volume can be found from the following relation, which utilizes the isothermal compressibility factor.

$$v = v^\circ + \int_{P=1}^{P} \left(\frac{\partial v}{\partial P}\right)_T dP = v^\circ + \int_{P=1}^{P} \frac{v}{v}\left(\frac{\partial v}{\partial P}\right)_T dP = v^\circ - \int_{P=1}^{P} v\beta_T\,dP \tag{b}$$

The superscript ° will be used in this example to indicate the properties at a pressure of 1 atm and a temperature of 25 C.

The specific volume changes only slightly with pressure, so that $v \approx v^\circ$. Also, we assume that β_T is constant and that we are considering a pressure of many atmospheres. With these assumptions this equation can be integrated to give

$$v = v^\circ - v^\circ\beta_T P = v^\circ(1 - \beta_T P) \tag{c}$$

We can now substitute this into Eq. a to give the relation

$$dg_T = [v^\circ(1 - \beta_T P)]\, dP_T \tag{d}$$

$$g - g^\circ = v^\circ(P - P^\circ) - v^\circ\beta_T \frac{(P^2 - P^{\circ 2})}{2}$$

If we assume that $P^\circ \ll P$ this reduces to

$$g - g^\circ = v^\circ\left[P - \frac{\beta_T P^2}{2}\right] \tag{e}$$

For the graphite, $g^\circ = 0$ and we can write

$$g_G = v_G{}^\circ\left[P - (\beta_T)_G \frac{P^2}{2}\right]$$

For the diamond, g° has a definite value and we have

$$g_D = g_D{}^\circ + v_D{}^\circ\left[P - (\beta_T)_D \frac{P^2}{2}\right]$$

But, at equilibrium the Gibbs function of the graphite and diamond are equal.

$$g_G = g_D$$

Therefore,

$$v_G{}^\circ\left[P - (\beta_T)_G \frac{P^2}{2}\right] = g_D{}^\circ + v_D{}^\circ\left[P - (\beta_T)_D \frac{P^2}{2}\right]$$

$$(v_G{}^\circ - v_D{}^\circ)P - [v_G{}^\circ(\beta_T)_G - v_D{}^\circ(\beta_T)_D]\frac{P^2}{2} = g_D{}^\circ$$

$$(0.00712 - 0.00456)P - (0.00712 \times 3.0 \times 10^{-6} - 0.00456 \times 0.16 \times 10^{-6})\frac{P^2}{2} = \frac{1233}{12} \times \frac{778}{14.7 \times 144}$$

Solving this for P we find

$$P = 15{,}750 \text{ atm}$$

That is, at 15,500 atm, 25 C, graphite and diamond can exist in equilibrium, and the possibility exists for conversion from graphite to diamonds.

16.3 Equilibrium of a Mixture of Ideal Gases Undergoing a Chemical Reaction

Consider the following chemical reaction.

$$aA + bB \rightleftarrows cC + dD \tag{16.3}$$

where A, B, C, D represent various substances
a, b, c, d represent the corresponding stoichiometric coefficients

In this section we will assume that all of the substances involved represent ideal gases. The double arrow indicates that when equilibrium is achieved each substance will be present in a certain amount, and each substance will exist at its partial pressure.

Let us assume that for each substance we know the Gibbs function for all temperatures at a pressure of 1 atm, and let us denote this value of the Gibbs function as $g°$. This is equivalent to saying that we know the specific heat as a function of temperature at 1 atm pressure for each substance. Then at any other pressure and a given temperature the value of the Gibbs function g is given by the relation

$$g - g° = RT \ln p \tag{16.4}$$

where p is the partial pressure in atmospheres
$g°$ is the value of the Gibbs function at the same temperature and a pressure of 1 atmosphere

This relation is derived as follows:

$$dg = v\,dP - s\,dT$$

But, we are always considering g and $g°$ at the same temperature. Therefore,

$$g - g° = \int_P^p v\,dP = \int_P^p RT\frac{dP}{P} = RT \ln \frac{p}{P}$$

However, if the pressures are expressed in atmospheres (i.e., the pressure at which we know $g°$),

$$P = 1 \text{ and therefore, } \ln P = 0$$

and it follows that

$$g - g° = RT \ln p$$

If we consider the mixture of gases given by Eq. 16.3, we can write Eq. 16.4 for each component on a mol basis:

$$\begin{aligned} \bar{g}_A - \bar{g}_A^{\circ} &= \bar{R}T \ln p_A \\ \bar{g}_B - \bar{g}_B^{\circ} &= \bar{R}T \ln p_B \\ \bar{g}_C - \bar{g}_C^{\circ} &= \bar{R}T \ln p_C \\ \bar{g}_D - \bar{g}_D^{\circ} &= \bar{R}T \ln p_D \end{aligned} \tag{16.5}$$

Now let us write an expression for the change in Gibbs function as the reaction given by Eq. 16.3 goes to completion; that is, we start with reactants A and B and we end with the products C and D.

$$\Delta G = G_{\text{products}} - G_{\text{reactants}}$$

$$\Delta G = c\bar{g}_C + d\bar{g}_D - a\bar{g}_A - b\bar{g}_B$$

Substituting Eq. 16.5 into this equation we have

$$\Delta G = c[\bar{g}_C^{\circ} + \bar{R}T \ln p_C] + d[\bar{g}_D^{\circ} + \bar{R}T \ln p_D] - a[\bar{g}_A^{\circ} + \bar{R}T \ln p_A] - b[\bar{g}_B^{\circ} + \bar{R}T \ln p_B] \tag{16.6}$$

But we define

$$\Delta G^{\circ} = c\bar{g}_C^{\circ} + d\bar{g}_D^{\circ} - a\bar{g}_A^{\circ} - b\bar{g}_B^{\circ} \tag{16.7}$$

In words this equation says that ΔG° is the change in Gibbs function during the chemical reaction if the reaction goes to completion, and if each constituent in the reactants is at a pressure of 1 atm (and the given temperature) before the reaction takes place, and if each constituent in the products is also at a pressure of 1 atm (and the given temperature) after the reaction takes place.

Substituting Eq. 16.7 into Eq. 16.6 we have

$$\Delta G = \Delta G^{\circ} + \bar{R}T \ln \frac{p_C{}^c p_D{}^d}{p_A{}^a p_B{}^b} \tag{16.8}$$

Let us now consider the significance regarding a chemical reaction of the fact that at equilibrium $\Delta G = 0$. Suppose we started the chemical reaction with a mols of substance A and b mols of substance B; that is, a stoichiometric mixture of the reactants. As the reaction proceeds, substances A and B disappear and substances C and D appear. If the reaction goes to completion, only C and D will finally be present. The term *degree of reaction* is used to indicate how far the reaction has proceeded to completion. Of course, at equilibrium all four substances will be present in some degree.

Let us schematically plot the sum of the Gibbs function for all constituents present as a function of the degree of reaction. We can do this by plotting the sum of the Gibbs functions of all constituents as a function of the number of moles of substance A. Before the reaction starts there are a moles of substance A present. When the reaction is complete, the number of moles of substance A will be zero. Such a plot is shown schematically in Fig. 16.2. The equilibrium point is indicated on the diagram as the point where the Gibbs function is a minimum. This is somewhat similar to a marble in a bowl; it is in equilibrium when its potential energy is minimum. Thus equilibrium in a chemical reaction is achieved when

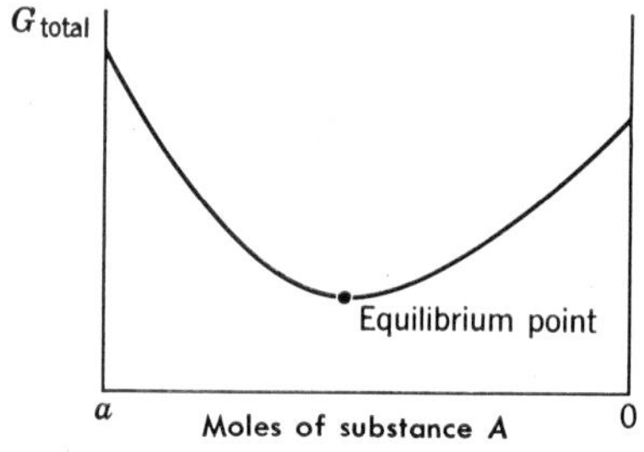

Fig. 16.2 Equilibrium point of a system undergoing a chemical reaction.

$$\left(\frac{\partial G_{\text{total}}}{\partial n_a}\right)_T = 0$$

This is what is implied when we say that the requirement for chemical equilibrium is that $\Delta G = 0$.

Returning now to Eq. 16.8, at equilibrium, $\Delta G = 0$. Therefore, at equilibrium Eq. 16.8 reduces to

$$\Delta G^\circ = -\bar{R}T \ln \frac{{p_C}^c {p_D}^d}{{p_A}^a {p_B}^b} \tag{16.9}$$

The equilibrium constant K_p, is defined as

$$K_P = \frac{{p_C}^c {p_D}^d}{{p_A}^a {p_B}^b} \tag{16.10}$$

Therefore, we can write

$$\Delta G^\circ = -\bar{R}T \ln K_p \tag{16.11}$$

We can make two important observations regarding K_P. First of all, ΔG° is independent of pressure and is a function of temperature only, and therefore, the equilibrium constant is a function of temperature only and is independent of pressure. Therefore, values of the equilibrium constant are given as a function of temperature, either as a graph, a table, or an equation. However, we noted in writing Eq. 16.4, that pressures were expressed in atmospheres. Therefore, the value given for K_P depends on the units for pressure, and most values given for K_P are based on pressures in atmospheres. Table 16.1 gives the value of the equilibrium constant for a number of chemical reactions.

Example 16.2 Determine the value of the equilibrium constant of the reaction $H_2O \rightleftarrows H_2 + \frac{1}{2}O_2$ at 298 K from the values for $\bar{g}°$ given in Table 15.4, and compare the value so calculated with the value given in Table 16.1.

The equilibrium constant is defined by Eq. 16.11

$$\ln K_P = -\frac{\Delta G°}{\bar{R}T}$$

For the reaction

$$H_2O\ (g) \rightleftarrows H_2 + \tfrac{1}{2}O_2$$

$\Delta G°$ is simply

$$\Delta G° = -\mathring{\bar{g}}_{H_2O(g)}$$

(since $\bar{g}°$ for the elements is assumed to be zero at 298 K and 1 atm pressure).

The value for $\mathring{\bar{g}}_{H_2O(g)}$ is found from Table 15.4.

$$\Delta G° = -\mathring{\bar{g}}_{H_2O(g)} = -(-98{,}344) = 98{,}344 \text{ Btu}$$

Therefore,

$$\ln K_P = -\frac{\Delta G°}{\bar{R}T} = -\frac{98{,}344}{1.986 \times 536.7} = -92.26$$

$$\log_{10} K_P = \frac{-92.26}{2.3026} = -40.05$$

This checks the value given in Table 16.1.

It should be pointed out that the subscript P in the equilibrium constant K_P is intended to indicate that the equilibrium constant is defined in terms of pressure and refers to ideal gases. Other equilibrium constants may be defined which involve nonideal gases, solutions, etc., and these are usually designated with other subscripts. These considerations however, are beyond the scope of this text.

It should also be noted that the equilibrium constant as defined by Eq. 16.10 has the products in the numerator and the reactants in the denominator. This is the usual procedure, but occasionally the equilibrium constant is defined as the reciprocal of this. Whenever one uses the value of the equilibrium constant from a table care should be exercised to note the exact way in which the equilibrium constant is defined.

TABLE 16.1

Logarithms to the Base 10 of the Equilibrium Constant K_p *

For the reaction $aA + bB \rightleftarrows cC + dD$ the equilibrium constant K_p is defined as

$$K_p = \frac{p_C{}^c p_D{}^d}{p_A{}^a p_B{}^b} \quad \text{where } p \text{ is the partial pressure in atmospheres}$$

Temp, °K	$H_2 \leftrightarrows 2H$	$O_2 \rightleftarrows 2O$	H_2O (g) $\rightleftarrows H_2 + \frac{1}{2}O_2$	H_2O (g) $\rightleftarrows OH + \frac{1}{2}H_2$	$CO_2 \rightleftarrows CO + \frac{1}{2}O_2$	$CO_2 + H_2 \leftrightarrows CO + H_2O$ (g)	$N_2 \rightleftarrows 2N$	$\frac{1}{2}O_2 + \frac{1}{2}N_2 \rightleftarrows NO$
298	−71.210	−80.620	−40.047	−46.593	−45.043	−4.996	−119.434	−15.187
400	−51.742	−58.513	−29.241	−33.910	−32.41	−3.169	− 87.473	−11.156
600	−32.667	−36.859	−18.633	−21.470	−20.07	−1.432	− 56.206	− 7.219
800	−23.074	−25.985	−13.288	−15.214	−13.90	−0.617	− 40.521	− 5.250
1000	−17.288	−19.440	−10.060	−11.444	−10.199	−0.139	− 31.084	− 4.068
1200	−13.410	−15.062	− 7.896	− 8.922	− 7.742	+0.154	− 24.619	− 3.279
1400	−10.627	−11.932	− 6.344	− 7.116	− 5.992	+0.352	− 20.262	− 2.717
1600	− 8.530	− 9.575	− 5.175	− 5.758	− 4.684	+0.490	− 16.869	− 2.294
1800	− 6.893	− 7.740	− 4.263	− 4.700	− 3.672	+0.591	− 14.225	− 1.966
2000	− 5.579	− 6.269	− 3.531	− 3.852	− 2.863	+0.668	− 12.106	− 1.703
2200	− 4.500	− 5.064	− 2.931	− 3.158	− 2.206	+0.725	− 10.370	− 1.488
2400	− 3.598	− 4.055	− 2.429	− 2.578	− 1.662	+0.767	− 8.922	− 1.309
2600	− 2.833	− 3.206	− 2.003	− 2.087	− 1.203	+0.800	− 7.694	− 1.157
2800	− 2.176	− 2.475	− 1.638	− 1.670	− 0.807	+0.831	− 6.640	− 1.028
3000	− 1.604	− 1.840	− 1.322	− 1.302	− 0.469	+0.853	− 5.726	− 0.915
3200	− 1.104	− 1.285	− 1.046	− 0.983	− 0.175	+0.871	− 4.925	− 0.817
3500	− 0.458	− 0.571	− 0.693	− 0.577	+ 0.201	+0.894	− 3.893	− 0.692
4000	+ 0.406	+ 0.382	− 0.221	− 0.035	+ 0.699	+0.920	− 2.514	− 0.526
4500	+ 1.078	+ 1.125	+ 0.153	+ 0.392	+ 1.081	+0.928	− 1.437	− 0.345
5000	+ 1.619	+ 1.719	+ 0.450	+ 0.799	+ 1.387	+0.937	− 0.570	− 0.298

* Values taken from or computed from "Selected Values of Chemical Thermodynamic Properties," Series III. National Bureau of Standards, Washington, D.C.

16.4 Use of the Equilibrium Constant to Determine Equilibrium Composition

In this section we consider how the equilibrium constant as defined by Eq. 16.10 can be used to determine the equilibrium composition for a chemical reaction involving ideal gases. We note first of all, that for a mixture of ideal gases, the mole fraction of a given constituent is equal to the ratio of the partial pressure of that constituent to the total pressure. Thus, for substance A

$$x_A P_T = p_A$$
$$x_A{}^a P_T{}^a = p_A{}^a \tag{16.12}$$

If we write similar relations for the other constituents and substitute these into Eq. 16.10, we have the following relation:

$$K_P = \frac{x_C{}^c x_D{}^d}{x_A{}^a x_B{}^b} P_T{}^{c+d-a-b} \tag{16.13}$$

This is called the law of mass action. It should be noted from this expression that the influence of temperature on the equilibrium composition is given by the value of K_P, since this is a function of temperature. The influence of pressure on the equilibrium composition is given by the term $P_T{}^{c+d-a-b}$. It is interesting to note that if $c + d = a + b$ (which means that the total number of moles does not change and therefore there is no change in volume during a chemical reaction at constant pressure and temperature) a change in pressure has no effect on the equilibrium composition. If there is a change in the number of moles, and therefore a change in volume, it is readily shown from Eq. 16.13 that the effect of an increase in pressure is to drive the reaction in the direction which has the smaller volume.

The matter of determining the equilibrium composition for a chemical reaction involving an ideal gas is illustrated by the following example.

Example 16.3 One mole of carbon dioxide at 77 F and 1 atm pressure is heated in a constant-pressure steady-flow process to 3000 K. Determine the equilibrium composition at 3000 K and the heat transfer for the process.

The equation for this reaction is initially written

$$CO_2 \rightarrow a\,CO_2 + b\,CO + d\,O_2$$

By using a carbon balance and an oxygen balance this equation can be simplified to

$$CO_2 \rightarrow a\,CO_2 + (1-a)CO + \frac{(1-a)}{2}\,O_2$$

In the solution of the problem the most convenient approach is to first find the partial pressures. Since there are three constituents, we must have three equations involving the partial pressures. Table 16.1 gives values for the equilibrium constant K_P for the reaction $CO_2 \rightleftarrows CO + \frac{1}{2}O_2$ as a function of temperature. Thus we can write the first equation involving partial pressures.

$$K_P = \frac{p_{CO} \times \sqrt{p_{O_2}}}{p_{CO_2}} \tag{a}$$

A second equation involving partial pressures can be written which simply states that the total pressure is the sum of the partial pressures.

$$P_T = p_{CO_2} + p_{CO} + p_{O_2} \tag{b}$$

A third equation can be written which involves the ratio of carbon atoms to oxygen atoms. Since initially only CO_2 was present, the ratio of carbon atoms to oxygen atoms is 1:2, and this ratio remains fixed. At equilibrium CO_2, CO, and O_2 are all present. Each CO_2 molecule contains one atom of carbon and two atoms of oxygen; each molecule of CO contains one carbon atom and one oxygen atom; each oxygen molecule contains two oxygen atoms. Thus we can write

$$\frac{\text{atoms carbon}}{\text{atoms oxygen}} = \frac{1}{2} = \frac{n_{CO_2} + n_{CO}}{2n_{CO_2} + n_{CO} + 2n_{O_2}} = \frac{a+b}{2a+b+2d} \tag{c}$$

This equation can also be written in terms of partial pressures because

$$\frac{n_{CO_2}}{n_T} = \frac{p_{CO_2}}{P_T} \qquad \frac{n_{CO}}{n_T} = \frac{p_{CO}}{P_T} \qquad \frac{n_{O_2}}{n_T} = \frac{p_{O_2}}{P_T}$$

Therefore,

$$\frac{\text{atoms carbon}}{\text{atoms oxygen}} = \frac{1}{2} = \frac{p_{CO_2} + p_{CO}}{2p_{CO_2} + p_{CO} + 2p_{O_2}} \tag{d}$$

Thus, we have three equations. (Equations a, b, and d) and three unknowns (p_{CO_2}, p_{CO}, p_{O_2}). We proceed by eliminating p_{CO_2} and p_{CO} and then solve for p_{O_2}. From Eqs. a, b, and d we can obtain the following equation in terms of K_P and P_T (which are known for a given problem) and p_{O_2}.

$$K_P^2 = \frac{4p_{O_2}^3}{(P_T - 3p_{O_2})^2} \tag{e}$$

From Table 16.1 we find the value of $\log_{10} K_P$ at 3000 K to be -0.469. Therefore, $K_P = 0.34$. From the statement of the problem, $P_T = 1$ atm. Thus, from Eq. (e)

$$(0.34)^2 = \frac{4p_{O_2}{}^3}{(1 - 3p_{O_2})^2}$$

Solving this by trial and error we determine that $p_{O_2} = 0.182$ atm, and substituting into the other equations we find $p_{CO} = 0.364$ atm and $p_{CO_2} = 0.454$ atm.

To find the number of mols of each constituent we note that

$$\frac{p_{CO_2}}{P_T} = \frac{n_{CO_2}}{n_{CO_2} + n_{CO} + n_{O_2}} = \frac{a}{a + (1 - a) + (1 - a)/2} = \frac{2a}{3 - a} = 0.454$$

Solving for a we find

$$a = 0.555$$

From a carbon balance we conclude that $a + b = 1$. Therefore,

$$b = 0.445$$

From an oxygen balance we find

$$d = 0.223$$

Therefore, for this process the chemical reaction is

$$CO_2 \rightarrow 0.555CO_2 + 0.445CO + 0.223O_2$$

The heat transfer for this process can be calculated by using the enthalpy of formation and enthalpy values from the *Gas Tables.* For this process

$$H_R + Q = H_P$$

$$H_R = \bar{h}^\circ_{CO_2} = -169{,}293 \text{ Btu/mole reactant}$$

$$\begin{aligned} H_P &= n_{CO_2}(\bar{h}^\circ + \bar{h}_{5400} - \bar{h}_{537})_{CO_2} + n_{CO}(\bar{h}^\circ + \bar{h}_{5400} - \bar{h}_{537})_{CO} \\ &\quad + n_{O_2}(\bar{h}_{5400} - \bar{h}_{537})_{O_2} \\ &= 0.555(-169{,}293 + 69{,}975 - 4030) + 0.445(-47{,}548 \\ &\quad + 43{,}953 - 3730) + 0.223(45{,}872 - 3725) \\ &= -51{,}219 \text{ Btu/mole reactant} \end{aligned}$$

$$\begin{aligned} Q &= H_P - H_R = -51{,}219 - (-169{,}293) \\ &= 118{,}074 \text{ Btu/mole reactant} \end{aligned}$$

Example 16.4 One mole of carbon monoxide at 77 F reacts with 1 mole of oxygen at 77 F in a steady-flow process. The final pressure is 1 atm and products consist of a mixture of CO, CO_2, and O_2 in equilibrium at the adiabatic flame temperature. Determine the adiabatic flame temperature and the equilibrium composition.

The chemical reaction in this case is

$$CO + O_2 \rightarrow a\, CO_2 + b\, CO + d\, O_2$$

By means of a carbon balance and oxygen balance this can be simplified to

$$CO + O_2 \rightarrow a\, CO_2 + (1 - a)CO + \left(\frac{2 - a}{2}\right) O_2$$

The coefficients a, b, and d must be determined. The approach to this problem is to assume a final temperature and determine the equilibrium composition and necessary heat transfer to achieve this temperature. By trial and error a temperature can be found for which the heat transfer is zero, and this is the adiabatic flame temperature.

We proceed by writing three equations involving the partial pressures. The equilibrium constant is given in Table 16.1 for the reaction $CO_2 \rightleftarrows CO + \frac{1}{2}O_2$. Therefore,

$$K_P = \frac{p_{CO}\sqrt{p_{O_2}}}{p_{CO_2}} \qquad \text{(a}$$

The second equation involves the sum of the partial pressures.

$$P_T = p_{CO_2} + p_{CO} + p_{O_2} \qquad \text{(b)}$$

The third equation involves the ratio of carbon atoms to oxygen atoms.

$$\frac{\text{atoms carbon}}{\text{atoms oxygen}} = \frac{1}{3} = \frac{n_{CO_2} + n_{CO}}{2n_{CO_2} + n_{CO} + 2n_{O_2}} \qquad \text{(c)}$$

Solving Eqs. a, b, and c simultaneously we have the relation

$$K_P = \frac{(3p_{O_2} - P_T)\sqrt{p_{O_2}}}{2P_T - 4p_{O_2}} \qquad \text{(d)}$$

Let us take as our first assumption a final temperature of 3000 K (5400 R). From Table 16.1 we find that at 3000 K, $\log_{10} K_P = -0.469$. Therefore,

$$K_P = 0.34$$

Substituting this into Eq. d we have, since $P_T = 1$,

$$0.34 = \frac{(3p_{O_2} - 1)\sqrt{p_{O_2}}}{2 - 4p_{O_2}}$$

By a trial-and-error solution we find

$$p_{O_2} = 0.402 \text{ atm} \qquad p_{CO} = 0.206 \text{ atm} \qquad p_{CO_2} = 0.392 \text{ atm}$$

The coefficients a, b, and d must be determined.

$$\frac{p_{CO_2}}{P_T} = \frac{n_{CO_2}}{n_T} = \frac{a}{a + (1 - a) + (2 - a)/2} = \frac{2a}{4 - a} = 0.392$$

$$a = 0.655 \qquad b = 0.345 \qquad d = 0.672$$

Thus, the equation for this reaction is

$$CO + O_2 \rightarrow 0.655CO_2 + 0.345CO + 0.672O_2$$

The heat transfer for this assumed temperature can now be calculated

$$Q = H_P - H_R$$

$$H_P = n_{CO_2}(\bar{h}^\circ + \bar{h}_{5400} - \bar{h}_{537})_{CO_2} + n_{CO}(\bar{h}^\circ + \bar{h}_{5400} - \bar{h}_{537})_{CO} + n_{O_2}(\bar{h}_{5400} - \bar{h}_{537})_{O_2}$$

$$= 0.655(-169{,}293 + 69{,}975 - 4030) + 0.345(-47{,}548 + 43{,}953 - 3730) + 0.672(45{,}872 - 3725)$$

$$= -41{,}897 \text{ Btu}$$

$$H_R = \bar{h}_{CO} = -47{,}548 \text{ Btu}$$

$$Q = -41{,}897 - (-47{,}548) = 5{,}651 \text{ Btu}$$

Since the heat transfer for a final temperature of 3000 K is positive, the adiabatic flame temperature is less than 3000 K.

As our second assumption we take a final temperature of 2900 K (5220 R). Following the same procedure we find the following for this temperature:

$$\log_{10} K_P = -0.638 \qquad K_P = 0.230$$

$$p_{O_2} = 0.389 \text{ atm} \qquad p_{CO} = 0.167 \text{ atm} \qquad p_{CO_2} = 0.444 \text{ atm}$$

$$n_{O_2} = 0.636 \qquad n_{CO} = 0.273 \qquad n_{CO_2} = 0.727$$

The chemical reaction is

$$CO + O_2 \rightarrow 0.727CO_2 + 0.273CO + 0.636O_2$$

$$H_P = -53{,}475 \text{ Btu}$$

$$H_R = -47{,}548 \text{ Btu}$$

$$Q = -53{,}475 - (-47{,}548) = -5927 \text{ Btu}$$

Thus, by linear interpolation we conclude that $Q = 0$ for this reaction at a temperature of 2951 K which is therefore the adiabatic flame temperature.

These two examples have involved very simple chemical reactions. Most actual combustion processes involve many different constituents. Thus, the combustion of methane with theoretical air might be as follows:

$$CH_4 + 2O_2 + 2(3.76)N_2 \rightarrow a\, CO_2 + b\, CO + c\, H_2O + d\, OH$$
$$+ e\, H_2 + f\, O_2 + g\, N_2 + h\, O + i\, N + j\, NO$$

The determination of the equilibrium composition would involve ten equations and ten unknowns. These ten equations would be obtained from six equilibrium constants, a carbon-hydrogen ratio, a carbon-oxygen ratio, a carbon-nitrogen ratio, and an equation involving the sum of the partial pressures and the total pressure. It is evident that the solution of such a problem is a tedious task, and modern tools for computation are being utilized in the solution of such problems. Reference should also be made to charts which have been prepared by H. C. Hottel and his associates.* These charts were prepared specifically for use in internal-combustion engines, and give the thermodynamic properties of the equilibrium mixture of the products of combustion on an internal energy-entropy chart. Charts are available for various air-fuel ratios. A detailed discussion of these charts is given in most textbooks dealing with internal-combustion engines.

16.5 Specific Impulse

An important parameter in the performance of rockets is the specific impulse. The specific impulse, I_s, is defined as the thrust F per unit mass rate of flow of propellant $\dot{m}$.

$$I_s = \frac{F}{\dot{m}} \tag{16.14}$$

* Hottel, H. C., *Thermodynamic Charts for Combustion Processes*, Wiley, New York, 1949.

The specific impulse of a rocket must be determined experimentally. The units frequently used are lbf-sec/lbm. (Sometimes the units for specific impulse are simply reported as sec, which is done by the dubious procedure of canceling the units of lbf against lbm.) The physical significance of this parameter is evident, because the range of a rocket is limited by its fuel carrying capacity, and the greater the thrust per unit mass rate of flow of propellant, the greater the range.

It is also the practice to determine the specific impulse of various propellants and propellant combinations (oxidizer and fuel). This is the specific impulse which would be achieved by a given propellant in an ideal rocket. An ideal rocket is one which operates in a steady-flow, adiabatic, frictionless process, and which has an axially directed exhaust of uniform velocity. It is further assumed in the ideal rocket that the products of combustion which leave the combustion chamber consist of a mixture of ideal gases in chemical equilibrium at the combustion chamber pressure and the adiabatic flame temperature.

When this mixture of gases leaves the combustion chamber and flows through the nozzle, a rapid decrease in pressure and temperature takes place. If chemical equilibrium is maintained during the expansion process, a change in composition must necessarily take place, the exact composition for a given propellant being determined by the pressure and temperature at each point in the nozzle. When this change in composition is assumed to take place, it is usually referred to as shifting equilibrium, and the specific impulse for a given propellant in an ideal rocket is referred to as the shifting equilibrium value of the specific impulse.

In an actual nozzle the gases remain in the nozzle a very short period of time, and equilibrium is not achieved as the gases flow through the nozzle. Therefore, in calculating the specific thrust for a given propellant in an ideal rocket, the assumption is frequently made that there is no change in the composition of the gases as they flow through the nozzle, and that the composition remains fixed and equal to the composition at the combustion chamber pressure and the adiabatic flame temperature. This is usually referred to as frozen equilibrium and the specific impulse so calculated is referred to as the specific impulse at frozen equilibrium. The composition at the exit of an actual nozzle is usually between the values obtained by these two assumptions.

The calculations involved in determining the equilibrium composition involve a rather tedious trial-and-error solution, and if shifting equilibrium is assumed this solution must be repeated at several points in the nozzle. The principles involved in this calculation have been covered in the previous section. The values of specific impulse for a few propellant combinations are given in Table 16.2. Note that the specific impulse multiplied by g_c gives the exhaust velocity.

TABLE 16.2 *

Propellant Combination Oxidizer and Fuel	Chamber Pressure, lbf/in.^{2}	Mixture Ratio, Oxidizer to Fuel	Exhaust Velocity, ft/sec	Specific Impulse, lbf-sec/lbm	Combustion Chamber Temp., F
Liquid oxygen and ammonia	300	1.4	8,220	255	4950
Liquid oxygen and liquid hydrogen	340	5.33	10,800	335	5430
Liquid oxygen and liquid hydrogen	500	8.0	10,120	317	5870
Liquid oxygen and liquid hydrogen	500	3.5	11,700	364	4500
Liquid oxygen and kerosene	300	2.2	7,970	248	5570
Fluorine and hydrazine	300	1.9	9,610	299	7530
Red fuming nitric acid and aniline	300	3.0	7,090	221	5020

* From G. P. Sutton, *Rocket Propulsion Elements*, p. 112, Wiley, New York, 1956.

When a given propellant combination is used in an actual rocket, the actual specific impulse is less than the ideal specific impulse by 5 to 10 per cent. However, the ideal specific impulse is an excellent parameter for evaluating various propellants and propellant combinations.

16.6 The Van't Hoff Isobar

An interesting and important relation between the equilibrium constant and the enthalpy of reaction may be derived. This relation is known as the Van't Hoff Isobar and is derived as follows.

From Eq. 16.11 and the definition of the Gibbs function we can write

$$\ln K_P = \frac{-\Delta G^\circ}{\bar{R}T} = \frac{-(\Delta H^\circ - T\,\Delta S^\circ)}{\bar{R}T} = -\frac{\Delta H^\circ}{\bar{R}T} + \frac{\Delta S^\circ}{\bar{R}}$$

In this relation ΔH° and ΔS° represent the change of enthalpy and entropy respectively when the reaction goes to completion with each constituent in the products at the given temperature and a pressure of 1 atm. When this relation is differentiated with respect to temperature we have

$$\frac{d \ln K_P}{dT} = \frac{\Delta H^\circ}{\overline{R}T^2}$$

$$\frac{d \ln K_P}{d\,(1/T)} = -\frac{\Delta H^\circ}{\overline{R}} \tag{16.15}$$

Thus, when the $\ln K_P$ is plotted against the reciprocal of the absolute temperature, the slope of the line is equal to $-\Delta H^\circ/\overline{R}$. This is shown

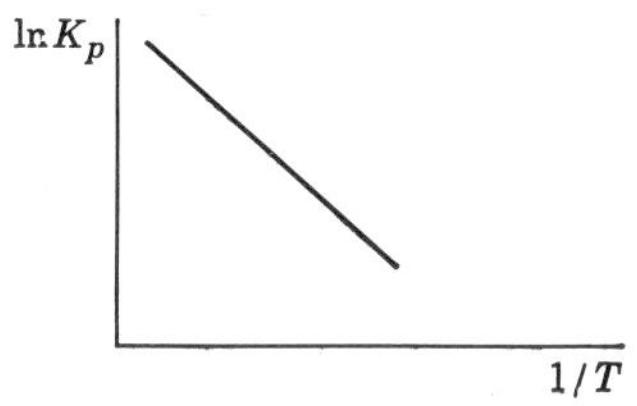

Fig. 16.3 Ln K_P vs. $1/T$ for a reacting system.

schematically in Fig. 16.3 for a chemical reaction which is endothermic. An exothermic reaction would have a positive slope.

16.7 The Gibbs Phase Rule

The matter of equilibrium between mixtures of different substances involving more than one phase is a complex problem, and in this last section of the book a few comments on this problem are given by way of introduction to the Gibbs phase rule. The Gibbs phase rule was derived by Professor J. Willard Gibbs of Yale University in 1875, and is one of the most important contributions to physical science that has ever been made.

A simple example of a mixture involving two phases and two components is a mixture of oxygen and nitrogen in which both liquid and vapor phases are present. In this case the liquid has one composition and the vapor another composition. Figure 16.4 shows the composition of the liquid and vapor phases which are in equilibrium as a function of temperature at a pressure of 1 atm. The upper line, marked "vapor line" gives the composition of the vapor phase, and the lower line gives the composition of the liquid phase. If we have pure nitrogen, the boiling point is 77.3 K. If we have pure oxygen, the boiling point is 90.2 K. If we have a mixture of nitrogen and oxygen with both phases present at a temperature of 84 K, the vapor will have a composition of 64 per cent N_2 and 36 per cent O_2. The liquid will have a composition of 30 per cent N_2 and 70 per cent O_2.

The requirement for equilibrium for such a mixture is best understood by introducing the chemical potential, which is given the symbol μ. If we consider only the gaseous phase of the mixture we can say that

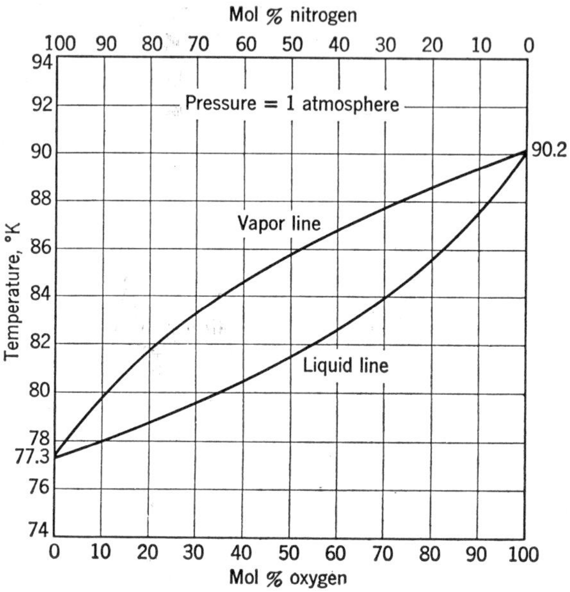

Fig. 16.4 Equilibrium diagram for liquid-vapor phases of the nitrogen-oxygen system at a pressure of 1 atm.

the Gibbs function of the gaseous phase depends on the temperature, pressure, and the number of mols of each substance in that phase. Thus,

$$G = f(T, P, n_1, n_2, n_3 \cdots) \tag{16.16}$$

Therefore,

$$dG = \left(\frac{\partial G}{\partial T}\right)_{P,n_1,n_2,n_3} dT + \left(\frac{\partial G}{\partial P}\right)_{T,n_1,n_2,n_3} dP + \left(\frac{\partial G}{\partial n_1}\right)_{T,P,n_2,n_3} dn_1 + \left(\frac{\partial G}{\partial n_2}\right)_{T,P,n_1,n_3} dn_2 + \cdots$$

But for constant composition we have the relation (Eq. 14.9)

$$dG = V\,dP - S\,dT$$

Therefore,

$$dG = V\,dP - S\,dT + \left(\frac{\partial G}{\partial n_1}\right)_{T,P,n_2,n_3} dn_1 + \left(\frac{\partial G}{\partial n_2}\right)_{T,P,n_1,n_3} dn_2 + \cdots \tag{16.17}$$

The chemical potential, μ, of each constituent is defined by the relation

$$\mu_1 = \left(\frac{\partial G}{\partial n_1}\right)_{T,P,n_2,n_3} \qquad \mu_2 = \left(\frac{\partial G}{\partial n_2}\right)_{T,P,n_1,n_3} \tag{16.18}$$

Therefore, we can write

$$dG = V\,dP - S\,dT + \mu_1\,dn_1 + \mu_2\,dn_2 + \mu_3\,dn_3 + \cdots \tag{16.19}$$

A similar equation could be written for the liquid phase.

The requirement for equilibrium in a mixture involving various components in various phases is that the chemical potential of each constituent is the same in all phases. Thus, in the mixture of nitrogen and oxygen which we have been considering, the requirement for equilibrium is that the chemical potential of the nitrogen be the same in the liquid and vapor phases and that the chemical potential of the oxygen be the same in each phase.

In the general case therefore we can write the requirement for equilibrium of various components in various phases as follows, where the numerical subscripts denote the various constituents and the letter superscripts denote the various phases.

$$\mu_1{}^A = \mu_1{}^B = \mu_1{}^C \cdots \text{ for all phases}$$

$$\mu_2{}^A = \mu_2{}^B = \mu_2{}^C \cdots \text{ for all phases}$$

Similar equations would be written for all the constituents.

By use of the Gibbs phase rule one can determine the number of intensive properties which can be varied in a mixture involving several phases. Frequently this number of intensive properties which can be varied is referred to as the degrees of freedom. One can also think of this as the number of intensive properties which determine the state of the system. The Gibbs phase rule is as follows

$$F = C + 2 - P \tag{16.20}$$

where F = degrees of freedom (the number of intensive properties that can be independently varied)

C = number of components

P = number of phases

Let us apply this first of all to a pure substance. In this case $C = 1$. Suppose we have one phase present, ($P = 1$) such as superheated vapor of a pure substance. In this case $F = 1 + 2 - 1 = 2$. That is, two intensive properties may be varied independently. Another way of saying this, is that two independent properties determine the state. Our familiarity with the superheated section of the Steam Tables indicates that we have already found this to be true.

Suppose we have two phases of a pure substance in equilibrium, such as saturated liquid and saturated vapor. In this case $C = 1$, $P = 2$, and $F = 1 + 2 - 2 = 1$. In this case temperature and pressure cannot be varied independently, but for each value of temperature the pressure is fixed, and such properties as specific volume and specific enthalpy of each phase are fixed.

Finally, consider also the triple point of a pure substance. In this case $C = 1$ and $P = 3$, and $F = 1 + 2 - 3 = 0$. That is, the temperature and pressure of the triple point are fixed, and if either is varied we are no longer at the triple point. This is called a nonvarient system.

Let us also apply the Gibbs phase rule to the mixture of nitrogen and oxygen we considered. Suppose we had only the vapor phase present. In this case $C = 2$ and $P = 1$, and $F = 2 + 2 - 1 = 3$. That is, in order to completely specify the system we would have to state the temperature, pressure, and say the mole fraction of one of the components. But suppose that we have both nitrogen and oxygen present and we have an equilibrium mixture of the vapor and liquid phases. In this case $C = 2$ and $P = 2$, and $F = 2 + 2 - 2 = 2$. It is evident from Fig. 16.4 that pressure alone is not sufficient to specify the state because at a pressure of one atmosphere we can have a variety of temperatures and compositions. However, if we specify both temperature and pressure, then the composition of both the liquid and vapor phases is fixed, and the state of the system is completely specified. Thus, in this case the pressure and temperature determine the state of the system.

These are of course simple applications of the Gibbs phase rule, but they serve to illustrate its use. The great utility of the phase rule is in more complex systems and in these cases it is a powerful tool for analysis.

PROBLEMS

16.1 Show that at the triple point of water all three phases have the same value of the Gibbs function.

16.2 One mole of oxygen at 77 F and 1 atm pressure is heated to 5000 K in a constant-pressure process. What is the composition at 5000 K and 1 atm pressure, assuming that O_2 and O are the constituents present?

16.3 One mole of CO_2 at a pressure of 10 atm and a temperature of 77 F is heated to a temperature of 3000 K in a steady-flow constant-pressure process. What is the equilibrium composition at 10 atm and 3000 K? Determine the heat transfer for this process. Compare these results with Example 16.3.

16.4 One mole of water vapor at 1 atm pressure and 400 K temperature is heated to 3000 K in a steady-flow constant-pressure process. Determine the final composition and heat transfer for this process, assuming that H_2, O_2, and H_2O are present in the final state.

16.5 One mole of water vapor at 1 atm pressure and 400 K temperature is heated to 3000 K in a steady-flow constant-pressure process. Determine the final composition, assuming that H_2, O_2, H_2O, and OH are present in the final state.

16.6 (*a*) One mole of hydrogen at 298 K, 1 atm pressure is heated to 5000 R while the pressure remains constant. What is the composition at 5000 R?

(*b*) Consider the reaction $H_2 \rightarrow 2H$ at 5000 R. Is this an endothermic or exothermic reaction?

16.7 One mole of hydrogen at 77 F and 1 atm pressure reacts with 2 moles of oxygen at 77 F and 1 atm pressure in a steady-flow adiabatic process. The products are at 1 atm pressure and consist of H_2, O_2, and H_2O. Determine the adiabatic flame temperature.

16.8 Many combustion processes involve products of combustion at temperatures not greater than 2000 F. A gas turbine is a typical example. Usually the effects of dissociation are ignored at these temperatures. Is this justified for CO_2 and H_2O?

16.9 Plot to scale the values of $\log_{10} K_P$ vs. $1/T$ for the reaction $CO_2 \rightleftarrows CO + \frac{1}{2}O_2$. Write an equation for the $\log_{10} K_P$ as a function of temperature.

16.10 (*a*) Liquid air at a pressure of 1 atm evaporates. What is the composition of the first bubble of vapor to form if this vapor is in equilibrium with the liquid? What is the temperature under these conditions?

(*b*) Air at a pressure of 1 atm condenses. What is the composition of the first drop of liquid to form if this drop is in equilibrium with the vapor? What is the temperature under these conditions?

APPENDIX I

THE CONTROL VOLUME

An alternate series of definitions is frequently used for what have been termed in this text the open and closed systems. The corresponding terms are system and control volume, and these terms are introduced here.*

In this alternate scheme a system is defined as a quantity of matter of fixed mass and identity. Everything external to the system is defined as the surroundings, and the boundary of the system separates the system from the surroundings. A system thus defined corresponds to the closed system as defined in this text.

A control volume is defined as a volume in space through which matter flows, and the control volume is bounded by a control surface. The control volume may change its position or size in a given process. The control volume thus defined corresponds to the open system as defined in this text.

In the paragraphs that follow, the law of the conservation of mass, the first law, the second law, and the momentum equation are developed for the control volume. These are all developed in relation to Fig. A.1.

The solid line in this diagram represents the control surface. The dotted line represents the boundary of the system. Initially, at time τ,

* The material in Appendix I has been drawn largely from J. C. Hunsaker and B. G. Rightmire, *Engineering Applications of Fluid Mechanics*, McGraw-Hill, 1947 and A. H. Shapiro, *The Dynamics and Thermodynamics of Compressible Fluid Flow*, Ronald Press, 1953.

the boundary of the system and the control surface coincide. Thus, the mass that constitutes the system is the mass within the control surface at time τ. After time $d\tau$ the system undergoes a process during which it moves to the position indicated. The system is divided into portions I and III at time τ, and into portions I and II at time $\tau + d\tau$. During this time $d\tau$ the quantity of heat δQ is transferred to the system,

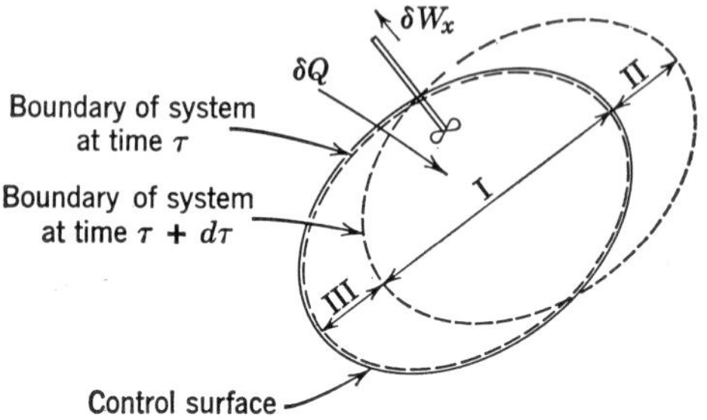

Fig. A.1

and the work δW_x is done by the system on the rotating shaft. There may also be some shear work, δW_{shear}, done by the system during this process.

Conservation of Mass

Because the mass of the system is conserved we can write

$$\begin{aligned} m_{\mathrm{I}\tau} + m_{\mathrm{III}\tau} &= m_{\mathrm{I}(\tau+d\tau)} + m_{\mathrm{II}(\tau+d\tau)} \\ m_{\mathrm{I}(\tau+d\tau)} - m_{\mathrm{I}\tau} &= m_{\mathrm{III}\tau} - m_{\mathrm{II}(\tau+d\tau)} \end{aligned} \tag{A.1}$$

But as the time $d\tau$ approaches 0, space I coincides with the control volume. Therefore, we can write

$$m_{\mathrm{I}(\tau+d\tau)} - m_{\mathrm{I}\tau} = \delta m_{\text{c.v.}} \tag{A.2}$$

where $\delta m_{\text{c.v.}}$ designates the change of mass within the control volume. We can also write

$$m_{\mathrm{III}\tau} = \int_{\text{c.s.}} \delta m_i \tag{A.3}$$

where the $\int_{\text{c.s.}}$ designates an integration over the entire area of the control surface across which matter flows into the control volume. In a similar manner we can write

$$m_{\mathrm{II}(\tau+d\tau)} = \int_{\text{c.s.}} \delta m_e \tag{A.4}$$

Therefore,

$$\int_V \delta m_{\text{c.v.}} = \int_{\text{c.s.}} \delta m_i - \int_{\text{c.s.}} \delta m_e \tag{A.5}$$

In a manner similar to the development in Section 5.6 we can introduce the velocity and density.

$$\frac{\delta m_{\text{c.v.}}}{d\tau} = \int_{\text{c.s.}} \frac{\delta m_i}{d\tau} - \int_{\text{c.s.}} \frac{\delta m_e}{d\tau} \tag{A.6}$$

$$\int_{\text{c.v.}} \frac{d\rho}{d\tau} dV = \int_{\text{c.s.}} \rho_i \overline{V}_i \cos\alpha\, dA - \int_{\text{c.s.}} \rho_e \overline{V}_e \cos\alpha\, dA \tag{A.7}$$

where $\int_{\text{c.v.}}$ designates an integration over the entire volume.

First Law

The first law was stated in Chapter 5 for a system (of fixed mass).

$$Q = E_2 - E_1 + W$$

$$\frac{\delta Q}{d\tau} = \frac{dE}{d\tau} + \frac{\delta W}{d\tau} \tag{A.8}$$

Consider now the system of Fig. A.1. The rate of heat transfer to the system is $\delta Q/d\tau$. But as $d\tau$ approaches 0 this is also the rate of heat transfer across the control surface,

$$\left(\frac{\delta Q}{d\tau}\right)_{\text{c.s.}} = \frac{\delta Q}{d\tau} \tag{A.9}$$

The change of the energy E of the system is now found

$$\frac{dE}{d\tau} = \frac{(E_{\text{I}(\tau+d\tau)} + E_{\text{II}(\tau+d\tau)}) - (E_{\text{I}\tau} + E_{\text{III}\tau})}{d\tau}$$

$$\frac{dE}{d\tau} = \frac{(E_{\text{I}(\tau+d\tau)} - E_{\text{I}\tau})}{d\tau} + \frac{E_{\text{II}(\tau+d\tau)}}{d\tau} - \frac{E_{\text{III}\tau}}{d\tau} \tag{A.10}$$

As the time $d\tau$ approaches zero, space I coincides with the control volume.

$$\frac{E_{\text{I}(\tau+d\tau)} - E_{\text{I}\tau}}{d\tau} = \left(\frac{dE}{d\tau}\right)_{\text{c.v.}} \tag{A.11}$$

Further,

$$\frac{E_{\text{II}(\tau+d\tau)}}{d\tau} = \int_{\text{c.s.}} e\ \delta\dot{m}_e \tag{A.12}$$

$$\frac{E_{\text{III}\tau}}{d\tau} = \int_{\text{c.s.}} e\ \delta\dot{m}_i$$

Thus, we can write

$$\frac{dE}{d\tau} = \left(\frac{dE}{d\tau}\right)_{\text{c.v.}} + \int_{\text{c.s.}} e_e\ \delta\dot{m}_e - \int_{\text{c.s.}} e_i\ \delta\dot{m}_i \tag{A.13}$$

Work can be done on the system in a variety of ways. In this analysis we consider three ways, namely, the work done by normal forces on the system as it undergoes the change of position indicated, the work done on the rotating shaft δW_x, and the work done by shearing forces.

$$\frac{\delta W}{d\tau} = \left(\frac{\delta W}{d\tau}\right)_{\text{normal}} + \frac{\delta W_x}{d\tau} + \left(\frac{\delta W}{d\tau}\right)_{\text{shear}} \tag{A.14}$$

Consider first the work done by normal forces. The work done against the normal forces during time $d\tau$ as the mass $m_{\text{III}\tau}$ enters the control volume (see Section 4.3) is $-\int_{\text{c.s.}} Pv\ \delta m_i$. Similarly the work done against normal forces during time $d\tau$ as the mass $m_{\text{II}(\tau+d\tau)}$ leaves the control volume is $\int_{\text{c.s.}} Pv\ \delta m_e$. Thus the total rate at which work is done by the system against normal forces is

$$\left(\frac{\delta W}{d\tau}\right)_{\text{normal}} = \int_{\text{c.s.}} Pv\ \delta\dot{m}_e - \int_{\text{c.s.}} Pv\ \delta\dot{m}_i \tag{A.15}$$

The rate at which shaft work is done is $\delta W_x/d\tau$, and the rate at which work is done against shearing forces is designated $(\delta W/d\tau)_{\text{shear}}$. Therefore the total work is

$$\frac{\delta W}{d\tau} = \int_{\text{c.s.}} Pv\ \delta\dot{m}_e - \int_{\text{c.s.}} Pv\ \delta\dot{m}_i + \frac{\delta W_x}{d\tau} + \left(\frac{\delta W}{d\tau}\right)_{\text{shear}} \tag{A.16}$$

Substituting Eqs. A.9, A.13, and A.16 into Eq. A.8, we have the first law for a control volume.

$$\frac{\delta Q}{d\tau} = \left(\frac{dE}{d\tau}\right)_{\text{c.v.}} + \int_{\text{c.s.}} e\ \delta\dot{m}_e - \int_{\text{c.s.}} e\ \delta\dot{m}_i + \int_{\text{c.s.}} Pv\ \delta\dot{m}_e - \int_{\text{c.s.}} Pv\ \delta\dot{m}_i + \frac{\delta W_x}{d\tau} + \left(\frac{\delta W}{d\tau}\right)_{\text{shear}} \tag{A.17}$$

We have previously defined the internal energy in terms of the relation

$$e = u + \frac{\overline{V}^2}{2g_c} + \frac{g}{g_c} Z$$

The definition of enthalpy, $h = u + Pv$ may also be introduced. With these substitutions, the first law for a control volume is

$$\frac{\delta Q}{d\tau} = \left(\frac{dE}{d\tau}\right)_{\text{c.v.}} + \int_{\text{c.s.}} \left(h + \frac{\overline{V}^2}{2g_c} + \frac{g}{g_c} Z\right) \delta \dot{m}_e - \int_{\text{c.s.}} \left(h + \frac{\overline{V}^2}{2g_c} + \frac{g}{g_c} Z\right) \delta \dot{m}_i + \frac{\delta W_x}{d\tau} + \left(\frac{\delta W}{d\tau}\right)_{\text{shear}} \quad \text{(A.18)}$$

Second Law

For a system of fixed mass, the second law of thermodynamics states that

$$dS \geq \frac{\delta Q}{T}$$

or

$$\frac{dS}{d\tau} \geq \frac{1}{T} \frac{\delta Q}{d\tau} \quad \text{(A.19)}$$

Consider the system of Fig. A.1. The rate of heat transfer to the system is $\delta Q/d\tau$. However, as the time $d\tau$ approaches 0, this is also the rate of heat transfer across the control surface $(\delta Q/d\tau)_{\text{c.s.}}$. Therefore,

$$\frac{1}{T} \frac{\delta Q}{d\tau} = \oint_{\text{c.s.}} \frac{1}{T} \frac{\delta Q}{d\tau} \quad \text{(A.20)}$$

Consider now the change in entropy of the system.

$$\frac{dS}{d\tau} = \frac{(S_{\text{I}(\tau+d\tau)} + S_{\text{II}(\tau+d\tau)}) - (S_{\text{I}\tau} + S_{\text{III}\tau})}{d\tau}$$

$$\frac{dS}{d\tau} = \frac{(S_{\text{I}(\tau+d\tau)} - S_{\text{I}\tau})}{d\tau} + \frac{S_{\text{II}(\tau+d\tau)}}{d\tau} - \frac{S_{\text{III}\tau}}{d\tau}$$

As the time $d\tau$ approaches 0, space I coincides with the control volume. Therefore,

$$\frac{S_{\text{I}(\tau+d\tau)} - S_{\text{I}\tau}}{d\tau} = \left(\frac{dS}{d\tau}\right)_{\text{c.v.}} \quad \text{(A.21)}$$

Also,

$$\frac{S_{\text{II}(\tau+d\tau)}}{d\tau} = \int_{\text{c.s.}} s\ \delta\dot{m}_e \tag{A.22}$$

$$\frac{S_{\text{III}\tau}}{d\tau} = \int_{\text{c.s.}} s\ \delta\dot{m}_i \tag{A.23}$$

Therefore,

$$\frac{dS}{d\tau} = \left(\frac{dS}{d\tau}\right)_{\text{c.v.}} + \int_{\text{c.s.}} s\ \delta\dot{m}_e - \int_{\text{c.s.}} s\ \delta\dot{m}_i \tag{A.24}$$

Therefore, substituting Eqs. A.20 and A.24 into Eq. A.19, we have the second law for a control volume.

$$\left[\left(\frac{dS}{d\tau}\right)_{\text{c.v.}} + \int_{\text{c.s.}} s\ \delta\dot{m}_e - \int_{\text{c.s.}} s\ \delta\dot{m}_i\right] \geq \int_{\text{c.s.}} \frac{1}{T}\frac{\delta Q}{d\tau}$$

Momentum Equation

Newton's second law for a system of fixed mass states that the force is equal to the rate of change of momentum.

$$\Sigma F_x = \frac{d(m\overline{V}_x)}{d\tau} \tag{A.25}$$

Considering the system of Fig. A.1,

$$\frac{d(m\overline{V}_x)}{d\tau} = \frac{(m\overline{V}_x)_{\text{I}(\tau+d\tau)} + (m\overline{V}_x)_{\text{II}(\tau+d\tau)} - (m\overline{V}_x)_{\text{I}\tau} - (m\overline{V}_x)_{\text{III}\tau}}{d\tau}$$

$$= \frac{(m\overline{V}_x)_{\text{I}(\tau+d\tau)} - (m\overline{V}_x)_{\text{I}\tau}}{d\tau} + \frac{(m\overline{V}_x)_{\text{II}(\tau+d\tau)}}{d\tau} - \frac{(m\overline{V}_x)_{\text{III}\tau}}{d\tau} \tag{A.26}$$

As time $d\tau$ approaches 0, space I coincides with the control volume, and therefore, the first term represents the rate of change of x momentum within the control volume.

$$\frac{(m\overline{V}_x)_{\text{I}(\tau+d\tau)} - (m\overline{V}_x)_{\text{I}\tau}}{d\tau} = \frac{d(m\overline{V}_x)_{\text{c.v.}}}{d\tau} \tag{A.27}$$

The second and third terms respectively represent the momentum flux leaving and entering the control volume.

$$\frac{(m\overline{V}_x)_{\mathrm{II}(\tau+d\tau)}}{d\tau} = \int_{\text{c.s.}} \overline{V}_{ex}\,\delta\dot{m}_e \tag{A.28}$$

$$\frac{(m\overline{V}_x)_{\mathrm{III}\tau}}{d\tau} = \int_{\text{c.s.}} \overline{V}_{ix}\,\delta\dot{m}_i \tag{A.29}$$

Therefore, for a control volume the momentum equation is

$$\Sigma F_x = \frac{d(m\overline{V}_x)_{\text{c.v.}}}{d\tau} + \int_{\text{c.s.}} \overline{V}_{ex}\,\delta\dot{m}_e - \int_{\text{c.s.}} \overline{V}_{ix}\,\delta\dot{m}_i \tag{A.30}$$

Introducing the velocity and density into Eq. A.30 we have

$$\Sigma F_x = \int_{\text{c.v.}} \frac{d(\rho\overline{V}_x)\,dV}{d\tau} + \int_{\text{c.s.}} \rho\overline{V}_e \cos\alpha\,\overline{V}_{ex}\,dA - \int_{\text{c.s.}} \rho\overline{V}_i \cos\alpha\,\overline{V}_{ix}\,dA \tag{A.31}$$

APPENDIX II

TABLES OF THERMODYNAMIC PROPERTIES

TABLE A.1

Steam Tables

TABLE A.1.1. DRY SATURATED STEAM: TEMPERATURE TABLE *

Temp., F t	Abs. Press., lbf/in.^{2} P	Specific Volume, ft^3/lbm: Sat. Liquid v_f	Specific Volume: Evap. v_{fg}	Specific Volume: Sat. Vapor v_g	Enthalpy, Btu/lbm: Sat. Liquid h_f	Enthalpy: Evap. h_{fg}	Enthalpy: Sat. Vapor h_g	Entropy, Btu/lbm R: Sat. Liquid s_f	Entropy: Evap. s_{fg}	Entropy: Sat. Vapor s_g
32	0.08854	0.01602	3306	3306	0.00	1075.8	1075.8	0.0000	2.1877	2.1877
35	0.09995	0.01602	2947	2947	3.02	1074.1	1077.1	0.0061	2.1709	2.1770
40	0.12170	0.01602	2444	2444	8.05	1071.3	1079.3	0.0162	2.1435	2.1597
45	0.14752	0.01602	2036.4	2036.4	13.06	1068.4	1081.5	0.0262	2.1167	2.1429
50	0.17811	0.01603	1703.2	1703.2	18.07	1065.6	1083.7	0.0361	2.0903	2.1264
60	0.2563	0.01604	1206.6	1206.7	28.06	1059.9	1088.0	0.0555	2.0393	2.0948
70	0.3631	0.01606	867.8	867.9	38.04	1054.3	1092.3	0.0745	1.9902	2.0647
80	0.5069	0.01608	633.1	633.1	48.02	1048.6	1096.6	0.0932	1.9428	2.0360
90	0.6982	0.01610	468.0	468.0	57.99	1042.9	1100.9	0.1115	1.8972	2.0087
100	0.9492	0.01613	350.3	350.4	67.97	1037.2	1105.2	0.1295	1.8531	1.9826
110	1.2748	0.01617	265.3	265.4	77.94	1031.6	1109.5	0.1471	1.8106	1.9577
120	1.6924	0.01620	203.25	203.27	87.92	1025.8	1113.7	0.1645	1.7694	1.9339
130	2.2225	0.01625	157.32	157.34	97.90	1020.0	1117.9	0.1816	1.7296	1.9112
140	2.8886	0.01629	122.99	123.01	107.89	1014.1	1122.0	0.1984	1.6910	1.8894
150	3.718	0.01634	97.06	97.07	117.89	1008.2	1126.1	0.2149	1.6537	1.8685
160	4.741	0.01639	77.27	77.29	127.89	1002.3	1130.2	0.2311	1.6174	1.8485
170	5.992	0.01645	62.04	62.06	137.90	996.3	1134.2	0.2472	1.5822	1.8293
180	7.510	0.01651	50.21	50.23	147.92	990.2	1138.1	0.2630	1.5480	1.8109
190	9.339	0.01657	40.94	40.96	157.95	984.1	1142.0	0.2785	1.5147	1.7932
200	11.526	0.01663	33.62	33.64	167.99	977.9	1145.9	0.2938	1.4824	1.7762

210	14.123	0.01670	27.80	27.82	178.05	971.6	1149.7	0.3090	1.4508	1.7598
212	14.696	0.01672	26.78	26.80	180.07	970.3	1150.4	0.3120	1.4446	1.7566
220	17.186	0.01677	23.13	23.15	188.13	965.2	1153.4	0.3239	1.4201	1.7440
230	20.780	0.01684	19.365	19.382	198.23	958.8	1157.0	0.3387	1.3901	1.7288
240	24.969	0.01692	16.306	16.323	208.34	952.2	1160.5	0.3531	1.3609	1.7140
250	29.825	0.01700	13.804	13.821	218.48	945.5	1164.0	0.3675	1.3323	1.6998
260	35.429	0.01709	11.746	11.763	228.64	938.7	1167.3	0.3817	1.3043	1.6860
270	41.858	0.01717	10.044	10.061	238.84	931.8	1170.6	0.3958	1.2769	1.6727
280	49.203	0.01726	8.628	8.645	249.06	924.7	1173.8	0.4096	1.2501	1.6597
290	57.556	0.01735	7.444	7.461	259.31	917.5	1176.8	0.4234	1.2238	1.6472
300	67.013	0.01745	6.449	6.466	269.59	910.1	1179.7	0.4369	1.1980	1.6350
310	77.68	0.01755	5.609	5.626	279.92	902.6	1182.5	0.4504	1.1727	1.6231
320	89.66	0.01765	4.896	4.914	290.28	894.9	1185.2	0.4637	1.1478	1.6115
330	103.06	0.01776	4.289	4.307	300.68	887.0	1187.7	0.4769	1.1233	1.6002
340	118.01	0.01787	3.770	3.788	311.13	879.0	1190.1	0.4900	1.0992	1.5891
350	134.63	0.01799	3.324	3.342	321.63	870.7	1192.3	0.5029	1.0754	1.5783
360	153.04	0.01811	2.939	2.957	332.18	862.2	1194.4	0.5158	1.0519	1.5677
370	173.37	0.01823	2.606	2.625	342.79	853.5	1196.3	0.5286	1.0287	1.5573
380	195.77	0.01836	2.317	2.335	353.45	844.6	1198.1	0.5413	1.0059	1.5471
390	220.37	0.01850	2.0651	2.0836	364.17	835.4	1199.6	0.5539	0.9832	1.5371
400	247.31	0.01864	1.8447	1.8633	374.97	826.0	1201.0	0.5664	0.9608	1.5272
410	276.75	0.01878	1.6512	1.6700	385.83	816.3	1202.1	0.5788	0.9386	1.5174
420	308.83	0.01894	1.4811	1.5000	396.77	806.3	1203.1	0.5912	0.9166	1.5078
430	343.72	0.01910	1.3308	1.3499	407.79	796.0	1203.8	0.6035	0.8947	1.4982
440	381.59	0.01926	1.1979	1.2171	418.90	785.4	1204.3	0.6158	0.8730	1.4887

* Abridged from *Thermodynamic Properties of Steam*, by Joseph H. Keenan and Frederick G. Keyes. Copyright 1936, by Joseph H. Keenan and Frederick G. Keyes. Published by John Wiley & Sons, Inc., New York.

TABLE A.1

Steam Tables (*Continued*)

TABLE A.1.1. DRY SATURATED STEAM: TEMPERATURE TABLE (*Continued*)

Temp., F t	Abs. Press., lbf/in.^{2} P	Specific Volume, ft^3/lbm Sat. Liquid v_f	Evap. v_{fg}	Sat. Vapor v_g	Enthalpy, Btu/lbm Sat. Liquid h_f	Evap. h_{fg}	Sat. Vapor h_g	Entropy, Btu/lbm R Sat. Liquid s_f	Evap. s_{fg}	Sat. Vapor s_g
450	422.6	0.0194	1.0799	1.0993	430.1	774.5	1204.6	0.6280	0.8513	1.4793
460	466.9	0.0196	0.9748	0.9944	441.4	763.2	1204.6	0.6402	0.8298	1.4700
470	514.7	0.0198	0.8811	0.9009	452.8	751.5	1204.3	0.6523	0.8083	1.4606
480	566.1	0.0200	0.7972	0.8172	464.4	739.4	1203.7	0.6645	0.7868	1.4513
490	621.4	0.0202	0.7221	0.7423	476.0	726.8	1202.8	0.6766	0.7653	1.4419
500	680.8	0.0204	0.6545	0.6749	487.8	713.9	1201.7	0.6887	0.7438	1.4325
520	812.4	0.0209	0.5385	0.5594	511.9	686.4	1198.2	0.7130	0.7006	1.4136
540	962.5	0.0215	0.4434	0.4649	536.6	656.6	1193.2	0.7374	0.6568	1.3942
560	1133.1	0.0221	0.3647	0.3868	562.2	624.2	1186.4	0.7621	0.6121	1.3742
580	1325.8	0.0228	0.2989	0.3217	588.9	588.4	1177.3	0.7872	0.5659	1.3532
600	1542.9	0.0236	0.2432	0.2668	617.0	548.5	1165.5	0.8131	0.5176	1.3307
620	1786.6	0.0247	0.1955	0.2201	646.7	503.6	1150.3	0.8398	0.4664	1.3062
640	2059.7	0.0260	0.1538	0.1798	678.6	452.0	1130.5	0.8679	0.4110	1.2789
660	2365.4	0.0278	0.1165	0.1442	714.2	390.2	1104.4	0.8987	0.3485	1.2472
680	2708.1	0.0305	0.0810	0.1115	757.3	309.9	1067.2	0.9351	0.2719	1.2071
700	3093.7	0.0369	0.0392	0.0761	823.3	172.1	995.4	0.9905	0.1484	1.1389
705.4	3206.2	0.0503	0	0.0503	902.7	0	902.7	1.0580	0	1.0580

TABLE A.1

Steam Tables (*Continued*)

TABLE A.1.2. DRY SATURATED STEAM: PRESSURE TABLE *

Abs. Press., lbf/in.^{2} P	Temp., F t	Specific Volume, ft^3/lbm: Sat. Liquid v_f	Sat. Vapor v_g	Enthalpy, Btu/lbm: Sat. Liquid h_f	Evap. h_{fg}	Sat. Vapor h_g	Entropy, Btu/lbm R: Sat. Liquid s_f	Evap. s_{fg}	Sat. Vapor s_g	Internal Energy, Btu/lbm: Sat. Liquid u_f	Sat. Vapor u_g
1.0	101.74	0.01614	333.6	69.70	1036.3	1106.0	0.1326	1.8456	1.9782	69.70	1044.3
2.0	126.08	0.01623	173.73	93.99	1022.2	1116.2	0.1749	1.7451	1.9200	93.98	1051.9
3.0	141.48	0.01630	118.71	109.37	1013.2	1122.6	0.2008	1.6855	1.8863	109.36	1056.7
4.0	152.97	0.01636	90.63	120.86	1006.4	1127.3	0.2198	1.6427	1.8625	120.85	1060.2
5.0	162.24	0.01640	73.52	130.13	1001.0	1131.1	0.2347	1.6094	1.8441	130.12	1063.1
6.0	170.06	0.01645	61.98	137.96	996.2	1134.2	0.2472	1.5820	1.8292	137.94	1065.4
7.0	176.85	0.01649	53.64	144.76	992.1	1136.9	0.2581	1.5586	1.8167	144.74	1067.4
8.0	182.86	0.01653	47.34	150.79	988.5	1139.3	0.2674	1.5383	1.8057	150.77	1069.2
9.0	188.28	0.01656	42.40	156.22	985.2	1141.4	0.2759	1.5203	1.7962	156.19	1070.8
10	193.21	0.01659	38.42	161.17	982.1	1143.3	0.2835	1.5041	1.7876	161.14	1072.2
14.696	212.00	0.01672	26.80	180.07	970.3	1150.4	0.3120	1.4446	1.7566	180.02	1077.5
15	213.03	0.01672	26.29	181.11	969.7	1150.8	0.3135	1.4415	1.7549	181.06	1077.8
20	227.96	0.01683	20.089	196.16	960.1	1156.3	0.3356	1.3962	1.7319	196.10	1081.9
25	240.07	0.01692	16.303	208.42	952.1	1160.6	0.3533	1.3606	1.7139	208.34	1085.1
30	250.33	0.01701	13.746	218.82	945.3	1164.1	0.3680	1.3313	1.6993	218.73	1087.8
35	259.28	0.01708	11.898	227.91	939.2	1167.1	0.3807	1.3063	1.6870	227.80	1090.1
40	267.25	0.01715	10.498	236.03	933.7	1169.7	0.3919	1.2844	1.6763	235.90	1092.0
45	274.44	0.01721	9.401	243.36	928.6	1172.0	0.4019	1.2650	1.6669	243.22	1093.7

* Abridged from *Thermodynamic Properties of Steam*, by Joseph H. Keenan and Frederick G. Keyes. Copyright 1936, by Joseph H. Keenan and Frederick G. Keyes. Published by John Wiley & Sons, Inc., New York.

TABLE A.1

Steam Tables (*Continued*)

TABLE A.1.2. DRY SATURATED STEAM: PRESSURE TABLE (*Continued*)

		Specific Volume, ft^3/lbm		Enthalpy, Btu/lbm			Entropy, Btu/lbm R			Internal Energy, Btu/lbm	
Abs. Press., $lbf/in.^2$ P	Temp., F t	Sat. Liquid v_f	Sat. Vapor v_g	Sat. Liquid h_f	Evap. h_{fg}	Sat. Vapor h_g	Sat. Liquid s_f	Evap. s_{fg}	Sat. Vapor s_g	Sat. Liquid u_f	Sat. Vapor u_g
50	281.01	0.01727	8.515	250.09	924.0	1174.1	0.4110	1.2474	1.6585	249.93	1095.3
55	287.07	0.01732	7.787	256.30	919.6	1175.9	0.4193	1.2316	1.6509	256.12	1096.7
60	292.71	0.01738	7.175	262.09	915.5	1177.6	0.4270	1.2168	1.6438	261.90	1097.9
65	297.97	0.01743	6.655	267.50	911.6	1179.1	0.4342	1.2032	1.6374	267.29	1099.1
70	302.92	0.01748	6.206	272.61	907.9	1180.6	0.4409	1.1906	1.6315	272.38	1100.2
75	307.60	0.01753	5.816	277.43	904.5	1181.9	0.4472	1.1787	1.6259	277.19	1101.2
80	312.03	0.01757	5.472	282.02	901.1	1183.1	0.4531	1.1676	1.6207	281.76	1102.1
85	316.25	0.01761	5.168	286.39	897.8	1184.2	0.4587	1.1571	1.6158	286.11	1102.9
90	320.27	0.01766	4.896	290.56	894.7	1185.3	0.4641	1.1471	1.6112	290.27	1103.7
95	324.12	0.01770	4.652	294.56	891.7	1186.2	0.4692	1.1376	1.6068	294.25	1104.5
100	327.81	0.01774	4.432	298.40	888.8	1187.2	0.4740	1.1286	1.6026	298.08	1105.2
110	334.77	0.01782	4.049	305.66	883.2	1188.9	0.4832	1.1117	1.5948	305.30	1106.5
120	341.25	0.01789	3.728	312.44	877.9	1190.4	0.4916	1.0962	1.5878	312.05	1107.6
130	347.32	0.01796	3.455	318.81	872.9	1191.7	0.4995	1.0817	1.5812	318.38	1108.6
140	353.02	0.01802	3.220	324.82	868.2	1193.0	0.5069	1.0682	1.5751	324.35	1109.6
150	358.42	0.01809	3.015	330.51	863.6	1194.1	0.5138	1.0556	1.5694	330.01	1110.5
160	363.53	0.01815	2.834	335.93	859.2	1195.1	0.5204	1.0436	1.5640	335.39	1111.2

170	368.41	0.01822	2.675	341.09	854.9	1196.0	0.5266	1.0324	1.5590	340.52	1111.9
180	373.06	0.01827	2.532	346.03	850.8	1196.9	0.5325	1.0217	1.5542	345.42	1112.5
190	377.51	0.01833	2.404	350.79	846.8	1197.6	0.5381	1.0116	1.5497	350.15	1113.1
200	381.79	0.01839	2.288	355.36	843.0	1198.4	0.5435	1.0018	1.5453	354.68	1113.7
250	400.95	0.01865	1.8438	376.00	825.1	1201.1	0.5675	0.9588	1.5263	375.14	1115.8
300	417.33	0.01890	1.5433	393.84	809.0	1202.8	0.5879	0.9225	1.5104	392.79	1117.1
350	431.72	0.01913	1.3260	409.69	794.2	1203.9	0.6056	0.8910	1.4966	408.45	1118.0
400	444.59	0.0193	1.1613	424.0	780.5	1204.5	0.6214	0.8630	1.4844	422.6	1118.5
450	456.28	0.0195	1.0320	437.2	767.4	1204.6	0.6356	0.8378	1.4734	435.5	1118.7
500	467.01	0.0197	0.9278	449.4	755.0	1204.4	0.6487	0.8147	1.4634	447.6	1118.6
550	476.94	0.0199	0.8424	460.8	743.1	1203.9	0.6608	0.7934	1.4542	458.8	1118.2
600	486.21	0.0201	0.7698	471.6	731.6	1203.2	0.6720	0.7734	1.4454	469.4	1117.7
650	494.90	0.0203	0.7083	481.8	720.5	1202.3	0.6826	0.7548	1.4374	479.4	1117.1
700	503.10	0.0205	0.6554	491.5	709.7	1201.2	0.6925	0.7371	1.4296	488.8	1116.3
750	510.86	0.0207	0.6092	500.8	699.2	1200.0	0.7019	0.7204	1.4223	498.0	1115.4
800	518.23	0.0209	0.5687	509.7	688.9	1198.6	0.7108	0.7045	1.4153	506.6	1114.4
850	525.26	0.0210	0.5327	518.3	678.8	1197.1	0.7194	0.6891	1.4085	515.0	1113.3
900	531.98	0.0212	0.5006	526.6	668.8	1195.4	0.7275	0.6744	1.4020	523.1	1112.1
950	538.43	0.0214	0.4717	534.6	659.1	1193.7	0.7355	0.6602	1.3957	530.9	1110.8
1000	544.61	0.0216	0.4456	542.4	649.4	1191.8	0.7430	0.6467	1.3897	538.4	1109.4
1100	556.31	0.0220	0.4001	557.4	630.4	1187.8	0.7575	0.6205	1.3780	552.9	1106.4
1200	567.22	0.0223	0.3619	571.7	611.7	1183.4	0.7711	0.5956	1.3667	566.7	1103.0
1300	577.46	0.0227	0.3293	585.4	593.2	1178.6	0.7840	0.5719	1.3559	580.0	1099.4
1400	587.10	0.0231	0.3012	598.7	574.7	1173.4	0.7963	0.5491	1.3454	592.7	1095.4
1500	596.23	0.0235	0.2765	611.6	556.3	1167.9	0.8082	0.5269	1.3351	605.1	1091.2
2000	635.82	0.0257	0.1878	671.7	463.4	1135.1	0.8619	0.4230	1.2849	662.2	1065.6
2500	668.13	0.0287	0.1307	730.6	360.5	1091.1	0.9126	0.3197	1.2322	717.3	1030.6
3000	695.36	0.0346	0.0858	802.5	217.8	1020.3	0.9731	0.1885	1.1615	783.4	972.7
3206.2	705.40	0.0503	0.0503	902.7	0	902.7	1.0580	0	1.0580	872.9	872.9

TABLE A.1

Steam Tables (*Continued*)

TABLE A.1.3. PROPERTIES OF SUPERHEATED STEAM *

Abs. Press., lbf/in.^{2} (Sat. Temp.)		Temperature, F 200	220	300	350	400	450	500	550	600	700	800	900	1000
	v	392.6	404.5	452.3	482.2	512.0	541.8	571.6	601.4	631.2	690.8	750.4	809.9	869.5
1	h	1150.4	1159.5	1195.8	1218.7	1241.7	1264.9	1288.3	1312.0	1335.7	1383.8	1432.8	1482.7	1533.5
(101.74)	s	2.0512	2.0647	2.1153	2.1444	2.1720	2.1983	2.2233	2.2468	2.2702	2.3137	2.3542	2.3923	2.4283
	v	78.16	80.59	90.25	96.26	102.26	108.24	114.22	120.19	126.16	138.10	150.03	161.95	173.87
5	h	1148.8	1158.1	1195.0	1218.1	1241.2	1264.5	1288.0	1311.7	1335.4	1383.6	1432.7	1482.6	1533.4
(162.24)	s	1.8718	1.8857	1.9370	1.9664	1.9942	2.0205	2.0456	2.0692	2.0927	2.1361	2.1767	2.2148	2.2509
	v	38.85	40.09	45.00	48.03	51.04	54.05	57.05	60.04	63.03	69.01	74.98	80.95	86.92
10	h	1146.6	1156.2	1193.9	1217.2	1240.6	1264.0	1287.5	1311.3	1335.1	1383.4	1432.5	1482.4	1533.2
(193.21)	s	1.7927	1.8071	1.8595	1.8892	1.9172	1.9436	1.9689	1.9924	2.0160	2.0596	2.1002	2.1383	2.1744
	v		27.15	30.53	32.62	34.68	36.73	38.78	40.82	42.86	46.94	51.00	55.07	59.13
14.696	h		1154.4	1192.8	1216.4	1239.9	1263.5	1287.1	1310.9	1334.8	1383.2	1432.3	1482.3	1533.1
(212.00)	s		1.7624	1.8160	1.8460	1.8743	1.9008	1.9261	1.9498	1.9734	2.0170	2.0576	2.0958	2.1319
	v			22.36	23.91	25.43	26.95	28.46	29.97	31.47	34.47	37.46	40.45	43.44
20	h			1191.6	1215.6	1239.2	1262.9	1286.6	1310.5	1334.4	1382.9	1432.1	1482.1	1533.0
(227.96)	s			1.7808	1.8112	1.8396	1.8664	1.8918	1.9160	1.9392	1.9829	2.0235	2.0618	2.0978
	v			11.040	11.843	12.628	13.401	14.168	14.93	15.688	17.198	18.702	20.20	21.70
40	h			1186.8	1211.9	1236.5	1260.7	1284.8	1308.9	1333.1	1381.9	1431.3	1481.4	1532.4
(267.25)	s			1.6994	1.7314	1.7608	1.7881	1.8140	1.8384	1.8619	1.9058	1.9467	1.9850	2.0214
	v			7.259	7.818	8.357	8.884	9.403	9.916	10.427	11.441	12.449	13.452	14.454
60	h			1181.6	1208.2	1233.6	1258.5	1283.0	1307.4	1331.8	1380.9	1430.5	1480.8	1531.9
(292.71)	s			1.6492	1.6830	1.7135	1.7416	1.7678	1.7926	1.8162	1.8605	1.9015	1.9400	1.9762

	v	5.803	6.220	6.624	7.020	7.410	7.797	8.562	9.322	10.077	10.830
80	h	1204.3	1230.7	1256.1	1281.1	1305.8	1330.5	1379.9	1429.7	1480.1	1531.3
(312.03)	s	1.6475	1.6791	1.7078	1.7346	1.7598	1.7836	1.8281	1.8694	1.9079	1.9442
	v	4.592	4.937	5.268	5.589	5.905	6.218	6.835	7.446	8.052	8.656
100	h	1200.1	1227.6	1253.7	1279.1	1304.2	1329.1	1378.9	1428.9	1479.5	1530.8
(327.81)	s	1.6188	1.6518	1.6813	1.7085	1.7339	1.7581	1.8029	1.8443	1.8829	1.9193
	v	3.783	4.081	4.363	4.636	4.902	5.165	5.683	6.195	6.702	7.207
120	h	1195.7	1224.4	1251.3	1277.2	1302.5	1327.7	1377.8	1428.1	1478.8	1530.2
(341.25)	a	1.5944	1.6287	1.6591	1.6869	1.7127	1.7370	1.7822	1.8237	1.8625	1.8990
	v		3.468	3.715	3.954	4.186	4.413	4.861	5.301	5.738	6.172
140	h		1221.1	1248.7	1275.2	1300.9	1326.4	1376.8	1427.3	1478.2	1529.7
(353.02)	s		1.6087	1.6399	1.6683	1.6945	1.7190	1.7645	1.8063	1.8451	1.8817
	v		3.008	3.230	3.443	3.648	3.849	4.244	4.631	5.015	5.396
160	h		1217.6	1246.1	1273.1	1299.3	1325.0	1375.7	1426.4	1477.5	1529.1
(363.53)	s		1.5908	1.6230	1.6519	1.6785	1.7033	1.7491	1.7911	1.8301	1.8667
	v		2.649	2.852	3.044	3.229	3.411	3.764	4.110	4.452	4.792
180	h		1214.0	1243.5	1271.0	1297.6	1323.5	1374.7	1425.6	1476.8	1528.6
(373.06)	s		1.5745	1.6077	1.6373	1.6642	1.6894	1.7355	1.7776	1.8167	1.8534
	v		2.361	2.549	2.726	2.895	3.060	3.380	3.693	4.002	4.309
200	h		1210.3	1240.7	1268.9	1295.8	1322.1	1373.6	1424.8	1476.2	1528.0
(381.79)	s		1.5594	1.5937	1.6240	1.6513	1.6767	1.7232	1.7655	1.8048	1.8415
	v		2.125	2.301	2.465	2.621	2.772	3.066	3.352	3.634	3.913
220	h		1206.5	1237.9	1266.7	1294.1	1320.7	1372.6	1424.0	1475.5	1527.5
(389.86)	s		1.5453	1.5808	1.6117	1.6395	1.6652	1.7120	1.7545	1.7939	1.8308
	v		1.9276	2.094	2.247	2.393	2.533	2.804	3.068	3.327	3.584
240	h		1202.5	1234.9	1264.5	1292.4	1319.2	1371.5	1423.2	1474.8	1526.9
(397.37)	s		1.5319	1.5686	1.6003	1.6286	1.6546	1.7017	1.7444	1.7839	1.8209

* Abridged from *Thermodynamic Properties of Steam*, by Joseph H. Keenan and Frederick G. Keyes. Copyright 1936, by Joseph H. Keenan and Frederick G. Keyes. Published by John Wiley & Sons, Inc., New York.

TABLE A.1

Steam Tables (*Continued*)

Table A.1.3. Properties of Superheated Steam (*Continued*)

Abs. Press., lbf/in.^{2} (Sat. Temp.)		Temperature, F: 200	220	300	350	400	450	500	550	600	700	800	900	1000
	v						1.9183	2.063	2.199	2.330	2.582	2.827	3.067	3.305
260	h						1232.0	1262.3	1290.5	1317.7	1370.4	1422.3	1474.2	1526.3
(404.42)	s						1.5573	1.5897	1.6184	1.6447	1.6922	1.7352	1.7748	1.8118
	v						1.7674	1.9047	2.033	2.156	2.392	2.621	2.845	3.066
280	h						1228.9	1260.0	1288.7	1316.2	1369.4	1421.5	1473.5	1525.8
(411.05)	s						1.5464	1.5796	1.6087	1.6354	1.6834	1.7265	1.7662	1.8033
	v						1.6364	1.7675	1.8891	2.005	2.227	2.442	2.652	2.859
300	h						1225.8	1257.6	1286.8	1314.7	1368.3	1420.6	1472.8	1525.2
(417.33)	s						1.5360	1.5701	1.5998	1.6268	1.6751	1.7184	1.7582	1.7954
	v						1.3734	1.4923	1.6010	1.7036	1.8980	2.084	2.266	2.445
350	h						1217.7	1251.5	1282.1	1310.9	1365.5	1418.5	1471.1	1523.8
(431.72)	s						1.5119	1.5481	1.5792	1.6070	1.6563	1.7002	1.7403	1.7777
	v						1.1744	1.2851	1.3843	1.4770	1.6508	1.8161	1.9767	2.134
400	h						1208.8	1245.1	1277.2	1306.9	1362.7	1416.4	1469.4	1522.4
(444.59)	s						1.4892	1.5281	1.5607	1.5894	1.6398	1.6842	1.7247	1.7623

		Temperature, F: 500	550	600	620	640	660	680	700	800	900	1000	1200	1400	1600
	v	1.1231	1.2155	1.3005	1.3332	1.3652	1.3967	1.4278	1.4584	1.6074	1.7516	1.8928	2.170	2.443	2.714
450	h	1238.4	1272.0	1302.8	1314.6	1326.2	1337.5	1348.8	1359.9	1414.3	1467.7	1521.0	1628.6	1738.7	1851.9
(456.28)	s	1.5095	1.5437	1.5735	1.5845	1.5951	1.6054	1.6153	1.6250	1.6699	1.7108	1.7486	1.8177	1.8803	1.9381

	v	0.9927	1.0800	1.1591	1.1893	1.2188	1.2478	1.2763	1.3044	1.4405	1.5715	1.6996	1.9504	2.197	2.442
500	h	1231.3	1266.8	1298.6	1310.7	1322.6	1334.2	1345.7	1357.0	1412.1	1466.0	1519.6	1627.6	1737.9	1851.3
(467.01)	s	1.4919	1.5280	1.5588	1.5701	1.5810	1.5915	1.6016	1.6115	1.6571	1.6982	1.7363	1.8056	1.8683	1.9262
	v	0.8852	0.9686	1.0431	1.0714	1.0989	1.1259	1.1523	1.1783	1.3038	1.4241	1.5414	1.7706	1.9957	2.219
550	h	1223.7	1261.2	1294.3	1306.8	1318.9	1330.8	1342.5	1354.0	1409.9	1464.3	1518.2	1626.6	1737.1	1850.6
(476.94)	s	1.4751	1.5131	1.5451	1.5568	1.5680	1.5787	1.5890	1.5991	1.6452	1.6868	1.7250	1.7946	1.8575	1.9155
	v	0.7947	0.8753	0.9463	0.9729	0.9988	1.0241	1.0489	1.0732	1.1899	1.3013	1.4096	1.6208	1.8279	2.033
600	h	1215.7	1255.5	1289.9	1302.7	1315.2	1327.4	1339.3	1351.1	1407.7	1462.5	1516.7	1625.5	1736.3	1850.0
(486.21)	s	1.4586	1.4990	1.5323	1.5443	1.5558	1.5667	1.5773	1.5875	1.6343	1.6762	1.7147	1.7846	1.8476	1.9056
	v		0.7277	0.7934	0.8177	0.8411	0.8639	0.8860	0.9077	1.0108	1.1082	1.2024	1.3853	1.5641	1.7405
700	h		1243.2	1280.6	1294.3	1307.5	1320.3	1332.8	1345.0	1403.2	1459.0	1513.9	1623.5	1734.8	1848.8
(503.10)	s		1.4722	1.5084	1.5212	1.5333	1.5449	1.5559	1.5665	1.6147	1.6573	1.6963	1.7666	1.8299	1.8881
	v		0.6154	0.6779	0.7006	0.7223	0.7433	0.7635	0.7833	0.8763	0.9633	1.0470	1.2088	1.3662	1.5214
800	h		1229.8	1270.7	1285.4	1299.4	1312.9	1325.9	1338.6	1398.6	1455.4	1511.0	1621.4	1733.2	1847.5
(518.23)	s		1.4467	1.4863	1.5000	1.5129	1.5250	1.5366	1.5476	1.5972	1.6407	1.6801	1.7510	1.8146	1.8729
	v		0.5264	0.5873	0.6089	0.6294	0.6491	0.6680	0.6863	0.7716	0.8506	0.9262	1.0714	1.2124	1.3509
900	h		1215.0	1260.1	1275.9	1290.9	1305.1	1318.8	1332.1	1393.9	1451.8	1508.1	1619.3	1731.6	1846.3
(531.98)	s		1.4216	1.4653	1.4800	1.4938	1.5066	1.5187	1.5303	1.5814	1.6257	1.6656	1.7371	1.8009	1.8595
	v		0.4533	0.5140	0.5350	0.5546	0.5733	0.5912	0.6084	0.6878	0.7604	0.8294	0.9615	1.0893	1.2146
1000	h		1198.3	1248.8	1265.9	1281.9	1297.0	1311.4	1325.3	1389.2	1448.2	1505.1	1617.3	1730.0	1845.0
(544.61)	s		1.3961	1.4450	1.4610	1.4757	1.4893	1.5021	1.5141	1.5670	1.6121	1.6525	1.7245	1.7886	1.8474
	v			0.4532	0.4738	0.4929	0.5110	0.5281	0.5445	0.6191	0.6866	0.7503	0.8716	0.9885	1.1031
1100	h			1236.7	1255.3	1272.4	1288.5	1303.7	1318.3	1384.3	1444.5	1502.2	1615.2	1728.4	1843.8
(556.31)	s			1.4251	1.4425	1.4583	1.4728	1.4862	1.4989	1.5535	1.5995	1.6405	1.7130	1.7775	1.8363
	v			0.4016	0.4222	0.4410	0.4586	0.4752	0.4909	0.5617	0.6250	0.6843	0.7967	0.9046	1.0101
1200	h			1223.5	1243.9	1262.4	1279.6	1295.7	1311.0	1379.3	1440.7	1499.2	1613.1	1726.9	1842.5
(567.22)	s			1.4052	1.4243	1.4413	1.4568	1.4710	1.4843	1.5409	1.5879	1.6293	1.7025	1.7672	1.8263
	v			0.3174	0.3390	0.3580	0.3753	0.3912	0.4062	0.4714	0.5281	0.5805	0.6789	0.7727	0.8640
1400	h			1193.0	1218.4	1240.4	1260.3	1278.5	1295.5	1369.1	1433.1	1493.2	1608.9	1723.7	1840.0
(587.10)	s			1.3639	1.3877	1.4079	1.4258	1.4419	1.4567	1.5177	1.5666	1.6093	1.6836	1.7489	1.8083

TABLE A.1

Steam Tables (*Continued*)

TABLE A.1.3. PROPERTIES OF SUPERHEATED STEAM (*Continued*)

Abs. Press., lbf/in.^{2} (Sat. Temp.)		Temperature, F													
		500	550	600	620	640	660	680	700	800	900	1000	1200	1400	1600
	v				0.2733	0.2936	0.3112	0.3271	0.3417	0.4034	0.4553	0.5027	0.5906	0.6738	0.7545
1600	h				1187.8	1215.2	1238.7	1259.6	1278.7	1358.4	1425.3	1487.0	1604.6	1720.5	1837.5
(604.90)	s				1.3489	1.3741	1.3952	1.4137	1.4303	1.4964	1.5476	1.5914	1.6669	1.7328	1.7926
	v					0.2407	0.2597	0.2760	0.2907	0.3502	0.3986	0.4421	0.5218	0.5968	0.6693
1800	h					1185.1	1214.0	1238.5	1260.3	1347.2	1417.4	1480.8	1600.4	1717.3	1835.0
(621.03)	s					1.3377	1.3638	1.3855	1.4044	1.4765	1.5301	1.5752	1.6520	1.7185	1.7786
	v					0.1936	0.2161	0.2337	0.2489	0.3074	0.3532	0.3935	0.4668	0.5352	0.6011
2000	h					1145.6	1184.9	1214.8	1240.0	1335.5	1409.2	1474.5	1596.1	1714.1	1832.5
(635.82)	s					1.2945	1.3300	1.3564	1.3783	1.4576	1.5139	1.5603	1.6384	1.7055	1.7660
	v							0.1484	0.1686	0.2294	0.2710	0.3061	0.3678	0.4244	0.4784
2500	h							1132.3	1176.8	1303.6	1387.8	1458.4	1585.3	1706.1	1826.2
(668.13)	s							1.2687	1.3073	1.4127	1.4772	1.5273	1.6088	1.6775	1.7389
	v								0.0984	0.1760	0.2159	0.2476	0.3018	0.3505	0.3966
3000	h								1060.7	1267.2	1365.0	1441.8	1574.3	1698.0	1819.9
(695.36)	s								1.1966	1.3690	1.4439	1.4984	1.5837	1.6540	1.7163
	v									0.1583	0.1981	0.2288	0.2806	0.3267	0.3703
3206.2	h									1250.5	1355.2	1434.7	1569.8	1694.6	1817.2
(705.40)	s									1.3508	1.4309	1.4874	1.5742	1.6452	1.7080
	v								0.0306	0.1364	0.1762	0.2058	0.2546	0.2977	0.3381
3500	h								780.5	1224.9	1340.7	1424.5	1563.3	1689.8	1813.6
	s								0.9515	1.3241	1.4127	1.4723	1.5615	1.6336	1.6968

4000	v	0.0287	0.1052	0.1462	0.1743	0.2192	0.2581	0.2943
	h	763.8	1174.8	1314.4	1406.8	1552.1	1681.7	1807.2
	s	0.9347	1.2757	1.3827	1.4482	1.5417	1.6154	1.6795
4500	v	0.0276	0.0798	0.1226	0.1500	0.1917	0.2273	0.2602
	h	753.5	1113.9	1286.5	1388.4	1540.8	1673.5	1800.9
	s	0.9235	1.2204	1.3529	1.4253	1.5235	1.5990	1.6640
5000	v	0.0268	0.0593	0.1036	0.1303	0.1696	0.2027	0.2329
	h	746.4	1047.1	1256.5	1369.5	1529.5	1665.3	1794.5
	s	0.9152	1.1622	1.3231	1.4034	1.5066	1.5839	1.6499
5500	v	0.0262	0.0463	0.0880	0.1143	0.1516	0.1825	0.2106
	h	741.3	985.0	1224.1	1349.3	1518.2	1657.0	1788.1
	s	0.9090	1.1093	1.2930	1.3821	1.4908	1.5699	1.6369

TABLE A.1

Steam Tables (*Continued*)

TABLE A.1.4. COMPRESSED LIQUID *

Abs. Press., lbf/in.^{2} (Sat. Temp.)		Temperature, F: 32	100	200	300	400	500	600	700
Saturated Liquid	P	0.08854	0.9492	11.526	67.013	247.31	680.8	1542.9	3093.7
	v_f	0.016022	0.016132	0.016634	0.017449	0.018639	0.020432	0.023629	0.03692
	h_f	0	67.97	167.99	269.59	374.97	487.82	617.0	823.3
	s_f	0	0.12948	0.29382	0.43694	0.56638	0.68871	0.8131	0.9905
200	$(v - v_f)\cdot 10^5$	−1.1	−1.1	−1.1	−1.1				
	$(h - h_f)$	+0.61	+0.54	+0.41	+0.23				
(381.79)	$(s - s_f)\cdot 10^3$	+0.03	−0.05	−0.21	−0.21				
400	$(v - v_f)\cdot 10^5$	−2.3	−2.1	−2.2	−2.8	−2.1			
	$(h - h_f)$	+1.21	+1.09	+0.88	+0.61	+0.16			
(444.59)	$(s - s_f)\cdot 10^3$	+0.04	−0.16	−0.47	−0.56	−0.40			
800	$(v - v_f)\cdot 10^5$	−4.6	−4.0	−4.4	−5.6	−6.5	−1.7		
	$(h - h_f)$	+2.39	+2.17	+1.78	+1.35	+0.61	−0.05		
(518.23)	$(s - s_f)\cdot 10^3$	+0.10	−0.40	−0.97	−1.27	−1.48	−0.53		
1000	$(v - v_f)\cdot 10^5$	−5.7	−5.1	−5.4	−6.9	−8.7	−6.4		
	$(h - h_f)$	+2.99	+2.70	+2.21	+1.75	+0.84	−0.14		
(544.61)	$(s - s_f)\cdot 10^3$	+0.15	−0.53	−1.20	−1.64	−2.00	−1.41		

	$(v - v_f) \cdot 10^5$	−8.4	−7.5	−8.1	−10.4	−14.1	−17.3		
1500	$(h - h_f)$	+4.48	+3.99	+3.36	+2.70	+1.44	−0.29		
(596.23)	$(s - s_f) \cdot 10^3$	+0.20	−0.86	−1.79	−2.53	−3.32	−3.56		
	$(v - v_f) \cdot 10^5$	−11.0	−9.9	−10.8	−13.8	−19.5	−27.8	−32.6	
2000	$(h - h_f)$	+5.97	+5.31	+4.51	+3.64	+2.03	−0.38	−2.5	
(635.82)	$(s - s_f) \cdot 10^3$	+0.22	−1.18	−2.39	−3.42	−4.57	−5.58	−4.3	
	$(v - v_f) \cdot 10^5$	−16.3	−14.7	−16.0	−20.7	−30.0	−47.1	−87.9	
3000	$(h - h_f)$	+9.00	+7.88	+6.76	+5.49	+3.33	−0.41	−6.9	
(695.36)	$(s - s_f) \cdot 10^3$	+0.28	−1.79	−3.56	−5.12	−7.03	−9.42	−12.4	
	$(v - v_f) \cdot 10^5$	−21.5	−19.2	−21.0	−27.5	−40.0	−64.5	−132.2	−821
4000	$(h - h_f)$	+11.88	+10.49	+9.03	+7.41	+4.71	−0.16	−10.0	−59.5
	$(s - s_f) \cdot 10^3$	+0.29	−2.42	−4.74	−6.77	−9.40	−13.03	−19.3	−55.8
	$(v - v_f) \cdot 10^5$	−26.7	−23.6	−26.0	−34.0	−49.6	−80.5	−169.3	−1017
5000	$(h - h_f)$	+14.75	+13.08	+11.30	+9.36	+6.08	+0.25	−12.1	−76.9
	$(s - s_f) \cdot 10^3$	+0.22	−3.07	−5.92	−8.40	−11.74	−16.47	−25.3	−75.3

* Abridged from *Thermodynamic Properties of Steam*, by Joseph H. Keenan and Frederick G. Keyes. Copyright 1936, by Joseph H. Keenan and Frederick G. Keyes. Published by John Wiley & Sons, Inc., New York.

TABLE A.1

Steam Tables (*Continued*)

TABLE A.1.5. SATURATION: SOLID-VAPOR *

Temp., F t	Abs. Press., lbf/in.² P	Specific Volume, ft³/lbm: Sat. Solid v_i	Sat. Vapor $v_g \times 10^{-3}$	Enthalpy, Btu/lbm: Sat. Solid h_i	Subl. h_{ig}	Sat. Vapor h_g	Entropy, Btu/lbm R: Sat. Solid s_i	Subl. s_{ig}	Sat. Vapor s_g
32	0.0885	0.01747	3.306	−143.35	1219.1	1075.8	−0.2916	2.4793	2.1877
30	0.0808	0.01747	3.609	−144.35	1219.3	1074.9	−0.2936	2.4897	2.1961
20	0.0505	0.01745	5.658	−149.31	1219.9	1070.6	−0.3038	2.5425	2.2387
10	0.0309	0.01744	9.05	−154.17	1220.4	1066.2	−0.3141	2.5977	2.2836
0	0.0185	0.01742	14.77	−158.93	1220.7	1061.8	−0.3241	2.6546	2.3305
−10	0.0108	0.01741	24.67	−163.59	1221.0	1057.4	−0.3346	2.7143	2.3797
−20	0.0062	0.01739	42.2	−168.16	1221.2	1053.0	−0.3448	2.7764	2.4316
−30	0.0035	0.01738	74.1	−172.63	1221.2	1048.6	−0.3551	2.8411	2.4860
−40	0.0019	0.01737	133.9	−177.00	1221.2	1044.2	−0.3654	2.9087	2.5433

* Abridged from *Thermodynamic Properties of Steam*, by Joseph H. Keenan and Frederick G. Keyes. Copyright 1936, by Joseph H. Keenan and Frederick G. Keyes. Published by John Wiley & Sons, Inc., New York.

TABLE A.2

Thermodynamic Properties of Ammonia *

TABLE A.2.1. SATURATED AMMONIA

Temp., F	Abs. Press., lbf/in.² P	Specific Volume, ft³/lbm Sat. Liquid v_f	Evap. v_{fg}	Sat. Vapor v_g	Enthalpy, Btu/lbm Sat. Liquid h_f	Evap. h_{fg}	Sat. Vapor h_g	Entropy, Btu/lbm R Sat. Liquid s_f	Evap. s_{fg}	Sat. Vapor s_g
−60	5.55	0.0228	44.707	44.73	−21.2	610.8	589.6	−0.0517	1.5286	1.4769
−55	6.54	0.0229	38.357	38.38	−15.9	607.5	591.6	−0.0386	1.5017	1.4631
−50	7.67	0.0230	33.057	33.08	−10.6	604.3	593.7	−0.0256	1.4753	1.4497
−45	8.95	0.0231	28.597	28.62	−5.3	600.9	595.6	−0.0127	1.4495	1.4368
−40	10.41	0.02322	24.837	24.86	0	597.6	597.6	0.000	1.4242	1.4242
−35	12.05	0.02333	21.657	21.68	5.3	594.2	599.5	0.0126	1.3994	1.4120
−30	13.90	0.0235	18.947	18.97	10.7	590.7	601.4	0.0250	1.3751	1.4001
−25	15.98	0.0236	16.636	16.66	16.0	587.2	603.2	0.0374	1.3512	1.3886
−20	18.30	0.0237	14.656	14.68	21.4	583.6	605.0	0.0497	1.3277	1.3774
−15	20.88	0.02381	12.946	12.97	26.7	580.0	606.7	0.0618	1.3044	1.3664
−10	23.74	0.02393	11.476	11.50	32.1	576.4	608.5	0.0738	1.2820	1.3558
−5	26.92	0.02406	10.206	10.23	37.5	572.6	610.1	0.0857	1.2597	1.3454
0	30.42	0.02419	9.092	9.116	42.9	568.9	611.8	0.0975	1.2377	1.3352
5	34.27	0.02432	8.1257	8.150	48.3	565.0	613.3	0.1092	1.2161	1.3253
10	38.51	0.02446	7.2795	7.304	53.8	561.1	614.9	0.1208	1.1949	1.3157
15	43.14	0.02460	6.5374	6.562	59.2	557.1	616.3	0.1323	1.1739	1.3062
20	48.21	0.02474	5.8853	5.910	64.7	553.1	617.8	0.1437	1.1532	1.2969
25	53.73	0.02488	5.3091	5.334	70.2	548.9	619.1	0.1551	1.1328	1.2879
30	59.74	0.02503	4.8000	4.825	75.7	544.8	620.5	0.1663	1.1127	1.2790
35	66.26	0.02518	4.3478	4.373	81.2	540.5	621.7	0.1775	1.0929	1.2704
40	73.32	0.02533	3.9457	3.971	86.8	536.2	623.0	0.1885	1.0733	1.2618
45	80.96	0.02548	3.5885	3.614	92.3	531.8	624.1	0.1996	1.0539	1.2535
50	89.19	0.02564	3.2684	3.294	97.9	527.3	625.2	0.2105	1.0348	1.2453
55	98.06	0.02581	2.9822	3.008	103.5	522.8	626.3	0.2214	1.0159	1.2373
60	107.6	0.02597	2.7250	2.751	109.2	518.1	627.3	0.2322	0.9972	1.2294
65	117.8	0.02614	2.4939	2.520	114.8	513.4	628.2	0.2430	0.9786	1.2216
70	128.8	0.02632	2.2857	2.312	120.5	508.6	629.1	0.2537	0.9603	1.2140
75	140.5	0.02650	2.0985	2.125	126.2	503.7	629.9	0.2643	0.9422	1.2065
80	153.0	0.02668	1.9283	1.955	132.0	498.7	630.7	0.2749	0.9242	1.1991
85	166.4	0.02687	1.7741	1.801	137.8	493.6	631.4	0.2854	0.9064	1.1918
90	180.6	0.02707	1.6339	1.661	143.5	488.5	632.0	0.2958	0.8888	1.1846
95	195.8	0.02727	1.5067	1.534	149.4	483.2	632.6	0.3062	0.8713	1.1775
100	211.9	0.02747	1.3915	1.419	155.2	477.8	633.0	0.3166	0.8539	1.1705
105	228.9	0.02769	1.2853	1.313	161.1	472.3	633.4	0.3269	0.8366	1.1635
110	247.0	0.02790	1.1891	1.217	167.0	466.7	633.7	0.3372	0.8194	1.1566
115	266.2	0.02813	1.0999	1.128	173.0	460.9	633.9	0.3474	0.8023	1.1497
120	286.4	0.02836	1.0186	1.047	179.0	455.0	634.0	0.3576	0.7851	1.1427
125	307.8	0.02860	0.9444	0.973	185.1	448.9	634.0	0.3679	0.7679	1.1358

* Reprinted by permission from National Bureau of Standards Circular No. 142, *Tables of Thermodynamic Properties of Ammonia.*

TABLE A.2

Thermodynamic Properties of Ammonia (*Continued*)

Table A.2.2. Superheated Ammonia

Abs. Press., lbf/in.² (Sat. Temp.)		Temperature, F 0	20	40	60	80	100	120	140	160	180	200	220
	v	28.58	29.90	31.20	32.49	33.78	35.07	36.35	37.62	38.90	40.17	41.45	
10	*h*	618.9	629.1	639.3	649.5	659.7	670.0	680.3	690.6	701.1	711.6	722.2	
(−41.34)	*s*	1.477	1.499	1.520	1.540	1.559	1.578	1.596	1.614	1.631	1.647	1.664	
	v	18.92	19.82	20.70	21.58	22.44	23.31	24.17	25.03	25.88	26.74	27.59	
15	*h*	617.2	627.8	638.2	648.5	658.9	669.2	679.6	690.0	700.5	711.1	721.7	
(−27.29)	*s*	1.427	1.450	1.471	1.491	1.511	1.529	1.548	1.566	1.583	1.599	1.616	
	v	14.09	14.78	15.45	16.12	16.78	17.43	18.08	18.73	19.37	20.02	20.66	21.3
20	*h*	615.5	626.4	637.0	647.5	658.0	668.5	678.9	689.4	700.0	710.6	721.2	732.0
(−16.64)	*s*	1.391	1.414	1.436	1.456	1.476	1.495	1.513	1.531	1.549	1.565	1.582	1.598
	v	11.19	11.75	12.30	12.84	13.37	13.90	14.43	14.95	15.47	15.99	16.50	17.02
25	*h*	613.8	625.0	635.8	646.5	657.1	667.7	678.2	688.8	699.4	710.1	720.8	731.6
(−7.96)	*s*	1.362	1.386	1.408	1.429	1.449	1.468	1.486	1.504	1.522	1.539	1.555	1.571
	v	9.25	9.731	10.20	10.65	11.10	11.55	11.99	12.43	12.87	13.30	13.73	14.16
30	*h*	611.9	623.5	634.6	645.5	656.2	666.9	677.5	688.2	698.8	709.6	720.3	731.1
(−.57)	*s*	1.337	1.362	1.385	1.406	1.426	1.446	1.464	1.482	1.500	1.517	1.533	1.550
	v		8.287	8.695	9.093	9.484	9.869	10.25	10.63	11.00	11.38	11.75	12.12
35	*h*		622.0	633.4	644.4	655.3	666.1	676.8	687.6	698.3	709.1	719.9	730.7
(5.89)	*s*		1.341	1.365	1.386	1.407	1.427	1.445	1.464	1.481	1.498	1.515	1.531
	v		7.203	7.568	7.922	8.268	8.609	8.945	9.278	9.609	9.938	10.27	10.59
40	*h*		620.4	632.1	643.4	654.4	665.3	676.1	686.9	697.7	708.5	719.4	730.3
(11.66)	*s*		1.323	1.347	1.369	1.390	1.410	1.429	1.447	1.465	1.482	1.499	1.515
	v		6.213	6.538	6.851	7.157	7.457	7.753	8.045	8.335	8.623	8.909	9.194
46	*h*		618.5	630.5	642.1	653.3	664.4	675.3	686.2	697.1	707.9	718.8	729.8
(17.87)	*s*		1.304	1.328	1.351	1.372	1.392	1.411	1.430	1.448	1.465	1.482	1.498
	v			5.988	6.280	6.564	6.843	7.117	7.387	7.655	7.921	8.185	8.448
50	*h*			629.5	641.2	652.6	663.7	674.7	685.7	696.6	707.5	718.5	729.4
(21.67)	*s*			1.317	1.340	1.361	1.382	1.401	1.420	1.437	1.455	1.472	1.488
	v			4.933	5.184	5.428	5.665	5.897	6.126	6.352	6.576	6.798	7.019
60	*h*			626.8	639.0	650.7	662.1	673.3	684.4	695.5	706.5	717.5	728.6
(30.21)	*s*			1.2913	1.3152	1.3373	1.3581	1.3778	1.3966	1.4148	1.4323	1.4493	1.4658

TABLE A.2

Thermodynamic Properties of Ammonia (*Continued*)

TABLE A.2.2. SUPERHEATED AMMONIA (*Continued*)

Abs. Press., lbf/in.^{2} (Sat. Temp.)		Temperature, F 60	80	100	120	140	160	180	200	240	280	320	360
	v	4.401	4.615	4.822	5.025	5.224	5.420	5.615	5.807	6.187	6.563		
70	*h*	636.6	648.7	660.4	671.8	683.1	694.3	705.5	716.6	738.9	761.4		
(37.7)	*s*	1.294	1.317	1.338	1.358	1.377	1.395	1.413	1.430	1.463	1.494		
	v	3.812	4.005	4.190	4.371	4.548	4.722	4.893	5.063	5.398	5.73		
80	*h*	634.3	646.7	658.7	670.4	681.8	693.2	704.4	715.6	738.1	760.7		
(44.4)	*s*	1.275	1.298	1.320	1.340	1.360	1.378	1.396	1.414	1.447	1.478		
	v	3.353	3.529	3.698	3.862	4.021	4.178	4.332	4.484	4.785	5.081		
90	*h*	631.8	644.7	657.0	668.9	680.5	692.0	703.4	714.7	737.3	760.0		
(50.47)	*s*	1.257	1.281	1.304	1.325	1.344	1.363	1.381	1.400	1.432	1.464		
	v	2.985	3.149	3.304	3.454	3.600	3.743	3.883	4.021	4.294	4.562		
100	*h*	629.3	642.6	655.2	667.3	679.2	690.8	702.3	713.7	736.5	759.4		
(56.05)	*s*	1.241	1.266	1.289	1.310	1.331	1.349	1.368	1.385	1.419	1.451		
	v		2.166	2.288	2.404	2.515	2.622	2.727	2.830	3.030	3.227	3.420	
140	*h*		633.8	647.8	661.1	673.7	686.0	698.0	709.9	733.3	756.7	780.0	
(74.79)	*s*		1.214	1.240	1.263	1.284	1.305	1.324	1.342	1.376	1.409	1.440	
	v			1.720	1.818	1.910	1.999	2.084	2.167	2.328	2.484	2.637	
180	*h*			639.9	654.4	668.0	681.0	693.6	705.9	730.1	753.9	777.7	
(89.78)	*s*			1.199	1.225	1.248	1.269	1.289	1.308	1.344	1.377	1.408	
	v				1.443	1.525	1.601	1.675	1.745	1.881	2.012	2.140	2.265
220	*h*				647.3	662.0	675.8	689.1	701.9	726.8	751.1	775.3	799.5
(102.42)	*s*				1.192	1.217	1.239	1.260	1.280	1.317	1.351	1.383	1.413
	v				1.302	1.380	1.452	1.521	1.587	1.714	1.835	1.954	2.069
240	*h*				643.5	658.8	673.1	686.7	699.8	725.1	749.8	774.1	798.4
(108.09)	*s*				1.176	1.203	1.226	1.248	1.268	1.305	1.339	1.371	1.402
	v				1.182	1.257	1.326	1.391	1.453	1.572	1.686	1.796	1.904
260	*h*				639.5	655.6	670.4	684.4	697.7	723.4	748.4	772.9	797.4
(113.42)	*s*				1.162	1.189	1.213	1.235	1.256	1.294	1.329	1.361	1.391
	v				1.078	1.151	1.217	1.279	1.339	1.451	1.558	1.661	1.762
280	*h*				635.4	652.2	667.6	681.9	695.6	721.8	747.0	771.7	796.3
(118.45)	*s*				1.147	1.176	1.201	1.224	1.245	1.283	1.318	1.351	1.382

TABLE A.3

Thermodynamic Properties of Freon-12 (Dichlorodifluoromethane) *

TABLE A.3.1. SATURATED FREON-12

Temp., F t	Abs. Press., lbf/in.^{2} P	Specific Volume, ft^3/lbm: Sat. Liquid v_f	Evap. v_{fg}	Sat. Vapor v_g	Enthalpy, Btu/lbm: Sat. Liquid h_f	Evap. h_{fg}	Sat. Vapor h_g	Entropy, Btu/lbm R: Sat. Liquid s_f	Evap. s_{fg}	Sat. Vapor s_g
−130	0.41224	0.009736	70.7203	70.730	−18.609	81.577	62.968	−0.04983	0.24743	0.19760
−120	0.64190	0.009816	46.7312	46.741	−16.565	80.617	64.052	−0.04372	0.23731	0.19359
−110	0.97034	0.009899	31.7671	31.777	−14.518	79.663	65.145	−0.03779	0.22780	0.19002
−100	1.4280	0.009985	21.1541	22.164	−12.466	78.714	66.248	−0.03200	0.21883	0.18683
−90	2.0509	0.010073	15.8109	15.821	−10.409	77.764	67.355	−0.02637	0.21034	0.18398
−80	2.8807	0.010164	11.5228	11.533	−8.3451	76.812	68.467	−0.02086	0.20229	0.18143
−70	3.9651	0.010259	8.5584	8.5687	−6.2730	75.853	69.580	−0.01548	0.19464	0.17916
−60	5.3575	0.010357	6.4670	6.4774	−4.1919	74.885	70.693	−0.01021	0.18716	0.17714
−50	7.1168	0.010459	4.9637	4.9742	−2.1011	73.906	71.805	−0.00506	0.18038	0.17533
−40	9.3076	0.010564	3.8644	3.8750	0	72.913	72.913	0	0.17373	0.17373
−30	11.999	0.010674	3.0478	3.0585	2.1120	71.903	74.015	0.00496	0.16733	0.17229
−20	15.267	0.010788	2.4321	2.4429	4.2357	70.874	75.110	0.00983	0.16119	0.17102
−10	19.189	0.010906	1.9628	1.9727	6.3716	69.824	76.196	0.01462	0.15527	0.16989
0	23.849	0.011030	1.5979	1.6089	8.5207	68.750	77.271	0.01932	0.14956	0.16888

10	29.335	0.011160	1.3129	1.3241	10.684	67.651	78.335	0.02395	0.14403	0.16798
20	35.736	0.011296	1.0875	1.0988	12.863	66.522	79.385	0.02852	0.13867	0.16719
30	43.148	0.011438	0.90736	0.91880	15.058	65.361	80.419	0.03301	0.13347	0.16648
40	51.667	0.011588	0.76198	0.77357	17.273	64.163	81.436	0.03745	0.12841	0.16586
50	61.394	0.011746	0.64362	0.65537	19.507	62.926	82.433	0.04184	0.12346	0.16530
60	72.433	0.011913	0.54648	0.55839	21.766	61.643	83.409	0.04618	0.11861	0.16479
70	84.888	0.012089	0.46609	0.47818	24.050	60.309	84.359	0.05048	0.11386	0.16434
80	98.870	0.012277	0.39907	0.41135	26.365	58.917	85.282	0.05475	0.10917	0.16392
90	114.49	0.012478	0.34281	0.35529	28.713	57.461	86.174	0.05900	0.10453	0.16353
100	131.86	0.012693	0.29525	0.30794	31.100	55.929	87.029	0.06323	0.09992	0.16315
110	151.11	0.012924	0.25577	0.26769	33.531	54.313	87.844	0.06745	0.09534	0.16279
120	172.35	0.013174	0.22019	0.23326	36.013	52.597	88.610	0.07168	0.09073	0.16241
130	195.71	0.013447	0.19019	0.20364	38.553	50.768	89.321	0.07583	0.08609	0.16202
140	221.32	0.013746	0.16424	0.17799	41.162	48.805	89.967	0.08021	0.08138	0.16159
150	249.31	0.014078	0.14156	0.15564	43.850	46.684	90.534	0.08453	0.07657	0.16110
160	279.82	0.014449	0.12159	0.13604	46.633	44.373	91.006	0.08893	0.07260	0.16053
170	313.00	0.014871	0.10386	0.11873	49.529	41.830	91.359	0.09342	0.06643	0.15985
180	349.00	0.015360	0.08794	0.10330	52.562	38.999	91.561	0.09804	0.06096	0.15900
190	387.98	0.015942	0.073476	0.089418	55.769	35.792	91.561	0.10284	0.05511	0.15793
200	430.09	0.016659	0.060069	0.076728	59.203	32.075	91.278	0.10789	0.04862	0.15651
210	475.52	0.017601	0.047242	0.064843	62.959	27.599	90.558	0.11332	0.03921	0.15453
220	524.43	0.018986	0.035154	0.053140	67.246	21.790	89.036	0.11943	0.03206	0.15149
230	577.03	0.021854	0.017581	0.039435	72.893	12.229	85.122	0.12739	0.01773	0.14512
233.6 (critical)	596.9	0.02870	0	0.02870	78.86	0	78.86	0.1359	0	0.1359

TABLE A.3

Thermodynamic Properties of Freon-12

Table A.3.2. Superheated Freon-12

Temp., F	v	h	s	v	h	s	v	h	s
		5 lbf/in.²			10 lbf/in.²			15 lbf/in.²	
0	8.0611	78.582	0.19663	3.9809	78.246	0.18471	2.6201	77.902	0.17751
20	8.4265	81.309	0.20244	4.1691	81.014	0.19061	2.7494	80.712	0.18349
40	8.7903	84.090	0.20812	4.3556	83.828	0.19635	2.8770	83.561	0.18931
60	9.1528	86.922	0.21367	4.5408	86.689	0.20197	3.0031	86.451	0.19498
80	9.5142	89.806	0.21912	4.7248	89.596	0.20746	3.1281	89.383	0.20051
100	9.8747	92.738	0.22445	4.9079	92.548	0.21283	3.2521	92.357	0.20593
120	10.234	95.717	0.22968	5.0903	95.546	0.21809	3.3754	95.373	0.21122
140	10.594	98.743	0.23481	5.2720	98.586	0.22325	3.4981	98.429	0.21640
160	10.952	101.812	0.23985	5.4533	101.669	0.22830	3.6202	101.525	0.22148
180	11.311	104.925	0.24479	5.6341	104.793	0.23326	3.7419	104.661	0.22646
200	11.668	108.079	0.24964	5.8145	107.957	0.23813	3.8632	107.835	0.23135
220	12.026	111.272	0.25441	5.9946	111.159	0.24291	3.9841	111.046	0.23614
		20 lbf/in.²			25 lbf/in.²			30 lbf/in.²	
20	2.0391	80.403	0.17829	1.6125	80.088	0.17414	1.3278	79.765	0.17065
40	2.1373	83.289	0.18419	1.6932	83.012	0.18012	1.3969	82.730	0.17671
60	2.2340	86.210	0.18992	1.7723	85.965	0.18591	1.4644	85.716	0.18257
80	2.3295	89.168	0.19550	1.8502	88.950	0.19155	1.5306	88.729	0.18826
100	2.4241	92.164	0.20095	1.9271	91.968	0.19704	1.5957	91.770	0.19379
120	2.5179	95.198	0.20628	2.0032	95.021	0.20240	1.6600	94.843	0.19918
140	2.6110	98.270	0.21149	2.0786	98.110	0.20763	1.7237	97.948	0.20445
160	2.7036	101.380	0.21659	2.1535	101.234	0.21276	1.7868	101.086	0.20960
180	2.7957	104.528	0.22159	2.2279	104.393	0.21778	1.8494	104.258	0.21463
200	2.8874	107.712	0.22649	2.3019	107.588	0.22269	1.9116	107.464	0.21957
220	2.9789	110.932	0.23130	2.3756	110.817	0.22752	1.9735	110.702	0.22440
240	3.0700	114.186	0.23602	2.4491	114.080	0.23225	2.0351	113.973	0.22915
		35 lbf/in.²			40 lbf/in.²			50 lbf/in.²	
40	1.1850	82.442	0.17375	1.0258	82.148	0.17112	0.80248	81.540	0.16655
60	1.2442	85.463	0.17968	1.0789	85.206	0.17712	0.84713	84.676	0.17271
80	1.3021	88.504	0.18542	1.1306	88.277	0.18292	0.89025	87.811	0.17862
100	1.3589	91.570	0.19100	1.1812	91.367	0.18854	0.93216	90.953	0.18434
120	1.4148	94.663	0.19643	1.2309	94.480	0.19401	0.97313	94.110	0.18988
140	1.4701	97.785	0.20172	1.2798	97.620	0.19933	1.0133	97.286	0.19527
160	1.5248	100.938	0.20689	1.3282	100.788	0.20453	1.0529	100.485	0.20051
180	1.5789	104.122	0.21195	1.3761	103.985	0.20961	1.0920	103.708	0.20563
200	1.6327	107.338	0.21690	1.4236	107.212	0.21457	1.1307	106.958	0.21064
220	1.6862	110.586	0.22175	1.4707	110.469	0.21944	1.1690	110.235	0.21553
240	1.7394	113.865	0.22651	1.5176	113.757	0.22420	1.2070	113.539	0.22032
260	1.7923	117.175	0.23117	1.5642	117.074	0.22888	1.2447	116.871	0.22502
		60 lbf/in.²			70 lbf/in.²			80 lbf/in.²	
60	0.69210	84.126	0.16892	0.58088	83.552	0.16556	...	...	...
80	0.72964	87.330	0.17497	0.61458	86.832	0.17175	0.52795	86.316	0.16885
100	0.76588	90.528	0.18079	0.64685	90.091	0.17768	0.55734	89.640	0.17489
120	0.80110	93.731	0.18641	0.67803	93.343	0.18339	0.58556	92.945	0.18070
140	0.83551	96.945	0.19186	0.70836	96.597	0.18891	0.61286	96.242	0.18629
160	0.86928	100.776	0.19716	0.73800	99.862	0.19427	0.63943	99.542	0.19170
180	0.90252	103.427	0.20233	0.76708	103.141	0.19948	0.66543	102.851	0.19696
200	0.93531	106.700	0.20736	0.79571	106.439	0.20455	0.69095	106.174	0.20207
220	0.96775	109.997	0.21229	0.82397	109.756	0.20951	0.71609	109.513	0.20706
240	0.99988	113.319	0.21710	0.85191	113.096	0.21435	0.74090	112.872	0.21193
260	1.0318	116.666	0.22182	0.87959	116.459	0.21909	0.76544	116.251	0.21669
280	1.0634	120.039	0.22644	0.90705	119.846	0.22373	0.78975	119.652	0.22135

TABLE A.3

Thermodynamic Properties of Freon-12 (*Continued*)

TABLE A.3.2. SUPERHEATED FREON-12 (*Continued*)

Temp., F	v	h	s	v	h	s	v	h	s
	90 lbf/in.^{2}			100 lbf/in.^{2}			125 lbf/in.^{2}		
100	0.48749	89.175	0.17234	0.43138	88.694	0.16996	0.32943	87.407	0.16455
120	0.51346	92.536	0.17824	0.45562	92.116	0.17597	0.35086	91.008	0.17087
140	0.53845	95.879	0.18391	0.47881	95.507	0.18172	0.37098	94.537	0.17686
160	0.56268	99.216	0.18938	0.50118	98.884	0.18726	0.39015	98.023	0.18258
180	0.58629	102.557	0.19469	0.52291	102.257	0.19262	0.40857	101.484	0.18807
200	0.60941	105.905	0.19984	0.54413	105.633	0.19782	0.42642	104.934	0.19338
220	0.63213	109.267	0.20486	0.56492	109.018	0.20287	0.44380	108.380	0.19853
240	0.65451	112.644	0.20976	0.58538	112.415	0.20780	0.46081	111.829	0.20353
260	0.67662	116.040	0.21455	0.60554	115.828	0.21261	0.47750	115.287	0.20840
280	0.69849	119.456	0.21923	0.62546	119.258	0.21731	0.49394	118.756	0.21316
300	0.72016	122.892	0.22381	0.64518	122.707	0.22191	0.51016	122.238	0.21780
320	0.74166	126.349	0.22830	0.66472	126.176	0.22641	0.52619	125.737	0.22235
	150 lbf/in.^{2}			175 lbf/in.^{2}			200 lbf/in.^{2}		
120	0.28007	89.800	0.16629	...	...	...	...	...	...
140	0.29845	93.498	0.17256	0.24595	92.373	0.16859	0.20579	91.137	0.16480
160	0.31566	97.112	0.17849	0.26198	96.142	0.17478	0.22121	95.100	0.17130
180	0.33200	100.675	0.18415	0.27697	99.823	0.18062	0.23535	98.921	0.17737
200	0.34769	104.206	0.18958	0.29120	103.447	0.18620	0.24860	102.652	0.18311
220	0.36285	107.720	0.19483	0.30485	107.036	0.19156	0.26117	106.325	0.18860
240	0.37761	111.226	0.19992	0.31804	110.605	0.19674	0.27323	109.962	0.19387
260	0.39203	114.732	0.20485	0.33087	114.162	0.20175	0.28489	113.576	0.19896
280	0.40617	118.242	0.20967	0.34339	117.717	0.20662	0.29623	117.178	0.20390
300	0.42008	121.761	0.21436	0.35567	121.273	0.21137	0.30730	120.775	0.20870
320	0.43379	125.290	0.21894	0.36773	124.835	0.21599	0.31815	124.373	0.21337
340	0.44733	128.833	0.22343	0.37963	128.407	0.22052	0.32881	127.974	0.21793
	250 lbf/in.^{2}			300 lbf/in.^{2}			400 lbf/in.^{2}		
160	0.16249	92.717	0.16462	...	...	...	...	...	...
180	0.17605	96.925	0.17130	0.13482	94.556	0.16537	...	...	...
200	0.18824	100.930	0.17747	0.14697	98.975	0.17217	0.091005	93.718	0.16092
220	0.19952	104.809	0.18326	0.15774	103.136	0.17838	0.10316	99.046	0.16888
240	0.21014	108.607	0.18877	0.16761	107.140	0.18419	0.11300	103.735	0.17568
260	0.22027	112.351	0.19404	0.17685	111.043	0.18969	0.12163	108.105	0.18183
280	0.23001	116.060	0.19913	0.18562	114.879	0.19495	0.12949	112.286	0.18756
300	0.23944	119.747	0.20405	0.19402	118.670	0.20000	0.13680	116.343	0.19298
320	0.24862	123.420	0.20882	0.20214	122.430	0.20489	0.14372	120.318	0.19814
340	0.25759	127.088	0.21346	0.21002	126.171	0.20963	0.15032	124.235	0.20310
360	0.26639	130.754	0.21799	0.21770	129.900	0.21423	0.15668	128.112	0.20789
380	0.27504	134.423	0.22241	0.22522	133.624	0.21872	0.16285	131.961	0.21253
	500 lbf/in.^{2}			600 lbf/in.^{2}					
220	0.064207	92.397	0.15683	...	...	...			
240	0.077620	99.218	0.16672	0.047488	91.024	0.15335			
260	0.087054	104.526	0.17421	0.061922	99.741	0.16566			
280	0.094923	109.277	0.18072	0.070859	105.637	0.17374			
300	0.10190	113.729	0.18666	0.078059	110.729	0.18053			
320	0.10829	117.997	0.19221	0.084333	115.420	0.18663			
340	0.11426	122.143	0.19746	0.090017	119.871	0.19227			
360	0.11992	126.205	0.20247	0.095289	124.167	0.19757			
380	0.12533	130.207	0.20730	0.10025	128.355	0.20262			
400	0.13054	134.166	0.21196	0.10498	132.466	0.20746			
420	0.13559	138.096	0.21648	0.10952	136.523	0.21213			
440	0.14051	142.004	0.22087	0.11391	140.539	0.21664			

TABLE A.4

Thermodynamic Properties of Saturated Mercury *

Press., lbf/in.^{2}	Temp., F	Enthalpy, Btu/lbm Sat. Liquid	Enthalpy, Btu/lbm Evap.	Enthalpy, Btu/lbm Sat. Vapor	Entropy, Btu/lbm R Sat. Liquid	Entropy, Btu/lbm R Evap.	Entropy, Btu/lbm R Sat. Vapor	Specific Volume Sat. Vapor, ft^3/lbf
0.020	259.88	7.532	127.614	135.146	0.01259	0.17735	0.18994	1893
0.040	288.32	8.463	127.486	135.949	0.01386	0.17044	0.18430	986
0.075	316.19	9.373	127.361	136.734	0.01504	0.16415	0.17919	545
0.100	329.73	9.814	127.300	137.114	0.01561	0.16126	0.17687	416
0.200	364.25	10.936	127.144	138.080	0.01699	0.15432	0.17131	217.3
0.400	401.98	12.159	126.975	139.134	0.01844	0.14736	0.16580	113.7
0.600	425.82	12.929	126.868	139.797	0.01932	0.14328	0.16260	77.84
0.800	443.50	13.500	126.788	140.288	0.01994	0.14038	0.16032	59.58
1.00	457.72	13.959	126.724	140.683	0.02045	0.13814	0.15859	48.42
2.00	504.93	15.476	126.512	141.988	0.02205	0.13116	0.15321	25.39
4.00	557.85	17.161	126.275	143.436	0.02373	0.12434	0.14787	13.38
6.00	591.2	18.233	126.124	144.357	0.02477	0.12002	0.14479	9.26
8.00	616.5	19.035	126.011	145.046	0.02551	0.11712	0.14262	7.12
10	637.0	19.685	125.919	145.604	0.02610	0.11483	0.14093	5.81
20	706.0	21.864	125.609	147.473	0.02800	0.10779	0.13579	3.09
40	784.4	24.345	125.255	149.600	0.03004	0.10068	0.13072	1.648
60	835.7	25.940	125.024	150.964	0.03127	0.09652	0.12779	1.144
80	874.8	27.159	124.849	152.008	0.03218	0.09356	0.12574	0.885
100	906.8	28.152	124.706	152.858	0.03290	0.09127	0.12417	0.725
120	934.3	29.005	124.582	153.587	0.03350	0.08938	0.12288	0.617
140	958.3	29.748	124.474	154.222	0.03401	0.08778	0.12179	0.538
160	979.9	30.415	124.376	154.791	0.03447	0.08640	0.12087	0.478
180	999.5	31.018	124.288	155.306	0.03488	0.08518	0.12006	0.431
200	1017.2	31.560	124.209	155.769	0.03523	0.08411	0.11934	0.392
225	1038.0	32.204	124.115	156.319	0.03565	0.08287	0.11852	0.354
250	1057.2	32.784	124.029	156.813	0.03603	0.08178	0.11781	0.322
275	1074.8	33.322	123.950	157.272	0.03637	0.08079	0.11716	0.297
300	1091.2	33.824	123.876	157.700	0.03669	0.07989	0.11658	0.276
350	1121.4	34.747	123.740	158.487	0.03725	0.07828	0.11553	0.241
400	1148.4	35.565	123.620	159.185	0.03775	0.07688	0.11463	0.215
500	1196.0	37.006	123.406	160.412	0.03861	0.07455	0.11316	0.177
600	1236.8	38.245	123.221	161.466	0.03932	0.07264	0.11196	0.151
800	1306.1	40.324	122.910	163.234	0.04047	0.06961	0.11008	0.118
1000	1364.0	42.056	122.649	164.705	0.04139	0.06726	0.10865	0.098
1100	1390.0	42.828	122.533	165.361	0.04179	0.06625	0.10804	0.090

* Abridged from *Thermodynamic Properties of Mercury Vapor*, by Lucian A. Sheldon. Courtesy of General Electric Company.

TABLE A.5

Critical Constants *

Substance	Formula	Molecular Weight	Temperature K	Temperature R	Pressure atm	Pressure lbf/in.^{2}	Volume, ft^3/lb-mole
Ammonia	NH_3	17.03	405.5	729.8	111.3	1636	1.16
Argon	A	39.944	151	272	48.0	705	1.20
Bromine	Br_2	159.832	584	1052	102	1500	2.17
Carbon dioxide	CO_2	44.01	304.2	547.5	72.9	1071	1.51
Carbon monoxide	CO	28.01	133	240	34.5	507	1.49
Chlorine	Cl_2	70.914	417	751	76.1	1120	1.99
Deuterium (Normal)	D_2	4.00	38.4	69.1	16.4	241	...
Helium	He	4.003	5.3	9.5	2.26	33.2	0.926
Helium^3	He	3.00	3.34	6.01	1.15	16.9	...
Hydrogen (Normal)	H_2	2.016	33.3	59.9	12.8	188.1	1.04
Krypton	Kr	83.7	209.4	376.9	54.3	798	1.48
Neon	Ne	20.183	44.5	80.1	26.9	395	0.668
Nitrogen	N_2	28.016	126.2	227.1	33.5	492	1.44
Nitrous oxide	N_2O	44.02	309.7	557.4	71.7	1054	1.54
Oxygen	O_2	32.00	154.8	278.6	50.1	736	1.25
Sulfur dioxide	SO_2	64.06	430.7	775.2	77.8	1143	1.95
Water	H_2O	18.016	647.4	1165.3	218.3	3208	0.90
Xenon	Xe	131.3	289.75	521.55	58.0	852	1.90
Benzene	C_6H_6	78.11	562	1012	48.6	714	4.17
n-Butane	C_4H_{10}	58.120	425.2	765.2	37.5	551	4.08
Carbon tetrachloride	CCl_4	153.84	556.4	1001.5	45.0	661	4.42
Chloroform	$CHCl_3$	119.39	536.6	965.8	54.0	794	3.85
Dichlorodifluoromethane	CCl_2F_2	120.92	384.7	692.4	39.6	582	3.49
Dichlorofluoromethane	$CHCl_2F$	102.93	451.7	813.0	51.0	749	3.16
Ethane	C_2H_6	30.068	305.5	549.8	48.2	708	2.37
Ethyl alcohol	C_2H_5OH	46.07	516.0	929.0	63.0	926	2.68
Ethylene	C_2H_4	28.052	282.4	508.3	50.5	742	1.99
n-Hexane	C_6H_{14}	86.172	507.9	914.2	29.9	439	5.89
Methane	CH_4	16.042	191.1	343.9	45.8	673	1.59
Methyl alcohol	CH_3OH	32.04	513.2	923.7	78.5	1154	1.89
Methyl chloride	CH_3Cl	50.49	416.3	749.3	65.9	968	2.29
Propane	C_3H_8	44.094	370.0	665.9	42.0	617	3.20
Propene	C_3H_6	42.078	365.0	656.9	45.6	670	2.90
Propyne	C_3H_4	40.062	401	722	52.8	776	...
Trichlorofluoromethane	CCl_3F	137.38	471.2	848.1	43.2	635	3.97

* K. A. Kobe and R. E. Lynn, Jr., *Chem. Rev.*, **52**, 117–236 (1953).

TABLE A.6

Zero-Pressure Properties of Gases

C_{po}, $C_{v\infty}$, and k are at 80 F

Gas	Chemical Formula	Molecular Weight	R ft-lbf/ lbm R	C_{po} Btu/ lbm R	$C_{v\infty}$ Btu/ lbm R	k
Air	...	28.96	53.34	0.240	0.171	1.400
Argon	A	39.94	38.66	0.1253	0.0756	1.668
Carbon Dioxide	CO_2	44.01	35.10	0.203	0.158	1.285
Carbon Monoxide	CO	28.01	55.16	0.249	0.178	1.399
Helium	He	4.003	386.0	1.25	0.753	1.66
Hydrogen	H_2	2.016	766.4	3.43	2.44	1.404
Methane	CH_4	16.04	96.35	0.532	0.403	1.32
Nitrogen	N_2	28.016	55.15	0.248	0.177	1.400
Oxygen	O_2	32.000	48.28	0.219	0.157	1.395
Steam	H_2O	18.016	85.76	0.445	0.335	1.329

TABLE A.7

Constant-Pressure Specific Heats of Various Substances at Zero Pressure *

Gas or Vapor	Equation, $\bar{C}_{po}$ in Btu/lb mol R T in degrees Rankine	Range, R	Max. Error, %
O_2	$\bar{C}_{po} = 11.515 - \frac{172}{\sqrt{T}} + \frac{1530}{T}$	540–5000	1.1
	$= 11.515 - \frac{172}{\sqrt{T}} + \frac{1530}{T} + \frac{0.05}{1000}(T - 4000)$	5000–9000	0.3
N_2	$\bar{C}_{po} = 9.47 - \frac{3.47 \times 10^3}{T} + \frac{1.16 \times 10^6}{T^2}$	540–9000	1.7
CO	$\bar{C}_{po} = 9.46 - \frac{3.29 \times 10^3}{T} + \frac{1.07 \times 10^6}{T^2}$	540–9000	1.1
H_2	$\bar{C}_{po} = 5.76 + \frac{0.578}{1000}T + \frac{20}{\sqrt{T}}$	540–4000	0.8
	$= 5.76 + \frac{0.578}{1000}T + \frac{20}{\sqrt{T}} - \frac{0.33}{1000}(T - 4000)$	4000–9000	1.4
H_2O	$\bar{C}_{po} = 19.86 - \frac{597}{\sqrt{T}} + \frac{7500}{T}$	540–5400	1.8
CO_2	$\bar{C}_{po} = 16.2 - \frac{6.53 \times 10^3}{T} + \frac{1.41 \times 10^6}{T^2}$	540–6300	0.8
CH_4	$\bar{C}_{po} = 4.52 + 0.00737T$	540–1500	1.2
C_2H_4	$\bar{C}_{po} = 4.23 + 0.01177T$	350–1100	1.5
C_2H_6	$\bar{C}_{po} = 4.01 + 0.01636T$	400–1100	1.5
C_8H_{18}	$\bar{C}_{po} = 7.92 + 0.0601T$	400–1100	est. 4
$C_{12}H_{26}$	$\bar{C}_{po} = 8.68 + 0.0889T$	400–1100	est. 4

* From *Bulletin No. 2*, Ga. School of Technology, by R. L. Sweigert and M. W. Beardsley, 1938.

TABLE A.8

Thermodynamic Properties of Air at Low Pressure *

T, R	h, Btu/lbm	P_r	u, Btu/lbm	v_r	ϕ, Btu/lbm R
200	47.67	0.04320	33.96	1714.9	0.36303
220	52.46	0.06026	37.38	1352.5	0.38584
240	57.25	0.08165	40.80	1088.8	0.40666
260	62.03	0.10797	44.21	892.0	0.42582
280	66.82	0.13986	47.63	741.6	0.44356
300	71.61	0.17795	51.04	624.5	0.46007
320	76.40	0.22290	54.46	531.8	0.47550
340	81.18	0.27545	57.87	457.2	0.49002
360	85.97	0.3363	61.29	396.6	0.50369
380	90.75	0.4061	64.70	346.6	0.51663
400	95.53	0.4858	68.11	305.0	0.52890
420	100.32	0.5760	71.52	270.1	0.54058
440	105.11	0.6776	74.93	240.6	0.55172
460	109.90	0.7913	78.36	215.33	0.56235
480	114.69	0.9182	81.77	193.65	0.57255
500	119.48	1.0590	85.20	174.90	0.58233
520	124.27	1.2147	88.62	158.58	0.59173
540	129.06	1.3860	92.04	144.32	0.60078
560	133.86	1.5742	95.47	131.78	0.60950
580	138.66	1.7800	98.90	120.70	0.61793
600	143.47	2.005	102.34	110.88	0.62607
620	148.28	2.249	105.78	102.12	0.63395
640	153.09	2.514	109.21	94.30	0.64159
660	157.92	2.801	112.67	87.27	0.64902
680	162.73	3.111	116.12	80.96	0.65621
700	167.56	3.446	119.58	75.25	0.66321
720	172.39	3.806	123.04	70.07	0.67002
740	177.23	4.193	126.51	65.38	0.67665
760	182.08	4.607	129.99	61.10	0.68312
780	186.94	5.051	133.47	57.20	0.68942
800	191.81	5.526	136.97	53.63	0.69558
820	196.69	6.033	140.47	50.35	0.70160
840	201.56	6.573	143.98	47.34	0.70747
860	206.46	7.149	147.50	44.57	0.71323

* Abridged from Table 1 in *Gas Tables*, by Joseph H. Keenan and Joseph Kaye. Copyright 1948, by Joseph H. Keenan and Joseph Kaye. Published by John Wiley & Sons, Inc., New York.

TABLE A.8

Thermodynamic Properties of Air at Low Pressure (*Continued*)

T, R	h, Btu/lbm	P_r	u, Btu/lbm	v_r	ϕ, Btu/lbm R
880	211.35	7.761	151.02	42.01	0.71886
900	216.26	8.411	154.57	39.64	0.72438
920	221.18	9.102	158.12	37.44	0.72979
940	226.11	9.834	161.68	35.41	0.73509
960	231.06	10.610	165.26	33.52	0.74030
980	236.02	11.430	168.83	31.76	0.74540
1000	240.98	12.298	172.43	30.12	0.75042
1020	245.97	13.215	176.04	28.59	0.75536
1040	250.95	14.182	179.66	27.17	0.76019
1060	255.96	15.203	183.29	25.82	0.76496
1080	260.97	16.278	186.93	24.58	0.76964
1100	265.99	17.413	190.58	23.40	0.77426
1120	271.03	18.604	194.25	22.30	0.77880
1140	276.08	19.858	197.94	21.27	0.78326
1160	281.14	21.18	201.63	20.293	0.78767
1180	286.21	22.56	205.33	19.377	0.79201
1200	291.30	24.01	209.05	18.514	0.79628
1220	296.41	25.53	212.78	17.700	0.80050
1240	301.52	27.13	216.53	16.932	0.80466
1260	306.65	28.80	220.28	16.205	0.80876
1280	311.79	30.55	224.05	15.518	0.81280
1300	316.94	32.39	227.83	14.868	0.81680
1320	322.11	34.31	231.63	14.253	0.82075
1340	327.29	36.31	235.43	13.670	0.82464
1360	332.48	38.41	239.25	13.118	0.82848
1380	337.68	40.59	243.08	12.593	0.83229
1400	342.90	42.88	246.93	12.095	0.83604
1420	348.14	45.26	250.79	11.622	0.83975
1440	353.37	47.75	254.66	11.172	0.84341
1460	358.63	50.34	258.54	10.743	0.84704
1480	363.89	53.04	262.44	10.336	0.85062
1500	369.17	55.86	266.34	9.948	0.85416
1520	374.47	58.78	270.26	9.578	0.85767
1540	379.77	61.83	274.20	9.226	0.86113
1560	385.08	65.00	278.13	8.890	0.86456
1580	390.40	68.30	282.09	8.569	0.86794
1600	395.74	71.73	286.06	8.263	0.87130
1620	401.09	75.29	290.04	7.971	0.87462

TABLE A.8

Thermodynamic Properties of Air at Low Pressure (*Continued*)

T, R	h, Btu/lbm	P_r	u, Btu/lbm	v_r	ϕ, Btu/lbm R
1640	406.45	78.99	294.03	7.691	0.87791
1660	411.82	82.83	298.02	7.424	0.88116
1680	417.20	86.82	302.04	7.168	0.88439
1700	422.59	90.95	306.06	6.924	0.88758
1720	428.00	95.24	310.09	6.690	0.89074
1740	433.41	99.69	314.13	6.465	0.89387
1760	438.83	104.30	318.18	6.251	0.89697
1780	444.26	109.08	322.24	6.045	0.90003
1800	449.71	114.03	326.32	5.847	0.90308
1820	455.17	119.16	330.40	5.658	0.90609
1840	460.63	124.47	334.50	5.476	0.90908
1860	466.12	129.95	338.61	5.302	0.91203
1880	471.60	135.64	342.73	5.134	0.91497
1900	477.09	141.51	346.85	4.974	0.91788
1920	482.60	147.59	350.98	4.819	0.92076
1940	488.12	153.87	355.12	4.670	0.92362
1960	493.64	160.37	359.28	4.527	0.92645
1980	499.17	167.07	363.43	4.390	0.92926
2000	504.71	174.00	367.61	4.258	0.93205
2020	510.26	181.16	371.79	4.130	0.93481
2040	515.82	188.54	375.98	4.008	0.93756
2060	521.39	196.16	380.18	3.890	0.94026
2080	526.97	204.02	384.39	3.777	0.94296
2100	532.55	212.1	388.60	3.667	0.94564
2120	538.15	220.5	392.83	3.561	0.94829
2140	543.74	229.1	397.05	3.460	0.95092
2160	549.35	238.0	401.29	3.362	0.95352
2180	554.97	247.2	405.53	3.267	0.95611
2200	560.59	256.6	409.78	3.176	0.95868
2220	566.23	266.3	414.05	3.088	0.96123
2240	571.86	276.3	418.31	3.003	0.96376
2260	577.51	286.6	422.59	2.921	0.96626
2280	583.16	297.2	426.87	2.841	0.96876
2300	588.82	308.1	431.16	2.765	0.97123
2320	594.49	319.4	435.46	2.691	0.97369
2340	600.16	330.9	439.76	2.619	0.97611
2360	605.84	342.8	444.07	2.550	0.97853
2380	611.53	355.0	448.38	2.483	0.98092
2400	617.22	367.6	452.70	2.419	0.98331

TABLE A.9

Properties of Some Gases at Low Pressure *

$\bar{h}$ = enthalpy, Btu/lb mole

$$\bar{\phi} = \int_{T=0}^{T} c_{p_0} \frac{dT}{T}, \text{ Btu/lb mole R}$$

$\bar{\phi}$ is essentially equal to the absolute entropy at 1 atm pressure, Btu/lb mole R

Temp R	Products of Combustion, 400% Theoretical Air		Products of Combustion, 200% Theoretical Air		Nitrogen		Oxygen	
	$\bar{h}$	$\bar{\phi}$	$\bar{h}$	$\bar{\phi}$	$\bar{h}$	$\bar{\phi}$	$\bar{h}$	$\bar{\phi}$
537	3746.8	46.318	3774.9	46.300	3729.5	45.755	3725.1	48.986
600	4191.9	47.101	4226.3	47.094	4167.9	46.514	4168.3	49.762
700	4901.7	48.195	4947.7	48.207	4864.9	47.588	4879.3	50.858
800	5617.5	49.150	5676.3	49.179	5564.4	48.522	5602.0	51.821
900	6340.3	50.002	6413.0	50.047	6268.1	49.352	6337.9	52.688
1000	7072.1	50.773	7159.8	50.833	6977.9	50.099	7087.5	53.477
1100	7812.9	51.479	7916.4	51.555	7695.0	50.783	7850.4	54.204
1200	8563.4	52.132	8683.6	52.222	8420.0	51.413	8625.8	54.879
1300	9324.1	52.741	9461.7	52.845	9153.9	52.001	9412.9	55.508
1400	10095.0	53.312	10250.7	53.430	9896.9	52.551	10210.4	56.099
1500	10875.6	53.851	11050.2	53.981	10648.9	53.071	11017.1	56.656
1600	11665.6	54.360	11859.6	54.504	11409.7	53.561	11832.5	57.182
1700	12464.3	54.844	12678.6	55.000	12178.9	54.028	12655.6	57.680
1800	13271.7	55.306	13507.0	55.473	12956.3	54.472	13485.8	58.155
1900	14087.2	55.747	14344.1	55.926	13741.6	54.896	14322.1	58.607
2000	14910.3	56.169	15189.3	56.360	14534.4	55.303	15164.0	59.039
2100	15740.5	56.574	16042.4	56.777	15334.0	55.694	16010.9	59.451
2200	16577.1	56.964	16902.5	57.177	16139.8	56.068	16862.6	59.848
2300	17419.8	57.338	17769.3	57.562	16951.2	56.429	17718.8	60.228
2400	18268.0	57.699	18642.1	57.933	17767.9	56.777	18579.2	60.594
2500	19121.4	58.048	19520.7	58.292	18589.5	57.112	19443.4	60.946
2600	19979.7	58.384	20404.6	58.639	19415.8	57.436	20311.4	61.287
2700	20842.8	58.710	21293.8	58.974	20246.4	57.750	21182.9	61.616
2800	21709.8	59.026	22187.5	59.300	21081.1	58.053	22057.8	61.934
2900	22581.4	59.331	23086.0	59.615	21919.5	58.348	22936.1	62.242
3000	23456.6	59.628	23988.5	59.921	22761.5	58.632	23817.7	62.540
3100	24335.5	59.916	24895.3	60.218	23606.8	58.910	24702.5	62.831
3200	25217.8	60.196	25805.6	60.507	24455.0	59.179	25590.5	63.113
3300	26102.9	60.469	26719.2	60.789	25306.0	59.442	26481.6	63.386
3400	26991.4	60.734	27636.4	61.063	26159.7	59.697	27375.9	63.654
3500			28556.8	61.329	27015.9	59.944	28273.3	63.914
3600			29479.9	61.590	27874.4	60.186	29173.9	64.168
3700			30406.0	61.843	28735.1	60.422	30077.5	64.415
3800			31334.8	62.091	29597.9	60.652	30984.1	64.657
3900			32266.2	62.333	30462.8	60.877	31893.6	64.893
4000					31329.4	61.097	32806.1	65.123
4100					32198.0	61.310	33721.6	65.350
4200					33068.1	61.520	34639.9	65.571
4300					33939.9	61.726	35561.1	65.788
4400					34813.1	61.927	36485.0	66.000
4500					35687.8	62.123	37411.8	66.208
4600					36563.8	62.316	38341.4	66.413
4700					37441.1	62.504	39273.6	66.613
4800					38319.5	62.689	40208.6	66.809
4900					39199.1	62.870	41146.1	67.003
5000					40079.8	63.049	42086.3	67.193
5100					40961.6	63.223	43029.1	67.380
5200					41844.4	63.395	43974.3	67.562
5300					42728.3	63.563	44922.2	67.743

* Abridged from Tables 4, 7, 11, 13, 15, 17, 19, and 21 in *Gas Tables*, by Joseph H. Keenan and Joseph Kaye. Copyright 1948, by Joseph H. Keenan and Joseph Kaye. Published by John Wiley & Sons, Inc., New York.

TABLE A.9

Properties of Some Gases at Low Pressure (*Continued*)

Temp R	Water Vapor $\bar{h}$	Water Vapor $\bar{\phi}$	Carbon Dioxide $\bar{h}$	Carbon Dioxide $\bar{\phi}$	Hydrogen $\bar{h}$	Hydrogen $\bar{\phi}$	Carbon Monoxide $\bar{h}$	Carbon Monoxide $\bar{\phi}$
537	4258.3	45.079	4030.2	51.032	3640.3	31.194	3729.5	47.272
600	4764.7	45.970	4600.9	52.038	4075.6	31.959	3168.0	48.044
700	5575.4	47.219	5552.0	53.503	4770.2	33.031	4866.0	49.120
800	6396.9	48.316	6552.9	54.839	5467.1	33.961	5568.2	50.058
900	7230.9	49.298	7597.6	56.070	6165.3	34.784	6276.4	50.892
1000	8078.9	50.191	8682.1	57.212	6864.5	35.520	6992.2	51.646
1100	8942.0	51.013	9802.6	58.281	7564.6	36.188	7716.8	52.337
1200	9820.4	51.777	10955.3	59.283	8265.8	36.798	8450.8	52.976
1300	10714.5	52.494	12136.9	60.229	8968.7	37.360	9194.6	53.571
1400	11624.8	53.168	13344.7	61.124	9673.8	37.883	9948.1	54.129
1500	12551.4	53.808	14576.0	61.974	10381.5	38.372	10711.1	54.655
1600	13494.9	54.418	15829.0	62.783	11092.5	38.830	11483.4	55.154
1700	14455.4	54.999	17101.4	63.555	11807.4	39.264	12264.3	55.628
1800	15433.0	55.559	18391.5	64.292	12526.8	39.675	13053.2	56.078
1900	16427.5	56.097	19697.8	64.999	13250.9	40.067	13849.8	56.509
2000	17439.0	56.617	21018.7	65.676	13980.1	40.441	14653.2	56.922
2100	18466.9	57.119	22352.7	66.327	14714.5	40.799	15463.3	57.317
2200	19510.8	57.605	23699.0	66.953	15454.4	41.143	16279.4	57.696
2300	20570.6	58.077	25056.3	67.557	16199.8	41.475	17101.0	58.062
2400	21645.7	58.535	26424.0	68.139	16950.6	41.794	17927.4	58.414
2500	22735.4	58.980	27801.2	68.702	17707.3	42.104	18758.8	58.754
2600	23839.5	59.414	29187.1	69.245	18469.7	42.403	19594.3	59.081
2700	24957.2	59.837	30581.2	69.771	19237.8	42.692	20434.0	59.398
2800	26088.0	60.248	31982.8	70.282	20011.8	42.973	21277.2	59.705
2900	27231.2	60.650	33391.5	70.776	20791.5	43.247	22123.8	60.002
3000	28386.3	61.043	34806.6	71.255	21576.9	43.514	22973.4	60.290
3100	29552.8	61.426	36227.9	71.722	22367.7	43.773	23826.0	60.569
3200	30730.2	61.801	37654.7	72.175	23164.1	44.026	24681.2	60.841
3300	31918.2	62.167	39086.7	72.616	23965.5	44.273	25539.0	61.105
3400	33116.0	62.526	40523.6	73.045	24771.9	44.513	26399.3	61.362
3500	34323.5	62.876	41965.2	73.462	25582.9	44.748	27261.8	61.612
3600	35540.1	63.221	43411.0	73.870	26398.5	44.978	28126.6	61.855
3700	36765.4	63.557	44860.6	74.267	27218.5	45.203	28993.5	62.093
3800	37998.9	63.887	46314.0	74.655	28042.8	45.423	29862.3	62.325
3900	39240.2	64.210	47771.0	75.033	28871.1	45.638	30732.9	62.551
4000	40489.1	64.528	49231.4	75.404	29703.5	45.849	31605.2	62.772
4100	41745.4	64.839	50695.1	75.765	30539.8	46.056	32479.1	62.988
4200	43008.4	65.144	52162.0	76.119	31379.8	46.257	33354.4	63.198
4300	44278.0	65.444	53632.1	76.464	32223.5	46.456	34231.2	63.405
4400	45553.9	65.738	55105.1	76.803	33070.9	46.651	35109.2	63.607
4500	46835.9	66.028	56581.0	77.135	33921.6	46.842	35988.6	63.805
4600	48123.6	66.312	58059.7	77.460	34775.7	47.030	36869.3	63.998
4700	49416.9	66.591	59541.1	77.779	35633.0	47.215	37751.0	64.188
4800	50715.5	66.866	61024.9	78.091	36493.4	47.396	38633.9	64.374
4900	52019.0	67.135	62511.3	78.398	37356.9	47.574	39517.8	64.556
5000	53327.4	67.401	64000.0	78.698	38223.3	47.749	40402.7	64.735
5100	54640.3	67.662	65490.9	78.994	39092.8	47.921	41288.6	64.910
5200	55957.4	67.918	66984.0	79.284	39965.1	48.090	42175.5	65.082
5300	57278.7	68.172	68479.1	79.569	40840.2	48.257	43063.2	65.252

TABLE A.10

One-Dimensional Isentropic Compressible-Flow Functions for an Ideal Gas with Constant Specific Heat and Molecular Weight and $k = 1.4$ *

M	M^*	$\frac{A}{A^*}$	$\frac{P}{P_o}$	$\frac{\rho}{\rho_o}$	$\frac{T}{T_o}$
0	0	∞	1.00000	1.00000	1.00000
0.10	0.10943	5.8218	0.99303	0.99502	0.99800
0.20	0.21822	2.9635	0.97250	0.98027	0.99206
0.30	0.32572	2.0351	0.93947	0.95638	0.98232
0.40	0.43133	1.5901	0.89562	0.92428	0.96899
0.50	0.53452	1.3398	0.84302	0.88517	0.95238
0.60	0.63480	1.1882	0.78400	0.84045	0.93284
0.70	0.73179	1.09437	0.72092	0.79158	0.91075
0.80	0.82514	1.03823	0.65602	0.74000	0.88652
0.90	0.91460	1.00886	0.59126	0.68704	0.86058
1.00	1.00000	1.00000	0.52828	0.63394	0.83333
1.10	1.08124	1.00793	0.46835	0.58169	0.80515
1.20	1.1583	1.03044	0.41238	0.53114	0.77640
1.30	1.2311	1.06631	0.36092	0.48291	0.74738
1.40	1.2999	1.1149	0.31424	0.43742	0.71839
1.50	1.3646	1.1762	0.27240	0.39498	0.68965
1.60	1.4254	1.2502	0.23527	0.35573	0.66138
1.70	1.4825	1.3376	0.20259	0.31969	0.63372
1.80	1.5360	1.4390	0.17404	0.28682	0.60680
1.90	1.5861	1.5552	0.14924	0.25699	0.58072
2.00	1.6330	1.6875	0.12780	0.23005	0.55556
2.10	1.6769	1.8369	0.10935	0.20580	0.53135
2.20	1.7179	2.0050	0.09352	0.18405	0.50813
2.30	1.7563	2.1931	0.07997	0.16458	0.48591
2.40	1.7922	2.4031	0.06840	0.14720	0.46468
2.50	1.8258	2.6367	0.05853	0.13169	0.44444
2.60	1.8572	2.8960	0.05012	0.11787	0.42517
2.70	1.8865	3.1830	0.04295	0.10557	0.40684
2.80	1.9140	3.5001	0.03685	0.09462	0.38941
2.90	1.9398	3.8498	0.03165	0.08489	0.37286
3.00	1.9640	4.2346	0.02722	0.07623	0.35714
3.50	2.0642	6.7896	0.01311	0.04523	0.28986
4.00	2.1381	10.719	0.00658	0.02766	0.23810
4.50	2.1936	16.562	0.00346	0.01745	0.19802
5.00	2.2361	25.000	$189(10)^{-5}$	0.01134	0.16667
6.00	2.2953	53.180	$633(10)^{-6}$	0.00519	0.12195
7.00	2.3333	104.143	$242(10)^{-6}$	0.00261	0.09259
8.00	2.3591	190.109	$102(10)^{-6}$	0.00141	0.07246
9.00	2.3772	327.189	$474(10)^{-7}$	0.000815	0.05814
10.00	2.3904	535.938	$236(10)^{-7}$	0.000495	0.04762
∞	2.4495	∞	0	0	0

* Abridged from Table 30 in *Gas Tables*, by Joseph H. Keenan and Joseph Kaye. Copyright 1948, by Joseph H. Keenan and Joseph Kaye. Published by John Wiley & Sons, Inc., New York.

TABLE A.11

One-Dimensional Normal-Shock Functions for an Ideal Gas with Constant Specific Heat and Molecular Weight and $k = 1.4$ *

M_x	M_y	$\frac{P_y}{P_x}$	$\frac{\rho_y}{\rho_x}$	$\frac{T_y}{T_x}$	$\frac{P_{oy}}{P_{ox}}$	$\frac{P_{oy}}{P_x}$
1.00	1.00000	1.00000	1.00000	1.00000	1.00000	1.8929
1.10	0.91177	1.2450	1.1691	1.06494	0.99892	2.1328
1.20	0.84217	1.5133	1.3416	1.1280	0.99280	2.4075
1.30	0.78596	1.8050	1.5157	1.1909	0.97935	2.7135
1.40	0.73971	2.1200	1.6896	1.2547	0.95819	3.0493
1.50	0.70109	2.4583	1.8621	1.3202	0.92978	3.4133
1.60	0.66844	2.8201	2.0317	1.3880	0.89520	3.8049
1.70	0.64055	3.2050	2.1977	1.4583	0.85573	4.2238
1.80	0.61650	3.6133	2.3592	1.5316	0.81268	4.6695
1.90	0.59562	4.0450	2.5157	1.6079	0.76735	5.1417
2.00	0.57735	4.5000	2.6666	1.6875	0.72088	5.6405
2.10	0.56128	4.9784	2.8119	1.7704	0.67422	6.1655
2.20	0.54706	5.4800	2.9512	1.8569	0.62812	6.7163
2.30	0.53441	6.0050	3.0846	1.9468	0.58331	7.2937
2.40	0.52312	6.5533	3.2119	2.0403	0.54015	7.8969
2.50	0.51299	7.1250	3.3333	2.1375	0.49902	8.5262
2.60	0.50387	7.7200	3.4489	2.2383	0.46012	9.1813
2.70	0.49563	8.3383	3.5590	2.3429	0.42359	9.8625
2.80	0.48817	8.9800	3.6635	2.4512	0.38946	10.569
2.90	0.48138	9.6450	3.7629	2.5632	0.35773	11.302
3.00	0.47519	10.333	3.8571	2.6790	0.32834	12.061
4.00	0.43496	18.500	4.5714	4.0469	0.13876	21.068
5.00	0.41523	29.000	5.0000	5.8000	0.06172	32.654
10.00	0.38757	116.50	5.7143	20.388	0.00304	129.217
∞	0.37796	∞	6.000	∞	0	∞

* Abridged from Table 48 in *Gas Tables*, by Joseph H. Keenan and Joseph Kaye. Copyright 1948, by Joseph H. Keenan and Joseph Kaye. Published by John Wiley & Sons, Inc., New York.

APPENDIX III

SOME SELECTED REFERENCES

Basic Thermodynamics

1. B. F. Dodge: *Chemical Engineering Thermodynamics*, McGraw-Hill Book Company, New York, 1944.
2. O. A. Hougen and K. M. Watson: *Industrial Chemical Calculations*, John Wiley & Sons, New York, 1936.
3. J. H. Keenan: *Thermodynamics*, John Wiley & Sons, New York, 1941.
4. J. F. Lee and F. W. Sears: *Thermodynamics*, Addison-Wesley Publishing Company, Cambridge, Mass., 1955.
5. D. A. Mooney: *Mechanical Engineering Thermodynamics*, Prentice-Hall, Englewood Cliffs, N. J., 1953.
6. E. F. Obert: *Thermodynamics*, McGraw-Hill Book Company, New York, 1948.
7. F. D. Rossini: *Chemical Thermodynamics*, John Wiley & Sons, New York, 1950.
8. E. Schmidt: *Thermodynamics, Principles and Applications to Engineering*, Clarendon Press, Oxford, 1949.
9. M. W. Zemansky: *Heat and Thermodynamics*, McGraw-Hill Book Company, New York, 1957.

Fluid Flow

1. N. A. Hall: *Thermodynamics of Fluid Flow*, Prentice-Hall, New York, 1951.
2. A. M. Kuethe and J. D. Schetzer: *Foundations of Aerodynamics*, John Wiley & Sons, New York, 1950.
3. H. W. Liepmann and A. Roshko: *Elements of Gasdynamics*, John Wiley & Sons, New York, 1957.
4. A. H. Shapiro: *The Dynamics and Thermodynamics of Compressible Fluid Flow*, The Ronald Press Company, New York, 1953.

ANSWERS TO SELECTED PROBLEMS

The answer to every third problem is given below. In problems involving ideal gases, slightly different answers are obtained when using variable specific heats (the gas tables) as compared to constant specific heats.

2.3 1440 lbm
2.6 45.3 lbf/in.^{2}, 3.08 atm
2.9 9.32 lbm
3.3 (*a*) 62.3%, 107 F
(*b*) 74.6%, 84.7 F
(*c*) 0.31%, 89.4%
3.6 (*a*) 0.1 lbm
(*b*) 1.58×10^{-4} ft^3, 0.0094 lbm
(*c*) 1.999 ft^3, 0.0906 lbm
3.9 40.2% liquid, 59.8% vapor
3.12 162 ft^3/min, 0.82%
4.3 84,000 ft-lbf
4.6 1,240,000 ft-lbf
5.3 (*a*) 100,000 Btu
(*b*) 0
5.6 10.9 Btu
5.9 −3505 Btu
5.12 −1648 Btu
5.15 (*a*) 300 lbf/in.^{2}
(*b*) 950 F
(*c*) 212.5 Btu
(*d*) 2132 Btu
5.18 (*a*) 1316 F
(*b*) 434.2 Btu
5.21 100%, 0.166 in.^{2}
5.24 13.6 hp
5.27 (*a*) 2160 lbm/hr
(*b*) 345.7 F
(*c*) 1,160,000 Btu/hr
5.30 187.75 Btu/lbm
5.33 0.339 lbm
5.36 55.5 lbm
5.39 97.7%
5.42 13.28%
5.45 (*a*) 6.1 lbm/hr
(*b*) 0.84
(*c*) 0.89
6.6 2270 Btu/min, 53.6 hp
6.9 2.02 hp
7.3 (*b*) 17.75%, 86.3%
(*c*) 4.58
7.6 −81.5 Btu/lbm
7.9 $q = 154$ Btu/lbm,
$w = 140$ Btu/lbm
7.12 434,000 lbm/hr
7.15 47 lbf/in.^{2}, 470 F
7.18 0.1226 Btu/lbm R
7.21 (*a*) Impossible
(*b*) Improbable
7.24 1.45 hp

7.27 705 Btu, 450 F
7.30 (*a*) 1.438 ft^3
(*b*) −1334 Btu
(*c*) −6736 Btu
(*d*) −701 Btu
(*e*) 1.789 Btu/R
8.3 $\Delta U = 698$ Btu, $\Delta H = 972$ Btu, $Q = 698$ Btu, $W = 0$
8.6 (*a*) 798 F
(*b*) 23.3 ft^3/lbm
(*c*) 202.7 Btu/lbm
(*d*) 271.4 Btu/lbm
(*e*) 271.4 Btu/lbm
8.15 577 hp
8.18 $w = -97$ Btu/lbm, $q = 0$
8.21 −101 hp
8.24 756 R
8.27 8.05×10^6 ft-lbf
8.30 $k = \dfrac{\ln P_1/P_o}{\ln P_1/P_2}$
9.3 569 R, 0.179 Btu/lbm R
9.6 (*a*) 0.01297
(*b*) 64.47 F
(*c*) 1.465 lbm
9.9 3.26×10^{-4}
9.12 (*a*) 1.715×10^{-2}, $\phi = 67\%$
(*b*) 1.74×10^{-2}, $\phi = 67\%$
9.15 −35.8 F, 57.7 F
10.3 −68.0 Btu/lbm, 11.8 Btu/lbm
10.6 2,120 hp, 77.3 Btu/lbm
10.9 1000 R, 500 Btu
10.12 6.3 ft^3
10.15 9.0 Btu/lbm
10.18 169,676 Btu, 169,676 Btu
11.3 10.2 hp
11.6 (*a*) 205 hp, 1310 ft^3/min
(*b*) 205 hp, 1780 ft^3/min
(*c*) 214 hp, 1375 ft^3/min
11.9 150.6 Btu/lbm
12.3 $\eta = 30.8\%, 31.6\%, 34.9\%, 38.7\%$
$(1 - x) = 24.9\%, 18.9\%, 9.0\%, 1.5\%$
12.6 342 Btu/lbm, 33.8%
12.12 (*a*) 5.2 Btu/lbm
(*b*) 1.18 tons
(*c*) 2.23
12.15 0.97 Btu
12.18 54.7%, 175 lbf/in.^{2}
12.21 60.6 Btu, 157.1 Btu/lbm, 30.4%
12.24 68.3 Btu/lbm, 145.2 Btu/lbm, 30.7%
12.27 (*a*) 46,200 hp
(*b*) 0.567
(*c*) 13,800 lbm/min, 190,000 ft^3/min
12.30 42.2%
13.3 88 lbf/in.^{2}, 466 F
13.6 (*b*) 0.0192 ft^2, 0.0371 ft^2
13.9 186.6 lbf/in.^{2}
13.12 1.405 ft^2
13.15 (*a*) 1.462 ft^2
(*b*) 1.522 ft^2
13.18 0.769 in.^{2}, 87.6%, 0.0256 Btu/lbm R
13.21 0.636
13.24 148.8 Btu/lbm
14.15 −105 Btu/lbm, −152 Btu/lbm
14.18 −22.7 Btu/lbm, −31.5 Btu/lbm (from Freon-12 tables)
14.21 8600 Btu, 3360 lbf/in.^{2}
14.24 −45,600 Btu/min
15.3 13.35, 23.5%
15.6 1182 Btu/ft^3
15.9 −30,700 Btu/lbm
15.12 2240 R
15.15 (*a*) 3.49 atm
(*b*) −95,440 Btu/lb mole
15.18 3220 R
15.21 3150 R
15.24 (*a*) −148,400 Btu/mole fuel
(*b*) 14.77 Btu/mole fuel R
(*c*) 291 Btu/mole fuel R
15.27 (*a*) 1180 R, 291 lbf/in.^{2}
(*b*) 4460 R, 1102 lbf/in.^{2}, 260.4 Btu/lb mole fuel R
(*c*) 139,800 Btu/lb-mole fuel
16.3 67.3% CO_2, 21.8% CO, 10.9% O_2
16.6 92.8% H_2, 7.2% H

Supplementary Problems

Chapter 2

2A.1 A cylinder containing a gas is fitted with a piston having a mass of 150 lbm. The cross-sectional area of the piston is 60 in.^{2} The atmospheric pressure is 14.2 lbf/in.^{2}, and the acceleration due to gravity at this location is 30.9 ft/sec^2. What is the absolute pressure of the gas?

2A.2 A mercury column is used to measure the pressure difference of 30 lbf/in.^{2} in a piece of apparatus which is located out of doors. The minimum temperature in the winter is 0 F and the maximum temperature in the summer is 100 F. What will be the difference in the height of the mercury column in the summer as compared to the winter when measuring this pressure difference of 30 lbf/in.^{2}? Assume standard gravitational acceleration. The following data are given for the density of mercury:

t °C	Density
−10	13.6198 gm/cm^3
0	13.5951
10	13.5704
20	13.5458
30	13.5213

2A.3 In space simulation chambers very low pressures are achieved by cryopumping. This involves maintaining certain surfaces at very low temperatures (as low as 5 K). Essentially all of the gas present (except helium) will freeze on these surfaces. Pressures of 1×10^{-8} mm Hg and lower are achieved in space chambers by this technique. What is this pressure of 10^{-8} mm Hg in atmospheres, lbf/in.^{2}, and dynes/cm^2? (Sometimes the name torr, after Torricelli, a pioneer worker in vacuum technology, is assigned to the unit of pressure of 1 mm Hg.)

Chapter 3

3A.1 Liquid nitrogen at a temperature of −240 F exists in a container, and both the liquid and vapor phases are present. The volume of the container is 3 ft^3, and it is determined that the mass of nitrogen in the container is 44.5 lbm. What is the mass of liquid and the mass of vapor present in the container?

3A.2 A refrigeration system is to be charged with Freon-12. The system, which has a volume of 0.85 ft^3, is first evacuated, and then slowly charged with Freon-12. The temperature of the Freon-12 remains constant at the ambient temperature of 80 F.

(*a*) What will be the mass of Freon-12 in the system when the pressure reaches 35 $lbf/in.^2$?

(*b*) What will be the mass of Freon-12 in the system when the system is filled with saturated vapor?

(*c*) What fraction of the Freon-12 will exist as a liquid when 3 lbm of Freon-12 have been placed in the system?

3A.3 One lbm of H_2O exists at the triple point. The volume of the liquid phase is equal to the volume of the solid phase, and the volume of the vapor phase is equal to ten times the volume of the liquid phase. What is the mass of H_2O in each phase?

3A.4 Compare the specific volume of nitrogen at 1000 $lbf/in.^2$, -200 F as reported in the nitrogen tables with the value calculated from the Beattie-Bridgman equation of state.

3A.5 A closed pressure vessel contains saturated liquid water at 30 $lbf/in.^2$ The liquid is heated until the temperature is 300 F. During this process the volume of the pressure vessel increases 1%. What is the final pressure in the tank?

3A.6 A tank contains Freon-12 at 100 F. The volume of the tank is 2 ft^3, and initially the volume of the liquid in the tank is equal to the volume of the vapor. Additional Freon-12 is forced into the tank until the mass of Freon-12 in the tank reaches 100 lbm. What is the final volume of liquid in the tank, assuming that the temperature is maintained at 100 F? How much mass enters the tank?

3A.7 A closed tank contains vapor and liquid H_2O in equilibrium at 400 F. The distance from the bottom of the tank to the liquid level is 10 ft. What is the pressure reading at the bottom of the tank as compared to the pressure reading at the top of the tank?

Chapter 4

4A.1 A sealed vessel having the shape of a rectangular prism with the area of the base A and height L_2, and of negligible mass, is initially floating on a liquid of density ρ, Fig. 4A.1.

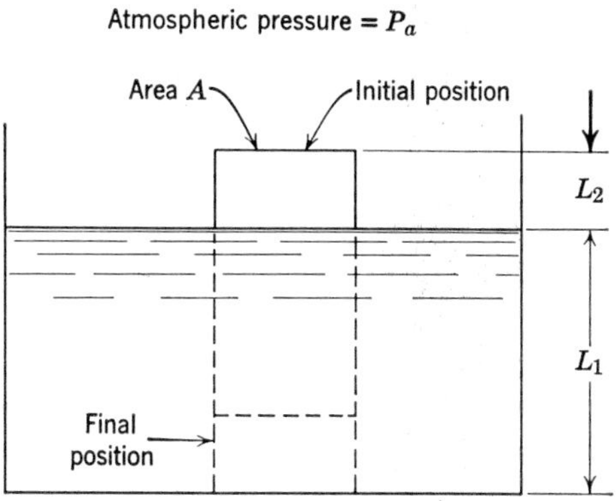

Fig. 4A.1

Derive an expression in terms of the given variables for the work required to move the vessel to the bottom of a very large tank in which depth of the liquid is L_1.

4A.2 Repeat Problem 4A.1 assuming that the sealed tank has a mass m.

4A.3 A metallic wire of initial length L_0 is stretched in the elastic region. Determine the work done in terms of the modulus of elasticity and the strain.

4A.4 The vertical cylinder shown in Fig. 4A.2 contains 0.185 lbm of H_2O at 100 F. The initial volume enclosed beneath the piston is 0.65 ft.^{3} The piston has an area of 60 in.^{2} and a mass of 125 lbm.

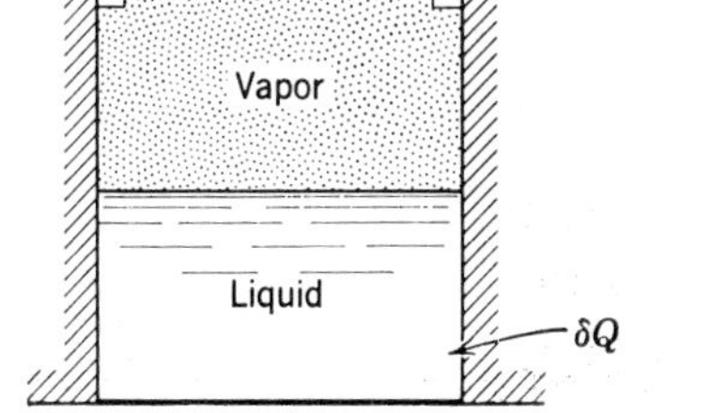

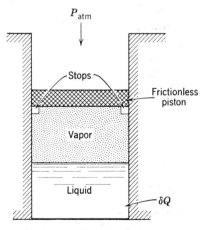

Fig. 4A.2

Initially the piston rests on the stops as shown. The atmospheric pressure is 14.0 and the gravitational acceleration is 30.9 ft/sec^2. Heat is then transferred to the steam until the cylinder contains saturated vapor.

(*a*) What is the temperature of the H_2O when the piston first rises from the stops?

(*b*) How much work is done by the steam during the entire process?

(*c*) Show the process on a T-V diagram.

4A.5 At 20 C methanol has a surface tension of 22.6 dynes/cm. Suppose that a film of methanol is maintained on the wire frame as shown in Fig. 4A.3, one side of which can be moved. The original dimensions of the wire frame are as shown. Determine the work done (consider the film to be the system), when the wire is moved 1 cm in the direction indicated.

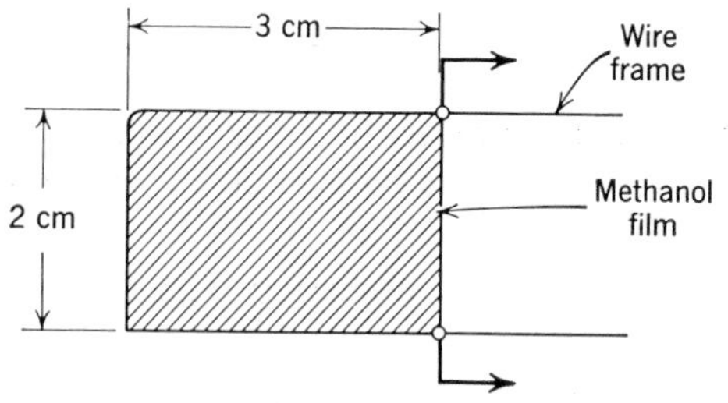

Fig 4A.3

4A.6 A storage battery is well insulated (thermally) while it is being charged. The charging voltage is 12.3 v and the current is 24.0 amp. Considering the storage battery as the system, what is heat transfer and work in a 15 min period?

Chapter 5

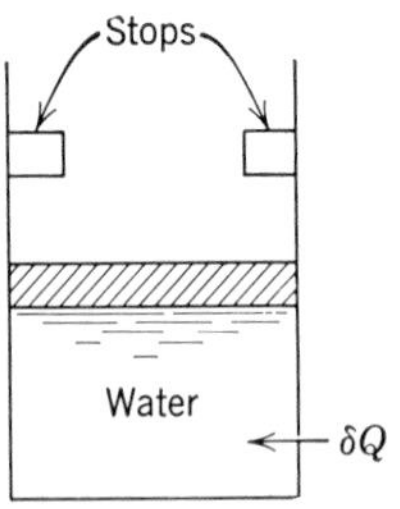

Fig. 5A.1

5A.1 Five lbm of water at 60 F is contained in a vertical cylinder by a frictionless piston of a mass such that the pressure of the water is 100 lbf/in.^{2} Heat is transferred slowly to the water, causing the piston to rise until it reaches the stops, at which point the volume inside the cylinder is 15 ft^3. More heat is transferred to the water until it exists as saturated vapor.

(*a*) Find the final pressure in the cylinder and the heat transfer and work done during the process.

(*b*) Show this process on a T-v diagram.

5A.2 A dewar vessel having a total volume of 100 gal contains liquid nitrogen at 1 atm pressure. The vessel is filled with 90% liquid and 10% vapor by volume. The dewar vessel is accidentally sealed off, so that as heat is transferred to the liquid nitrogen across the vacuum space, the pressure will increase. It is anticipated that the inner vessel will rupture when the pressure reaches 60 lbf/in.² How long will it take to reach this pressure if the heat leak into the dewar vessel is 90 Btu/hr?

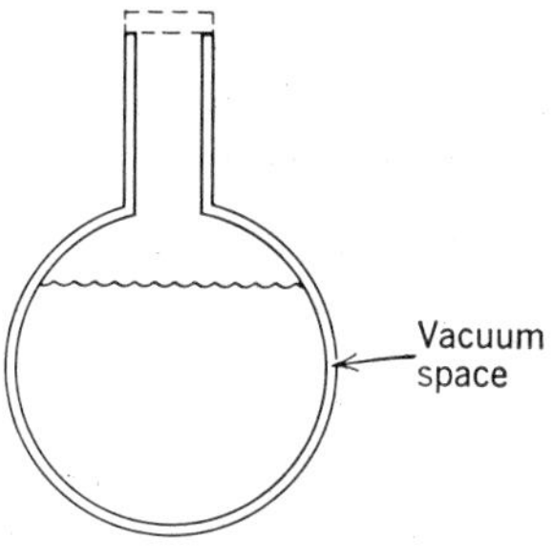

Fig. 5A.2

5A.3 0.285 lbm of H_2O at 100 F is contained in a vertical cylinder fitted with a frictionless piston, which is initially resting on two small stops, as shown in the diagram. The piston has a face area of 60 in.², and a mass of 225 lbm. The initial volume enclosed beneath the piston is 0.95 ft³.

At the location of this cylinder, the gravitational acceleration is 31.9 ft/sec², and the atmospheric pressure, 14.0 lbm/in.²

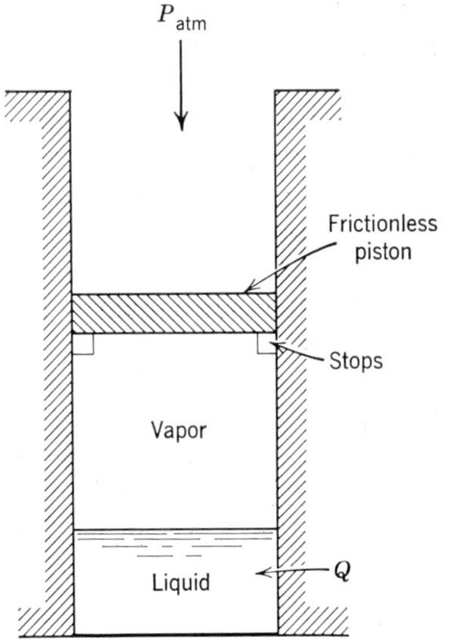

Fig. 5A.3

If heat is slowly transferred to the water, raising its temperature

(*a*) Find the temperature of the H_2O at which the piston begins to rise from the stops.

(*b*) Find the temperature of the H_2O and also the distance that the piston has risen when 300 Btu of heat has been transferred.

(*c*) Show the entire process on T-v coordinates.

5A.4 A steam boiler is fired up with all the valves closed. Initially the boiler contains 40% liquid and 60% vapor by volume. The total volume of the enclosed H_2O is 300 ft³. The relief valve lifts when the pressure reaches 1100 lbf/in.² How much heat will be transferred to the boiler before the relief valve lifts? The initial temperature is 100 F.

5A.5 A missile propellant tank which will be filled with liquid hydrogen is cooled down from the ambient temperature (80 F) to −300 F by means of liquid nitrogen. The liquid nitrogen enters as saturated liquid at 1 atm pressure, and it leaves at 1 atm pressure and a temperature of −100 F. If this process requires a heat transfer of 25,000 Btu from the tank, what amount of nitrogen must be provided to accomplish this?

5A.6 A system is to be charged with ammonia. It is initially evacuated. A charging bottle filled with saturated liquid ammonia at 80 F is then connected to the system, and the ammonia flows into the system. The charging bottle remains connected and open to the system. The final temperature in the system is 80 F. The charging bottle has a volume of 0.4 ft^3, and the system has volume of 10 ft^3. What is the heat transfer to the ammonia?

5A.7 Repeat Problem 5A.6, but assume that there is no heat transfer to the ammonia in the process. What is the final pressure?

5A.8 One method of producing liquid nitrogen is to use the arrangement shown in Fig. 5A.4. Nitrogen gas at 1500 $lbf/in.^2$, 80 F, flows through the heat exchanger, where it is cooled. As it flows across the expansion valve the pressure drops from 1500 $lbf/in.^2$ to 15 $lbf/in.^2$, and during this process some liquid is formed. The vapor flows out through the counterflow heat exchanger, and leaves at a temperature of 75 F. Assume no heat transfer between the surroundings and the nitrogen.

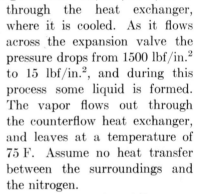

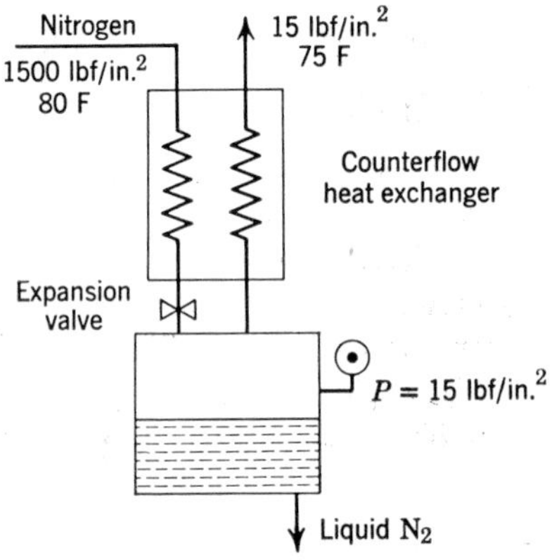

Fig. 5A.4

(*a*) What fraction of the entering gas stream can be liquefied?

(*b*) What is the temperature difference between the two streams at that point in the heat exchanger where the high pressure stream is at a temperature of −200 F?

5A.9 Steam enters a constant area tube at 80 $lbf/in.^2$, saturated vapor, with a velocity of 400 ft/sec. Heat is transferred to the steam in a steady flow process, until at the exit of the tube the steam leaves at 78 lbf/in.^{2}, 650 F. Find the heat transfer per lbm of steam passing through the tube.

5A.10 A schematic arrangement for a proposed procedure for producing fresh water from salt water which would operate in conjunction with a large steam power plant and utilize a flash evaporator is shown in Fig. 5A.5. Cooling water at the rate of 2,500,000 lbm/hr

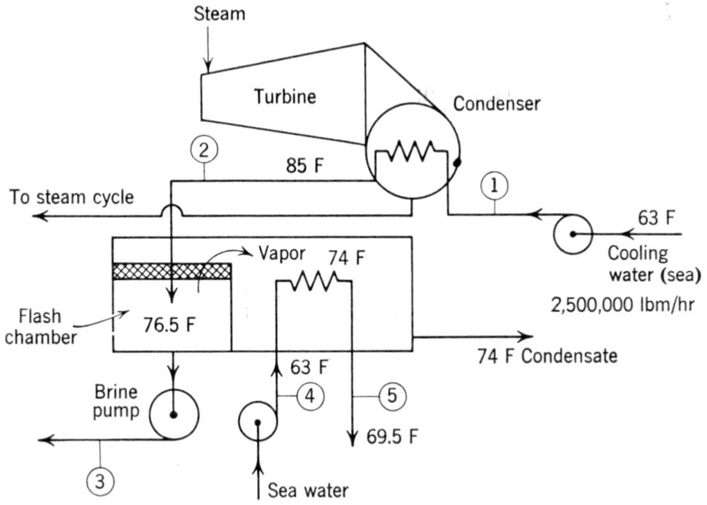

Fig. 5A.5

enters the condenser, where its temperature is increased from 63 F to 85 F. It then enters a flash evaporator, where the pressure is reduced to that corresponding to a saturation temperature of 76.5 F. During this process some of the liquid flashes into vapor, and the remainder is pumped back to the sea. The vapor is then condensed at 74 F to form the desired fresh water. This takes place by utilizing the sea water which enters at 63 F and leaves at 69.5 F. Calculate:

(*a*) The amount of fresh water which is produced per hour. Assume that the mixture which is formed when the liquid at 80 F is throttled as it enters the flash evaporator is in equilibrium at 76.5 F, and that it is perfectly separated.

(*b*) The amount of cooling water which enters at 4.

5A.11 Usually a cryogenic fluid is stored in a vacuum insulated tank, Fig. 5A.6. One way of discharging the fluid from such a tank is to withdraw some of the liquid from the storage tank and vaporize it in a heat exchanger, as shown in the sketch, and then feed the vapor to the top of the tank at such a rate as to maintain constant pressure. The heat transferred to the fluid is from the surroundings which are at temperature T_0. Suppose the fluid in the tank is saturated at a given pressure, and it is desired to withdraw fluid from the tank at this pressure at the rate of $\dot{m}_e$. Assume that the vapor leaving the heat exchanger and entering the storage tank is saturated vapor at the given pressure.

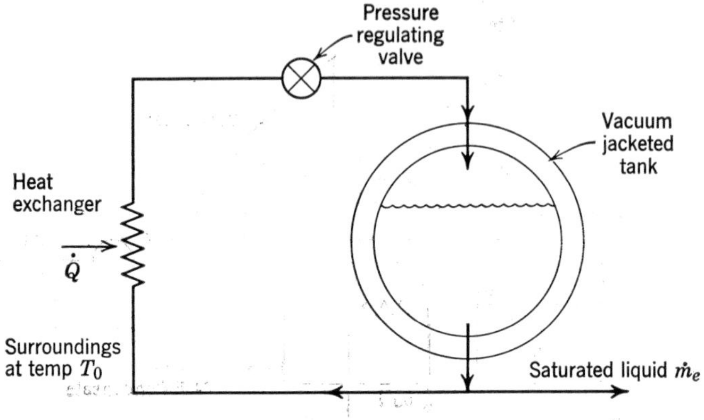

Fig. 5A.6

Determine heat transfer rate to the heat exchanger, $\dot{Q}$, as a function of the mass rate of liquid flow $\dot{m}_e$, and the thermodynamic properties of the fluid.

5A.12 When a steam boiler is used in conjunction with a reactor in a nuclear power plant, the time required for a rapid blowdown is an important factor in safety considerations. In making this calculation it is necessary to know the pressure in the boiler as a function of the amount of liquid in the tank. To illustrate the procedure involved consider the following situation.

The boiler has a volume of 100 ft^3. Initially it is half filled with liquid and half with vapor. The initial pressure is 200 lbf/in.^{2} Assume that there is no heat transfer to the liquid or vapor during this process, and that the liquid and vapor are always in equilibrium.

Determine the pressure in the tank when the volume of the liquid is:

(*a*) $\frac{3}{8}$ of the volume of the boiler
(*b*) $\frac{1}{4}$ of the volume of the boiler
(*c*) $\frac{1}{8}$ of the volume of the boiler

5A.13 Heat is transferred at a given rate to a mixture of liquid and vapor in equilibrium in a closed container. Determine the rate of change of temperature as a function of the thermodynamic properties of the liquid and vapor and the mass of liquid and the mass of vapor.

Chapter 6

6A.1 Helium has the lowest normal boiling point of any of the elements, namely 4.2 K. At this temperature it has an enthalpy of evaporation of 19.7 cal/g mole.

A Carnot refrigeration cycle is to be used in the production of 1 g mole of liquid helium at 4.2 K from saturation vapor at the same temperature. What is the work input to the refrigerator and the coefficient of performance of this refrigeration cycle? Assume an ambient temperature of 300 K.

6A.2 Temperatures of 0.01 K can be achieved by a technique known as magnetic cooling. In this process a strong magnetic field is imposed on a paramagnetic salt which is maintained at 1 K by transferring heat to liquid helium which is boiling at very low pressure. The salt is then thermally isolated from the helium, the magnetic field is removed, and the temperature drops.

Assume that 1×10^{-3} calories are to be removed from the paramagnetic salt at an average temperature of 0.1 K, and that the necessary refrigeration is produced by a Carnot refrigeration cycle. What is the work input to the refrigerator and the coefficient of performance of this refrigeration cycle? Assume an ambient temperature of 300 K.

6A.3 The lowest temperature which has been achieved at the present time (1962) is about 1×10^{-6} K. Achieving this temperature involved an additional stage to that described in Problem 6A.2, namely nuclear cooling. This is similar to magnetic cooling, but involves the magnetic moment associated with the nucleus rather than that associated with certain ions in the paramagnetic salt.

Suppose that 10^{-8} Btu were to be removed from a specimen at an average temperature of 10^{-5} K (10^{-8} Btu is about the amount of energy associated with the dropping of a pin through a distance of $\frac{1}{8}$ in., and is about equal to the energy transferred as heat at this temperature in some experiments). If this amount of refrigeration at an average temperature of 1×10^{-5} K is produced by a Carnot refrigeration cycle, determine the work input and the coefficient of performance of the refrigeration cycle. Assume an ambient temperature of 300 K.

6A.4 Consider an engine in outer space which operates on the Carnot cycle. The only way in which heat can be transferred from the engine is by radiation. The energy radiated is proportional to the fourth power of the absolute temperature and the area of the radiating surface. Show that for a given amount of work and a given T_H, the area of the radiator will be a minimum when $T_L/T_H = \frac{3}{4}$.

6A.5 It is desired to produce refrigeration at −10 F.

A reservoir is available at a temperature of 300 F. The ambient temperature is 90 F. Determine the ratio of the heat transferred from the high temperature reservoir to the heat transferred from the refrigerator, assuming all processes to be reversible.

Chapter 7

7A.1 Two tanks, *A* and *B*, contain steam in the amounts and at the condition indicated in Fig. 7A.1. The tanks are connected to a

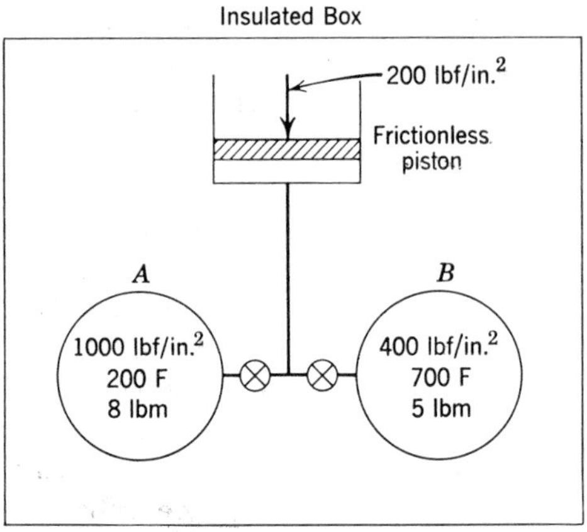

Fig. 7A.1

common cylinder equipped with a piston of such a weight that a pressure of 200 $lbf/in.^2$ is required to support it. Initially the piston is at the bottom of the cylinder.

Calculate the change of enthalpy, internal energy, and entropy of all the steam if the valves are opened and all the steam comes to a uniform pressure and temperature. Assume negligible volume in the pipes and valves and assume no heat transfer between the containing walls and the steam.

7A.2 Consider the scheme shown for raising a weight by heat transfer from the reservoir at 300 F to the Freon-12. The pressure on the Freon-12 due to the weight plus the atmosphere is 200 $lbf/in.^2$ Initially the temperature of the Freon-12 is 160 F and its volume is 0.4 ft^3. Heat is transferred until the temperature of the Freon-12 is 300 F.

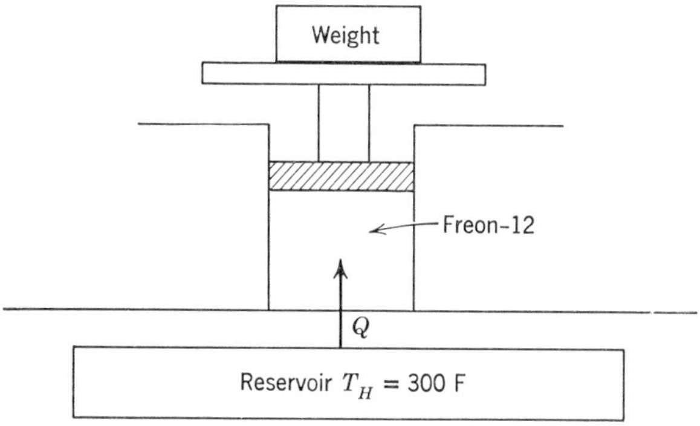

Fig. 7A.2

(*a*) Determine the heat transfer and work for this process.
(*b*) Determine the net change of entropy.
(*c*) Devise a scheme for accomplishing this with no net change of entropy.

7A.3 Steam enters a turbine at 100 lbf/in.^{2}, 800 F, and exhausts at 20 lbf/in.^{2}, 400 F. Any heat transfer is with the surroundings at 80 F, and changes in kinetic and potential energy are negligible. It is claimed that this turbine produces 100 hp with a mass flow of 1800 lbm/hr. Does this process violate the second law of thermodynamics?

7A.4 Steam flowing steadily in a line at 1000 lbf/in.^{2}, 700 F, 20 ft/sec, is expanded in a nozzle to 200 lbf/in.^{2} in an irreversible, isothermal process during which 200 Btu of heat are transferred per lbm of steam from a large reservoir at 800 F. The surroundings are at 14.7 lbf/in.^{2}, 77 F. Calculate:
(*a*) Exit velocity from the nozzle
(*b*) The net increase of entropy

7A.5 The cylinder of a refrigeration compressor has a bore of 3 in. When the piston is at the bottom position the total volume in the cylinder is 40 in.^{3}, and it is filled with Freon-12 at 25 lbf/in.^{2}, 80 F. As the piston moves upward it first compresses the Freon-12, which is the refrigerant, until the pressure reaches 125 lbf/in.^{2}, and then the discharge valve opens and the refrigerant is forced out the cylinder during the remainder of the upward stroke. Assume that the compression process is reversible and adiabatic.
(*a*) How far will the piston travel before the discharge valve opens?
(*b*) What is the work required for this portion of the upward stroke?

7A.6 Repeat Problem 7A.5(*a*) assuming that the leakage past the piston is 3% of the initial mass in the cylinder.

7A.7 A spherical balloon initially 6 in. in diameter and containing Freon-12 at 15 lbf/in.^{2} is connected to an uninsulated 1 ft^3 tank containing Freon-12 at 80 lbf/in.^{2} Both are at 80 F, the temperature of the surroundings.

The valve connecting the two is then opened very slightly and left open until the pressures become equal. During the process, heat is transferred with the surroundings such that the temperature of all the Freon remains at 80 F. It may also be assumed that the balloon diameter is proportional to the pressure inside the balloon at any point during the process. Calculate:

(*a*) The final pressure
(*b*) Work done by the Freon during the process
(*c*) Heat transfer to the Freon during the process
(*d*) Net entropy change for the process

7A.8 Determine the increase of entropy associated with the production of 1 lb of liquid nitrogen by the process and under the conditions given in Problem 5A.8.

7A.9 It is desired to quickly cool a given quantity of material to 40 F. The heat transfer is 1000 Btu. One possibility is to immerse it in a mixture of ice and water, in which case the heat transfer from the material results in the melting of ice. Another possibility is to cool the material by evaporating Freon-12 at 0 F, in which case the heat transfer results in changing the Freon-12 from saturated liquid to saturated vapor. A third possibility is to do the same with liquid nitrogen at one atmosphere pressure.

(*a*) Calculate the change of entropy of the cooling medium in each of the three cases.

(*b*) What is the significance of these results?

Chapter 8

8A.1 Consider the system shown in Fig. 8A.1. Tank A has a volume of 10 ft^3 and initially contains air at 100 lbf/in.^{2}, 100 F. Cylinder B is fitted with a frictionless piston resting on the bottom, at which position the spring is fully extended. The piston has a cross-sectional area of 100 in.^{2} and mass of 100 lbm, and the spring constant K_s is 100 lbf/in. Atmospheric pressure is 14.7 lbf/in.^{2}

The valve is opened and air flows into the cylinder until the pressures in A and B become equal, after which the valve is closed. During this process, the air finally remaining in A may be considered to have undergone a reversible process, and the entire process is adiabatic. The spring force is proportional to the displacement.

Determine the final pressure in the system and the temperature in cylinder B.

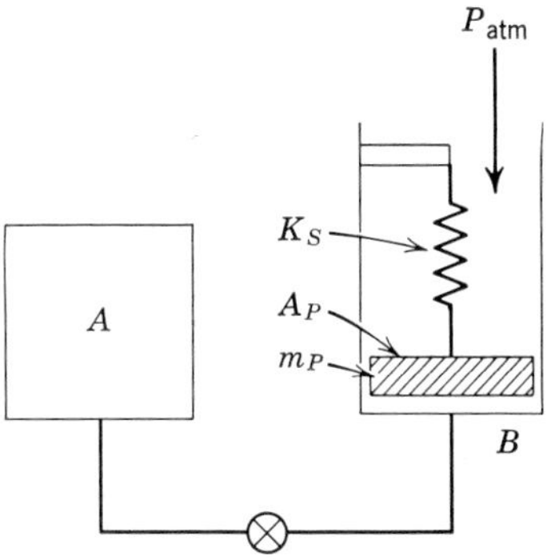

Fig. 8A.1

8A.2 Air flows in a pipe in which frictional effects are present. At one point in the pipe the pressure is 100 lbf/in.^{2}, 200 F, and the velocity is 300 ft/sec. At a certain point down stream the pressure is 70 lbf/in.^{2} Assuming no heat transfer during the process, determine the final velocity and temperature of the air.

8A.3 An ideal gas flows into an evacuated vessel. For gas A the specific heat ratio $k = 1.67$ and for gas B, $k = 1.4$. The initial pressure and temperature is the same for both gases, and the final pressure is equal to the initial pressure in both cases. For gas A, as compared to gas B, the final temperature is: greater______, the same______, less______.

8A.4 An incompressible liquid and an ideal gas each flow through a pipe in an adiabatic process. As a result of friction the pressure decreases in both cases. For the incompressible liquid, as compared with the ideal gas,

	Greater	The Same	Less
The increase in velocity is	______	______	______
The increase in internal energy is	______	______	______
The work is	______	______	______

8A.5 A neophyte engineer has a problem to solve which involves the entropy of superheated steam at pressures below 1 lbf/in.^{2}, but he finds that the steam table which he has does not give the properties of superheated steam below this pressure.

(*a*) How would you advise him to proceed?

(*b*) What assumptions does this procedure involve?

(*c*) Set up, in accordance with the assumptions made, a line similar to the 1 lbf/in.^{2} line in Table A.1.3 which gives v, h, and s at 0.5 lbf/in.^{2} at temperatures of 200 F, 500 F, 1000 F.

8A.6 One proposal for a nuclear-powered rocket engine involves liquid hydrogen (which has a saturation temperature at one atmosphere pressure of 20.4 K). The liquid would be pumped to a high pressure, and then vaporized and heated to a high temperature in the reactor.

(*a*) How much heat is required to raise the temperature of the hydrogen from 500 to 5000 R, assuming the hydrogen to be an ideal gas with no dissociation during this process?

(*b*) If the hydrogen is at 800 lbf/in.^{2}, 5000 R when it leaves the reactor and enters the nozzle, what will be the exit velocity if it expands in a reversible adiabatic process to 10 lbf/in.^{2} at the nozzle exit?

(*c*) What would be the exit area of the nozzle under these conditions if the flow is 1 lbm/sec?

8A.7 A cylinder contains air at 1 atm pressure, -40 F. The volume of the cylinder is 1 ft^3. Air at 10 atm pressure and a temperature of 100 F enters the cylinder until the pressure in the cylinder reaches 10 atm. There is no heat transfer to the air during this process. What is the net change of entropy?

8A.8 A cylinder shown in Fig. 8A.2 is so constructed that compartments A and B are separated by a metal partition, and compart-

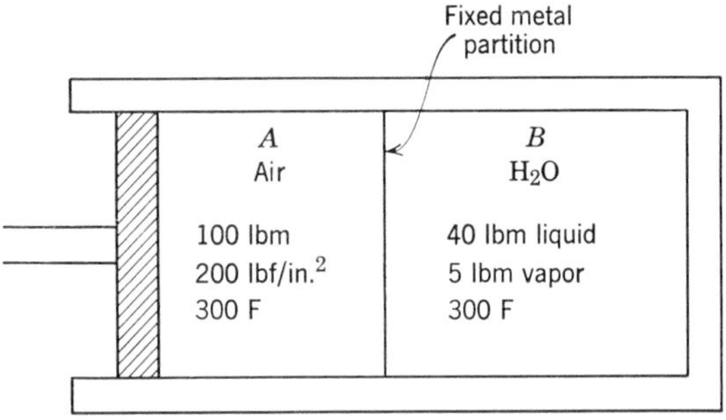

Fig. 8A.2

ment A is fitted with a frictionless piston. Assume that the temperatures in compartments A and B differ only by an infinitesimal. Compartment A initially contains 100 lbm of air at 200 lbf/in.^{2}, 300 F. Compartment B contains 40 lbm of liquid H_2O and 5 lbm of vapor H_2O at 300 F. The piston is moved to the right until the pressure in compartment A reaches 300 lbf/in.^{2} Assume no heat transfer except across the metal partition. Determine the total change of entropy and internal energy.

8A.9 A tank of volume V_T is filled with air at the initial condition P_1 and T_1. It is to be evacuated with a vacuum pump. Atmospheric Pressure $= P_a$.

(*a*) Calculate the work of evacuation as a function of the pressure in the tank, and plot a curve of work of evacuation vs pressure.

(*b*) Considering the pump to handle a constant volume per unit time, determine horsepower as a function of pressure in the tank and horsepower as a function of time.

Assume both the tank and the pump to operate under isothermal conditions, and assume that the pumping process is reversible.

Chapter 9

9A.1 A vessel contains one lb mole of nitrogen and one lb mole of helium, each at 1 atm pressure and 77 F and separated by a membrane.

Determine the change of entropy which occurs when the membrane ruptures and a homogeneous mixture fills the vessel. Assume no heat transfer during the process.

9A.2 The vacuum space in the dewar shown in Fig. 9A.1 contained air under the initial conditions when the dewar was empty. However, when the liquid hydrogen was placed in the dewar, essentially all of the CO_2, O_2, N_2, and H_2O froze out on the inner walls, leaving only hydrogen and helium in the vacuum space. If the air initially contained 0.01% helium and 0.01% hydrogen by volume, what would be the pressure in the vacuum space if the average temperature of the gas is considered to be 50 K?

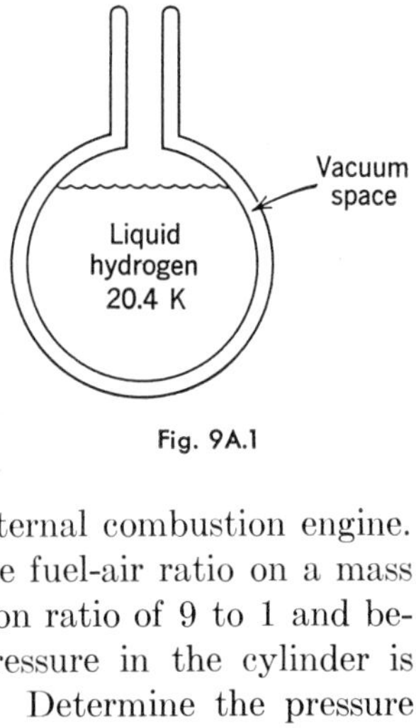

Fig. 9A.1

9A.3 Consider the compression of a fuel-air mixture which takes place in an internal combustion engine. Assume that the fuel is ethane and that the fuel-air ratio on a mass basis is 15 to 1. The engine has a compression ratio of 9 to 1 and before the compression stroke begins the pressure in the cylinder is 13.0 lbf/in.^{2} and the temperature is 100 F. Determine the pressure and temperature after compression and the work of compression per pound of mixture.

9A.4 Two mixtures, one consisting of helium and water vapor and the other consisting of air and water vapor are each at the same total pressure and temperature. In each mixture the partial pressure of the water vapor is the same. For the helium-water vapor mixture, as compared to the air-water vapor mixture the

	Greater	The Same	Less
relative humidity is	____	____	____
humidity ratio is	____	____	____
mole fraction of water vapor is	____	____	____

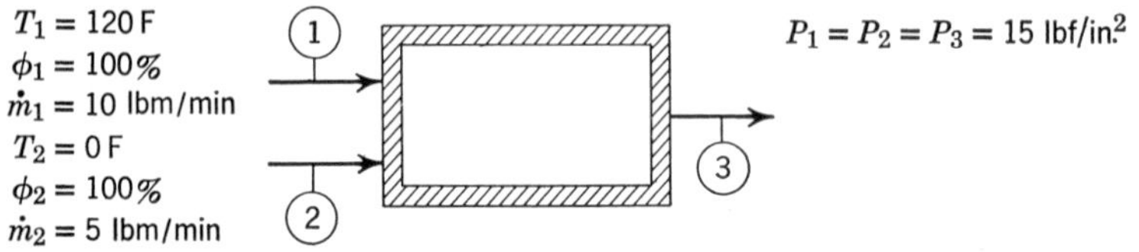

Fig. 9A.2

9A.5 Two streams of air, each saturated with water vapor, and at the pressure and temperature shown in Fig. 9A.2 are mixed together in an adiabatic, steady-flow process. Determine the exit temperature, T_3.

9A.6 Frequently a source of dry air is needed for such purposes as pressurizing telephone cables and similar applications. One scheme for providing dry air is shown in Fig. 9A.3. Atmospheric air is compressed to 165 lbf/in.^{2} It is cooled to 70 F in an after cooler and a counterflow heat exchanger. Finally it is cooled to 35 F by heat transfer to the refrigerant in the evaporator of the refrigeration cycle. The water condensed in these processes is separated from the air and leaves through an automatic water ejector. The air-water vapor mixture

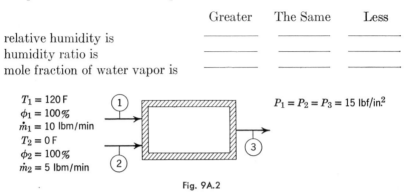

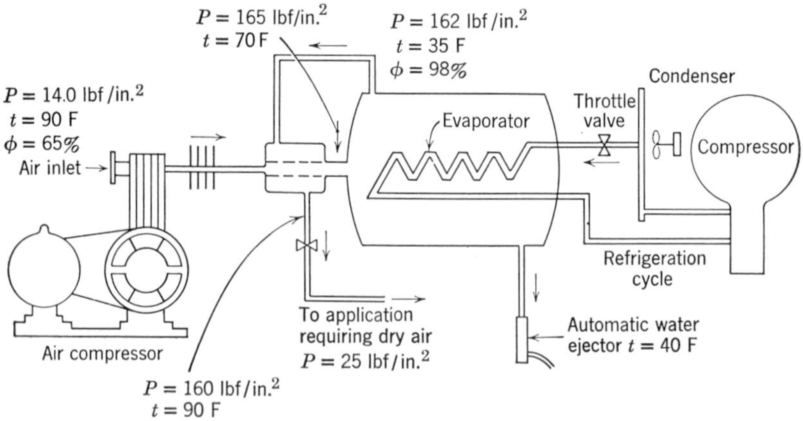

Fig. 9A.3

remaining is used as the cooling medium in the heat exchanger and is then throttled to 25 lbf/in.^{2} and is used for the intended application.

If the volume to the compressor is 200 ft^3/hr, determine:

(*a*) The amount of moisture condensed per hour.

(*b*) The heat transfer to the refrigerant in the evaporator per hour.

(*c*) The relative humidity of the air-vapor mixture entering the apparatus which requires the dry air.

9A.7 In an air liquefaction and separation plant, air is compressed from the ambient pressure to 3000 lbf/in.^{2}, after which the air enters a heat exchanger. The ambient conditions of the air are 14.7 lbf/in.^{2}, 85 F, 65% relative humidity. After compression and before entering the heat exchanger the air is cooled to 90 F. In the heat exchanger the air is cooled to -150 F, at which temperature the vapor pressure of the H_2O is very low, and it may be assumed that all of the water vapor which enters the heat exchanger freezes out on the tubes of the heat exchanger. If the compressor handles 3000 ft^3/min of air, how much moisture will freeze out in the heat exchanger in 1 hr of operation? Would you recommend operating a plant in this manner?

9A.8 The process of injection cooling may be demonstrated by the following problem: Consider a column of water 10 ft high which is insulated from the surroundings. Let dry air be bubbled through the water. During this process some of the water will evaporate and be carried away by the air and the temperature of the remaining liquid will decrease.

Assume that the water is initially at a temperature of 90 F, and that the dry air enters at 90 F and a pressure slightly above 1 atm. Assume that the air vapor mixture leaves the surface of the water at 1 atm pressure and at the same temperature as the water and at a relative humidity of 100%.

(*a*) What is the heat transfer from the water for each pound of air entering under the initial conditions?

(*b*) What will be the ratio of the mass of air bubbled through the liquid to the mass of water cooled if the temperature of the water is decreased 10 F?

9A.9 Consider the compressor and aftercooler shown in the sketch below. Atmospheric air at a pressure of 14.7 lbf/in.^{2}, a temperature of 90 F, and a relative humidity of 75% enter the compressor. Under these conditions the volume flow into the compressor is 1000 ft^3/min. The rate of heat transfer from the compressor is 800 Btu/min, and the air-vapor mixture leaves the compressor at 100 lbf/in.^{2}, 400 F. It then enters the aftercooler, where it is cooled to 100 F in a constant

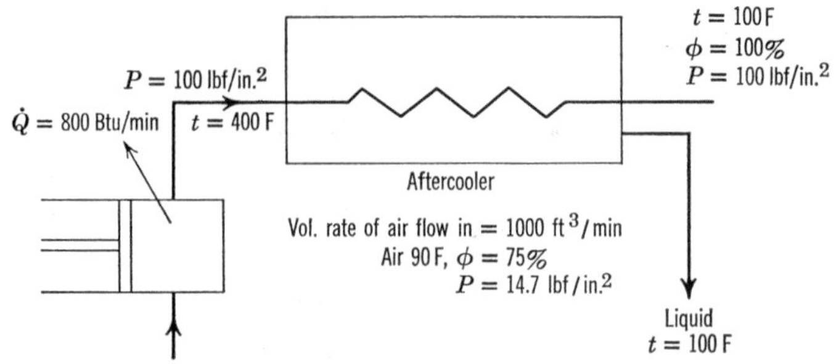

Fig. 9A.4

pressure process. The air-vapor mixture leaving the aftercooler has a relative humidity of 100%, and the condensate leaves at 100 F.

(*a*) What is the power input to the compressor?

(*b*) Determine the amount of condensate per hour.

Chapter 10

10A.1 Compressed air is used to start a diesel engine. The air is stored at 300 lbf/in.^{2}, 77 F in a tank having a volume of 20 ft^3, and it can be used for starting until the pressure is 40 lbf/in.^{2} It can be assumed that during the starting process that the temperature of the air remains at 77 F. What is the maximum work which is available for starting the engine?

10A.2 What is the minimum work required to change 1 lbm of nitrogen from 15 lbf/in.^{2}, 80 F to saturated liquid at 1 atm pressure? Sketch an apparatus which would be used in such a process and show the process on a temperature entropy diagram.

10A.3 Suppose nitrogen is to be liquefied by the process of Problem 5A.8. Assume that a reversible isothermal compressor is used to compress the air from 15.0 lbf/in.^{2}, 80 F to 1500 lbf/in.^{2} before the high pressure nitrogen enters the heat exchanger.

(*a*) What is the required work input to liquefy 1 lb of nitrogen?

(*b*) What is the irreversibility per pound of nitrogen liquefied?

10A.4 Liquid nitrogen at 1 atm pressure is to be vaporized and delivered to a pipeline at 500 lbf/in.^{2}, 0 F. Three possible schemes for doing this are shown in Fig. 10A.1. The first scheme involves pumping the liquid and then vaporizing it. The second involves vaporizing the liquid and superheating it just enough so that it would leave a reversible adiabatic compressor at 500 lbf/in.^{2}, 0 F. The third involves

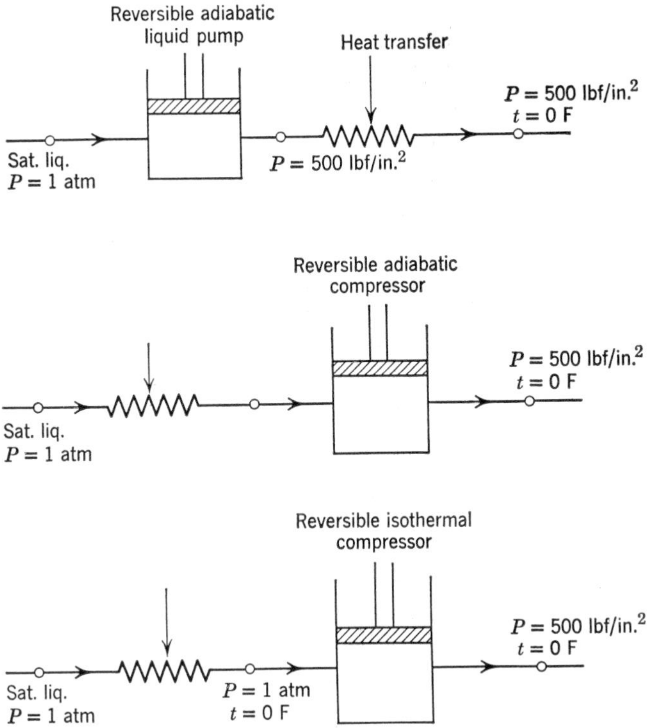

Fig. 10A.1

vaporizing it and superheating it to 0 F, followed by a reversible isothermal compressor.

(*a*) What is the availability of liquid nitrogen at 1 atm pressure?

(*b*) What is the minimum work necessary to deliver 1 lb of nitrogen at 500 lbf/in.^{2}, 0 F.

Show a schematic arrangement of how you would do this with this minimum work input.

(*c*) Determine the work and the irreversibility per pound of nitrogen delivered for each of the three suggested ways of accomplishing this.

10A.5 Consider the proposed scheme for making fresh water from salt water shown in Fig. 10A.2. Salt water enters the apparatus at 35 F and 1 atm pressure. It flows into a vacuum chamber where a low pressure is maintained by a reversible adiabatic vacuum pump. 10% of the entering liquid leaves as a brine solution at 32 F. The solid, which is sufficiently free of salt, also leaves at 32 F. (Some "pump" for removing the solid must be utilized, but this will be ignored in this

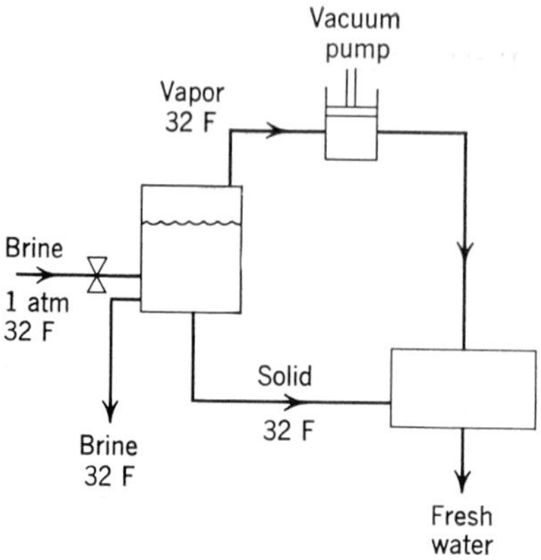

Fig. 10A.2

analysis.) The vapor which leaves the vacuum chamber at 32 F is compressed to 1 atm pressure in a reversible adiabatic compressor (vacuum pump) and then flows through a heat exchanger where it is condensed by melting the ice which was formed.

(*a*) Determine the fraction of the incoming stream which leaves the vacuum chamber as vapor and as solid.

(*b*) What is the work input to the vacuum pump per pound of pure water?

(*c*) What is the irreversibility per pound of pure water?

Assume that the salt water entering and the brine leaving have the properties of water.

10A.6 A rather simplified thermodynamic model of how rain is caused is shown in Fig. 10A.3. Air at 60 F, 50% relative humidity

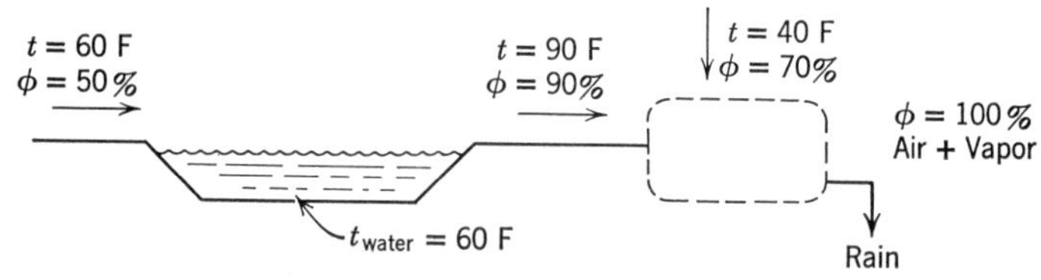

Fig. 10A.3

moves over a large body of water. As the result of heat transfer through the radiant energy of the sun the air leaves the lake at 90 F, 90% relative humidity. Later, it mixes with cold air at 40 F, 70% relative humidity, with the result that rain forms. Assume that the mass rates of flow of the hot and cold air streams are equal, and that the pressure remains 1 atm throughout the process.

(*a*) Determine the heat transfer per pound of rain.

(*b*) Determine the irreversibility per pound of rain.

10A.7 Consider the process shown in Fig. 5A.6 in conjunction with Problem 5A.11. Assume that the tank initially contains saturated liquid nitrogen at −300 F, and that the liquid is to be delivered at the saturation pressure corresponding to this temperature. What is the irreversibility per pound of liquid delivered at this pressure?

10A.8 The composition of air on a volumetric basis is essentially 78% nitrogen, 21% oxygen, and 1% argon. Determine the minimum work required to separate 1 mole of air at 77 F, 1 atm pressure into nitrogen, oxygen, and argon, each at one pressure and 77 F. Compare this result with Problem 10.5(*b*).

10A.9 The general principle of a multiple effect evaporator is shown in Fig. 10A.4. One purpose of such a device is to make fresh water from salt water, although the sketch shown is somewhat simplified. Saturated steam enters the first evaporator at a pressure of 20 lbf/in.^{2} As this condenses, it evaporates water at a lower pressure, in this case 10 lbf/in.^{2} To simplify this calculation, it may be assumed that the water being evaporated enters as saturated liquid at 10 lbf/in.^{2} and leaves as saturated vapor at 10 lbf/in.^{2} This vapor is in turn con-

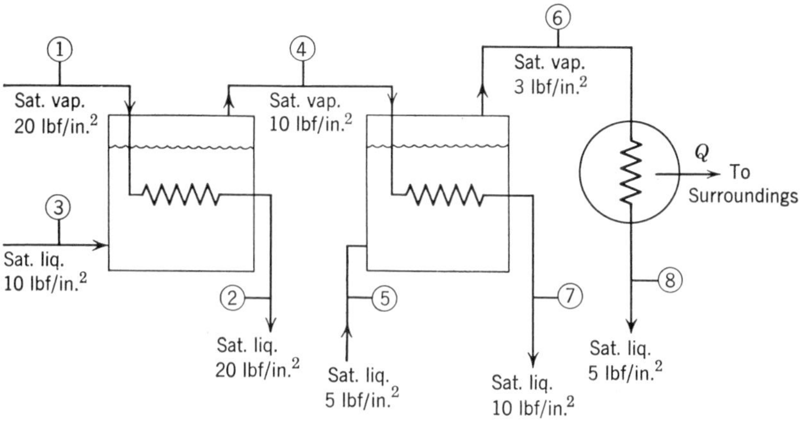

Fig. 10A.4

densed by evaporating liquid at 5 lbf/in.^{2} This latter vapor is condensed by heat transfer to the surroundings at 537 R.

(*a*) Determine the total amount of condensed liquid obtained from 1 lbm of saturated vapor entering at *A* at 20 lbf/in.^{2}

(*b*) Determine the availability of the steam entering the first evaporator.

(*c*) Determine the irreversibility of this process.

10A.10 Consider the expansion of gases in a gas turbine having four stages. For purposes of this calculation assume that the gas is air, which enters the turbine at 60 lbf/in.^{2}, 1400 F, and leaves at 15.0 lbf/in.^{2} The pressure ratio across each stage is the same, and the efficiency (based on a reversible adiabatic process) for each stage is 78%. Determine the efficiency of the turbine and the reheat factor.

Chapter 11

11A.1 One device which is used in the production of low temperatures is the reciprocating expansion engine. The general schematic arrangement is shown in Fig. 11A.1. For purposes of this analysis assume that nitrogen enters at 180 lbf/in.^{2}, -100 F and exhausts at 15 lbf/in.^{2} Assume an ideal cycle for the expansion engine, which is the reverse of the air compressor cycle shown in Fig. 11.4.

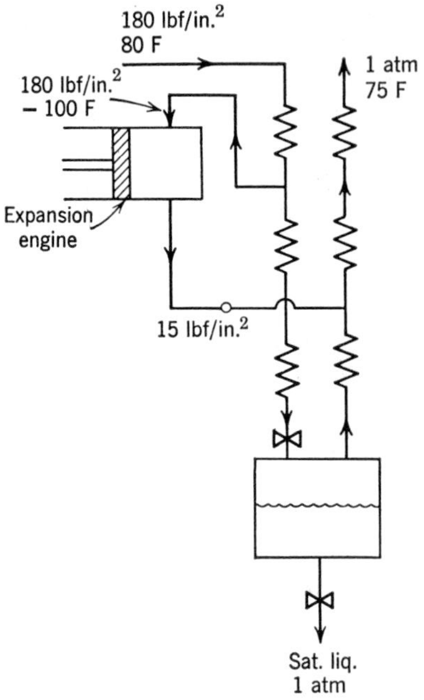

Fig. 11A.1

Consider an expansion engine which has a 4-in. bore, a 5-in. stroke, a 6% clearance volume, and operates at 400 rpm.

(*a*) At what point of the inlet stroke should the inlet valve close?

(*b*) At what point in the exhaust stroke should the discharge valve close?

(*c*) What is the power output of the expansion engine?

(*d*) What fraction of the inlet air can be liquefied, if 60% of the incoming air passes through the expansion engine?

11A.2 At the inlet of a reciprocating ammonia compressor the pressure is 20 lbf/in.^{2} and the temperature is 40 F. The compressor has a bore of 5 in. and a stroke of 6 in., and the volumetric efficiency is estimated to be 78%. At what speed must the compressor run if it is required to handle 400 lbm of ammonia per hour?

11A.3 Consider one stage of a reciprocating helium compressor at 15 lbf/in.^{2}, 80 F. The pressure ratio across the compressor is 3.5, and during the compression and expansion processes the polytropic exponent is 1.5. The clearance volume is 8%.

(*a*) What is volumetric efficiency of an ideal compressor under these conditions?

(*b*) What is the work per pound of helium delivered?

(*c*) What is the heat transfer from the compressor per pound of helium delivered?

11A.4 A compressor is to be designed which will receive air at 14.7 lbf/in.^{2}, 80 F, and deliver it at 2800 lbf/in.^{2} It has been decided that four stages will be used, with equal pressure ratio across each stage. The volumetric efficiency of each stage is estimated at 75%, and the temperature at the exit of all the intercoolers will be 110 F. Each stage is single acting, and is driven off a common shaft.

Determine the ratio of the piston displacements for the four stages.

Chapter 12

12A.1 A nuclear reactor to produce power is so designed that the maximum temperature in the steam cycle is 800 F. The minimum temperature of the steam cycle is 100 F. It is planned to build a steam power plant which has one open feedwater heater.

Select what you consider to be a reasonable cycle within these specifications, determine the ideal cycle efficiency, and explain why the particular pressures involved were selected.

12A.2 Consider the preliminary design for a supercritical steam power plant cycle. The maximum pressure will be 4000 lbf/in.^{2}, and the maximum temperature will be 1100 F. The cooling water temperature is such that 1 lbf/in.^{2} pressure can be maintained in the condensers. Turbine efficiencies of at least 85% can be expected.

(*a*) Do you recommend any reheat for this cycle? If so, how many stages of reheat and at what pressures? Give reasons for your decisions.

(*b*) Do you recommend feedwater heaters? If so, how many and at what pressures would they operate? Would they be open or closed feedwater heaters?

(*c*) Estimate the thermal efficiency of the cycle which you recommend.

12A.3 It is desired to study the influence of the number of feedwater heaters on thermal efficiency for a cycle in which the steam leaves the steam generator at 2400 lbf/in.^{2}, 1000 F, and which has a condenser pressure of 2 in. of mercury. Assume that open feedwater heaters are used. Determine the thermal efficiency for each of the following cycles.

(*a*) No feedwater heater

(*b*) One feedwater heater which operates at 140 lbf/in.^{2}

(*c*) Two feedwater heaters, one operating at 450 lbf/in.^{2} and the other at 35 lbf/in.^{2}

12A.4 Repeat problem 12A.3 assuming a turbine efficiency of 85%.

12A.5 In a nuclear power plant heat is transferred in the reactor to liquid sodium. The liquid sodium is then pumped to a heat exchanger where heat is transferred to steam. The steam leaves this heat exchanger as saturated vapor at 800 lbf/in.^{2}, and is then superheated in an external gas fired superheater to 1100 F. The steam then enters the turbine, which has one extraction point at 60 lbf/in.^{2}, where steam flows to an open feedwater heater. The turbine efficiency is 70% and the condenser pressure is 1 lbf/in.^{2}

Determine the heat transfer in the reactor and in the superheater to produce a power output of 60,000 kw.

12A.6 Consider an ideal regenerative steam cycle with a single closed feedwater heater. Steam enters the turbine at 600 lbf/in.^{2}, 600 F, and the condenser pressure is 2 lbf/in.^{2} Steam is extracted at 100 lbf/in.^{2} to heat the feedwater, which leaves the heater at the temperature of the condensing steam. Condensate from the heater drains through a trap to the condenser.

(*a*) Show the various states on a temperature-entropy diagram.

(*b*) Calculate the fraction of mass extracted, the net work per lbm of steam entering the turbine and the thermal efficiency of the cycle.

12A.7 A freezer is designed to maintain a temperature of -10 F in the storage compartment while placed in an ambient of 90 F. Under these conditions it is necessary to transfer 4000 Btu/day from the compartment which is maintained at -10 F.

(*a*) What is the absolute minimum power rating (in horsepower) for the motor which drives the refrigeration compressor?

(*b*) Suppose that Freon-12 is used as the refrigerant, and that the temperature of the Freon-12 in the condenser is 30 F above the ambient, and in the evaporator it is 20 F below the temperature in the freezing compartment. What is the minimum power rating for the motor driving the compressor under these conditions? What is the pressure of the Freon-12 in the condenser and in the evaporator?

(*c*) Suppose cycle shown in Fig. 12.18 is used with the temperatures in the condenser and evaporator as given in part (*b*). What is the minimum power under these conditions?

12A.8 In a conventional refrigerant cycle which uses Freon-12 as the refrigerant, the temperature of the evaporating fluid is -10 F. It leaves the evaporator as saturated vapor at -10 F and enters the

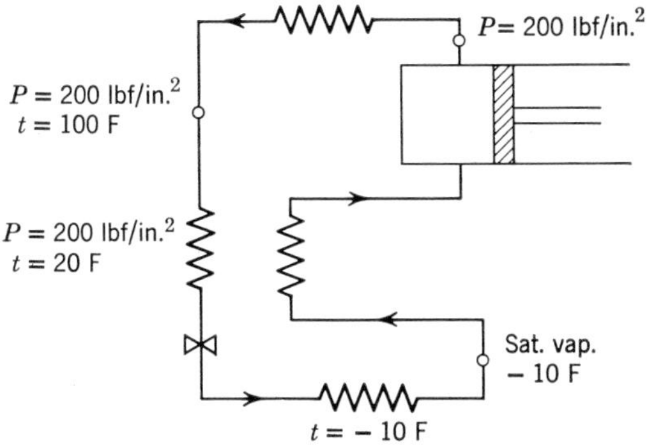

Fig. 12A.1

compressor. The pressure in the condenser is 200 $lbf/in.^2$ The liquid leaves the condenser and enters the expansion valve at a temperature of 100 F.

It is proposed to modify this cycle as shown in Fig. 12A.1. In this case an additional heat exchanger is supplied in which the cold vapor leaving the evaporator cools the liquid which enters the expansion valve.

Compare the coefficient of performance of these two cycles.

12A.9 Consider a simple open gas turbine cycle in which the minimum cycle temperature is 40 F and the minimum pressure is 15 $lbf/in.^2$ The maximum temperature in the cycle is 1640 F and the maximum pressure is 60 $lbf/in.^2$ There is a pressure drop of 2 $lbf/in.^2$ between the compressor and the turbine. The efficiency of the compressor is 86% and the efficiency of the turbine is 90%.

Determine the over-all cycle efficiency.

12A.10 A gas turbine is to be used for pumping natural gas through a cross country pipeline. The required power input to the natural gas compressor is 1000 hp. It has been decided to use a gas turbine to provide the necessary power input. A simple open cycle will be used with a regenerator of 50% efficiency. Since a supply of fuel is readily available, and since some of these units may be located in relatively isolated places, low maintenance costs are more important than a high efficiency.

Select a cycle which you would recommend, making appropriate assumptions for compressor and turbine efficiencies, and determine (for the gas turbine unit) the power output of the turbine, the power input to the compressor, and the efficiency.

Chapter 14

14A.1 Derive an expression for each of the following quantities which does not contain h, u, or s:

$$\left(\frac{\partial u}{\partial T}\right)_P ; \quad \left(\frac{\partial h}{\partial s}\right)_v$$

14A.2 An insulated tank having a volume of 2 ft^3 contains carbon dioxide at 200 lbf/in.^{2}, 100 F. It is pressurized from a line in which carbon dioxide is flowing at 1200 lbf/in.^{2}, 100 F. The valve is closed when the pressure in the tank reaches 1200 lbf/in.^{2}

Using the generalized charts determine the amount of carbon dioxide which flows into the tank.

14A.3 Carbon dioxide is flowing at low velocity in a line at 700 R, 1000 lbf/in.^{2} The velocity of the carbon dioxide is to be increased to 800 ft/sec by having it flow through an appropriate nozzle. Assuming the flow to be isentropic, determine the final pressure and temperature by use of the generalized charts.

14A.4 Consider Fig. 3.4, p. 35, and show that for water the slope of the sublimation line is greater than the slope of the vaporization line at the triple point.

14A.5 An uninsulated cylinder with a volume of 2 ft^3 contains ethylene at 1000 lbf/in.^{2}, 80 F, the temperature of the surroundings. The cylinder valve leaks very slightly so that after a considerable length of time the cylinder pressure drops to 500 lbf/in.^{2} Determine:

(*a*) The mass of ethylene which escaped from the cylinder
(*b*) The heat transferred during the process

14A.6 Determine, using the various generalized charts,

(*a*) The change of enthalpy
(*b*) The change of entropy

when the state of methane is changed from 800 lbf/in.^{2}, 350 R to 14.7 lbf/in.^{2}, 520 R.

It may be assumed that the zero pressure specific heat for methane given in Table A.7 is also valid at 1 atm pressure.

14A.7 Freon-12, which has a molecular weight of 121, expands in a cylinder from 300 lbf/in.^{2}, 180 F to 35 lbf/in.^{2} in an isothermal process. During this process the heat transfer to the Freon-12 is 20 Btu/lbm. The process is not reversible.

Determine, for this process, the work per pound of Freon-12 using the following sources for thermodynamic data:

(*a*) The Freon-12 tables
(*b*) The generalized charts
(*c*) The relation (which should first be derived)

$$\left(\frac{\partial u}{\partial v}\right)_T = T\left(\frac{\partial P}{\partial T}\right)_v - P$$

and Van der Waals' equation of state.

14A.8 A transfer line for liquid nitrogen is so designed that it has a double walled pipe as indicated in Fig. 14A.1. The annular space is so

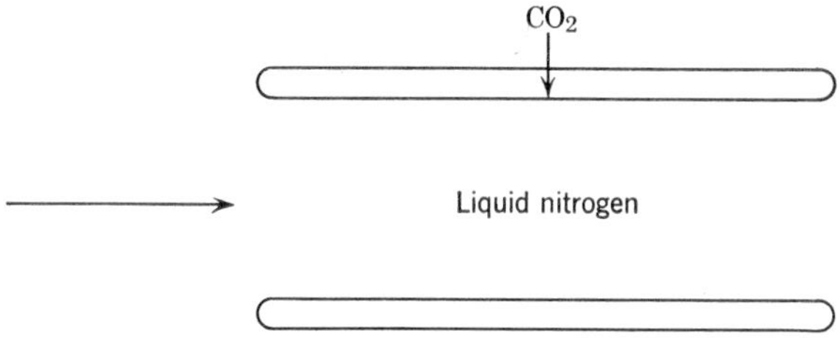

Fig. 14A.1

constructed that when the pipeline is at ambient temperature it contains carbon dioxide at 1 atm pressure. When liquid nitrogen at 1 atm pressure flows through the pipe the carbon dioxide will freeze on the inner walls. The pressure of the carbon dioxide in the annular space will essentially be equal to the vapor pressure of carbon dioxide at the saturation temperature of nitrogen at 1 atm pressure. Suppose that the tables of thermodynamic properties which are available do not go down to this temperature. However, at 5 lbf/in.^{2} pressure the saturation temperature is -13 F and the latent heat of sublimation is -247 Btu/lbm. Estimate the pressure of the carbon dioxide in the annular space when liquid nitrogen is flowing through the tube.

14A.9 Ethane is stored in a tank having a volume of 4 ft^3 at a pressure of 1000 lbf/in.^{2} and a temperature of 100 F. Ethane escapes slowly from the tank until the pressure is reduced to 500 lbf/in.^{2} During this process the temperature remains constant.

(*a*) What fraction of the original mass escapes from the tank?

(*b*) What is the decrease of internal energy per unit mass within the tank?

14A.10 Propane gas expands in a turbine from 400 lbf/in.^{2}, 300 F to 50 lbf/in.^{2}, 200 F. Assume the process to be adiabatic. Over this temperature range the average specific heat of propane may be taken as 0.57 Btu/lbm R. Using the generalized charts determine:

(*a*) The work done per pound of propane entering the turbine

(*b*) The increase of entropy per pound of propane as it flows through the turbine.

Chapter 15

15A.1 A natural gas consisting of 90% methane (CH_4) and 10% ethane (C_2H_6) (on a volume analysis) is burned with 150% theoretical air in a steady flow process. Heat is transferred from the products of combustion until the temperature reaches 600 F. The fuel enters the combustion chamber at 77 F and the air at 300 F.

Determine the heat transfer per mole of fuel.

15A.2 An internal combustion engine burns liquid octane (C_8H_{18}) and uses 120% theoretical air. The air and fuel enter at 77 F, the products leave the engine exhaust ports at 1200 F. In the engine 80% of the carbon burns to CO_2 and the remainder burns to CO. The heat transfer from this engine is just equal to the work done by the engine. Determine:

(*a*) Power output of the engine if the engine burns 20 lbm fuel/hr

(*b*) The Orsat analysis and the dew point of the products of combustion.

15A.3 Consider the analysis of a rocket which uses C_8H_{18}(l) and liquid oxygen for propellants. 100% excess oxygen is used. In a static firing the octane enters the rocket at 14.7 lbf/in.^{2} and 77 F and the liquid oxygen at 14.7 lbf/in.^{2} and its normal boiling point of 162 R. (The difference in enthalpy of oxygen in this state and gaseous oxygen at 14.7 lbf/in.^{2} and 77 F is 5560 Btu/lb mole.)

If complete combustion is achieved in this static firing, and the products leave at 5000 R, what is the velocity of the products of combustion?

15A.4 A certain fuel consisting of a mixture of various compounds has a chemical analysis of 86% carbon, 13% hydrogen, and 1% nitrogen, on a mass basis, and has a higher heating value of 19,200 Btu/lbm. This fuel, at 77 F, is burned with 180% theoretical air at 77 F in a steady flow process. Determine the adiabatic flame temperature.

15A.5 A mixture of butane and 100% theoretical air enters a combustor at 537 R, 1 atm, and products of combustion leave at 1500 R, 1 atm. Assuming complete combustion, calculate the heat transfer from the combustor and the irreversibility for the process, both per mole of butane entering.

15A.6 Consider the cylinder of an internal combustion engine. The contents of the cylinder at top dead center (after compression) and at bottom dead center (after expansion) are as follows:

Constituent	Number of lb moles at T.D.C.	Number of lb moles at B.D.C.
Fuel (C_8H_{18})	0.728×10^{-6}	0
O_2	0.838×10^{-5}	0
N_2	0.326×10^{-5}	0.326×10^{-4}
CO_2	0.166×10^{-6}	0.500×10^{-5}
H_2O	0.210×10^{-6}	0.608×10^{-5}
CO	0.342×10^{-7}	0.102×10^{-5}
H_2	0.161×10^{-7}	0.483×10^{-6}

The specific heat of C_8H_{18} is given by the following relation (T = R)

$$\bar{C}_p = a + bT + cT^2 + dT^3$$

$$a = -0.141264 \times 10^1$$

$$b = 0.10221$$

$$c = -0.308597 \times 10^{-4}$$

$$d = 0.361327 \times 10^{-8}$$

The enthalpy of formation of C_8H_{18} (g) is $-89{,}617$ Btu/lb mole.

At T.D.C. the pressure is 16.1 atm and the temperature is 1220 R. At bottom dead center the pressure is 4.42 atm and the temperature is 2560 R. From a pressure volume diagram it is found that the work of expansion done during the process is 0.705 Btu.

Determine the heat transfer during this process.

15A.7 The following data are taken from the test of a gas turbine on a test stand.

Fuel—C_4H_{10} (g) at 77 F and 1 atm
Air—300% theoretical air at 77 F and 1 atm
Velocity of inlet air = 200 ft/sec
Velocity of products at exit = 2200 ft/sec
Temperature and pressure of products = 1200 F and 1 atm

Assuming complete combustion determine,

(*a*) The net heat transfer per mole of fuel.

(*b*) The net increase of entropy (including the surroundings) per mole of fuel.

Chapter 16

16A.1 Using Table 16.1, determine the value of K_p at 5400 R for the reaction

$$N + O \rightleftarrows NO$$

16A.2 A gas mixture at 77 F, 15 lbf/in.^{2}, which consists of 50% CO_2 and 50% CO (by volume) is fed to a chemical reactor at the rate of 2 moles/min. Steam from a line at 400 F, 200 lbf/in.^{2} is also fed to the reactor at the rate of 2 moles/min. In the reactor the gases are heated in the presence of a catalyst such that the mixture leaving the reactor at 1340 F, 15 lbf/in.^{2} may be assumed to be in equilibrium according to the reaction

$$CO + H_2O \rightleftharpoons CO_2 + H_2$$

Assume that all the constituents are ideal gases except for the inlet steam. Calculate:

(*a*) Enthalpy of the inlet steam relative to the elements at 77 F, 1 atm.

(*b*) Composition (on a mole basis) of the mixture leaving the reactor.

(*c*) Heat transfer to the reactor per minute.

16A.3 Methyl alcohol (CH_3OH) is burned with 140% theoretical air at 20 lbf/in.^{2}, after which the products of combustion are passed through a heat exchanger and cooled to 140 F. Considering the process to be steady-flow, calculate the entropy of the products leaving the heat exchanger per mole of alcohol burned.

16A.4 A mixture of 2 moles CO_2, 2 moles O_2, and b moles N_2 at room temperature and 50 lbf/in.^{2} is heated to 3500 F in a steady-flow process. Consider that at the outlet the mixture consists of CO_2, O_2, N_2, CO in equilibrium. Determine the minimum value for b if the mole fraction of CO at the outlet is to be kept below 0.001.

16A.5 A mixture of 1 mole CO_2, $\frac{1}{2}$ mole CO, and 2 moles O_2 at 77 F and 1 atm pressure is heated in a constant pressure steady-flow process to 3000 K. Determine the equilibrium composition at 3000 K on a percent basis, assuming the reaction equation to be

$$CO_2 \rightleftarrows CO + \tfrac{1}{2}O_2$$

16A.6 Repeat problem 16A.5 for the case where the initial mixture also includes 2 moles N_2, which does not dissociate during the process.

16A.7 Calculate the equilibrium constant at 3000 K for the reaction

$$2CO_2 \rightleftarrows 2CO + O_2$$

Repeat problem 16A.5 using this reaction equation instead of the one specified in Problem 16A.5.

16A.8 Oxygen is heated from room temperature in a steady flow process at a constant pressure of 10 atm. At what temperature will the mole fraction of O_2 in the mixture of O_2 and O be 30%? (That is, the mole fraction of the O_2 is 30% and of the O, 70%.)

16A.9 One mole of carbon monoxide at 1 atm pressure, 1000 K, reacts with one mole of oxygen at 1 atm pressure, 1000 K to form an equilibrium mixture at 2800 K and a total pressure of 1 atm.

Determine the heat transfer for this process, assuming the reaction takes place in a steady-flow process.

16A.10 One mole of carbon at 2 atm pressure and 77 F reacts with 1.5 moles of oxygen at 2 atm pressure and 77 F to form an equilibrium mixture of CO_2, CO, and O_2 at a temperature of 2800 K and a pressure of 2 atm.

Determine the composition of the products of combustion.

Thermodynamic Properties of Superheated Nitrogen

Temp. F	v	h	s	v	h	s	v	h	s
	P = 15 lbf/in.^{2} (−320.1 F)			P = 25 lbf/in.^{2} (−312.3 F)			P = 50 lbf/in.^{2} (−298.8 F)		
−320	3.410	85.73	0.6149						
−300	3.947	91.25	0.6508	2.314	90.50	0.6109			
−280	4.492	96.49	0.6824	2.654	95.89	0.6437	1.272	94.39	0.5891
−260	5.023	101.60	0.7092	2.983	101.13	0.6717	1.452	99.95	0.6187
−240	5.548	106.68	0.7334	3.305	106.30	0.6961	1.617	105.35	0.6442
−220	6.078	111.75	0.7555	3.617	111.45	0.7186	1.785	110.52	0.6673
−200	6.598	116.79	0.7758	3.927	116.53	0.7389	1.946	115.75	0.6881
−150	7.885	129.31	0.8199	4.698	129.13	0.7829	2.344	128.63	0.7329
−100	9.167	141.79	0.8572	5.326	141.65	0.8206	2.739	141.30	0.7706
−50	10.453	154.24	0.8892	6.259	154.13	0.8530	3.126	153.86	0.8032
0	11.729	166.68	0.9178	7.037	166.60	0.8815	3.575	166.36	0.8320
100	14.295	191.56	0.9667	8.577	191.45	0.9304	4.288	191.30	0.8817

Temp. F	v	h	s	v	h	s	v	h	s
	P = 100 lbf/in.^{2} (−282.7 F)			P = 200 lbf/in.^{2} (−263.7 F)			P = 300 lbf/in.^{2} (−250.4 F)		
−280	0.7581	91.04	0.5257						
−260	0.6846	97.51	0.5607	0.2879	91.29	0.4875			
−240	0.7742	103.40	0.5885	0.3480	98.93	0.5249	0.2000	93.27	0.4760
−220	0.8649	109.08	0.6132	0.4021	105.53	0.5536	0.2451	101.48	0.5124
−200	0.9510	114.54	0.6352	0.4516	111.72	0.5783	0.2838	108.50	0.5410
−150	1.1566	127.74	0.6815	0.5635	125.77	0.6278	0.3664	123.73	0.5943
−100	1.3586	140.60	0.7202	0.6690	139.08	0.6679	0.4391	137.65	0.6360
−50	1.554	153.33	0.7530	0.7706	152.18	0.7017	0.5101	151.05	0.6711
0	1.752	165.90	0.7821	0.8726	165.05	0.7310	0.5794	164.10	0.7007
100	2.144	191.00	0.8313	1.0721	190.46	0.7816	0.7147	189.84	0.7515

Temp. F	v	h	s	v	h	s	v	h	s
	P = 500 lbf/in.^{2}			P = 1000 lbf/in.^{2}			P = 1500 lbf/in.^{2}		
−220	0.1120	90.32	0.4395	0.03259	56.98	0.2807	0.02876	53.40	0.2540
−200	0.1470	101.09	0.4831	0.04725	76.54	0.3588	0.03375	66.37	0.3061
−150	0.2072	119.50	0.5481	0.08933	107.97	0.4704	0.05536	97.49	0.4163
−50	0.3016	148.78	0.6297	0.14659	143.14	0.5699	0.09605	138.29	0.5318
0	0.3452	162.30	0.6612	0.1706	157.97	0.6039	0.11317	154.23	0.5684
100	0.4288	188.65	0.7133	0.2153	185.97	0.6593	0.14480	183.74	0.6262

Thermodynamic Properties of Saturated Nitrogen* Liquid-Vapor Equilibrium

Temp. F t	Abs. Press. lbf/in.^{2} P	Specific Volume, ft^3/lbm			Enthalpy, Btu/lbm			Entropy, Btu/lbm R		
		Sat. Liquid v_f	Evap. v_{fg}	Sat. Vapor v_g	Sat. Liquid h_f	Evap. h_{fg}	Sat. Vapor h_g	Sat. Liquid s_f	Evap. s_{fg}	Sat. Vapor s_g
−320.4	14.696	0.01985	3.45315	3.473	0.000	85.641	85.64	0.0000	0.6152	0.6152
−310	27.539	0.02053	1.91047	1.931	5.076	82.319	87.40	0.0350	0.5502	0.5851
−300	46.532	0.02129	1.16671	1.188	10.18	78.59	88.77	0.0673	0.4923	0.5596
−290	73.852	0.02215	0.73385	0.7560	15.41	74.36	89.77	0.0985	0.4384	0.5369
−280	111.27	0.02316	0.48454	0.5077	20.80	69.45	90.25	0.1287	0.3863	0.5150
−270	160.66	0.02444	0.32256	0.3470	26.56	63.37	89.93	0.1589	0.3341	0.4930
−260	224.14	0.02604	0.21376	0.2398	32.75	56.01	88.76	0.1889	0.2806	0.4695
−250	303.97	0.02820	0.13700	0.1652	39.80	46.82	86.62	0.2198	0.2229	0.4427
−240	402.82	0.03238	0.07562	0.1080	48.09	33.75	81.84	0.2543	0.1537	0.4080
−232.42	492.90	0.05092	0.0000	0.05092	66.19	0.00	66.19	0.3357	0.0000	0.3357

* Abridged from "Thermodynamic Properties of Nitrogen". O. T. Bloomer and K. N. Rao, Institute of Gas Technology Research Bulletin No. 18, Chicago (1952).

Index